Impacts of Point Polluters on Terrestrial Biota

AF181519

ENVIRONMENTAL POLLUTION

VOLUME 15

Editors

Brian J. Alloway, *Department of Soil Science, The University of Reading, U.K.*
Jack. T. Trevors, *Department of Environmental Biology, University of Guelph, Ontario, Canada*

Editorial Board

I. Colbeck, *Institute for Environmental Research, Department of Biological Sciences, University of Essex, Colchester, U.K.*
R.L. Crawford, *Food Research Center (FRC) 204, University of Idaho, Moscow, Idaho, U.S.A.*
W. Salomons, *GKSS Research Center, Geesthacht, Germany*

For other titles published in this series, go to
www.springer.com/series/5929

Impacts of Point Polluters on Terrestrial Biota

Comparative Analysis of 18 Contaminated Areas

Mikhail V. Kozlov
Section of Ecology, Department of Biology,
University of Turku, Finland

Elena L. Zvereva
Section of Ecology, Department of Biology,
University of Turku, Finland

Vitali E. Zverev
Section of Ecology, Department of Biology,
University of Turku, Finland

Springer

Mikhail V. Kozlov
Section of Ecology
Department of Biology
University of Turku
Finland

Elena L. Zvereva
Section of Ecology
Department of Biology
University of Turku
Finland

Vitali E. Zverev
Section of Ecology
Department of Biology
University of Turku
Finland

ISBN 978-94-007-3671-9 e-ISBN 978-90-481-2467-1
DOI 10.1007/978-90-481-2467-1
Springer Dordrecht Heidelberg London New York

© Springer Science+Business Media B.V. 2009
Softcover reprint of the hardcover 1st edition 2009
No part of this work may be reproduced, stored in a retrieval system, or transmitted in any form or by any means, electronic, mechanical, photocopying, microfilming, recording or otherwise, without written permission from the Publisher, with the exception of any material supplied specifically for the purpose of being entered and executed on a computer system, for exclusive use by the purchaser of the work.

Cover illustration:
Main photo (part of the color plate 36): industrial barren located 2 km W of smelter at Monchegorsk. Photo by V. Zverev.

Top right-hand photo (part of the color plate 30): nickel-copper smelter at Monchegorsk, Russia. Photo by V. Zverev.

Printed on acid-free paper

Springer is part of Springer Science+Business Media (www.springer.com)

Executive Summary

1 Background

The adverse consequences of pollution impact on terrestrial ecosystems have been under careful investigation since the beginning of the twentieth century. Several thousand case studies have documented the biotic effects occurring in contaminated areas. However, after more than a century of research, ecologists are still far from understanding the effects of pollution on biota. Only a few generalisations have been made on the basis of extensive monitoring programs and numerous experiments with industrial contaminants.

The need to reveal general patterns in the responses of terrestrial biota to industrial pollution and to identify the sources of variation in these responses became obvious more than a decade ago. At about that time, our team initiated a quantitative research synthesis of the biotic effects caused by industrial pollution, based on a meta-analysis[1] of published data. All meta-analyses conducted so far (covering diversity and abundance of soil microfungi, diversity of vascular plants, diversity and abundance of terrestrial arthropods, and plant growth and reproduction) consistently showed high heterogeneity in the responses of terrestrial biota to industrial pollution. At the same time, they demonstrated an unexpected shortage of information suitable for meta-analyses, as well as a considerable influence of methodology of primary studies on the outcome of the research syntheses.

To overcome the identified problems, we designed a comparative study, the results of which are reported in this book. We used a properly replicated sampling design, and applied identical techniques for collecting and processing the data on the same characteristics of organisms and ecosystems around different polluters. In this way we substantially increased the amount of primary data available for research synthesis and minimised the effects of both methodological differences and the quality of primary studies on the outcomes of our analyses. A comparison of our original, uniformly collected data with data from published studies allowed

[1] Meta-analysis is a quantitative synthetic research method that statistically integrates results from individual studies to find common trends and differences.

for the identification of biases and the mitigation of their impacts on conclusions based on published data.

In this book, we summarise our observations on a number of characteristics reflecting the structure and functions of terrestrial ecosystems, which were conducted in the impact zones of 18 polluters.

2 Materials and Methods

We explored environmental effects imposed by eight non-ferrous smelters, six aluminium plants, one iron pellet plant, one fertilising plant, and two power stations. These polluters differ in the type and amount of emissions and in pollution history, and are situated in localities that differ in climate and vegetation type. Considered together, these polluters are sufficiently representative to detect the effects of industrial pollutants on terrestrial ecosystems of temperate to northern regions of the Northern hemisphere. We briefly describe the history of each polluter, summarise available emission data, and review the results of earlier environmental research.

We selected ten study sites around each polluter. These sites were grouped in two blocks (transects) of five, located in two different directions from the polluter. At each site, we measured environmental contamination, soil quality (stoniness, topsoil depth, pH, and electrical conductivity), and functional (soil respiration and consumption of plant foliage by herbivorous insects) and structural parameters of terrestrial ecosystems. Organism-level effects (two parameters of photosynthesis, fluctuating asymmetry, leaf/needle size, shoot length, radial increment, and needle longevity) were explored in 43 species of vascular plants. Community-level effects included cover and diversity of vegetation (separately for different functional groups), stand characteristics, regeneration of woody plants, and abundance of herbivorous insects.

We summarised our data by means of meta-analyses. We also linked the observed effects with some characteristics of both polluters and the affected communities. The site-specific values of all measured parameters, reported in this book (about 7,570 data lines), may provide a basis for future comparative studies.

3 Key Findings

1. All measured parameters generally demonstrated substantial intersite variation around all investigated polluters. However, only a small part of this variation was associated with pollution load. The Pearson linear correlation coefficient, calculated between site-specific values of measured parameters and pollution loads, and averaged across all parameters and all impact zones, was around 0.4, suggesting that the effects of pollution on many parameters are weak.

2. Soil quality generally decreased with pollution. Changes in soil chemical properties (measured by pH and electrical conductivity) were industry-specific, whereas changes in soil morphology or physical properties (measured by topsoil depth and stoniness) were similar across the studied polluters. Both losses of topsoil and overall decreases in soil respiration indicated adverse effects of industrial polluters on ecosystem services.
3. Plant vitality indices responded differently to the impacts of polluters. Radial increment and needle longevity in conifers strongly decreased with pollution. Photosynthesis and shoot length demonstrated slight, albeit significant, adverse effects, while leaf/needle size did not change. No differences in responses were detected between species, life forms, or between evergreen and deciduous plants.
4. It is generally accepted that an increase in fluctuating asymmetry[2] (FA) reflects the development instability and can serve a measure of environmental stress (in particular stress imposed by pollution). However, we did not detect the expected increase of plant FA with increasing levels of pollution. We suggest that either plants persisting in polluted habitats do not experience environmental stress (due to acclimation or selection for pollution resistance), or the hypothesis linking FA and environmental stress requires reformulation and restriction of its scope.
5. Although some point polluters drastically change both the structure and diversity of surrounding vegetation, pollution effects were not uniform across the polluters. Vegetation cover, stand basal area, proportion of conifers among top-canopy plants and diversity of vascular plants significantly decreased near non-ferrous smelters, while the effects of other polluters (including aluminium smelters) on these characteristics were neutral or even positive.
6. Overall decreases in stand basal area and stand height, decline of field layer vegetation, and overall reduction in plant growth suggest that a decrease in net primary productivity with pollution is likely to be a general ecosystem-level phenomenon.
7. Foliar damage to woody plants by insect herbivores showed a slight but significant decrease with pollution, while the densities of selected groups of herbivorous insects did not change. The dome-shaped patterns in herbivory along pollution gradients, earlier considered to be a typical population response to pollution, were rare. We conclude that increases in herbivory at polluted sites are an infrequent phenomenon rather than a general regularity.
8. The direction and magnitude of changes in ecosystem parameters around industrial polluters generally depended on both composition and amount of emissions but were only rarely associated with the duration of the impact. Polluters that caused acidifying effects on soil imposed a stronger impact on biota than polluters that caused alkalifying effects. The impacts of metal-emitting industries (non-ferrous smelters) were generally more detrimental than the impacts of polluters emitting only sulphur dioxide or fluorine.

[2] Fluctuating asymmetry represents small, random deviations from symmetry of a bilaterally symmetrical trait.

9. Pollution-induced changes in some vegetation characteristics were smaller in low diversity plant communities. The adverse effects of pollution on community- and organism-level parameters were stronger in southern ecosystems and increased with mean July temperatures. Variations in responses of the investigated components of terrestrial ecosystems to pollution were not explained by critical loads of sulphur, nickel, or copper.
10. Comparison of outcomes of meta-analyses of our data with the outcomes of meta-analyses of published data allowed us to conclude that published results generally overestimated (by an average factor of 5) the magnitude of the effects of industrial pollution on terrestrial biota. This effect can be explained by both research and publication biases frequently observed in pollution studies.

4 Conclusions

1. Effects of pollution should be considered within the scope of basic ecology and have to be studied in the same manner as any natural ecological phenomena. They require proper documentation, explanation, generalisation through building of theoretical models, and experimental verification of these models followed by practical application. This research requires an understanding of the ecological principles governing ecosystem-level processes in both pristine and contaminated habitats.
2. Short-term experiments with non-adapted organisms in over-simplified laboratory environments are likely to overestimate the adverse effects of pollutants relative to the effects observed in polluted ecosystems.
3. Since the links between the organism-level and ecosystem-level effects appeared weak, pollution impacts on ecosystem-level processes should be explored directly rather than deducted from organism-level studies. Similarly, the importance of indirect and secondary effects in shaping community structure prevents the deduction of community-level responses to pollution from the results of single-species studies.
4. Impact zones of point polluters can be seen as 'unintentional experiments' and can serve as suitable models to explore biotic consequences of environmental contamination. Investigation of the effects imposed by point polluters lacks some of the benefits of controlled experiments, but instead allows for the consideration of: (a) the consequences of long-term impacts on abiotic environment and biota, (b) multiple interactions among different organisms and processes, (c) large-scale, landscape- and ecosystem-level effects, and (d) secondary effects, such as modifications of the microclimate.
5. Non-significant overall effects of pollution on many of the explored characteristics primarily resulted from a high diversity in responses, which is linked with variations in characteristics of both polluters and impacted ecosystems. Identification of factors underlying this variation is crucial for understanding and predicting pollution-induced changes in terrestrial ecosystems.

6. Geographical variation in ecosystem responses to pollution is likely to be a general phenomenon. This variation deserves thorough investigation, since our results support the hypothesis that under a warmer climate the existing pollution loads may become more harmful.

5 Data Gaps and Research Needs

Study objects

- Geography: Africa, South-Eastern Asia, South America, Oceania
- Biomes and ecosystems: tropical forests, deserts, grasslands, savannah, tundra
- Industries: power plants, chemical industries, small polluters

Research topics

- Belowground processes
- Changes in species interactions
- Functional diversity of communities
- Spatial structure and demography of populations
- Ecosystem-level effects

Research approaches

- Comparisons between polluters, organisms, communities, and ecosystems
- Long-term studies
- Exploration of dose-and-effect and cause-and-effect relationships
- Research synthesis

6 Methodological Recommendations

1. A high (35%) proportion of studies based on only two or three study sites and a low (25% on average) power of the majority of statistical tests in published studies, reporting effects of industrial polluters on terrestrial biota, emphasise the need to substantially improve the design of impact research. Special attention should be paid to proper replication of sampling on all hierarchical levels. In particular, the impact versus reference sites design should include at least two polluted and two control plots. However, more than 30 study sites are necessary to achieve sufficient statistical power of correlation analysis. Choosing study sites in several directions from the polluter is likely to minimise the impact of confounding variables on the data analysis outcomes.
2. We suggest that more attention be paid to identification of non-linear dose–effect relationships. This requires improvements in research methodology, including an increase in the number of study sites to allow for: (a) an accurate

approximation of data using non-linear functions and (b) statistical comparison of the fit of non-linear and linear models to the data. Exploration of non-linear effects is particularly important to enhance the understanding of causal relationships behind the observed phenomena.

3. Selection of the pollution load measure in studies conducted around point polluters (pollutant concentrations or distance from the pollution source) depends primarily on the research goals. If the study aims at a demonstration of the effect, then the distance from the polluter may appear to be the best proxy of pollution load. Exploration of dose–response relationships requires the use of pollutant concentration (usually log-transformed) instead of distance.
4. Primary studies that detected pollution-induced changes in biotic parameters by calculating correlations with distance, correlations with concentrations of selected pollutant(s), or contrasting polluted and control sites, generally yield similar estimates of effect sizes, and the results of all these studies can therefore be combined in research synthesis.
5. We strongly encourage researchers and editors to publish the quality results that are unexpected or seem strange. Elimination of these results at the pre-publication stage (decision not to submit the manuscript) or by reviewers (frequently due to disagreement with the prevailing paradigm) is likely to bias estimates of overall effects. This process may lead to the dominance of incorrect or exaggerated opinions. 'Outliers' are especially important for exploration of the sources of variation in responses of organisms and ecosystems to pollution.

Acknowledgements

The creation of this book was made possible by the consolidated support of several funding bodies, organisations, friends and colleagues. At the data collection stage, we benefited from advice provided by Peter Beckett, Vincas Buda, Tatiana Chernenkova, John Gunn, Jan Kulfan, Pekka Niemelä, Lukasz Przybylowicz, Elena Runova, Vitaly Savchenko, Tatiana Vlasova, Eugene Vorobeichik and Peter Zach. Among the persons who participated fieldwork we are most grateful to Sarah Bogart, Jan Kulfan, Eugene Melnikov, Irina Mikhailova, Anna Sandulskaya, Erik Szkokan, Andrey Vassiliev, Brian Wesolek, Peter Zach, Anna Zvereva and Ljubov Zvereva. Identifications of vascular plants were provided by Sarah Bogart, Viktor Chepinoga, Anna Doronina, Alexandr Egorov, Elena Khantemirova, Galina Konechnaya, Irina Sokolova, Marina Trubina, and Nikolay Tzvelev; plant nomenclature was checked by Anna Doronina. Chemical analyses were conducted under the supervision of Andrey Bakhtiarov, Tamara Gorbacheva, Leena Kuusisto, Katarina Kvapilova, Natalia Lukina and Hannu Raitio. A substantial amount of the measurements for the analyses of fluctuating asymmetry were conducted by Tatjana Koskello and Andrey Stekolshchikov. Assistance in the search for information and the translation of information published in national languages was kindly provided by Valery Barcan, Eduardas Budrys, Janne Eränen, Gottskálk Friðgeirsson, Gísli Már Gíslason, Tatyana Karapetyan, Andrey Kozlovich, Jan Kulfan, Yngvild Lorentzen, Sigurður Magnússon, Tatiana Mikhailova, Marja Roitto, Anna Liisa Ruotsalainen, Nadia Solovieva, and Hans Tømmervik. Andrey Maisov, Astkhik Vinokurova and Ljubov Zvereva contributed to creation of the list of references. We are pleased to thank Valery Barcan, John Gunn, Julia Koricheva, Jan Kulfan, Annamari Markkola, Tatiana Mikhailova, Anders Pape Møller, Pekka Niemelä, Anna Liisa Ruotsalainen, and Eugene Vorobeichik for inspiring discussions, as well as for their perceptive reading and comments on selected chapters. The language was improved by highly qualified editors from 'American Journal Experts'.

The study was primarily supported by several research grants provided by Maj and Tor Nessling Foundation (Finland) to the co-authors. These grants covered both fieldwork and book production, especially publication of the colour section showing both polluted and pristine landscapes. Additional financial support was provided by the Academy of Finland (projects 124152, 209219, and 211734, and several researcher exchange grants to M.V.K. and E.L.Z.), and by EC through the

BASIS and BALANCE projects carried out under contracts ENV4-CT97-0637 and EVK2-2002-00169, respectively. Work on this book for one of us (M.V.K.) was partially supported by NordForsk through the visiting professorship programme from 2000 to 2007.

We are grateful to Ulf Molau and Brian Huntley for their permission to refer to the unpublished report of the DART project, as well as to Elsevier and Natturfræðingurinn for their permissions to reproduce, in a modified form, Tables 2.7 and 2.19, respectively.

Finally, we thank the University of Turku, and the staff of Ecology and Biodiversity Sections in particular, for providing an excellent environment for conducting this research. Grateful thanks are given to Erkki Haukioja who, by inviting us to work in Turku, created the basis for the development of our research careers.

Contents

Chapter 1
Introduction

1.1 'Pollution Science' – Applied or Basic Ecology?

We are living in a rapidly changing world. Human domination of the Earth alters the composition, structure and function of ecosystems, emphasising an urgent need to consider ecological principles on a global scale. Knowledge of these principles is necessary to understand ecosystem development and to manage ecosystem services crucial to human survival (Kremen & Ostfeld 2005; Mokany et al. 2006; Grimm et al. 2008).

Pollution, the introduction of contaminants into an environment that causes instability, disorder, harm or discomfort to the physical systems or living organisms, is just one of many actors behind global changes. Along with direct toxic impacts, pollution often triggers numerous secondary effects, such as modifications of the microclimate, leading to further disturbance of the contaminated environments. Disturbance-induced changes in ecosystems are of central concern in ecology, and a challenge for ecologists is to understand the factors that affect the resilience of community structures and ecosystem functions (Moretti et al. 2006).

The domain of ecology is defined as the spatial and temporal patterns of the distribution and abundance of organisms, including causes and consequences (Scheiner & Willig 2008). This domain obviously incorporates the effects of pollution, which have frequently been reported to affect the distribution and abundance of living beings, both directly, through toxicity of pollutants, and indirectly, through habitat degradation and loss. The effects of pollution on biota were, for example, considered by E. Odum (1969a), who included a chapter on pollution in his book *Fundamentals of Ecology* and further developed the theoretical basis of 'stress ecology' (as defined by Barrett et al. 1976) in several influential papers (Odum et al. 1979; Odum 1985). However, the attitude of 'basic' ecologists towards the impacts of pollution on ecosystems has changed since then (Filser 2008), and pollution, at best, is now mentioned in only a few lines of basic ecological textbooks (Dodson et al. 1998; Krebs 2001).

Freedman (1989) considered pollution effects among other topics of 'environmental ecology', a somewhat odd expression since ecology, by definition, is the study

M.V. Kozlov et al. *Impacts of Point Polluters on Terrestrial Biota*,
DOI 10.1007/978-90-481-2467-1_1, © Springer Science+Business Media B.V. 2009

of interactions between organisms and their natural environment. In the Russian literature, some authors discuss 'ecology of impacted areas' (Vorobeichik 2002) or even 'dirty ecology' (Vorobeichik 2005). Several universities offer course on 'pollution ecology' (www.uvm.edu, www.csuchico.edu). Direct effects of pollution are often seen as a topic of ecotoxicology, which aims to study how organisms are affected by chemicals released into the environment due to human activities. For example, the effects of zinc smelters at Palmerton, Pennsylvania, are considered, among other case histories, in the *Handbook of Ecotoxicology* (Hoffman et al. 1995). However, the toxicological approach is too narrow to adequately describe the complex effects of pollution, since pollution alters not only the fate of individuals but also biotic interactions within and among species. In particular, while changes on a larger scale arise from interactions on a finer scale, it is difficult to predict the effects of chemical contamination across scales. Several ecotoxicologists have argued for putting more 'eco' into ecotoxicology (Calow et al. 1997; Chapman 2002), and several new terms have been suggested. These include, for example, 'community ecotoxicology', addressing effects of chemicals on species abundance, diversity, and interactions (Clements & Newman 2002), 'syntropic ecotoxicology', aimed at multilevel conceptualisation beyond the simple organism-level effects of poisons (Downs & Ambrose 2001), and 'disturbance ecology', bridging ecotoxicological studies and ecological theories (Filser et al. 2008). An attempt to properly account for toxicity effects acting on multiple spatial scales resulted in the development of landscape ecotoxicology (Cairns & Niederlehner 1996; Johnson 2002). Chapters on the pollution effects on wildlife also appeared in textbooks on applied ecology (Newman 1993) and conservation biology (Hunter & Gibbs 2007). Development of heavy metal tolerance in plants provides a well-documented example of rapid evolutionary adaptation (Bradshaw & McNeilly 1981), a topic that that belongs to evolutionary ecology. Thus, in spite of several attempts to integrate this research field, different aspects of the pollution impact on biota are still being considered by different scientific disciplines.

It is difficult to discern the reasons behind a recent reluctance of 'basic' ecologists to consider pollution effects on biota as a part of general ecology, as was common in the 1960–1970s (Odum 1969b; Barrett et al. 1976; Odum et al. 1979). Instead, this ecological approach to pollution-mediated effects is becoming increasingly popular among ecotoxicologists (Clements & Newman 2002; Filser 2008). We think that data from contaminated environments are well suited to advance basic ecology, in particular by exploring the limits of ecological 'rules' or 'laws' that are usually based on studies of unpolluted habitats. This opinion is also in line with the recent position of the British Ecological Society, which arranged the 'Ecology of Industrial Pollution' meeting (2008). This meeting stressed the importance of basic ecology in solving environmental problems. In particular, ecologists were called to answer several vital questions, namely: How do we monitor and assess the ecological status of environments? What constitutes 'good ecological status'? Does industrial pollution always result in lower biodiversity? How can we use ecology in remediation technologies? (www.britishecologicalsociety.org).

1.2 Pollution, Polluters, and Pollutants

It is problematic to define either pollution or a pollutant. Any foreign substance, primarily waste from human activities, that renders the air, soil, water, or other natural resource harmful or unsuitable for a specific purpose, can be classified as a pollutant, especially when it has detrimental effects on living organisms. Harmless or even useful substances become pollutants when they appear where they should not be, for example gasoline in drinking water.

Aerial emissions from any polluter consist of an endless number of substances, dozens of which are known to be toxic. In the following text we will only briefly mention the most common groups of pollutants that (a) appear in ambient air due to industrial activities, (b) are common and widespread, and (c) are considered primary drivers of environmental deterioration near industrial enterprises.

Historically, sulphur dioxide was the very first pollutant that caused local but severe environmental deterioration several centuries ago (for an early history of air pollution, consult Brimblecombe & Makra 2005). This is the best-studied pollutant, both in experimental and natural conditions. In high concentrations, it causes acute plant damage, which weakens and then kills the trees (please see color plates 32, 33 and 53 in Appendix II); in low concentrations, it contributes to regional acidification. Animals, especially invertebrates, are less susceptible to SO_2 than plants.

Over the past century, the importance of different sources of sulphur dioxide emissions changed substantially. While in 1940 both smelters and power plants contributed approximately equal parts to global SO_2 emissions, by 1990 the contribution of smelters had declined to 5% (Dudka & Adriano 1997).

Fluorine emissions into the atmosphere started to increase in the late 1930s, reaching peak values in the late 1960s. These emissions, primarily associated with aluminium production, resulted in severe localised damage (please see color plate 6 in Appendix II). Fluoride causes severe health problems in vertebrates, including humans. However, fluorine emissions strongly decreased between 1970 and 1980 due to effective precautions taken to minimise the release of fluoride from aluminium smelters into the atmosphere.

Heavy metals have been common pollutants since ancient times (for an early history of metal pollution, consult Makra & Brimblecombe 2004); luckily, heavy metals generally do not spread far from smelters. Only a few of the largest polluters caused detectable contamination of soils and vegetation at distances exceeding 50 km from the emission source. Although most heavy metals are extremely toxic, they were only rarely reported as proximate causes of the deterioration of landscapes adjacent to polluters. However, while mature trees can withstand extreme soil concentrations of metal pollutants, heavy metals efficiently prevent seedling establishment, thus hampering natural revegetation of contaminated areas long after pollution decline (please see color plate 23 in Appendix II). They also pose toxicity risks to vertebrates, including humans.

Increased deposition of nitrogen started to play an important role in European forests several decades ago. Although this pollutant does not create dramatic landscapes, its effects are insidious and long lasting. In some countries, such as the Netherlands, annual deposition of nitrogen by the late 1980s reached 200 kg/ha, making eutrophication (an increase in chemical nutrients, mostly compounds containing nitrogen and phosphorous, in an ecosystem) more important than the impact of traditional acidifying pollutants. The importance of nitrogen deposition has also been increasing in some parts of North America. It has recently been argued that atmospheric nitrogen deposition may represent a threat to floristic diversity on a global scale (Phoenix et al. 2006).

Emissions of nitrogen oxides and volatile organic compounds have significantly increased levels of ozone over large regions of the globe. Ozone has recently been identified as a probable contributor to the observed forest decline and reduction of crop production in Europe and North America. Although unequivocal evidence for O_3-induced foliar injury on woody species under field conditions has only been found in a few places, ozone obviously weakens plants, making them vulnerable to other assaults and stresses. Overall, the quantitative risk assessment of the ozone impact on natural ecosystems is vague on a European scale.

Finally, carbon dioxide is sometimes listed among pollutants, even as the number one pollutant (Birdsey 2003). However, although CO_2 fits the definition of a pollutant and is emitted by industry, its concentrations, distribution, and mode of action differ substantially from other pollutants. In our opinion, the effects of CO_2 elevation are outside the scope of pollution ecology.

1.3 Extent and Severity of Impacts

For more than a century, pollution has been recognised as one of the most severe, albeit local, environmental stressors associated with large industrial towns, primarily those housing non-ferrous smelters and iron factories (Haywood 1907; National Research Council of Canada 1939; Gordon & Gorham 1963). However, by the 1960s, both the extent and impact of air pollution were already much greater than previously thought (Likens & Bormann 1974), and air pollution came to represent one of the major factors of environmental disturbance (Freedman 1989; Soulé 1991; Lee 1998). Now pollution affects every part of our planet, and its impacts on ecosystem structure and services have considerable implications for human welfare (Emberson et al. 2003).

1.3.1 Local Scale

Extensive forest mortality around large sources of pollution has sometimes created barren 'moonscapes' (please see color plates 9–11, 20–23, 29, 34–39, 45–47, 54, 59 in Appendix II) that, in spite of their relatively small areas, have attracted considerable public and scientific attention over the past decades (reviewed by Kozlov & Zvereva 2007a). The most striking examples of severe local pollution have long been associated with Canadian smelters (in Trail, Sudbury, and Wawa). However, after implementation of strong emission controls in most industrial countries in the 1970s, the largest individual polluters remain in Eastern Europe and Russia. Among them, the Norilsk smelter in Northern Siberia (please see color plate 50 in Appendix II) is the largest on a global scale: it causes damage to vegetation over 100–150 km distances. The largest sources of fluorine-containing emissions (Bratsk, Shelekhov) are also situated in Siberia. The most extensive scientific information concerning impacts imposed by an individual polluter, with more than 1,000 published data sources (Kozlov 2006), has been collected around the Monchegorsk nickel-copper smelter (Kola Peninsula, NW Russia).

Importantly, local problems do not disappear with emission decline or even with the closure of polluters since natural leaching of some toxic pollutants, e.g., metals from contaminated soils, may last for centuries (Tyler 1978; Barcan 2002a). Therefore, there exists a time lag between emission decline and the start of ecosystem recovery (Tarko et al. 1995), which in the case of the smelter at Monchegorsk, Russia obviously exceeds 20 years (Zverev 2009). This time lag can be shortened only by applying costly remediation efforts (Gunn 1995; Winterhalder 2000).

1.3.2 Regional Scale

While the majority of aerial emissions are deposited near the polluter, some substances are spread over huge areas, leading to increases in pollutant deposition on a regional scale. The best known example of a regional pollution problem is acidification, which is caused mostly by emissions of sulphur and nitrogen compounds that react in the atmosphere to produce acids (Likens & Bormann 1974).

Mapping of precipitation pH in the late 1980s identified two regions most affected by acid rain: Western Europe and the Atlantic coast of North America (Newson 1995). Acidification is still widespread and occurs even in remote parts of the Arctic (AMAP 2006). It is of special concern in areas with sensitive geology when deposition exceeds the system's acid neutralising capacity. Model calculations (Fowler et al. 1999) demonstrate that by 2050, severe regional problems may be expected in rapidly developing regions of South-Eastern Asia, South Africa, Central America, and along the Atlantic coast of South America.

Large, primarily forested, areas require rehabilitation as a result of the impacts of acidic deposition and other forms of pollution, including ozone. The problem is particularly apparent in Central Europe, with the most striking example being the 'Black Triangle', an area along the German-Czech-Polish border. This region has been heavily affected by industrial pollution for the past 50 years, with severe consequences for forests, landscapes, the environment and public health (Markert et al. 1996; Renner 2002).

1.3.3 *Global Scale*

In 1985, 8% of the forested areas of the world received >1 kg H^+ ha^{-1} annually due to sulphur deposition, and it is estimated that 17% of the forested areas of the world will reach this pollution load by 2050 (Fowler et al. 1999). Similarly, 24% of global forest was exposed to ozone concentrations exceeding 60 ppb by 1990, and this proportion is expected to increase to 50% of global forest by 2100 (Fowler et al. 1999). Polymetallic dust emitted by large smelters, such as Norilsk and Monchegorsk, can be detected more than 2,000 km away from the emission source (Shevchenko et al. 2003), and significant increases in metal concentrations relative to a pre-industrial period have been reported even from Greenland (Boutron et al. 1995; Johansen et al. 2000) and Antarctica (Wolff et al. 1999). Thus, more and more extensive areas will require specific management, which will account for pollution impacts. Still, current projections of ecosystem responses to global climate change only rarely consider important physiological changes induced by air pollutants that may amplify climatic stress.

1.4 The State of Pollution-Oriented Studies and the Need for Generalization

Adverse consequences of the pollution impact on terrestrial ecosystems have been under careful investigation since the very beginning of the twentieth century, and several thousand case studies have documented the biotic effects occurring in contaminated areas. The overall number of studies containing the key word 'pollut*' in the title showed steady growth from the beginning of the century until the 1980s (Fig. 1.1a). However, as can be seen from the number of papers that appeared in 'Nature' (Fig. 1.1b) and 'Science' (Fig. 1.1c), this problem attracted attention of multidisciplinary scientists in the 1970s–1980s, and then the general interest in pollution started to decline.

Industrial development, environmental legislation and public attitude have influenced the focal research topics of pollution ecology. Sulphur dioxide is historically the first of the well-known pollutants that was reported to cause health problems and deteriorate the natural environment. Fluorine emissions associated with aluminium production became important in the middle of the last century, and the importance of ozone and nitrogen was appreciated in the 1980s. This process, along with the reversal of the trend in global anthropogenic sulphur emissions (the peak value of 73.2 million tons was achieved in 1989; Stern 2006), naturally shifted research priorities.

The proportion of presentations devoted to industrial polluters at bi-annual meetings for specialists in pollution impact on forest ecosystems, conducted by the International Union of Forestry Research Organisations (IUFRO), declined from 20–40% in the early 1990s to 3–7% in 2004–2008. At the same meetings, reported studies of ozone increased from approximately 30% to 70% at the expense of studies on the effects of sulphur dioxide (35–55% in the early 1990s and less than 10% in 2004–2008).

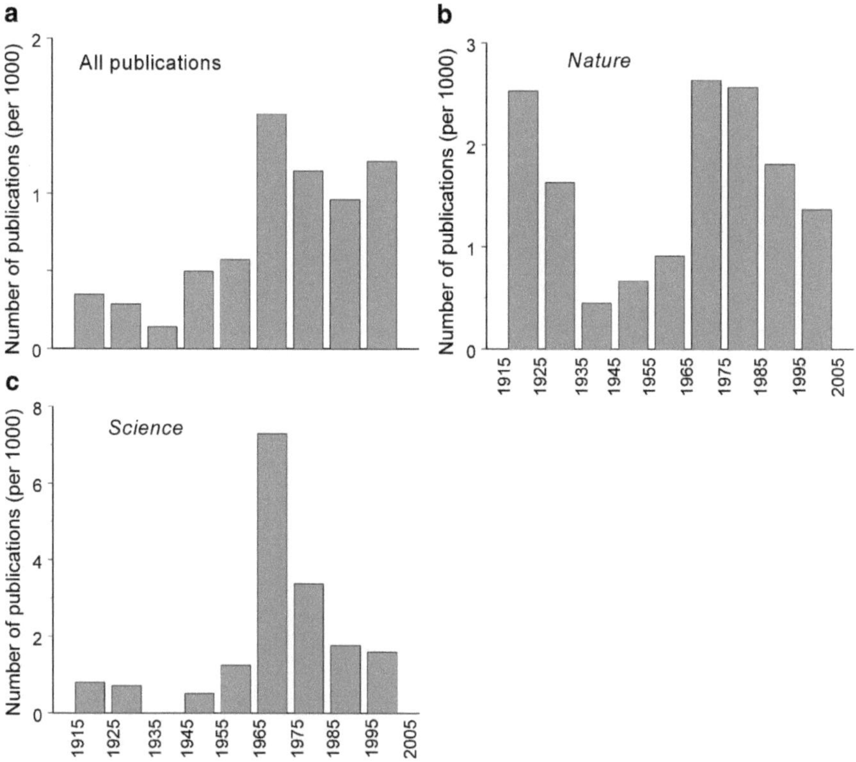

Fig. 1.1 Temporal trends in proportion (×1,000) of publications containing key word 'pollut*' in the title relative to all publications. a, all journals listed in the database of the Institute of Scientific Information; b, '*Nature*'; c, '*Science*'

Thus, pollution ecology, which started from the exploration of acute damage caused by traditional pollutants (such as SO_2, fluorine, and metals) associated with large industrial enterprises (Freedman 1989; Treshow 1984), turned towards studies focused on the regional effects of ozone and, to a lesser extent, nitrogen deposition. In line with this pattern, we detected a recent decrease in the number of descriptive studies, providing quantitative information on biotic effects of industrial polluters in a form suitable for meta-analysis (Fig. 1.2).

However, 'traditional' pollutants should not be forgotten; although the environmental impact of large polluters is gradually declining, especially in high-income OECD member states, overall emission rates of sulphur dioxide and heavy metals still remain high on a global scale, particularly due to the contributions of rapidly developing countries (Fowler et al. 1999; Stern 2006). As a result, large-scale deterioration and the decline of natural ecosystems remain a serious concern both globally and regionally, in particular in the European Community, which recently 'inherited' environmental problems from new member states (Freer-Smith 1997; Fowler et al. 1999; Innes & Oleksyn 2000; Innes & Haron 2000; Kozlov & Zvereva 2007a).

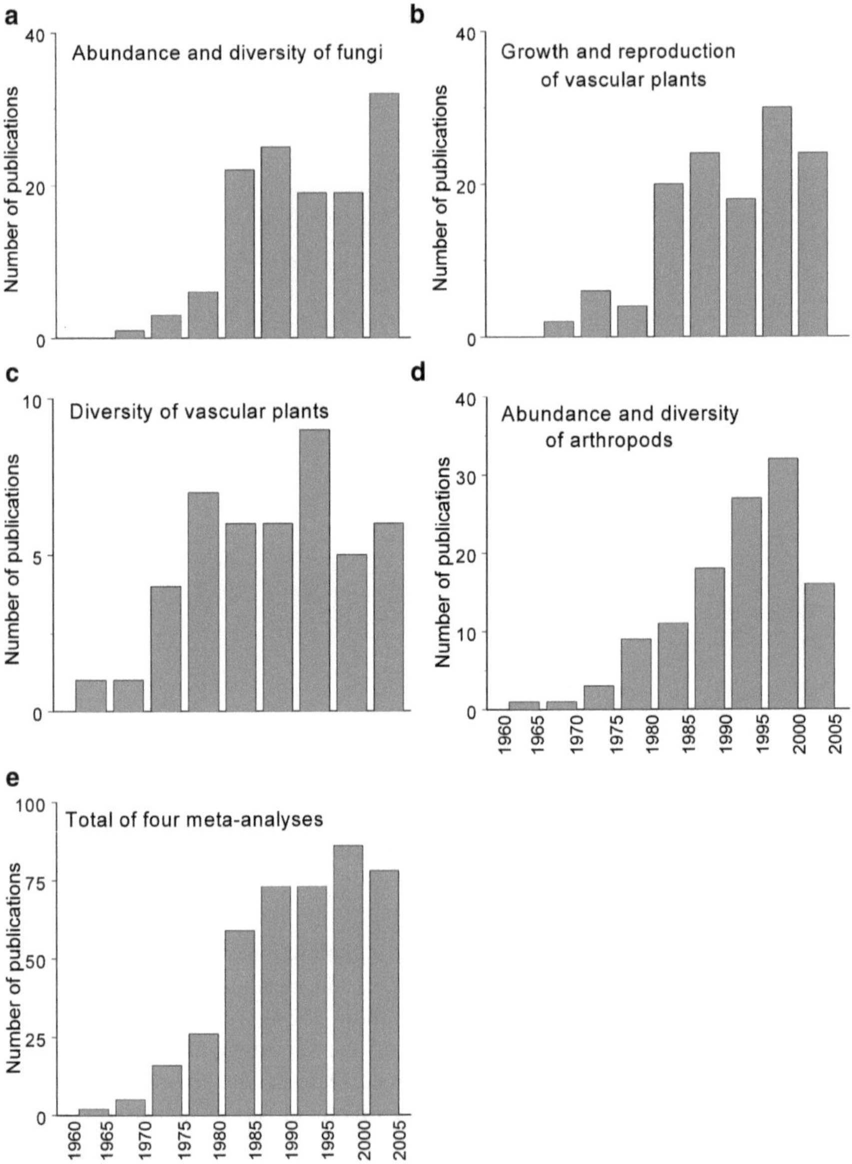

Fig. 1.2 Numbers of publications that reported quantitative information suitable for meta-analysis of biotic effects caused by industrial polluters. For selection criteria, consult Zvereva et al. (2008). a, diversity and abundance of soil microfungi (after Ruotsalainen & Kozlov 2006); b, growth and reproduction of vascular plants (after Roitto et al. 2009); c, diversity of vascular plants (after Zvereva et al. 2009); d, abundance and diversity of terrestrial arthropods (after Zvereva & Kozlov 2009); e, four meta-analyses pooled

After more than a century of research on the biotic effects of industrial pollution, ecologists are still far from understanding pollution effects on biota. Even the eternal questions, such as 'Does industrial pollution always result in lower biodiversity?', are not answered yet. This is in particular due to surprisingly weak links between pollution-oriented environmental studies and development of ecological theories (see above, Section 1.1). Only a few generalizations have been made on the basis of extensive monitoring programs and numerous experiments with industrial contaminants (Odum 1985; Rapport et al. 1985; Freedman 1989; Genoni 1997; Clements & Newman 2002). Environmental scientists clearly prefer a descriptive approach to the testing of research hypotheses (Ormerod et al. 1999), although several models in the field of population and community ecology (Andrén et al. 1997; Doak et al. 1998; Tilman 1999; Scheffer & Carpenter 2003; Orwin et al. 2006; Didham & Norton 2006) are readily available to predict ecosystem behaviour under pollution impacts. On the other hand, researchers addressing basic ecological problems reluctantly use the results of 'unintentional pollution experiments' (Lee 1998; Zvereva & Kozlov 2000b, c, 2001, 2004; Clements & Newman 2002).

Scientists working around industrial polluters have only rarely addressed the question of causal links between pollution and the recorded effects; therefore, the published case studies are generally narrative and often fail to deliver quantitative information. We were surprised by an acute shortage of information suitable for meta-analyses; although any search for relevant publications brings several hundred 'hits', only a small fraction of these studies allow calculation of effect sizes (Fig. 1.2). Moreover, following several trials that have carefully investigated the evidence of causal links between emissions and environmental damage (National Research Council of Canada 1939; Fitzgerald 1980; MacMillan 2000), ordinary people, as well as many scientists, tend to routinely attribute all adverse effects observed in polluted areas to the impact of the largest local polluter (usually a smelter or power plant). Although this approach seems logical, we suggest that a presumption of innocence needs to be accepted by environmental researchers dealing with pollution effects; any environmental change should be considered natural until proof of causal links between pollution and effect have been collected.

The majority of studies conducted around point polluters are still aimed at logical rationalisation and an historical explanation of patterns and processes occurring in these contaminated environments. Narrative reviews (we surveyed approximately 50) offer low levels of generalization, frequently just listing selected case studies as examples supporting intuitive conclusions (see, for example, Smith 1974; Linzon 1986; Bååth 1989; Freer-Smith 1997; Klumpp et al. 1999; Rusek & Marshall 2000; Padgett & Kee 2004). Moreover, the accuracy of the conclusions made in these reviews may suffer from the research bias in original studies, i.e., the tendency to perform experiments on organisms or under conditions in which one has a reasonable expectation of detecting significant effects (Gurevitch & Hedges 1999). It was recently found that some historically formed views concerning pollution effects on biota, such as a higher sensitivity of Northern ecosystems, overall decline in biodiversity, or increase in herbivory, lack proper statistical support (Zvereva et al. 2008; Zvereva & Kozlov 2009). Many problems vital for understanding and predicting pollution effects on biota, such as dose–response relationships, separation between specific responses to individual pollutants and general response to pollution, synergic effects of pollutants, effects of climate on bioavailability and toxicity of pollutants, and the role of catastrophic events in the transformation of contaminated communities, remain largely unexplored (Vorobeichik et al. 1994; Holmstrup et al. 1998; Clements & Newman 2002; Settele et al. 2005; Kozlov & Zvereva 2007a; Zvereva et al. 2008). This may partly explain why the biotic impacts of traditional pollutants, such as sulphur dioxide, fluorine and heavy metals, are not considered in global change scenarios.

The need to reveal general patterns in the responses of terrestrial biota to industrial pollution and to identify the sources of variation in these responses became obvious more than a decade ago. At about that time, our research team initiated a quantitative research synthesis of biotic responses to industrial pollution, based on a meta-analysis of published data (Kozlov & Zvereva 2003). This task is extremely time-consuming, and so far we have only been able to summarise the data on the diversity and abundance of soil microfungi (Ruotsalainen & Kozlov 2006), diversity of vascular plants (Zvereva et al. 2008), diversity and abundance of terrestrial arthropods (Zvereva & Kozlov 2009), and plant growth and reproduction (Roitto & Kozlov 2007; Roitto et al. 2009). All these meta-analyses consistently showed high heterogeneity in the responses of terrestrial biota to industrial polluters.

Problems with the heterogeneity of primary studies summarised by meta-analyses have been discussed widely, in particular in terms of the contrast between 'real' differences in the target phenomena and 'artefactual' differences due to the way these phenomena have been studied (Glasziou & Sanders 2002). One continuing criticism is that, because no two studies can ever be perfectly identical, meta-analysts are comparing apples and oranges (Hunt 1997; Glass 2000), thereby calling into question the generalizability of conclusions made by meta-analyses (Markow & Clarke 1997; Matt 2003). However, the same arguments are applicable to comparisons of study objects (or subjects) *within* a study, since they are never identical. Of course, narrative reviews also compare apples and oranges, although in a different manner than meta-analyses. Thus, all studies differ, and the problem lies in formulating an adequate research question: how study outcomes vary across the factors that we consider important (Glass 2000). Other concerns include the diverse quality of primary studies (Conn & Rantz 2003) and the comparison of results arising from different research designs. The latter source of variation deserves specific attention.

First and most importantly, many of the studies reporting pollution effects on biota suffer from pseudoreplication; i.e., statistical conclusions on the effects of pollution are based on comparisons between samples collected from one polluted and one unpolluted site, while true replication requires comparison between several (at least two) polluted and several unpolluted sites. Conclusions based on these two sampling designs may well differ (Zvereva & Kozlov 2009), yet we cannot disregard pseudoreplicated data since the number of correctly designed studies is too small to allow in-depth analyses.

Second, the spatial arrangement of study sites varies greatly between studies (Ruotsalainen & Kozlov 2006). The majority of researchers prefer a gradient approach, with several study sites selected at different distances from an emission source. With this design, effects are usually quantified by calculating correlation coefficients, either with the concentration of one 'major' or 'representative' pollutant, or with the distance from polluter. Whether or not the magnitude of the effect depends on the choice of explanatory variable, had not, to our knowledge, ever been tested statistically.

Neither is there any information on whether an ANOVA-oriented design (several polluted vs. several control sites) yields the same conclusions as a gradient design. Last but not least, some researchers, to be on the safe side, select control sites very far from polluters. However, this increases the importance of confounding factors, in particular climatic differences between the impact zone and arbitrarily selected distant control(s).

Although many strategies have been proposed to address the heterogeneity of primary studies, no consensus exists on any one set of methods. That is why we decided not only to restrict our research synthesis to meta-analysis of the published data but also to design a comparative study, the results of which are reported in this book. Using a uniform, properly replicated sampling design, and applying identical techniques for data collection and processing, we attempted to minimise the effects of both methodological differences and the quality of primary studies. Of course our data remain biased; as with any data produced by the same team, they share problems and errors. However, we believe that comparison of our original, uniformly collected data with data from published studies would favour distinguishing artefactual differences from real variation in biotic responses to pollution.

In this book, we aimed to reveal general patterns in the responses of terrestrial biota to industrial pollution and to identify the sources of variation in these responses. We also intended to provide some numerical estimates that, in terms of building phenomenological models, would allow linking of the observed effects with some characteristics of both the polluter and the affected community. We address the following questions:

1. What are the relative sensitivities of key ecosystem components and services to pollution impacts?
2. Which characteristics of polluters and of impacted ecosystems explain variation in biotic responses among the explored impact zones?
3. Which operational indicators are best suited to detect pollution-induced changes in basic ecosystem properties and services?
4. What are the minimum levels of pollution that cause changes in basic ecosystem properties/services?

Answering these questions is necessary to predict the possible fates of the impacted ecosystems under different economic and climatic scenarios.

1.5 Impact Zones of Point Polluters as Models for Ecological and Environmental Research

A point source of pollution is a single identifiable *localised* source that, relative to the impacted territory, has negligible extent. The sources are called *point sources* because in modelling they can be approximated as mathematical points to simplify analysis. This is in contrast to non-point pollution sources, the largest of which are agriculture, with huge areas treated by various chemicals, and big cities, with plenty of small, stationary and mobile sources, the individual impacts of which are impossible to distinguish.

We would like to specifically stress the fact that our focus on the data collected around point polluters does not mean that we overestimate the importance of local effects. There is no doubt that relatively minor regional increases in the deposition of pollutants has much larger ecological and economic consequences than the acute impacts of the largest individual emitters, such as Norilsk in Siberia or (earlier) Sudbury in Canada. However, local impacts on biota are well suited to reveal directions and magnitudes of biotic effects and to investigate dose–response relationships, as well as to explore general patterns and sources of variation (including relative sensitivities) in the responses of organisms, communities and ecosystems to pollution. In particular, histories of point polluter impacts are usually documented, effects are well pronounced, environmental gradients are sharp and short, and therefore the unavoidable influence of confounding factors can be accounted for more easily than on a regional scale (Ruotsalainen & Kozlov 2006; Kozlov & Zvereva 2007a; Zvereva et al. 2008).

On the other hand, the exploration of impact zones around point polluters allows us to bridge 'observational' and 'experimental' studies. From a scientific perspective, ecosystems transformed by pollution, for example, unusual landscapes of denuded barren land with lifeless lakes, can be considered opportunistic macrocosms (unique laboratories) for integrated research on the effects of harsh environmental conditions on ecosystem structure and function (Nriagu et al. 1998). This opinion is in line with the perception that ecological effects caused by industrial pollutants are the result of unintentional experiments with terrestrial ecosystems, and this result can be utilised by ecologists (Lee 1998). Environmental gradients created by point polluters share many characteristics with urban-rural gradients (simply because polluters are usually associated with large settlements), which were already considered 'an unexploited opportunity for ecology' long ago (McDonnell & Pickett 1990).

When studying the impact zone of a given polluter, we know where and when the 'experimental treatment' started. We can estimate the amount of pollutants that have been 'applied' to study sites. We can compare 'controls' and 'treatments' after long-term exposure and thus evaluate the effects. Of course, we cannot randomise 'treatments' among our study sites, but this is a reasonable cost of harvesting the data from long-term and incredibly expensive experiments established by industries many years ago. Neglect of these field-collected data in favour of simplified short-term experiments will obviously have detrimental consequences for understanding, predicting, and mitigating consequences of pollution impacts on biota. Instead, ecologists involved in stress-effect evaluation should combine the results obtained from microcosms and seminatural and natural ecosystems in order to properly extrapolate the results of laboratory-type studies to nature itself (Barrett et al. 1976; Filser et al. 2008).

It has sometimes been argued that since each impacted area has developed in its own way due to a unique history of events, experience with one system is rarely directly applicable to a prediction of outcome in another (Cairns & Niederlehner 1996; Matthews et al. 1996). However, the same argument can be applied, for example, to forest fires; each fire event is unique, but this is not seen as an obstacle for studying and predicting the ecological effects of fires. The difference is that fire ecologists never restrict themselves to the study of a single fire event, while pollution ecologists devote many years of their lives to exploration of the impact zone of a single polluter. Examples of this concentration are endless, while comparisons between at least two polluted areas have only rarely been performed. In a random sample of 1,000 publications reporting the effects of point polluters on terrestrial biota (a history of creation of the collection of primary studies was briefly described by Kozlov & Zvereva 2003; Zvereva et al. 2008; criteria for selecting publications are listed in Section 5.1.2), only 9% of publications reported the effects of two or more polluters; even among these studies, only rarely was the question asked whether some effects were shared among the studied impact zones. We are aware of only two publications (except for those produced by our team) that explicitly address comparative analysis. One of these papers compared the data from two independent projects addressing pollution issues (Grodziński & Yorks 1981), and another reported the results of similarly designed studies conducted in impact zones of four Russian polluters (Chernenkova 2002).

Understanding the mechanisms of pollution impacts on biota and predicting the long-term consequences of these impacts require building integrative phenomenological models, i.e., providing some numerical estimates that would allow researchers to link changes in biota with some characteristics of both the polluter and the impacted community (Zvereva et al. 2008). These models would allow prediction of the biotic effects of pollution at local, regional and (to a certain extent) global scales, when they are built on the basis of representative samples of contaminated territories. We appreciate that we are still far from reaching this goal, which was determined to be a major end result of any stress-response analysis decades ago (Barrett et al. 1976). This is due to both low integration between different levels of research of toxic effects (Filser 2008) and geographical biases; most pollution studies originated in Europe and North America, while almost no information is available from the tropics and from the Southern hemisphere (Innes & Oleksyn 2000; Innes & Haron 2000; Padgett & Kee 2004; Ruotsalainen & Kozlov 2006; Zvereva et al. 2008; Zvereva & Kozlov 2009). In this book we did our best to make our sample of impacted areas as representative as possible, given time and resource constrains.

Last but not least, a comparison of data on environmental contamination of areas adjacent to selected point polluters (Tables 2.26–2.43) with modelled global deposition of pollutants (Fowler et al. 1999) suggests that an *average* local effect detected by our study should be higher than any regional effect of pollution. Therefore, our conclusions on overall effects caused by point polluters can be seen as an upper estimate of possible regional effects of pollution on biota of northern and temperate regions of the Northern hemisphere.

1.6 Summary

Effects of pollution should be considered within the scope of basic ecology and need to be studied in the same manner as any natural ecological phenomena. They require proper documentation, explanation, generalization through building theoretical models, and experimental verification of these models followed by practical applications. This requires an understanding of the ecological principles governing ecosystem-level processes in both pristine and contaminated habitats. We argue that impact zones of point polluters can be seen as 'unintentional experiments' and thus serve as suitable models to explore biotic consequences of environmental contamination. By studying changes in structure and functions of ecosystems around point polluters, we aim to reveal general patterns in the responses of terrestrial biota to industrial pollution and to identify the sources of variation in these responses. We also intended to provide some numerical estimates that would allow linking of the observed effects with some characteristics of both the polluter and the impacted community. These models can be further used to predict the biotic effects of pollution on different spatial scales.

Chapter 2
Methodology of the Research and Description of Polluters

2.1 Selection of Polluters

In order to obtain an unbiased estimate of the overall effect of point polluters on terrestrial biota, we needed to explore a random sample from the entire population of polluters. Unfortunately, this task is hardly feasible, keeping in mind obvious financial and time constraints. Therefore we selected a representative sample of point polluters according to the following basic criteria:

1. The explored polluters should represent the entire range of pollution loads, from highest to reasonably low.
2. Basic types of polluters (identified in earlier meta-analyses; Ruotsalainen & Kozlov 2006; Zvereva et al. 2008) should be included.
3. Impact zones should represent the basic vegetation types of the northern and temperate regions of the Northern hemisphere.

When other selection criteria were met, the preference was given to impact zones that were best studied either by our team or by other researchers.

We explored environmental effects imposed by 18 point polluters (Fig. 2.1; Table 2.1) representing five of six classes that were used in earlier meta-analyses (Ruotsalainen & Kozlov 2006; Zvereva et al. 2008): non-ferrous industries, emitting heavy metals and SO_2; aluminium plants, emitting fluorine; iron and steel producing factories, emitting both SO_2 and alkaline dust; fertilising and chemical plants, emitting various nitrogen and sulphur containing compounds; and power stations, emitting mostly SO_2. Detailed information on these polluters is given below (Section 2.2). Table 2.1 shows classifications of these polluters, listing the variables that are used in categorical meta-analysis (Section 2.5).

M.V. Kozlov et al. *Impacts of Point Polluters on Terrestrial Biota*,
DOI 10.1007/978-90-481-2467-1_2, © Springer Science+Business Media B.V. 2009

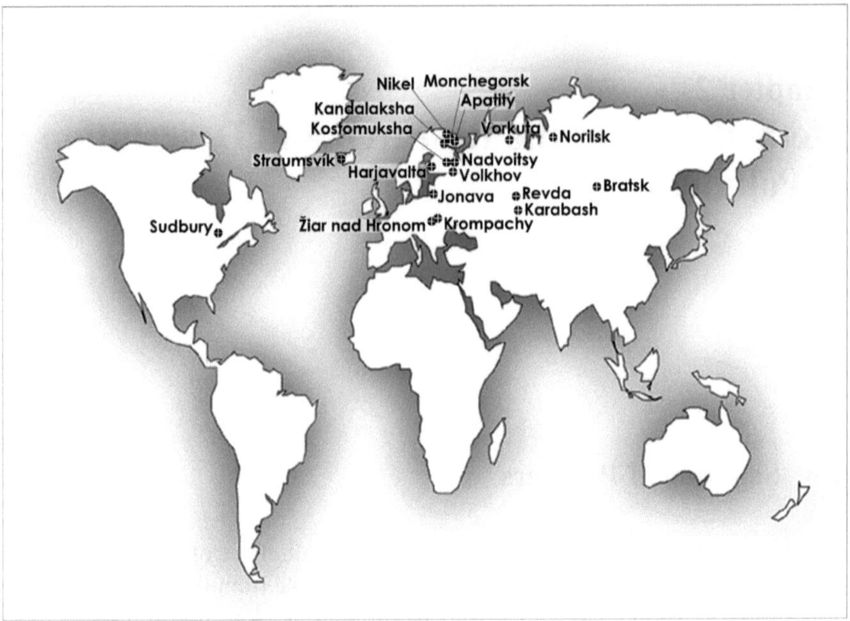

Fig. 2.1 Location of the investigated polluters

Table 2.1 Values of classificatory variables used in meta-analysis

	Classificatory variables		
Locality and polluter	Polluter type[a]	Effect on soil pH[b]	Geographic position[c]
Apatity, Russia: power plant	Pow	B	N
Bratsk, Russia: aluminium smelter	Al	B	S
Harjavalta, Finland: nickel-copper smelter	Nf	N	S
Jonava, Lithuania: fertiliser factory	Fert	N	S
Kandalaksha, Russia: aluminium smelter	Al	B	N
Karabash, Russia: copper smelter	Nf	A	S
Kostomuksha, Russia: iron pellet plant	Fe	B	S
Krompachy, Slovakia: copper smelter	Nf	A	S
Monchegorsk, Russia: nickel-copper smelter	Nf	A	N
Nadvoitsy, Russia: aluminium smelter	Al	B	N
Nikel, Russia: nickel-copper smelter	Nf	A	N
Norilsk, Russia: nickel-copper smelters	Nf	N	N
Revda, Russia: copper smelter	Nf	A	S
Straumsvík, Iceland: aluminium smelter	Al	N	N
Sudbury, Canada: nickel-copper smelter	Nf	N	S
Volkhov, Russia: aluminium smelter	Al	B	S
Vorkuta, Russia: power plant	Pow	B	N
Žiar nad Hronom, Slovakia: aluminium smelter	Al	B	S

[a] Classification follows Zvereva et al. (2008).
[b] A, acidifying; B, alkalysing; N, neutral; classification is primarily based on data reported in Section 3.3.
[c] Classification is based on mean July temperature: below 14°C – northern polluters (N); above 14°C – southern polluters (S). For a map showing location of polluters, consult Fig. 2.1; coordinates of polluters are given in Tables 2.25–2.42; climatic data are summarised in Table 2.2.

2.2 History of the Selected Polluters and Their Environmental Impacts

2.2.1 *Introduction*

In this chapter, we provide background information on both polluters selected for our study and their environmental impacts. We also briefly describe the surrounding landscapes (climatic data are summarised in Table 2.2), a history of landscape deterioration under the impacts of pollution, and a history of environmental research conducted around the selected polluters. The amount of information included in this chapter for each polluter depends on its availability to the international scientific community. We do not repeat details easily available from review papers (e.g., on Harjavalta, Monchegorsk, and Sudbury), but we do include more data for less well-known polluters, especially from sources published in national languages. Historical information is referenced only in cases of discrepancies between different data sources.

Emission data are of vital importance for understanding changes in the polluted environment, conducting comparative analyses and exploring dose-response relationships (Ruotsalainen & Kozlov 2006; Zvereva et al. 2008). These data are also necessary for re-assessment of regional and global emissions from anthropogenic sources (Nriagu & Pacyna 1988; Pacyna & Pacyna 2001). Therefore, every effort was made to obtain emission data from both published and unpublished sources for

Table 2.2 Temperature and precipitation at the study areas

Polluter	Mean temperatures, °C		Annual precipitation, mm
	January	July	
Apatity	−13.2	13.6	640
Bratsk	−20.5	17.4	357
Harjavalta	−6.8	16.4	545
Jonava	−5.2	16.8	630
Kandalaksha	−13.7	14.3	519
Karabash	−16.4	15.4	666
Kostomuksha	−12.8	15.0	580
Krompachy	−3.6	19.2	613
Monchegorsk	−13.8	14.1	561
Nadvoitsy	−11.7	14.1	566
Nikel/Zapolyarnyy	−10.9	11.7	499
Norilsk	−28.9	13.8	528
Revda	−15.7	17.4	546
Straumsvík	−11.7	14.1	569
Sudbury	−11.9	19.1	811
Volkhov	−10.8	16.7	652
Vorkuta	−24.8	14.3	437
Žiar nad Hronom	−3.3	18.7	644

Climatic data were estimated using *New_LocClim* (FAO 2006).

all the polluters involved in our analysis. Since emission data are very diverse due to a variety of methods for their estimation and reporting, we referenced each value in Tables 2.4–2.24. If values that differ from each other by more than 7% were reported for the same year, they were all included in our tables. If the values differ less than by 7%, then the value reported with higher accuracy was included in the table (i.e., 0.83 was preferred over 0.8).

Importantly, the data on both ambient concentrations of pollutants and, later on, emission of pollutants in industrial cities of the USSR (before 1990) and of the Russian Federation (since 1991) have been summarised in annual reports by the Voeikov Main Geophysical Observatory (St. Petersburg) from 1966 until 1995, and by the Research Institute of Ambient Air Protection (St. Petersburg) since 1996. These low-circulation reports (100–250 copies, published in Russian) were, until 1987, 'for the restricted use only', with numbered copies deposited into specially authorised departments of the main libraries, and public access to old reports was allowed only in 2007. However, even the accessible reports (for the years 1988–2005) remain unknown to the majority of environmental scientists, leading to large uncertainties in estimation of local and regional depositions of pollutants (Boyd et al. 2009). We were lucky to obtain access to most of these reports (Berlyand 1967, 1968, 1969, 1970, 1971, 1972, 1973, 1974, 1975, 1976, 1979, 1980, 1981, 1982, 1983, 1984, 1985, 1986, 1987, 1988, 1989, 1990, 1991, 1992, 1994; Milyaev & Nikolaev 1996; Milyaev et al. 1997a, b, 1998, 1999, 2000, 2001, 2002, 2003; Milyaev & Yasenskij 2004, 2005, 2006). For reasons of brevity, these reports are later referred to as 'Emissions of pollutants in Russia (1966–2005)'. Since these reports combine emissions of all polluters within a city, the data should be used with caution, especially for sulphur dioxide and nitrogen oxides that are emitted by many industries. Still, these values are good approximations for the giant polluters (such as Monchegorsk, Nikel, or Karabash, that produce over 95% of a town's emissions), as well as for pollutants associated with specific industrial processes (such as fluorine-containing substances emitted by aluminium industries). Amounts of both production and emission are reported in metric tons (i.e., 1,000 kg; abbreviated 't' hereafter).

The data on pollution loads and the extent of the contaminated territory are even less uniform than the emission data. Therefore, we have chosen to report them in their original form. We paid special attention to the information on the spatial extent of the contaminated territory, i.e., the area with statistically significant increases in pollution load relative to the regional background level. Since this area was often measured from published maps showing 'pollution zones', criteria for defining these zones vary across the selected polluters.

We provide several estimates reflecting (to a certain extent) the level of knowledge on both environmental contamination around the selected polluters and biotic effects attributable to pollution (Table 2.3). The number of publications known to us is counted from an extensive collection of reprints accumulated by M. Kozlov and E. Zvereva over 25 years of research on pollution ecology; it recently included approximately 5,000 publications reporting pollution effects on terrestrial biota. Potential publications were searched using the ISI Web of Science and several more databases. The publications in Russian were identified by surveying the abstract

Table 2.3 Amount and quality of available information on environmental impact of the selected polluters

	Number of publications		Contamination[a]		Biotic effects[a]		Number of effect sizes used in meta-analyses[b]			
Polluter	Total[c]	ISI[d]	A	Q	A	Q	1	2	3	4
Apatity, Russia: power plant	3	1	1	3	1	3	0	0	1	2
Bratsk, Russia: aluminium smelter	50	0	2	3	2	2	1	0	29	5
Harjavalta, Finland: nickel-copper smelter	200	57	3	3	3	3	3	5	46	13
Jonava, Lithuania: fertiliser factory	60	9	2	3	2	3	1	0	0	3
Kandalaksha, Russia: aluminium smelter	30	6	1	2	1	1	1	1	5	9
Karabash, Russia: copper smelter	80	8	2	3	2	3	0	5	5	2
Kostomuksha, Russia: iron pellet plant	100	7	3	3	2	2	4	5	1	13
Krompachy, Slovakia: copper smelter	20	8	2	2	1	2	0	0	0	0
Monchegorsk, Russia: nickel-copper smelter	1,000+	91	3	3	3	3	10	5	58	80
Nadvoitsy, Russia: aluminium smelter	0	0	0	–	0	–	0	0	0	0
Nikel, Russia: nickel-copper smelter (and ore-roasting plant in Zapolyarnyy)	100	40	2	3	2	3	6	3	6	24
Norilsk, Russia: nickel-copper smelters	80	17	2	3	1	1	4	0	0	0
Revda, Russia: copper smelter	300	12	3	3	3	3	4	5	17	21
Straumsvík, Iceland: aluminium smelter	10	0	1	3	1	1	0	0	0	0
Sudbury, Canada: nickel-copper smelter	100	20	3	3	2	3	1	5	3	0
Volkhov, Russia: aluminium smelter	5	1	1	2	1	3	0	0	10	0
Vorkuta, Russia: power plant	40	8	2	2	2	2	0	0	0	4
Žiar nad Hronom, Slovakia: aluminium smelter	80	11	2	2	2	2	0	5	18	3

Numbers of publications may overestimate the level of knowledge on environmental effects due to repeated publication of the same results by several Russian research groups. This especially concerns studies conducted near Kandalaksha, Karabash, Kostomuksha and Monchegorsk.

[a] Evaluation is based on publications available to the authors. Amount of published information (A) scored as: 0 – absent, 1 – insufficient; 2 – sufficient to outline the general picture; 3 – sufficient to represent the detailed picture. Quality of published information (Q), in terms of sampling design (number and distribution of study sites), applied methodology, statistical analysis and presentation form, scored as: 1 – unsatisfactory; 2 – moderate; 3 – high (an overall impression based on all available data sources).

[b] Publication coded as follows: 1 – abundance and diversity of soil fungi (Ruotsalainen & Kozlov 2006), 2 – diversity of vascular plants (Zvereva et al. 2008), 3 – abundance and diversity of arthropods (Zvereva & Kozlov 2009), 4 – growth and reproduction of vascular plants (Roitto et al. 2009).

[c] Only scientific publications were counted (grey literature included, but newspaper reports and web-based data in HTML format excluded); all values exceeding ten are approximate; for the efforts taken to obtain relevant publications, consult text (Section 2.2.1).

[d] ISI Web of Knowledge was searched for location of the polluter, its name, and the type of the industry; all searches were conducted between 25 December 2007 and 10 January 2008.

journal *Nature protection* (1990–2007; code 72 in the series of monthly abstract journals published by the All-Russian Institute of Scientific and Technical Information in Moscow). We also checked reference sections of all obtained data sources, and asked colleagues from many countries for assistance in locating grey literature and publications in national languages. The review of publications describing the effects of each of the polluters on terrestrial biota are by no means exhaustive; additional publications can be located using review papers published by our team (Ruotsalainen & Kozlov 2006; Kozlov & Zvereva 2007a; Zvereva et al. 2008; Zvereva & Kozlov 2009; Roitto et al. 2009).

Although a detailed investigation of the perception of environmental and health problems caused by industrial pollution is outside the scope of our study, we still found it interesting to give a brief overview of materials reflecting the degree of public awareness of local environmental problems. For reasons of simplicity, in this review we used a dual referencing system: all individual data sources, such as newspaper reports or individual web-based documents, are included in the list of references, However, overall impressions obtained by surveying some web sites, or small pieces of web-based information that cannot be attributed to an individual author, are referred to by short URLs inserted into the text. All websites mentioned in this chapter were visited between 20 December 2007 and 19 January 2008. An absence of information can be seen as an absence of public interest in local environmental problems.

2.2.2 Descriptions of the Polluters

2.2.2.1 Apatity, Russia: Power Plant (Color Plates 1–3 in Appendix II)

Location and Environment

Apatity (population 62,600; data from 2007) is a town in the Murmansk Region, located between Imandra Lake and the Khibiny Mountains 160 km south of Murmansk (Fig. 2.2). Originally it was surrounded by mixed forests of Scots pine (*Pinus sylvestris*) and mountain birch[1] (*Betula pubescens* subsp. *czerepanovii*), usually with dwarf shrub and moss cover; stands with ground lichens were relatively infrequent.

Pre-industrial History

The earliest settlements on Imandra Lake, 15 km east of the current location of Apatity, appeared 7–8,000 years ago. In 1574, a country churchyard, with six huts and 40 inhabitants subsisting on fishing and hunting, existed at about the same locality.

[1]Jonsell (2000) considered mountain birch, *Betula pubescens* subsp. *czerepanovii* (Orlova) Hämet–Ahti, an ecological form of white birch (*Betula pubescens*), not a distinct taxon.

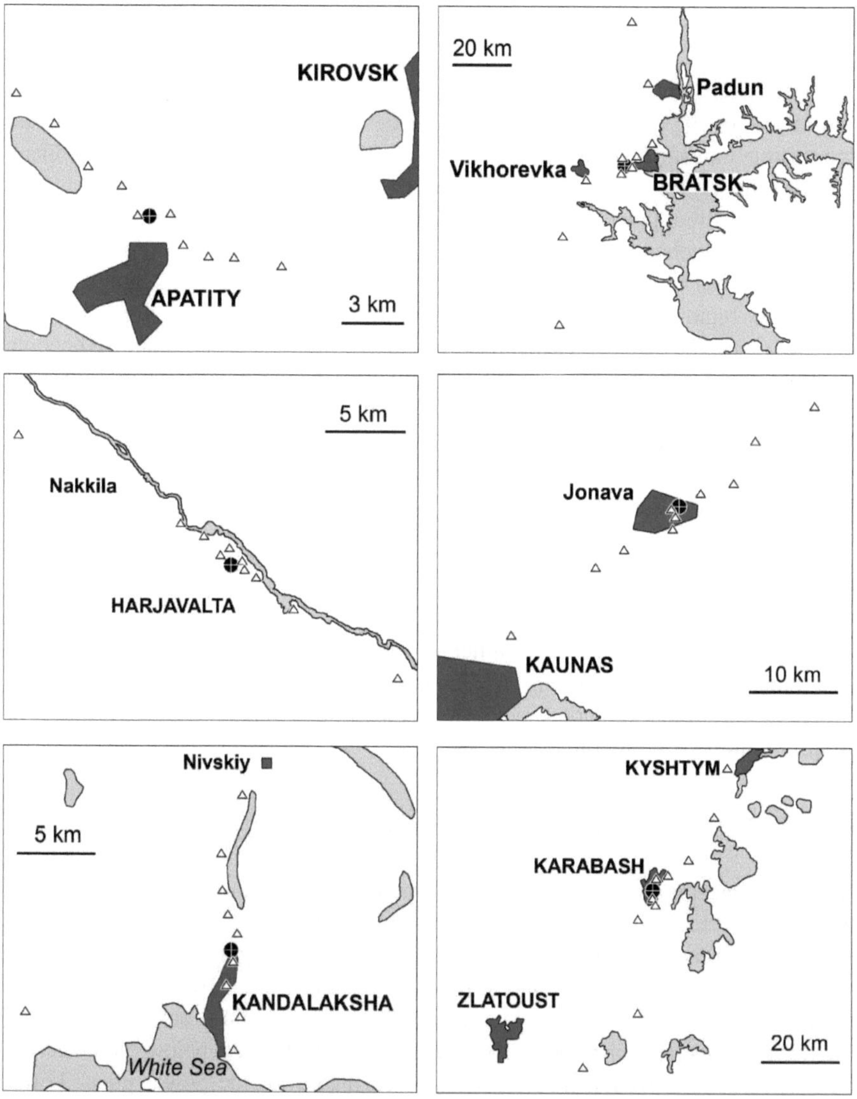

Figs. 2.2–2.7 Distribution of study sites (triangles) around investigated polluters (black circles with white cross inside). Names of larger settlements (usually towns) are capitalised. **2.2**, power plant in Apatity, Russia; **2.3**, aluminium smelter in Bratsk, Russia; **2.4**, nickel-copper smelter in Harjavalta, Finland; **2.5**, fertilising factory in Jonava, Lithuania; **2.6**, aluminium smelter in Kandalaksha, Russia; **2.7**, copper smelter in Karabash, Russia

The first permanent settlement at the current location of the city appeared in 1916 in connection with the building of a railway from St. Petersburg to Murmansk. Colonisation started in 1924 when the village of Byelorechenskij was established close to the place where the power plant was later built. Apatity was granted town status in 1966.

Industrial History

The intensive growth of Apatity (from seven inhabitants in 1926 to 4,000 in 1939 and 15,200 in 1959) was associated with the development of the apatite mining and processing industry. The main employer of Apatity is the joint stock company 'Apatit', the largest mining and concentrating enterprise in Europe.

The coal-fired power plant ('Kirovskaya GRES', now named 'Apatitskaya TEC') was launched in 1959 and reached its full capacity in 1961. For the first 2 decades, it mostly produced electricity (up to 37% of the total output of the Murmansk region), but since the late 1970s, it is used almost exclusively for heating the town. The power plant employed 983 people in 2003, but only 520 people in 2005; it achieves peak capacity (80 MWt) during the winter, when daily coal consumption reaches 2,800–3,000 t. In the summer, the daily coal consumption is around 650 t and the capacity is 10 MWt. The station mainly uses coal from Inta, Northern Ural (sulphur content 1.5–1.9%), but sometimes from Spitsbergen or Khakassia (sulphur content 0.7–1.0%).

Emissions Data

The power plant is the only local emitter of sulphur dioxide and some metals, such as iron, zinc, chromium, cadmium and lead (Golubeva 1991). The station contributed 99.0% of the sulphur dioxide and 75.4% of the mineral dust to overall emissions of Apatity (summarised in Table 2.4).

Pollution Loads and the Extent of the Contaminated Territory

Ambient concentrations of pollutants in Apatity have been monitored since 1972; the peak reported values were 1.4 mg/m^3 of dust in 1972 and 1.43 mg/m^3 of SO_2 in 1982 (Emissions of pollutants in Russia 1966–2006). During the period 1987–1989, annual depositions of sulphur and nitrogen in Apatity were 3,100 and 400 kg/km^2, respectively (Vasilenko et al. 1991). Concentrations of chromium, arsenic and lead in the leaves of the speckled alder (*Alnus incana*) in 1992 were below the detection limit (10 mg/kg), even within a few hundred meters of the power plant. Foliar concentrations of nickel, copper and zinc showed no relationship to the distance from the polluter. The peak concentrations of strontium were detected 4–7 km northwest of the power plant and were obviously associated with dust contamination from tailing ponds of the apatite and nepheline-processing factory (located 400–800 m from these sites). Foliar concentrations of iron appeared to be a good measure of pollution load; they exceeded the background level by a factor of 3 within 2 km of the power plant and approached regional background levels at about 5 km (Kozlov 2003). Consistently, soil contamination was detected only within approximately 2 km^2 east of the power plant (Kalabin 1999). On the other hand, from 1986 to 1991 a doubling of pollutant deposition (relative to the regional background) was recorded for the 800 km^2 around Apatity (Prokacheva et al. 1992).

Table 2.4 Aerial emissions (t) from industrial enterprises at Apatity, Russia

Year	Dust	SO_2	CO	NO_x
1972	14,600	51,800	–	–
1973	14,600	51,100	–	12,800
1974	14,600	51,100	–	12,800
1975	18,250	51,100	–	12,800
1978	19,500	25,500	–	2,100
1979	17,400	31,300	–	3,300
1980	17,700	31,200	10	3,200
1981	14,400	21,500	200	2,900
1982	13,900	31,800	100	2,500
1983	15,800	30,800	5,000	2,700
1984	19,000	36,900	4,900	2,800
1985	19,700	35,300	5,500	3,300
1986	15,200	31,000	8,300	4,500
1987	16,300	36,400	800	4,500
1988	14,400	30,700	900	4,100
1989	13,800	29,400	600	2,700
			4,900[a]	3,200[a]
1990	16,100	31,900	700	3,500
			9,200[a]	4,400[a]
1991	17,600	28,200	300	3,200
			8,800[a]	4,100[a]
1993	10,200	21,400	700	6,900
1994	7,100	14,600	300	5,600
1995	7,400	11,800	400	5,500
1996	7,300	14,700	200	5,100
1997	8,900	16,200	1,100	5,300
1998	8,600	17,900	700	4,500
1999	8,100	16,800	500	4,200
2000	7,300	16,200	300	4,000
2001	7,200	14,300	300	4,200
2002	5,800	12,000	200	3,900
2003	5,300	7,300	200	3,600
2004	6,000	10,300	100	3,900
2005	5,100	8,200	100	3,300

Non-referenced values were extracted from Emissions of Pollutants in Russia (1966–2006).
Other data sources:
[a] Kryuchkov (1993b).

Habitat Transformation due to Human Activity

Forests located 1–3 km from the power plant (please see color plate 3 in Appendix II) are dominated by mountain birch and speckled alder due to selective logging of conifers during the expansion of the settlement (between the 1930s and 1970s). Field layer vegetation is sparse, most likely due to both recreational activities (picking of berries and mushrooms) and dust contamination from the apatite-processing plant.

Brief History of Environmental Research

The data on the environmental impact of the power plant at Apatity are restricted to a study modelling the spatial distribution of sulphur dioxide (Baklanov et al. 1993), a report mentioning (but not quantifying) the contribution of the power plant to soil contamination by some metals (Golubeva 1991), and a paper reporting the results of long-term (1991–2001) monitoring of the population density of a tiny moth, *Phyllonorycter strigulatella* (Lienig et Zeller) (Lepidoptera, Gracillariidae), whose larvae develop in the leaves of the speckled alder (Kozlov 2003).

Perception of the Environmental Situation

The local population is most worried by the health effects of dust from tailing ponds of the apatite-processing factory, which affects the quality of ambient air in the city (especially on windy days) and contaminates drinking water (www.liga-rf.ru).

2.2.2.2 Bratsk, Russia: Aluminium Smelter (Color Plates 4–6 in Appendix II)

Location and Environment

Bratsk is a large (population 253,200; data from 2007) industrial city in the Irkutsk Region, located on the Angara River near the vast Bratsk Reservoir, 490 km North of Irkutsk (Fig. 2.3). Bratsk is surrounded by southern taiga forests (please see color plate 5 in Appendix II) consisting mostly of Scots pine (56%) and common birch (*Betula pendula*) (22%); other abundant trees are Siberian larch (*Larix sibirica*), Siberian spruce (*Picea abies* ssp. *obovata*), and Siberian fir (*Abies sibirica*).

Pre-industrial History

The area surrounding Bratsk has long been populated by Buryats who subsisted mostly from hunting. Bratsk was founded in 1631, when the stockaded fort was built at the confluence of the Oka and Angara rivers (now submerged by the Bratsk reservoir). In 1702, 128 homesteads were recorded in the Bratsk area. In 1805, the newly established Bratsk administrative unit had 5,210 inhabitants. In 1948, the regional centre, the old village of Bratsk, had 2,247 inhabitants. Town status was granted in 1955.

Industrial History

The railway reached Bratsk in 1947. The construction of the Bratsk dam and the hydroelectric station was completed in 1961. Construction of the aluminium smelter started in 1961. Production of aluminium was launched in 1966, but extension of the facilities continued until 1976; annual production reached 850,000 t in the late 1990s (US Geological Survey 1999). In 2006, the smelter employed approximately 5,000 people and produced 983,000 t of primary aluminium (30% of the aluminium production in Russia and 4% of the world's output), totalling 30,000,000 t of aluminium since its launch. In the early 1990s, the smelter implemented an ecologically oriented modernisation program, which will continue until 2012 (Ditrikh 2001). The city also houses large-scale timber processing industries and several coal-fired power plants.

Emissions Data

Although the emission of pollutants from all industries in Bratsk has been reported since 1970 (Table 2.5), emissions from the aluminium smelter were first estimated in 1980 (Ditrikh 2001). All emissions of fluorine-containing gases reported for the city of Bratsk (Emissions of pollutants in Russia 1966–2006) were attributed to the aluminium smelter; contributions from other sources to HF emissions is less than 0.1%, and to fluorine-containing dust emissions is approximately 6% (Lyubashevsky et al. 1996). Data on other pollutants are scarce. Bezuglaya et al. (1991, fig. 2.104) indicated that during the period 1984–1998, emissions of nitrogen oxides from the aluminium smelter were approximately 300 t, i.e., 5% of the overall emissions in Bratsk. Lyubashevsky et al. (1996), referring to local authorities, reported (for an unknown year, presumably 1992 or 1993) emissions of 25,358 t of dust, 3,968 t of SO_2, 2,448 t of HF, 2,808 t of fluorine-containing dust (it remains unclear, whether this value is included in an overall estimate of dust emission given above), 274 t of NO_x, and 107,199 t of CO. Importantly, these values (totalling 142,055 t) suggest substantially higher emissions than reported by Ditrikh (2001).

Pollution Loads and the Extent of the Contaminated Territory

Ambient concentrations of pollutants in Bratsk have been monitored since 1969; the peak reported values were 3.18 mg/m^3 of SO_2 in 1975, 1.39 mg/m^3 of NO_x in 1971, 7.0 mg/m^3 of dust in 1970, and 0.44 mg/m^3 of HF in 1978 (Emissions of pollutants in Russia 1966–2006).

The peak annual depositions of fluorine near the polluter were 7,200 kg/km^2 in the early 1980s and 5,000 kg/km^2 in the early 1990s (Morshina 1986; Davydova 2001), i.e., 300–1,000 times as high as the regional background. The increase in sulphur deposition was less extreme, with peak values exceeding the background by a factor of 10–30 (Davydova 2001). In 1987–1989, annual depositions of sulphur and nitrogen in Bratsk were 1,100 and 500 kg/km^2, respectively (Vasilenko et al. 1991).

The deposition of airborne pollutants in the 1980s exceeded the regional background within approximately 100 km of Bratsk (Morshina 1986). These data suggest that an area of 25,000–30,000 km^2 was contaminated by smelter emissions; however, the larger part of this territory (with 3–25 times excess fluorine deposition and 2–10 times excess sulphur deposition over the regional background) did not

Table 2.5 Aerial emissions (t) from industrial enterprises at Bratsk, Russia

Year	Total[a]	Dust	SO_2	CO	NO_x	F-containing dust	HF
1970	–	–	–	–	–	–	730
1971	–	5,500	8,400	–	–	–	2,200
1972	–	16,400	25,600	–	–	–	7,300
1973	–	51,100	40,150	–	9,100	3,500	2,800
1974	–	73,000	43,800	186,200	14,600	–	4,000
1975	–	73,000	43,800	175,200	14,600	–	4,000
1978	–	51,200	17,500	183,100	8,300	–	5,490
1979	–	49,900	21,600	188,000	9,100	–	5,100
1980	181,598[a]	45,500	18,200	165,500	7,400	–	6,100
1981	172,779[a]	57,800	17,500	152,700	5,000	–	2,600
1982	168,981[a]	60,200	22,400	147,100	5,400	4,000	3,200
1983	161,995[a]	57,600	23,900	148,200	5,600	3,455	2,862
1984	149,925[a]	51,600	23,800	136,800	5,400	3,355	2,960
1985	132,596[a]	51,000	22,300	118,200	5,100	3,061	2,448
1986	125,999[a]	44,400	21,100	114,800	4,600	2,970	2,240
1987	111,572[a]	41,900	22,400	101,700	4,700	2,652	2,020
1988	99,465[a]	41,300	20,700	85,400	5,700	–	1,960[b]
1989	55,171[a]	42,400	22,700	50,900	6,100	2,652	2,020
1990	52,947[a]	41,300	16,700	54,700	7,500	–	2,020
1991	73,386[a]	50,800	18,500	59,300	7,600	3,295	2,155
1992	92,998[a]	–	–	–	–	–	–
1993	78,297[a]	41,600	15,900	56,300	5,800	–	2,463
1994	73,843[a]	36,700	13,800	49,000	4,800	–	2,069
1995	64,487[a]	–	–	–	–	–	2,070
1996	58,609[a]	23,700	10,300	46,600	8,300	–	1,714
1997	46,642[a]	20,200	8,400	45,300	7,200	–	1,453
1998	43,556[a]	22,800	9,700	45,800	7,200	–	1,324
1999	42,680[a]	20,800	8,900	45,300	7,500	–	1,299
2000	82,426[a]	25,900	10,300	75,000	6,400	–	1,456
2001	–	21,300	9,000	62,500	5,600	–	1,280
2002	–	22,200	7,600	52,000	6,100	–	1,070
2004	–	16,300	5,900	49,300	7,300	–	1,037
2005	–	19,400	7,200	55,200	8,500	–	1,110

Non-referenced values were extracted from Emissions of Pollutants in Russia (1966–2006); emissions of SO_2 from the aluminium smelter comprise approximately 18% of the total emissions reported for Bratsk (proportion based on data of Lyubashevsky et al. 1996).
Other data sources:
[a] Ditrikh (2001), values refer to the aluminium smelter only.
[b] Bezuglaya et al. (1991).
Note added in proof: Detailed information on the emissions of Bratsk aluminium smelter during 1980–2001 is published by Pavlov (2006).

exhibit visible signs of environmental damage. During the period 1986–1991, a doubling of pollutant deposition (relative to the regional background) was recorded for 3,000 km^2 around Bratsk (Prokacheva et al. 1992). Detectable soil contamination by fluorine extended up to 15 km (Morshina 1986), with a peak value of 3,400 mg/kg (Sataeva 1991). On the basis of pollutant deposition and changes in vegetation structure, Davydova (2001) outlined zones of extreme (13.9 km^2), strong (82.5 km^2), moderate (200 km^2), and modest (316 km^2) impact.

Habitat Transformation due to Human Activity

Forests have been cut around Bratsk for centuries, while the impact of agriculture was negligible until the 1950s. Recently, pollution has been the leading factor in habitat transformation, although the contribution of recreational activities is also detectable (Alpatov et al. 2001).

The first signs of forest damage became apparent in 1968, when 1.36 km^2 of Scots pine stands died on the slopes of Mogrudon hill, situated near the smelter. Forest dieback increased to 15–20 km^2 by 1976, 50–75 km^2 by 1985, and 100–150 km^2 by the early 2000s. The extent of damaged forests reached 800 km^2 by 1985 (Ugrjumov et al. 1996) and 1,000–1,500 km^2 by the late 1990s (Runova 1999); another data source reported damage to 1,400 km^2 of forests by the early 1980s (Mikhailova 2003) and 2,000 km^2 by the mid-1990s (Mikhailova 1997). Davydova (2001) estimated that landscapes within approximately 70 km were affected by the emissions.

The aluminium smelter shares the responsibility for adverse environmental effects with power plants and timber-processing industries, emitting 82% of the sulphur dioxide released in Bratsk. Although fluorine contamination may cause more pronounced environmental effects than sulphur dioxide, it is estimated that only 41.5% of forest damage within 50 km of Bratsk is due to emissions from the aluminium plant (Lyubashevsky et al. 1996).

In spite of the extreme pollution loads, we have not found industrial barrens (bleak open landscapes evolved due to deposition of airborne pollutants, with only small patches of vegetation surrounded by bare land) around the Bratsk smelter (Kozlov & Zvereva 2007a). However, trees are nearly absent within the zone of extreme impact, which is mostly covered by herbaceous vegetation and low-stature alder and willow bushes, exhibiting clear signs of pollution-induced damage, i.e., dwarfed growth forms and chlorosis (please see color plate 6 in Appendix II). The surrounding zone of strong impact is mostly covered by white birch (*Betula pubescens*) and European aspen (*Populus tremula*); surviving Scots pine and Siberian spruce often have dead upper canopies and stunted growth forms (Davydova 2001).

Brief History of Environmental Research

All publications on the environmental impact of the Bratsk aluminium smelter, including five monographs (Rozhkov & Mikhailova 1989; Lyubashevsky et al. 1996; Ugrjumov et al. 1996; Runova 1999; Pavlov 2006), are in Russian. The book by Rozhkov and Mikhailova (1992) was translated to English. The vast majority of publications do not report quantitative data in a form suitable for meta-analysis (Table 2.3). Especially disappointing is a book by Lyubashevsky et al. (1996), which lists a large number of biotic effects, such as smaller size and germinability of the seeds of white and common birches, a decline in species richness of lichens, changed growth of several herbaceous species, and changes in abundances of different groups of insects. However, this book did not provide either supporting numerical information or the results of statistical tests. Majority of these raw data have never been published (N. Lyubashevsky 2006 and M. Trubina, personal communication, 2006) and are, in fact, lost to science.

The very first studies of the environmental impacts of the aluminium smelter on forest ecosystems were conducted in the early 1970s by researchers from the Siberian Institute of Plant Physiology and Biochemistry, Irkutsk. In particular, they documented growth of conifers (Sokov & Rozhkov 1975) and listed xylophagous insects in damaged forests (Anisimova 1989). Since 1977, integrated research of the forests affected by emissions from both the aluminium smelter and the timber-processing industry has been conducted (for nearly 20 years) by scientists from the Forest Technical Academy in St. Petersburg. They have documented changes in forest vitality, including decreases in the vertical growth and radial increment of conifers, suppression of regeneration of conifers with simultaneous increases in the density of birch and aspen seedlings, and quantified damage to woody plants by insect herbivores (Kataev et al. 1981; Golutvin et al. 1983; Popovichev & Golutvin 1983; Selikhovkin 1992, 1995). Decreases in the increment of Scots pine near Bratsk (detected by dendrochronological analysis) started in the 1960s and were therefore associated with the beginning of the pollution impact (Lairand et al. 1979). Since the mid-1990s, a research team from Bratsk University has monitored forest damage and developed silvicultural measures for impacted forests (Ugrjumov et al. 1996; Alpatov et al. 2001).

Perception of the Environmental Situation

Bratsk was declared an ecological disaster zone in 1993, and the government of the Russian Federation developed plans to improve the situation from 1994–2000. In spite of a substantial decrease in emissions (Table 2.5), the Chekanovskiy settlement (located 1.5 km northeast of the smelter) was evacuated in 2001 due to repeated health emergencies. Bratsk is listed by the 2004 report of the Central Geophysical Laboratory in St. Petersburg among the ten most polluted Russian cities (english.pravda.ru), and by the report of the Blacksmith Institute (2007) among the 30 most polluted places in the world in the category 'petrochemicals'. The most acute environmental problems have recently been identified as contamination of drinking water by mercury, and not with air pollution (www.pollutedplaces.org).

2.2.2.3 Harjavalta, Finland: Nickel-Copper Smelter (Color Plates 7–11 in Appendix II)

Location and Environment

Harjavalta is a small (population 7,700; data from 2007) town in Western Finland, 170 km northwest of Helsinki (Fig. 2.4). Harjavalta is situated on an esker that runs to the southeast of the smelter. The region is part of the southern boreal coniferous zone, with the forest consisting mainly of Scots pine (please see color plate 8 in Appendix II). According to the Finnish classification, the forest along the esker varies from the *Vaccinium* to the *Calluna* type.

Pre-industrial History

The Harjavalta region has been populated since the Iron Age, with the first signs of human settlement dating back to about 1500 BC. Written records mentioning the village of Harjavalta begin in the AD 1400s. Harjavalta was established as a municipality in 1869 and an independent parish in 1878, when it included 1,600 inhabitants. Telephones and railroads reached Harjavalta in the late nineteenth century, but Harjavalta remained a typical countryside village until 1944. By 1950, Harjavalta had a population of 6,000. Town status was granted in 1977.

Industrial History

The Outokumpu copper plant was moved to Harjavalta from Imatra in 1944 (at the end of World War II) because of the proximity of Imatra to the Russian border; it started operations in 1945. The Kemira sulphuric acid plant was launched in 1947 and the nickel smelter in 1960; recently, this industrial area has hosted 13 different firms (Heino & Koskenkari 2004). Originally part of Outokumpu, a Finnish company, the copper business is now owned by Boliden (since 2003) and the nickel business by Norilsk Nickel (since 2007). The copper smelter has a nominal annual capacity of 160,000 t. Sulphur is recovered and sold as sulphuric products. The nickel smelter, operating on a tolling basis, produces nickel matte.

Emissions Data

Accurate emissions data from the Harjavalta smelters (Table 2.6) have been available through Outokumpu reports and the web-site, as well as from the documents of the Finnish Ministry of Environment, since the mid-1980s; they were partially published in many research reports, including those by Nieminen et al. (2002) and Kiikkilä (2003). Estimates of dust emissions for 5-year periods starting with the beginning of smelting have been published by Nieminen et al. (2002) and are included in Table 2.6; for a detailed history of emissions and for annual production-based estimates of dust emissions (not included in Table 2.6), consult Nieminen (2005). Data on annual emissions of sulphur dioxide in the early 1970s are contradictory: 20,000 t was reported by Levula (1993), while Laaksovirta and Silvola (1975) indicated 4,800 t.

Data from 1990 demonstrated that smelters contributed 98.7% of the sulphur dioxide, 90% of the nitrogen oxides (with annual emissions of 180 t), 92.7% of the dust

Table 2.6 Annual emissions (t) from Harjavalta smelters, Finland

Year	Dust	SO_2	Cu	Ni	Pb	Zn	Cd	As
1945	–	30,000[a]	–	–	–	–	–	–
1947	–	35,000[b]	–	–	–	–	–	–
1945–1949	478[c]	–	–	–	–	–	–	–
1950	–	10,000[a]	–	–	–	–	–	–
1950–1954	534[c]	–	–	–	–	–	–	–
1955–1959	784[c]	–	–	–	–	–	–	–
1960–1964	1,195[c]	–	–	–	–	–	–	–
1965–1969	632[c]	–	–	–	–	–	–	–
1970	–	20,000[a]	–	–	–	–	–	–
1970–1974	548[c]	–	–	–	–	–	–	–
1975–1979	778[c]	–	–	–	–	–	–	–
1980–1984	903[c]	–	–	–	–	–	–	–
1984	1,100	11,500	300	70	50	200	–	–
1985	1,100	7,800	98	47	55	216	1.7[d]	15[d]
1986	1,200	7,530	126	46	60	232	7.1[d]	17[d]
1987	1,800	6,860	140	90	94	162	3.9[d]	19[d]
1988	1,000	8,500	104	45	48	103	3.2[d]	19[d]
1989	1,000	9,500	80	33	70	190	3.6[d]	22[d]
1990	960	8,800	80	31	80	160	4.2[d]	28[d]
1991	640	5,200	80	15	45	90	1.6[e]	–
1992	280	4,800	60	10	9.0	12	1.0[e]	–
1993	250	4,700	50	7.0	6.0	11	0.9[e]	10[e]
1994	190	5,000	40	6.0	3.0	6.0	0.6[e]	5.0[e]
1995	70	3,270	17	1.4	0.5	1.7	0.0[e]	0.2[e]
1996	195	3,243	49	1.2	1.9	5.3	0.2[e]	4.2[e]
1997	355	2,980	69	2.9	3.9	14	0.3[e]	10[e]
1998	132	3,041	23	1.7	2.4	6.1	0.4	10
1999	48	3,397	5.9	0.8	1.0	4.2	0.3	1.8
2000	36	3,002	6.6	1.2	0.2	1.1	0.1	0.8
2001	50	3,387	7.4	0.8	0.7	3.0	0.4	1.6
2002	55	3,300	11.6	0.6	0.4	1.5	0.1	0.5
2003	–	3,000	6.0[f]	0.6[f]	0.3[f]	0.9[f]	–	0.4[f]
2004	–	2,900	–	–	–	–	–	–
2005	–	2,850	–	–	–	–	–	–

Non-referenced values were obtained from Outokumpu Harjavalta Metals Ltd. Other data sources:
[a] Levula (1993).
[b] Poutanen and Kuisma (1998).
[c] Nieminen et al. (2002).
[d] Harjavallan Kaupunki (1991).
[e] Helmisaari (2000).
[f] Nieminen (2005).

and 99.5% of the heavy metals emissions at Harjavalta (Harjavallan Kaupunki 1991). However, before the closure of the fertiliser factory in 1989, the emissions share of the smelter for nitrogen oxides was much lower; 125 t emitted in 1978 constituted only 17.2% of the total emissions at Harjavalta (Helmisaari 2000). Emissions of NO_2 in 2002 were 250 t (Outokumpu Harjavalta Metals Ltd.). In 2002, Outokumpu claimed the smelter to be 'among the cleanest in the world'.

Pollution Loads and the Extent of the Contaminated Territory

Long-term pollution impacts resulted in soil contamination by metals, with peak recorded values of 49,000 mg/kg of copper, 913 mg/kg of nickel, 58 mg/kg of arsenic, 620 mg/kg of zinc, and 204 mg/kg of lead (Helmisaari et al. 1995; Derome & Nieminen 1998; Uhlig et al. 2001; Salemaa et al. 2001; Nieminen et al. 2002; Nieminen 2004). However, due to a substantial decline in emissions, both deposition of pollutants and ambient concentrations of sulphur dioxide have been very low recently. For example, average ambient concentrations of sulphur dioxide near the smelter in 2001–2002 were 0.004–0.006 mg/m^3 (Outokumpu Harjavalta Metals Ltd.), i.e., well below sanitary limits.

The polluted area, defined by the accumulation of copper in moss bags exceeding the background level by a factor of 5 or more (Hynninen 1986), in 1981–1982 was 70 km^2. In 1995, the area contaminated by copper (over 8 mg/kg in green mosses) was 107 km^2 (Kubin et al. 2000).

Habitat Transformation due to Human Activity

The heathland Scots pine forests in the immediate vicinity of the smelter died during the first years of industrial activity, but they were left uncut until the mid-1950s (Poutanen & Kuisma 1998). The very first evaluation, conducted at the beginning of the 1970s, classified 8.8 km^2 as a lichen desert, i.e., the area where epiphytic lichens disappeared due to the pollution impact. Visible damage to Scots pine was detected within approximately the same area, and another 52 km^2 area was classified as a transitional zone (Laaksovirta & Silvola 1975).

In the mid- to late 1950s, the dying forest around the smelter was restored by planting Scots pine seedlings, which had a high rate of survival but stunted growth. They have formed a kind of forest with unusually open canopies and dead or nearly dead ground layer vegetation (please see color plates 9 and 11 in Appendix II). The overall extent of barren and semi-barren habitats around Harjavalta is now approximately 0.5 km^2 (Kozlov & Zvereva 2007a).

Brief History of Environmental Research

The first scientific report known to us about the environmental effects of the Harjavalta smelters is the paper by Laaksovirta and Silvola (1975). Many of the studies conducted near Harjavalta were based on a poorly replicated sampling design, with four study sites, each representing a different pollution load. The studies cover soil microbiota, invertebrates (mostly epigeic fauna and insect herbivores), the structure of plant communities, the growth and vitality of Scots pine, and the performance of insectivorous birds (for a review and references, consult Kiikkilä 2003). Recent declines in emissions have already resulted in ecosystem recovery, which was detected by monitoring bird communities (Eeva & Lehikoinen 2000).

Perception of the Environmental Situation

The Harjavalta smelter was removed from a 'hot spot' list of top Baltic polluters in 2003. A census by the Association of Finnish Local and Regional Authorities in 2004 clearly showed that Harjavalta's image has changed greatly due to a significant decrease in emissions since 1990; it is no longer perceived as a polluted little town (Rintakoski 2004). In spite of several accidents reported in newspapers (www.turunsanomat.fi), local inhabitants pay little attention to environmental contamination.

2.2.2.4 Jonava, Lithuania: Fertiliser Factory (Color Plates 12–14 in Appendix II)

Location and Environment

Jonava (population 34,500; data from 2007) is located in the central part of Lithuania, 30 km northeast of Kaunas (Fig. 2.5). Before the onset of pollution, Scots pine (50%), Norway spruce (*Picea abies* ssp. *abies*) (23%), common birch and black alder (*Alnus glutinosa*) prevailed in the local forests (Armolaitis 1998).

Pre-industrial History

Jonava (founded in 1740) was an estate until the mid-eighteenth century, owned by the Kosakowski family. In 1750, Jonava received town and market rights, but it did not develop into a city. In 1923, proper city rights were granted, but no substantial growth had occurred prior to World War II.

Industrial History

Construction of the nitrogen fertiliser factory 'Achema' (formerly 'Azotas') started near Jonava, at the confluence of the rivers Neris and Sventoji, in 1962. Production of synthetic ammonia was launched in 1965; development of the factory and the launching new departments continued until 1973. The factory became the 'Achema' stock company following privatisation in 1994. Recently, the plant has become the biggest producer of nitrogen fertilisers in Lithuania, Latvia and Estonia. It employs 1,600 workers and produces ammonia, ammonium nitrate, urea ammonium nitrate solution, and many other chemicals.

Emissions Data

In the 1980s, the factory was one of the biggest air pollution sources in Lithuania (Table 2.7), emitting the larger part of all pollutants reported for Jonava. Along with the pollutants included in Table 2.7, emissions of chlorine have been occasionally

Table 2.7 Annual emissions (t) from all industrial enterprises at Jonava, Lithuania (1974–1989) and from the Achema plant (1980–2005)

Year	Dust	SO_2	CO	NO_x	NH_4
1974	7,300	11,000	25,600	–	–
1975	7,300	11,000	25,600	–	–
1978	4,800	7,500	19,600	2,200	2,420
1979	14,600	4,200	20,500	5,600	4,100
1980	16,400	5,400	28,100	6,400	5,100
	12,995[a]	3,901[a]	8,548[a]	4,148[a]	3,622[a]
1981	14,100	5,200	10,000	3,900	3,734[a]
	13,860[a]	4,630[a]	9,874[a]	3,862[a]	
1982	12,400	4,300	10,100	3,900	3,700
	11,476[a]	4,079[a]	10,541[a]	3,896[a]	3,633[a]
1983	7,300	4,400	10,400	4,500	3,720
	8,098[a]	3,699[a]	9,711[a]	3,595[a]	3,034[a]
1984	7,000	1,700	9,800	3,600	3,840
	6,328[a]	3,312[a]	9,928[a]	2,777[a]	2,591[a]
1985	7,400	1,600	9,800	3,000	3,870
	4,477[a]	2,635[a]	9,456[a]	2,886[a]	2,426[a]
1986	6,700	1,700	9,100	2,600	4,010
	3,238[a]	2,975[a]	10,291[a]	2,730[a]	2,559[a]
1987	5,500	3,000	9,400	2,500	2,495[a]
	1,705[a]	2,416[a]	9,682[a]	2,358[a]	
1988	4,300	2,300	8,700	2,400	1,551
	629[a]	1,991[a]	8,356[a]	2,206[a]	
1989	1,300	1,600	8,300	2,000	3,658
	271[a]	1,295[a]	6,608[a]	1,824[a]	
1990	305[a]	716[a]	6,143[a]	908[a]	2,249[a]
1991	351[a]	450[a]	7,318[a]	1,133[a]	2,249[a]
1992	246[a]	438[a]	6,904[a]	690[a]	2,055[a]
1993	212[a]	630[a]	3,924[a]	243[a]	678[a]
1994	274[a]	379[a]	3,013[a]	359[a]	720[a]
1995	292[a]	370[a]	3,475[a]	324[a]	1,287[a]
1996	285[a]	24[a]	3,450[a]	122[a]	645[a]
1997	265[a]	541[a]	3,510[a]	389[a]	382[a]
1998	203[a]	68[a]	5,704[a]	390[a]	257[a]
1999	301[a]	83[a]	5,362[a]	381[a]	197[a]
2000	323[a]	8[a]	5,770[a]	416[a]	286[a]
2001	–	50[b]	4,282[b]	430[b]	420[b]
2002	–	34[b]	4,192[b]	428[b]	446[b]
2003	–	24[b]	3,527[b]	478[b]	423[b]
2004	–	7[b]	3,639[b]	408[b]	416[b]
2005	–	0[b]	2,598[b]	514[b]	426[b]

Non-referenced values extracted from Emissions of Pollutants in Russia (1966–2006).
Other data sources:
[a] Juknys et al. (2003, Table 2.1; reprinted, in a modified form, with permission from Elsevier).
[b] Šujetovinė and Stakėnas (2007, and personal communication 2009).

reported (2,030 t in 1981, 1,561 t in 1984, 1,166 in 1987) (Emissions of pollutants in Russia 1966–2006). Small quantities of metals (Zn, Cu, Mn, Cr, Ni, Cd) detected in snow samples (Slavenene et al. 1987) presumably resulted from burning organic fuel.

Pollution Loads and the Extent of the Contaminated Territory

Ambient concentrations of pollutants in Jonava have been monitored since 1974; the peak reported values were 4.2 mg/m^3 of dust in 1975, 4.63 mg/m^3 of SO_2 in 1979, 0.34 mg/m^3 of NO_x in 1975, 0.70 mg/m^3 of HF in 1979, and 5.00 mg/m^3 of NH_4 in 1974–1975 (Emissions of pollutants in Russia 1966–2006).

In 1987–1989, annual depositions of sulphur and nitrogen in Jonava were 1,900 and 1,100 kg/km^2, respectively (Vasilenko et al. 1991). However, according to another data source (Armolaitis 1998), total (wet and dry) annual deposition of sulphur at a distance of 1–2 km from the factory in the mid-1980s comprised 5,000 kg/km^2 but was reduced to 1,500 kg/km^2 by the mid-1990s. To our knowledge, pollution loads have never been mapped; at the time of peak emissions (in the mid-1980s), the contaminated zone most likely ranged from 500 to 1,000 km^2.

Habitat Transformation due to Human Activity

Local forest damage was recognised in 1972 and has become extremely acute since 1979, when crown defoliation in conifers was recorded up to 10–12 km in the direction of prevailing winds, and complete dieback occurred within 2–3 km from the polluter (Juknys et al. 2003). By 1983, 20 km^2 of forest had been damaged by pollution (Dauskevicius 1984). Despite essential reduction of emissions (Table 2.7), at the end of the 1980s the forest damage expanded to 20–25 km in the direction of prevailing winds. Some signs of recovery appeared only in the 1990s (Juknys et al. 2003).

Brief History of Environmental Research

The studies of biotic effects caused by emissions of Achema are generally forestry-oriented, with detailed descriptions of the growth and vitality of Scots pine (Augustaitis 1989; Armolaitis 1998; Juknys et al. 2003) and fragmentary data on soil microbiota (Lebedeva & Lugauskas 1985), insect pests (Mastauskis 1987), and birds (Knistautas 1982, 1983).

Perception of the Environmental Situation

According to the web-based reports of the mass media, Achema was the first factory in Lithuania to adopt the EU requirement for an environmental management system, and was recently assessed as one of the top ten least polluting big chemical plants of the EU (E. Budrys, personal communication 2008).

2.2.2.5 Kandalaksha, Russia: Aluminium Smelter (Color Plates 15–17 in Appendix II)

Location and Environment

Kandalaksha is a small (population 38,100; data from 2007) town in the Murmansk Region, located on the Kola Peninsula on the coast of the Kandalaksha Gulf on the White Sea, 210 km south of Murmansk (Fig. 2.6). Originally, it was surrounded by mixed forests dominated either by Scots pine or, less frequently, by Norway spruce. Ground vegetation typically consisted of dwarf shrubs and green mosses (please see color plate 16 in Appendix II).

Pre-industrial History

Fishing has been the primary business of the local population for millennia. Although Kandalaksha has existed since eleventh century, only 806 inhabitants were recorded there in 1914. Town status was granted in 1938.

Industrial History

The industrial development of Kandalaksha started in 1915 with the building of the railway connecting St. Petersburg and Murmansk. The next important events were construction of the Niva hydropower station (completed in 1934) and the mechanical factory (launched in 1936). Building of the aluminium plant started in 1939 but was interrupted by World War II, and the smelter was only commissioned in 1951. Its main products are primary aluminium ingots, aluminium wire rod and billets. Annual production of aluminium in the late 1990s was 63,000 t (US Geological Survey 1999). During the period 2001–2007, the smelter employed 1,400 people and produced 70–75,000 t of aluminium annually (Boltramovich et al. 2003; www.rusal.ru).

Emissions Data

The aluminium smelter contributed 70–80% of all aerial emissions reported for the town of Kandalaksha in 1989–1991; in 1993–2000, its share declined to 53–64% (Emissions of pollutants in Russia 1966–2006), in particular due to 'severe pressure from nature protection agencies, up to demand to reduce production of aluminium by 12%' (Kruglyashov 2001). All emissions of both aluminium oxide (around 5,000 t in the mid-1990s; Evdokimova et al. 2007) and fluorine-containing substances (Table 2.8) originate from the smelter. In 2005, the new gas-cleaning system was put into operation, and a further decrease in emissions is expected in the coming years.

Table 2.8 Aerial emissions (t) from industrial enterprises at Kandalaksha, Russia

Year	Dust	SO_2	CO	NO_x	F-containing dust	HF
1970	–	–	–	–	–	>1,300
1971	–	–	3,100	–	–	1,300
1973	7,300	5,500	1,800	–	–	1,100
1974	7,300	5,500	1,800	–	–	1,100
1975	7,300	5,500	–	–	–	1,100
1978	20,100	5,500	8,300	–	–	1,425
1979	19,400	5,500	8,300	800	–	550
1980	20,100	4,200	7,600	300	–	790
1981	18,000	4,100	7,200	1,500	–	800
1982	23,200	3,700	7,300	300	–	800
1983	23,300	5,200	8,000	500	836	795
1984	23,700	6,600	8,900	200	833	790
1985	9,100	5,500	7,800	300	830	790
1986	8,600	5,200	7,000	200	824	800
1987	8,400	5,000	6,900	900	821	800
1988	8,100	4,700	6,800	200	816	784
1989	8,500	5,200	7,100	300	814[a]	780
			15,500[a]	600[a]		
1990	7,600	4,600	6,600	200	806	660
			8,800[a]	300[a]		775[a]
1991	5,244[b]	3,800	7,500	200	936	659[a]
	9,500		19,400[a]	1,400[a]		
1992	5,193[b]	–	–	–	–	–
1993	5,198[b]	5,800	9,300	600	848	705
	10.800					
1994	5,158[b]	8,700	11,300	800	–	694[b]
	9,400					848
1995	5,126[b]	4,800	7,800	340	–	693[b]
	8,900					848
1996	5,031[b]	5,300	8,100	300	–	688[b]
	9,000					839
1997	4,992[b]	5,100	8,500	300	–	688[b]
	9,700					839
1998	4,946[b]	5,700	8,700	300	–	687
	9,700					
1999	4,950[b]	5,800	8,200	300	–	686
	9,300					
2000	8,400	6,100	7,100	600	–	686
2001	8,400	6,200	7,300	600	–	672
2002	8,200	5,400	7,200	600	–	670
2003	5,400	5,400	7,100	600	–	383
2004	5,000	5,800	7,400	600	–	349
2005	3,800	6,000	6,700	500	–	274

Non-referenced values extracted from Emissions of Pollutants in Russia (1966–2006); emission of CO in 1983 corrected from the reported 800 to 8,000 t – the typing error presumed.

Other data sources:

[a] Kryuchkov (1993b).

[b] Kruglyashov (2001), data refer to the aluminium smelter only.

Pollution Loads and the Extent of the Contaminated Territory

Ambient concentrations of pollutants in Kandalaksha have been monitored since 1967; the peak reported values were 1.60 mg/m^3 of dust in 1967, 2.21 mg/m^3 of SO_2 in 1982, 0.43 mg/m^3 of NO_x in 1971, and 1.30 mg/m^3 of HF in 1970. Importantly, high concentrations of HF (0.30–0.50 mg/m^3) were recorded up to 2 km from the smelter in 1971–1973 (Emissions of pollutants in Russia 1966–2006).

A local increase in sulphate deposition from 1979–1983 was detected within approximately 300 km^2 (Kryuchkov & Makarova 1989). However, by the late 1980s–early 1990s, the sulphur-contaminated territories around major industrial centres of the Murmansk region had merged, and no local increase in sulphur deposition near Kandalaksha was detected in the 1990s (Kalabin 1999). In 1987–1989, annual depositions of sulphur and nitrogen in Kandalaksha were 1,000 and 300 kg/km^2, respectively (Vasilenko et al. 1991). The impact of the aluminium smelter on the local geochemistry in the mid-1990s was characterised as 'surprisingly small' (Reimann et al. 1998); however, emissions caused a substantial increase in the pH of both snow precipitation and forest litter (Evdokimova et al. 2005). The peak concentrations of fluorine in soils exceeded the regional background by a factor of 25–30 (Evdokimova et al. 2005).

During the period 1986–1991, a doubling of pollutant deposition (relative to the regional background) was recorded for 250 km^2 around Kandalaksha (Prokacheva et al. 1992). The content of fluorine in forest litter suggests that approximately 500 km^2 have been recently contaminated, with the severely contaminated zone (fluorine content increased by a factor of 6–25 relative to the regional background) limited to 2.5 km from the smelter (Evdokimova et al. 1997, 2007).

Habitat Transformation due to Human Activity

Severe damage to conifers and depression of field layer vegetation were apparent in the early 1980s, when Syroid (1987) classified the habitats next to the smelter as a zone of 'complete destruction of ecosystems', with 70–90% of soils void of vegetation. However, felling had been quite intensive around Kandalaksha until at least the 1970s; thus, initial deforestation near the smelter was most likely the result of clearcutting. Moreover, the presence of young Scots pines near the smelter (Syroid 1987) suggests that forest recovery was not prevented even at periods with the highest pollution pressure.

Kryuchkov and Makarova (1989, Fig. 18) claimed the presence of 25 km^2 of industrial barrens around the aluminium smelter at Kandalaksha; however, this information seems exaggerated. In particular, it contradicts the data by Georgievskij

(1990), who reported a decline of epiphytic lichens within 5 km^2 only. We repeatedly visited Kandalaksha in the early 1980s and observed Scots pine stands growing 1–3 km from the smelter (please see color plate 17 in Appendix II). We suggest that the acute environmental effects were local and mostly resulted from deposition of fluorine-containing dust in combination with other kinds of human-induced disturbances.

Visible forest damage in the early 1980s was observed within 450 km^2 (Syroid 1987); detectable lichen damage was reported within 1,300 km^2 (Georgievskij 1990).

Brief History of Environmental Research

The first reports of the environmental situation around Kandalaksha (Syroid 1987; Georgievskij 1990) were narrative and contained neither numerical estimates of the effects nor statistical analysis. Quantitative information is available only for soil micromycetes (Evdokimova et al. 1997, 2007) and several groups of soil invertebrates (Zainagutdinova 2003). An integrated environmental study (Evdokimova et al. 2005) produced mostly inconclusive results due to methodological flaws, including a poorly replicated sampling design, selective reporting of information (e.g., data on individual taxa of soil mesofauna are given for the most polluted site, but data for the control site are missing), and an absence of proper statistical support for the majority of the conclusions. In particular, the contrast between one polluted and one unpolluted site (e.g., Evdokimova et al. 2007) does not allow to attribute the detected differences to the impact of pollution. Our team recently published information on growth of three species of herbs (Kozlov & Zvereva 2007b) and phenology of mountain birch (Kozlov et al. 2007) in the impact zone of the smelter at Kandalaksha.

Perception of the Environmental Situation

Recent publications in mass media indicate that the situation in this town is reasonably good in comparison to other towns of the Murmansk Region, and suggest that the existing sanitary zone around the smelter (it is now 1 km wide) is too large (Anonymous 2007).

2.2.2.6 Karabash, Russia: Copper Smelter (Color Plates 18–23 in Appendix II)

Location and Environment

Karabash (population 15,400; data from 2007) is an industrial town of the Chelyabinsk Region, located on the eastern slopes of the South Ural Mountains 90 km northwest of Chelyabinsk (Fig. 2.7). The town lies within a flat-bottomed valley, roughly parallel to the prevailing wind direction and surrounded by hills (please see color plate 21 in Appendix II) having an altitude up to 610 m. This valley was originally covered by south taiga forests formed by Scots pine and common birch, with well-developed herbaceous vegetation (please see color plate 19 in Appendix II).

Pre-industrial History

The region has hosted mining and metal production for over 3,000 years. The settlement Sak-Elginskij (later renamed Soimanovskij) was founded in 1822 close to the current location of Karabash by gold-miners; in the early 1900s, the population was around 400.

Industrial History

For unknown reasons, many publications have reported contradictory dates for the most important events in the industrial history, including in particular the opening of a concentrating mill and a period of temporary closure of the smelter. All the dates below follow the book by Novoselov (2002).

The building of the first copper smelter started in 1834 at Sak-Elginskij; from 1837 to 1842, it produced 22 t of copper from local ores, and it was then developed into an iron factory. The second copper smelter was launched at another locality in 1907 and operated for 3 years. The third smelter was built in 1910, again at a new site (approximately between the two earlier works), and the settlement grew around it. This smelter utilised modern technology to produce blister copper, and by 1915, one third of the Russian copper output originated from Karabash. Copper production from 1907 to 1918 totalled 56,000 t.

Both mining and smelting operations were discontinued between 1918 and 1925. In 1930, a concentrating mill was built in the eastern part of the settlement. Town status was granted in 1933. Copper production in 1940 reached 22,000 t but declined during the World War II to 6,750 t in 1945 due to a shortage of electricity and workforce. The peak population of 50,000 occurred between the late 1950s and the 1960s, when mining, ore beneficiation and smelting operations were at their peaks.

In 1987, Soviet authorities decided to modify the smelter for recycling of non-ferrous metals. The concentrating mill and one of the copper furnaces were closed in November 1989. Closures occurred gradually until 1997, when the smelting operations were discontinued totally. The ownership of the plant and of the nearby mine changed in 1998, and a new enterprise, 'Karabashmed' (with 1,400 employees), was launched using the existing facilities. Annual copper production reached 30,000 t in 2000 and further increased to 40,000 t in the mid-2000s. Since 2005, the plant has also produced sulphuric acid (31,600 t from May to December of 2005).

Emissions Data

The emissions history of the smelter is well documented (Table 2.9), although official data (Emissions of pollutants in Russia 1966–2006) have only been available since 1984. During the period 1989–2003, the smelter contributed 98.3–99.9% of all emissions at Karabash (Emissions of pollutants in Russia 1966–2006). It is expected that ongoing modernisation will reduce the annual emissions of sulphur dioxide to less than 5,000 t (www.karmed.ru).

Table 2.9 Aerial emissions (t) from industrial enterprises at Karabash, Russia

Year	Dust	SO_2	CO	NO_x	Cu	Pb	Zn	As
1956	–	–	–	–	540[a]	1,760[a]	1,940[a]	–
1957	–	–	–	–	410[a]	990[a]	1,660[a]	–
1958	12,100[a]	–	–	–	540[a]	1,280[a]	1,830[a]	–
1959	–	–	–	–	1,030[a]	1,060[a]	2,130[a]	–
1960	20,400[a]	–	–	–	1,100[a]	1,300[a]	3,290[a]	1,900[b]
1961	23,800[a]	–	–	–	1,200[a]	2,250[a]	4,020[a]	–
1962	10,600[a]	–	–	–	1,780[a]	1,790[a]	3,560[a]	–
1965	29,400[a]	–	–	–	1,600[a]	2,300[a]	3,820[a]	1,610[a]
1966	25,600[a]	–	–	–	1,350[a]	1,990[a]	3,410[a]	1,850[a]
1967	28,600[a]	–	–	–	1,340[a]	2,880[a]	4,410[a]	2,070[a]
1968	31,700[a]	–	–	–	1,360[a]	3,280[a]	5,370[a]	2,420[a]
1969	29,900[a]	–	–	–	1,440[a]	3,820[a]	5,140[a]	2,600[a]
1970	28,800[a]	364,500[a]	–	–	1,530[a]	2,570[a]	4,550[a]	1,920[a]
1971	25,800[a]	281,200[a]	–	–	1,250[a]	3,350[a]	4,680[a]	1,680[a]
1972	18,700[a]	–	–	–	830[a]	3,110[a]	3,150[a]	1,070[a]
1973	27,900[a]	273,100[a]	–	–	860[a]	3,300[a]	3,430[a]	1,720[a]
1974	27,400[a]	259,800[a]	–	–	740[a]	2,870[a]	3,470[a]	1,480[a]
1975	27,100[a]	252,000[a]	–	–	600[a]	2,720[a]	2,960[a]	1,380[a]
1976	25,900[a]	233,800[a]	–	–	1,070[a]	3,400[a]	4,740[a]	2,300[a]
1977	29,000[a]	267,200[a]	–	–	1,370[a]	3,930[a]	5,140[a]	2,400[a]
1978	26,000[a]	181,100[a]	–	–	–	3,410[a]	–	1,870[a]
1979	23,900[a]	183,700[a]	–	–	–	3,760[a]	–	2,330[a]
1980	20,300[a]	184,000[a]	–	–	1,140[a]	3,160[a]	4,100[b]	1,990[a]
1984	23,600	137,300	–	100	1,600	2,532	3,200	1,803
1985	23,300	153,500	–	100	1,400	2,130	3,300	1,700
1986	21,300	153,000	–	100	1,200	1,960	3,000	1,670
1987	18,700	148,100	–	100	1,300	1,650	2,600	1,300
1988	20,500	142,500	–	100	–	–	–	–
1989	20,500	142,500	–	100	–	483	–	–
1990	3,400	46,900	500	–	435	254	1,363[b]	75
	3,900[b]				485[b]			
1991	2,200	6,000	100	100	88	119	–	31
1994	200[b]	600[b]	–	–	21[b]	9[b]	63[b]	4[b]
1996	200	40	200	100	–	–	–	–
1997	1,200[b]	4,500[b]	–	–	100[b]	30[b]	200[b]	20[b]
1998	200	5,000	200	100	6	3	–	–
2000	10,100	79,000	15,800	400	–	72	–	99
2001	13,800	71,900	12,200	300	–	92	–	110
2002	14,800	67,700	14,800	400	–	63	–	83
2003	–	–	–	–	340[c]	–	–	–
2004	2,300	66,300	800	100	–	26	–	7
2005	1,300	38,100	1,600	90	–	20	–	7

Non-referenced values were extracted from Emissions of Pollutants in Russia (1966–2006); note that data for the years 1993–1995, 1997, 1999 and 2003 have not been reported in these publications.
Other data sources:
[a] Stepanov et al. (1992).
[b] Chernenkova et al. (2001).
[c] E. Vorobeichik (personal communication, 2008).

Pollution Loads and the Extent of the Contaminated Territory

Ambient concentrations of pollutants in Karabash have been monitored since 1972; the peak reported values were 4.00 mg/m^3 of dust in 1973 and 6.20 mg/m^3 of SO_2 in 1974 (Emissions of pollutants in Russia 1966–2006). The peak ambient SO_2 level in 2000 (i.e., after reopening the smelter) reached 20 mg/m^3 at a 1 km distance from the smokestack (Udachin et al. 2003). A detailed investigation of total airborne suspended particulates performed in 2000–2001 demonstrated that a large (83–100%) fraction was respirable; these particulates contain high levels of potentially toxic metals and are thought to pose a severe risk to human health (Williamson et al. 2004).

From 1987 to 1989, annual depositions of sulphur and nitrogen in Karabash were 16,200 and 1,100 kg/km^2, respectively (Vasilenko et al. 1991). In the early 1990s, deposition of heavy metals near the smelter exceeded the regional background by a factor of 45 for copper, 20 for lead, and 5 for cadmium (Bolshakov et al. 2001). In 1981, soils near the smelter contained 2,500 mg/kg of copper and 1,000 mg/kg of zinc (Chernenkova 1986); later, a peak value of 6,744 mg/kg of copper in forest litter was recorded 3 km northwest of the polluter (Stepanov et al. 1992). The peak concentrations of other elements in soils of the industrial barrens were 10,000 mg/kg of zinc, 3,000 mg/kg of lead, and 1,500 mg/kg of arsenic (Makunina 2002). In the mid-1990s, copper concentrations in soils exceeded the background level by a factor of 128 or more for 5.5 km^2 (Nesterenko 1997).

During the period 1986–1991, a doubling of pollutant deposition (relative to the regional background) was recorded for 250 km^2 around Karabash (Prokacheva et al. 1992). In 1998, depositions of arsenic and copper reached the regional background levels between 30 and 40 km from the smelter (Frontasyeva et al. 2004). Metal accumulation in lichens demonstrated a similar extent of the contaminated territory (Purvis et al. 2004). Thus, although the distribution of pollutants has not been properly mapped, we estimate that emissions from the Karabash smelter have contaminated 3–4,000 km^2.

Habitat Transformation due to Human Activity

The occurrence of industrial barrens around Karabash is well documented (Stepanov et al. 1992; Kozlov & Zvereva 2007a). Development of the barrens presumably started in the 1940s, following intensive felling near the town of Karabash. Clearcutting continued until the 1950s, leaving no stands within 5 km of the smelter (Stepanov et al. 1992). The barren sites (please see color plates 20, 21 and 23 in Appendix II) are now surrounded by semi-barren landscapes – birch forests with missing field layer vegetation (please see color plate 22 in Appendix II). The absence of birch seedlings in these stands suggests that their state is transient; these birch forests will likely turn into industrial barrens following either fire or felling (Kozlov & Zvereva 2007a), or when birches will reach their upper age limit (Zverev 2009). In 1983, dead and declining stands occupied 80 km^2, surrounded by 210 km^2 of damaged forests; in 1992, these areas were 35 and 160 km^2, respectively (Butusov et al. 1998). These data are in line with the conclusions of Chernenkova et al. (1989) that birch forests located 10–12 km from the polluter were not modified by pollution.

Brief History of Environmental Research

Research on pollution impacts of the Karabash smelter started in the early 1980s, when a team from the Severtsev Institute of Evolutionary Morphology and Ecology of Animals (Moscow) collected data on the contamination of soils, water, and plants, as well as on the composition and productivity of plant communities (Chernenkova 1986). This team continued an integrated monitoring (with an emphasis on vegetation) until the mid-1990s (Chernenkova et al. 1999; Chernenkova 2002), involving remote sensing data for outlining the extent of pollution damage (Bugrovskii et al. 1985; Butusov et al. 1998). The results have been summarised in two monographs (Stepanov et al. 1992; Chernenkova 2002). Unfortunately, T. Chernenkova and her colleagues published many of their data sets and conclusions repeatedly. An absence of cross-references and missing sampling dates in these publications create numerous problems for using these data in meta-analyses.

Although published data are diverse, changes in plant growth and community structure were much better documented than the effects of pollution on animals. Several data sources (Stepanov et al. 1992, and references therein; Kucherov & Muldashev 2003) reported a decline in the growth of Scots pine, as well as an absence of changes in the radial increment of common birch; however, non-replicated (or poorly replicated) study designs and an absence of statistical analyses diminish the value of this information. Changes in species richness and biomass (i.e., productivity) were reported for soil microbiota (Mukatanov & Shigapov 1997), soil algae (Stepanov et al. 1992), epiphytic lichens (Williamson et al. 2004), vascular plants (Chernenkova et al. 1989), and soil macrofauna (Nekrasova 1993). Studies of vertebrates seem to be restricted to short reports on changes in vole reproduction (Lukyanova 1987; Chernousova 1990) and information on the absence of small mammals in the most polluted habitats (Stepanov et al. 1992).

A decrease in emissions due to the closure of the plant in the 1990s resulted in a slight improvement of the environmental situation. The initial stages of the recovery of field layer vegetation were observed in slightly to moderately polluted forests (5–9 km from the smelter), while the state of the most polluted forests (3 km northeast of the smelter) did not change (Chernenkova et al. 1999). Later on, grasses started to colonise barren, heavily eroded slopes of the Zolotaya Mountain next to the smelter (Chernenkova et al. 2001), and the radial increment of Scots pine increased near the polluter (Kucherov & Muldashev 2003).

Perception of the Environmental Situation

In 1992, the United Nations Environmental Programme designated Karabash as one of the world's most polluted towns (Anonymous 1999). Newspaper's descriptions of Karabash resemble apocalyptic nightmares: 'The sky turned black.

The snow became black. The withered branches of ailing trees got a new coating of black' (Filipov 2004). The mountains near Karabash are called 'bald' by the locals, as their slopes are almost void of vegetation. The other end of town looks like a lunar landscape, with nothing but dust for hundreds of metres (Anonymous 1999). A plan to hold a referendum on shutting down the smelter faltered after unidentified assailants badly beat one of the initiators (Filipov 2004).

Karabash was visited by a team from the Blacksmith Institute in February of 2005. The adverse health effects caused by the smelter were partly acknowledged by local health and municipal authorities. The priorities here are relocating people living in the buffer zone elsewhere and installation of pollution control devices in the smelter (www.pollutedplaces.org).

Intriguingly, in 2005 the smelter at Karabash was awarded a medal 'For achievements in nature protection'. At the ceremony held in the Kremlin, Moscow, it was announced that 'already now, Karabash can be removed from the list of ecologically unfavourable localities' (www.elisprom.ru). However, this conclusion seems premature, since in 2007 we have not observed an improvement of the polluted landscapes around Karabash (please see color plates 18–23 in Appendix II).

2.2.2.7 Kostomuksha, Russia: Iron Pellet Plant (Color Plates 24–26 in Appendix II)

Location and Environment

Kostomuksha (population 30,000; data from 2007) is a newly built town. It was established in 1977 as an urban-type settlement and populated by people from various regions of the Soviet Union; town status was granted in 1983. Kostomuksha is located in the north-western part of the Republic of Karelia on the shore of Kontoki Lake, 40 km from the border with Finland (Fig. 2.8).

In the mid-1970s, two thirds of the territory surrounding Kostomuksha was covered by forest, and about one-quarter by bogs. Some stands were still in an almost virgin state, representing old-growth mid-taiga forests dominated by Scots pine or Norway spruce, with a ground layer formed by dwarf shrubs and green mosses (please see color plate 25 in Appendix II).

Pre-industrial History

Kostomuksha area is best known for poems of the Finnish epos 'Kalevala'. When Elias Lönnrot visited Kostomuksha in 1837 to record this epos, the village had only ten houses. Before the Winter War (1939), Kostomuksha (or Kostamus) belonged to the sparsely populated Finnish municipality of Kontokki, with 1,872 inhabitants in 1907. During the war, the village was ruined, and in 1948, only five persons who were capable of working, a few horses, a cow and a sheep were living in Kostomuksha. By the late 1950s, Kostamuksha had become unpopulated, and only a few people were still living in the entire area.

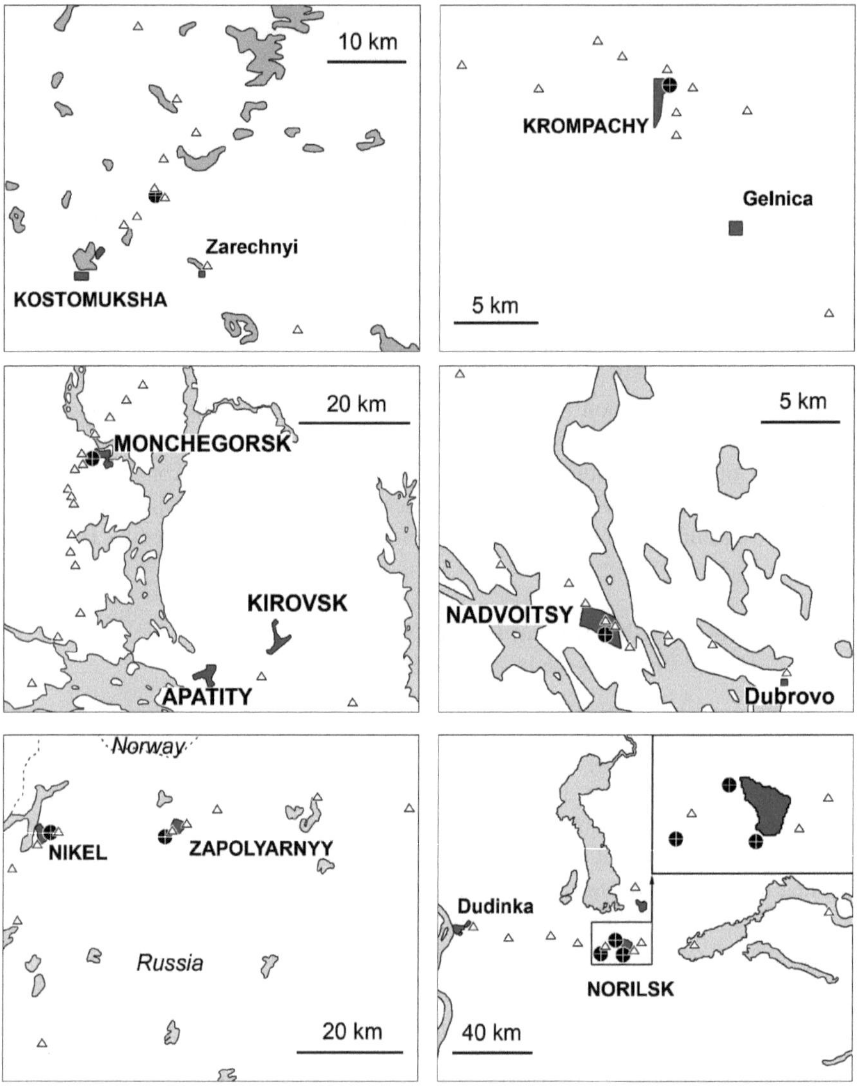

Figs. 2.8–2.13 Distribution of study sites (triangles) around investigated polluters (black circles with white cross inside). Names of larger settlements (usually towns) are capitalised. **2.8**, iron pellet plant in Kostomuksha, Russia; **2.9**, copper smelter in Krompachy, Slovakia; **2.10**, nickel-copper smelter in Monchegorsk, Russia; **2.11**, aluminium smelter in Nadvoitsy, Russia; **2.12**, nickel-copper smelter in Nikel, and ore-roasting factory in Zapolyarnyy, Russia **2.13**, nickel-copper smelters in Norilsk, Russia

The presence of iron ores around Kostomuksha has been known for a long time, at least since the beginning of the nineteenth century. In the 1850s, some local inhabitants dug mud from Kostamus Lake, smelted it and made axes and other iron items. Intensive geological exploration of this region started in 1946.

Industrial History

The decision to build the Kostomuksha ore mining and processing enterprise was made in 1969, and a contract with Finnish companies that conducted a substantial part of the work was signed in 1977. Mining of depleted ferruginous quartzite began in 1978, and production of iron-ore pellets (high-grade metallurgical stock) was launched in 1982. Development continued until 1985, when the full capacity had been achieved. In 1993, the enterprise was reorganised into an open joint-stock company, Karelsky Okatysh, which is the principal employer (8,600 persons in 2001) of the town (okatysh.home.spb.ru). In 2004, the company mined 22,320,000 t of ore and produced 7,584,000 t of iron pellets (ru.wikipedia.org).

Emissions Data

Karelsky Okatysh contributed 99.4–99.8% to the total emissions of pollutants at Kostomuksha (Emissions of pollutants in Russia 1966–2006). Only the amounts of gaseous pollutants have been reported (Table 2.10); no data are available on emissions of either metals (with dust) or organic pollutants. However, metal emissions can be roughly estimated from the amount of dust using concentrations reported by Shiltsova and Lastochkina (2004); 6,900 t of dust (emitted in 2000) is likely to contain approximately 350 t of iron, approximately 27 t of zinc, and approximately 0.6 t of copper.

Table 2.10 Aerial emissions (t) from industrial enterprises at Kostomuksha, Russia

Year	Dust	SO_2	CO	NO_x
1987	5,400	61,300	400	2,500
1988	5,100	61,900	400	2,700
1989	4,800	62,500	400	2,700
1990	4,700	62,200	2,200	3,400
1991	5,200	54,700	600	2,500
1992	6,000[a]	60,400[a]	1,300[a]	2,200[a]
1993	5,300	53,100	600	1,600
1994	6,400	48,100	600	1,500
1995	5,000	44,600	600	1,300
1996	6,400	47,400	500	1,200
1997	6,000	32,200	900	1,500
1998	6,400	37,700	900	1,600
1999	6,900	34,400	1,000	1,300
2000	6,900	30,200	1,100	1,300
2001	6,500	30,000	1,200	1,400
2002	6,400	32,700	1,300	1,400
2003	6,600	33,500	1,300	1,400
2004	6,400	35,400	1,200	1,400
2005	6,100	36,500	1,400	2,100

Non-referenced values extracted from Emissions of Pollutants in Russia (1966–2006).
Other data sources:
[a] Krutov et al. (1998).

Pollution Loads and the Extent of the Contaminated Territory

During the mid-1990s, sulphur deposition within 2–3 km from the plant was 2,100 kg/km^2, i.e., nearly ten times as high as the regional background of 260 kg/km^2 (Shiltsova & Lastochkina 2004). In 1986–1987, the extent of the most contaminated zone (with an approximately 100-fold increase in dust deposition) was 4–5 km, and the extent of the moderately contaminated zone (with an approximately 10–30-fold increase in dust deposition) was 25–30 km (Lazareva et al. 1988). Monitoring of wet precipitation from 1992 to 1994 indicated that to the west of the smelter (nearly opposite the direction of prevailing winds), emissions extend to about 30–40 km. However, a substantial (tenfold) increase in annual deposition of iron was only detected within 5 km, while the deposition at localities 22–38 km from the polluter was only 1.5 times as high as at more distant (63–114 km) plots (Poikolainen & Lippo 1995). The iron content in soils near the polluter in the late 1990s–early 2000s exceeded the regional background by a factor of 60 (Opekunova & Arestova 2005). Levels of Li, B, Al, Fe, Ni, Cu, Zn, Hg, Mn, and Mo, as well as hydrocarbons, phthalates and phenols in snow samples collected near the polluter in 2001, were found to exceed the maximum allowable concentrations (Lebedev et al. 2003). High concentrations of calcium-containing dust in aerial emissions resulted in increased pH values of precipitation near the factory (Potapova & Markkanen 2003). However, at distances exceeding 2 km, a slight acidification of forest litter was detected after 6 years of pollution impact (Lazareva et al. 1988).

During the period 1986–1991, a doubling of pollutant deposition (relative to the regional background) was recorded for 390 km^2 around Kostomuksha (Prokacheva et al. 1992). However, mapping of the entire territory of Karelia for sulphur and metal contamination in the mid-1990s revealed increases in sulphur and iron content in mosses (by a factor of 2 or more) for approximately 10.1 and 12.9 km^2 near Kostomuksha, respectively (Fedorets et al. 1998). On the other hand, the extent of the territory contaminated by manganese, chromium and vanadium was recently estimated as approximately 1,200 km^2 (Gulyaeva & Kalieva 2004).

Habitat Transformation due to Human Activity

Three years of pollution impact appeared sufficient to enhance accumulation of forest litter and reduce the biomass of mosses and field layer vegetation near the factory (Chernenkova 1991). The appearance of dust on plants and litter in the immediate vicinity of the factory, a partial replacement of dwarf shrubs by herbaceous vegetation, and a decline of epiphytic lichens are directly attributable to pollution (please see color plate 26 in Appendix II). However, Scots pine stands at distances exceeding 5 km showed no signs of pollution damage, and no effects on growth of Scots pine had been detected until very recently (Krutov et al. 1998; Sinkevich 2001).

Brief History of Environmental Research

Environmental and ecological research conducted from 1972 to 1975, i.e., several years prior to launching the factory, provided detailed information on the pre-industrial state of many components of the terrestrial and aquatic ecosystems (Biske et al. 1977). Since the launch, environmental monitoring has continued almost uninterrupted until very recently, providing a unique opportunity to follow environmental changes during the first years of pollution impact. The largest body of information is collected on pollution-induced changes in soil quality and microbial activity (Zaguralskaya 1997; Germanova & Medvedeva 2001), soil nematoda (Matveeva et al. 2001), epiphytic lichens (Fadeeva 1999), growth, physiology and vitality of Scots pine (Fuksman et al. 1997; Kaibiyainen et al. 2001; Lamppu & Huttunen 2003; Sazonova et al. 2005), and on the structure and productivity of plant communities (Chernenkova 1991, 2002). Since researchers from the Forest Institute at Petrozavodsk tend to publish the same results repeatedly, less than half of the data sources available to us (Table 2.3) contain original information.

Perception of the Environmental Situation

Due to the proximity to the Finnish border, emissions from Kostomuksha soon became an international concern. Several research projects established in the early 1990s concluded that the effects of contamination on the Finnish environment and population are relatively minor (Lumme et al. 1997), and the recent report by the Finnish Ministry of the Environment (Ympäristöministeriö 2007) says that there are no 'direct, continuous, significant environmental threats' to the Finnish environment from the Karelian side.

2.2.2.8 Krompachy, Slovakia: Copper Smelter (Color Plates 27–29 in Appendix II)

Location and Environment

Krompachy (population 8,800; data from 2001) is a town in Slovakia, situated in the central Spiš region, in the valley of the Hornád River, surrounded by three mountain ranges with summits 900–1,100 m above sea level (Fig. 2.9). Norway spruce is the prevailing tree in coniferous and mixed forests between 400 and 800 m above sea level (please see color plate 28 in Appendix II). At lower altitudes, the forests are mostly composed of European beech (*Fagus sylvatica*) with contributions of Silver fir (*Abies alba*), Norway spruce, and common birch.

Pre-industrial History

Metallurgy ranks among the most ancient industries within the territory of Slovakia; local polymetallic ores have been exploited since the Bronze Age. Iron and non-ferrous (copper- and silver-containing) ores have been mined and processed in the area surrounding Krompachy for about 700 years. The first written record of the existence of the settlement is dated 1282, and the town of Krompachy was founded in the mid-fourteenth century.

Industrial History

During the nineteenth century, the Krompachy Iron Mining Company was gradually formed from smaller iron-mills; iron and copper rolling mills were also built. In the early 1900s, the Krompachy Ironworks, with 3,500 employees (of 6,500 inhabitants of the town), produced 85,000 t of iron annually and was the biggest ironworks of the former Hungary. It was closed after World War I. Production of copper started in 1843. A copperworks built in 1937 on the site of the old ironworks produced 1,500 t of copper annually and operated until early 1945, when it was destroyed by the withdrawing German troops. The present smelter was launched in 1951 for processing ore concentrates and recycled copper materials, with an annual capacity of 20,000 t of copper; 29,000 t was produced in 1995 and 25,000 in both 1996 and 1997. For a long time Krompachy was a flourishing town; however, production was ceased in 1999 due to falling world copper prices, and 1,000 workers were dismissed. The smelter was reopened in September 2000 by the newly established joint-stock company, Kovohuty (Barecz 2000). The town of Krompachy also houses several other industries, including production of alloys, electronic goods, and electromechanical equipment for both industrial and domestic use.

Emissions Data

While emissions of sulphur dioxide are more or less properly documented (Table 2.11), data on metal emissions are scarce. The composition of dust arising from different technological processes was reported by Hronec (1996); however, contributions of these processes to the total amount of emitted dust remain unknown. Occasional information suggests that annual copper emissions in the 1990s were between 40 and 55 t, lead emissions decreased during the first half of the 1990s from 50 to 20 t (Kalač et al. 1996), and arsenic emissions in 1986 were as high as 90 t (Andersen 2000).

Pollution Loads and the Extent of the Contaminated Territory

The highest reported concentrations of metals in soil are 8,437 mg/kg of copper, 1,482 mg/kg of zinc, and 3,343 mg/kg of lead (Maňkovská 1984; Banásová & Lackovičová 2004); the regional background levels are reached at 7–12 km from the smelter (Wilcke et al. 1999). Soil acidification was only detected within 3–5 km from the smelter (Hronec 1996; Wilcke et al. 1999). Importantly, a decrease in emissions in early the 1990s did not result in lower metal contamination of edible mushrooms, and concentrations of Hg, Cd, Pb and Cu in many samples collected 3–5 km east of the smelter exceeded the statutory limits (Svoboda et al. 2000).

Maňkovská (1984) outlined three zones of environmental contamination around Krompachy. The first zone, with pollutant concentrations exceeding the regional background by a factor of 150 or more, covered 28.4 km^2; the second zone (pollution loads 6 to 150 times higher than the background) covered 44.7 km^2. The borders of the third, less contaminated zone, were incompletely shown on the map by Maňkovská (1984), but its area still exceeded 150 km^2.

Table 2.11 Aerial emissions (t) from copper smelter at Krompachy, Slovakia

Year	SO_2	NO_x	Dust
1952	5,600	–	11,900
1955	6,800	–	12,220
1960	7,400	–	12,670
1965	7,032	–	10,096
1971	3,040	–	1,387
1975	15,616	–	1,086
1980	19,654	–	1,869
1985	24,341	–	1,724
1986	20,000[a]	–	1,400[a]
1989	13,700[b]	–	–
1990	23,000	–	1,100
1992	10,156	–	73
1995	5,348[b]	43[b]	86[b]
1996	10,187[c] 9,008[d]	90[d]	188[c]
1997	7,236[d]	99[d]	296[e]
1998	2,543[e]	98[e]	151[e]
2002	0.5[f]	20[f]	3.3[f]

Non-referenced values were extracted from Hronec (1996, Table 2.8).
Other data sources:
[a] Andersen (2000).
[b] Lackovičová et al. (1994).
[c] Slovak Environmental Agency (1997).
[d] Slovak Environmental Agency (1998).
[e] Maňkovská (2003).
[f] Ministry of the Environment of the Slovak Republic (2004).

Habitat Transformation due to Human Activity

Development of the industrial barrens presumably began in the 1980s, and recently they covered less than 0.5 km^2. Extensive soil erosion (please see color plates 27 and 29 in Appendix II) occurs on the north-facing slope of a hill adjacent to the smelter (Maňkovská 1984; Banásová & Lackovičová 2004; Kozlov & Zvereva 2007a). Although Krompachy is surrounded by a 'lichen desert' (Lackovičová 1995), a rare lichen species, *Cladonia rei* Schaer, is abundant on bare, acid soils near the smelter (Hajdúk & Lisiká 1999).

Brief History of Environmental Research

Information on the environmental effects caused by aerial emissions of the Krompachy smelter is scarce, with contamination patterns documented much better than the biotic effects of pollution. Data on soil pH, as well as concentrations of both macronutrients and pollutants, were published by Hronec (1996) and Wilcke et al. (1999). Deposition of metals with dust was measured at 15 sites between Krompachy and Košice in 1999 (Bobro et al. 2000). The effects of emissions on vegetation were first explored by Kaleta (1982). A detailed description of field layer vegetation at the most polluted, nearly barren, site was provided by Banásová and Lackovičová (2004). Data on animals are restricted to reports on soil nematodes (Sabová & Valocká 1996).

2.2.2.9 Monchegorsk, Russia: Nickel-Copper Smelter (Color Plates 30–39 in Appendix II)

Location and Environment

Monchegorsk (population 49,400; data from 2007) is an industrial town in the Murmansk Region, located 115 km south of Murmansk, between Imandra Lake and the Monche-tundra mountain ridge (reaching 965 m above sea level) (Fig. 2.10). Monchegorsk is only 150 km south of the tree line formed by mountain birch; up to the mid–1930s, virgin Scots pine stands and an impenetrable Norway spruce forest with beards of epiphytic lichens covered the area close to the recent position of Monchegorsk (Bobrova & Kachurin 1936).

Pre-industrial History

Although the Kola Peninsula has been populated since approximately 6000 BC, in 1913 there were only 13,200 inhabitants, mostly in seashore settlements. Until the railway was built in 1916, central parts of the Kola Peninsula were accessible only by foot or by boat. Descriptions of the landscape and vegetation made in the 1920s–1930s reflect a virgin nature, slightly affected by traditional forms of land use such as reindeer herding, fishing and hunting (Kozlov & Barcan 2000).

The discovery of apatite ores in the Khibiny Mountains (1920–1926) and of nickel-copper ores in the Monche-tundra (1929–1932), followed by the decision of the Soviet government to develop mining and ore processing factories in the central part of the Kola Peninsula (1929–1935), resulted in rapid changes. The Severonikel smelter and the town of Monchegorsk were built in the mid-1930s on a previously unpopulated territory using forced labour. The population of the Monchegorsk area grew from a single Saami family in 1930 to 200 persons in 1933 and 34,190 persons in 1938, and town status was granted to Monchegorsk in 1937. Intensive development of the town required the clear-cutting of large forest areas, and the rapid population growth resulted in a drastic increase in forest fires.

Industrial History

Construction of the smelter started in 1934, and it was officially opened in 1937. Operations were discontinued the week after the German attack on the USSR in 1941. The equipment was partially sent to Norilsk, where it served as the basis for the recently operating nickel smelter; another portion of the equipment was moved to Orsk.

Restoration of the factory started in 1942, the mine was restored in 1943, and nickel production began before the end of World War II. However, regular operations started only in 1946–1947 with ores from the local Nittis-Kumuzhje deposit, which was exhausted in 1977. Later, sinter-roasted ores from Zapolyarnyy and both ores and nickel matte from Norilsk (since 1968) were processed. Recently, the smelter has used a breakage, waste products and raw material from both domestic and foreign suppliers.

The smelter had no air-cleaning facilities until 1968, when production of sulphuric acid was launched to utilise converter and roaster gases. The second line of sulphuric acid production was installed in 1979. The last substantial expansion of the smelter took place in the 1980s. One of the most polluting smelting departments was closed in 1998, due to both economic and ecological reasons (Pozniakov 1993, 1999; Alexeyev 1993; Barcan 2002b).

Until 1950, nickel and copper production did not exceed 10,000 and 6,300 t, respectively. In the 1980s, annual production reached 140,000 t of nickel and 100–120,000 t of copper (Pozniakov 1993). Production decreased in the 1990s, but it increased in the 2000s to 100–122,000 t of nickel and 74–106,000 t of copper. Other products include concentrates of precious metals and sulphuric acid (www.nornik.ru). Between 2002 and 2007, the number of employees decreased from 16,601 to 11,374.

Emissions Data

During the first years of smelting, losses were as high as 10–15% of the metal content of the ores, which, in combination with the data on metal production and on sulphur content in ores, allows us to estimate annual emissions for the late 1940s as 60,000 t of sulphur dioxide, 1,000–1,500 t of nickel and 500–1,000 t of copper (Kozlov & Barcan 2000). During the period 1989–1999, the smelter produced 99.6–99.9% of all aerial emissions reported from the town of Monchegorsk (Table 2.12). Along with the pollutants included in this table, emissions of H_2SO_4 (255 t in 1979, 170 t in 1987) and chlorine (438 t in 1987) have been occasionally reported (Emissions of pollutants in Russia 1966–2006). Both the nature and origin of emissions of the smelter at Monchegorsk are described in details by Barcan (2002b), a professional chemist who worked at this smelter for many years. Recently, Boyd et al. (2009) estimated emissions of As, Cd, Cr, Pb, Sb, V and Zn on the basis of ore chemistry and annual production data.

Pollution Loads and the Extent of the Contaminated Territory

Ambient concentrations of pollutants in Monchegorsk have been monitored since 1967; the peak reported values were 2.90 mg/m^3 of dust in 1970, 6.84 mg/m^3 of SO_2 in 1973, and 1.12 mg/m^3 of NO_x in 1967 (Emissions of pollutants in Russia 1966–2006). In 1987–1989, annual depositions of sulphur and nitrogen in Monchegorsk were 2,100 and 400 kg/km^2, respectively (Vasilenko et al. 1991). The long-term emissions

Table 2.12 Aerial emissions (t) from industrial enterprises at Monchegorsk, Russia

Year	Dust	SO_2	CO	NO_x	Ni	Cu	Co
1960	–	134,000[a]	–	–	–	–	–
1961	–	139,000[a]	–	–	–	–	–
1962	–	136,000[a]	–	–	–	–	–
1963	–	147,000[a]	–	–	–	–	–
1964	–	148,000[a]	–	–	–	–	–
1965	–	200,000[a]	–	–	–	–	–
1966	–	191,000[a]	–	–	–	–	–
1967	–	199,000[a]	–	–	–	–	–
1968	–	76,400[a]	–	–	–	–	–
1969	–	94,000[b]	–	–	–	–	–
1970	–	101,000[b]	–	–	–	–	–
1971	–	117,200	–	–	–	–	–
1972	–	117,200	–	–	–	–	–
1973	7,300	138,700 215,000[b]	–	–	–	–	–
1974	7,300	138,700 259,000[b]	–	–	–	–	–
1975	9,100	259,200	–	–	–	–	–
1976	–	268,000[b] 307,100[c]	–	–	–	–	–
1977	–	253,000[c]	–	–	–	–	–
1978	8,600	254,200 237,000[c]	–	–	–	–	–
1979	8,600	254,200 189,000[b]	–	–	–	–	–
1980	11,800	206,100	200	1,300	3,420[c] 5,600[d]	1,370[c]	–
1981	11,700	204,900 187,000[b]	400	2,900	–	–	–
1982	13,300	239,200	400	3,500	2,970[c] 3,200[d]	1,600[c]	–
1983	15,500	278,200	400	4,000	1,560[d]	2,130[c]	–
1984	19,900	257,200	400	4,000	3,110	2,490	116[b]
1985	17,800	254,500 236,000[b]	400	4,000	3,013	2,420	82[b]
1986	17,600	243,600	1,200	5,100	6,770[c]	5,000[c]	–
1987	17,500	224,400	1,200	5,100	7,480[c]	5,670[c]	–
1988	17,100	212,200	1,200	5,100	2,710[c]	4,110[c]	–
1989	16,600	200,400	1,300 5,400[e]	5,100 5,500[e]	3,100	2,200	100
1990	15,786[f]	232,600	1,200 6,600[e]	5,100 5,600[e]	2,712	1,813	97
1991	15,422[f]	196,200	1,400 9,700[e]	5,100 6,100[e]	2,660	1,739	97
1992	14,147[f]	182,000[b,c]	–	–	2,118[b]	1,456[b]	91[b]
1993	12,528[f]	136,900 145,000[f]	1,300	5,100	1,960	1,049	89
1994	10,206[f]	97,700	900	1,300	1,619	933	82
1995	8,445[f]	129,400	1,000	1,000	1,366	726	56

(continued)

Table 2.12 (continued)

Year	Dust	SO_2	CO	NO_x	Ni	Cu	Co
1996	7,729[f] 9,800	110,500	2,300	1,000	1,309	699	41
1997	8,092[f]	140,200	1,500	700	1,348	761	37
1998	7,246[f] 8,500	88,600	1,400	700	1,304	873	35
1999	6,016[f] 7,200	46,100	2,000	800	1,128	856	32
2000	9,900	45,800	4,000	1,500	1,216	873	34
2001	9,100	43,900	3,600	1,300	1,200	827	44[b]
2002	7,700	43,900	3,800	1,200	818[f] 910	696	–
2003	7,200	42,100	3,800	800	734[f]	699	10
2004	6,500	37,900	2,900	800	687	580	12
2005	6,500	41,100	2,400	700	501[f]	608	12

Non-referenced values were extracted from Emissions of Pollutants in Russia (1966–2006).
Other data sources:
[a] Alexeyev (1993).
[b] Barcan (2002b).
[c] Mokrotovarova (2003).
[d] V. Nikonov (personal communication, 1998).
[e] Kryuchkov (1993b).
[f] Miroevskij et al. (2001), the data refer to the smelter only.
[f] Severonikel smelter (statistical forms TP-2).

impact resulted in slight soil acidification and severe contamination by metals. The reported peak concentrations in soils were 4,622 mg/kg of copper, 9,288 mg/kg of nickel, and 210 mg/kg of zinc (Barcan & Kovnatsky 1998; V. Barcan, personal communication, 2006).

By the time of the first survey (1966), the contaminated area covered approximately 6,000 km (Kozlov & Barcan 2000). In 1983, sulphur deposition exceeded the regional background over 60,000 km^2, forming a continuous polluted zone centred at Monchegorsk and extending over Kovdor, Olenegorsk, Kirovsk and Kandalaksha (Kryuchkov & Makarova 1989); however, there still was a gap between this area and the contaminated area around Nikel and Zapolyarnyy. These polluted areas merged by the late 1980s–early 1990s, and a potentially critical deposition of sulphur was exceeded over an area of 150,000 km^2, 32,000 km^2 of which was in Finland and 19,000 km^2 in Norway (Tuovinen et al. 1993). However, according to official reports, a doubling of pollutant deposition (relative to the regional background) during the period 1986–1991 was recorded only for 800 km^2 (Prokacheva et al. 1992).

Habitat Transformation due to Human Activity

Visible signs of forest damage around Monchegorsk appeared immediately after the beginning of smelting. In the early 1950s, clear changes in forest vegetation were detected within 2–3 km from the smelter (Kozlov & Barcan 2000). Industrial barrens (please see color plates 34–39 in Appendix II) appeared in the early 1960s, and by the early 1990s, they were estimated to cover 200–250 km^2, surrounded by 400–500 km^2 of dead forests (please see color plates 32 and 33 in Appendix II) (Kryuchkov 1993a, c; Tikkanen & Niemelä 1995; Rigina & Kozlov 2000; Kozlov & Zvereva 2007a).

Brief History of Environmental Research

The effects of emissions of the Monchegorsk smelter on terrestrial ecosystems have been systematically studied since 1971, and the impact zone of this polluter is explored most thoroughly on a global scale. Some 30 research teams have been working in the area affected by the Severonikel smelter since 1970s; the results of the research are reported in over 1,000 publications, including approximately 20 monographs. For the history of environmental research and references for the most important studies of Russian scientists, consult Kozlov et al. (1993) and Kozlov and Barcan (2000); the results of the Finnish project were summarised by Tikkanen and Niemelä (1995). A somewhat outdated list of relevant publications is available at www.ngu.no/Kola/bibliodb.html. In spite of the high number of studies, there remain a number of knowledge gaps (identified by Rigina & Kozlov 2000). The extensive overview of recent findings is given by the AMAP (2006) assessment.

Perception of the Environmental Situation

Although severe damage to forests around Monchegorsk has been clearly visible for decades, neither the Soviet nor the Russian governments invested money in searching for solutions to the ecological problems of the Kola Peninsula. Even publication of the alarming letter 'Smoke over the reserve' by O. I. Semenov-Tian-Schanskij and A. B. Bragin in the central communist newspaper 'Pravda' in 1979 had no effect (Kozlov & Barcan 2000). Later, the distribution of 'negative' ecological information was prohibited until 1989, when the appearance of publications warning of the threat to nature led to mass actions by local residents calling for an end to the poisoning of the environment. However, since 1990 the smelter has been privately owned, leaving no hope for governmental funding of nature protection measures. Furthermore, on 12 May 1990, the director of the Severonikel smelter declared in the local newspaper that reducing production is the only realistic way to reduce emissions, but that this would lead to substantial cutting the workforce. Thus, in a town inhabited mainly by smelter workers, every family would be affected (Kryuchkov 1993c).

There is no doubt that the inhabitants of Monchegorsk were and continue to be chronically exposed to exceptionally high levels of pollutants. The concentrations of nickel and copper in local vegetables, berries and mushrooms greatly exceed the maximum tolerable limits, and the extensive health problems reported for the residents of these cities are obviously related to the emissions of non-ferrous metallurgy. However, local residents, although regularly informed through the press and other mass media of the possible adverse effects of contaminated air and food, continue to collect berries and mushrooms near the smelter and to grow potatoes and vegetables within severely contaminated areas (Kozlov & Barcan 2000). The situation started to change in the late 1990s, with a decrease in smelter emissions and implementation of a municipal re-greening program widely advertised by mass media (Petrov 2004; Timofeeva 2005). The environmental situation at Monchegorsk was recently debated on one of the local web forums (www.hibiny.ru).

2.2.2.10 Nadvoitsy, Russia: Aluminium Smelter (Color Plates 40–42 in Appendix II)

Location and Environment

Nadvoitsy is a small (population 10,600; data from 2006) urban-type settlement in the Segezhsky district of the Republic of Karelia. It is located on the shore of Vygozero Lake, 240 km north of Petrozavodsk (Fig. 2.11). Forests around Nadvoitsy (please see color plate 40 in Appendix II) were originally dominated by Scots pine, with a mixture of Norway spruce and frequent occurrences of *Sphagnum* and aapa mires.

Pre-industrial History

The oldest settlements in the area surrounding Nadvoitsy date to the Stone Age. Prior to building of the railway between St. Petersburg and Murmansk, the area was only sparsely populated. The railway station at Nadvoitsy was established in 1915. Neither agriculture nor forestry had affected the surrounding forest prior to the 1930s, when a pulp and paper factory was built in the nearby town of Segezha.

Industrial History

The aluminium smelter, located on the southern outskirts of Nadvoitsy, was commissioned in 1954. Its capacities were expanded in 1961, allowing production of 60,000 t of aluminium (in pigs, silumin, powder, and alloys) annually. Annual production of aluminium increased to 68,000 t in the late 1990s (US Geological Survey 1999). The new electrolysis production line, using technology developed by Kaiser (USA), was launched in 1999, leading to both an increase of aluminium production (80,000 t in 2006) and a decrease in emissions. Building of new emission control systems (dry gas-washing) started in 2006. Following implementation of the new technologies, the number of employees decreased from 3,000 in the early 1990s to 2,000 in 2001 (Boltramovich et al. 2003) and 1,458 in 2006 (www.sual.ru). The town's economy is fully dependent on the smelter.

Emissions Data

Since there are no other industries in the town, the aluminium smelter is responsible for a vast majority of the total emissions of sulphur dioxide and nitrogen oxides at Nadvoitsy, and for the entire emission of fluorine (Table 2.13).

Pollution Loads and the Extent of the Contaminated Territory

Ambient concentrations of pollutants in Nadvoitsy have been monitored since 1978; the peak reported values were 0.30 mg/m^3 of SO_2 in 1981, and 0.27 mg/m^3 of HF in 1982 (Emissions of pollutants in Russia 1966–2006). In 1990, ambient air monitored 1,200 m north–northwest of the smelter contained on average 0.037 mg/m^3 of SO_2, 0.026 mg/m^3 of NO_x and 0.0084 mg/m^3 of HF, with peak values of 0.15, 0.08 and 0.027 mg/m^3, respectively (Ministry of the Environment of Finland 1991b). On the other hand, it was reported (Kozlovich 2006) that ambient fluorine

Table 2.13 Aerial emissions (t) from industrial enterprises at Nadvoitsy, Russia

Year	Dust	SO_2	CO	NO_x	F-containing dust	HF
1978	1,700	12,900	7,200	200	–	360
1979	1,700	12,900	7,200	200	–	360
1980	6,200	–	7,200	200	–	400
1981	4,700	1,500	16,200	200	–	400
1984	5,000	1,500	7,500	100	–	410
1985	4,600	1,500	7,500	100	3,330	400
1986	4,700	1,400	7,500	100	3,330	400
1987	6,100	2,300	5,200	100	4,704	375
1988	5,900	2,300	5,100	100	4,700	373
1989	6,000	2,200	3,700	100	4,487	437
1990	5,700 3,450[a]	2,200	3,500	100	690[a]	410[a]
1991	4,300	1,800	3,300	100	–	–
1993	3,900	1,900	3,300	100	445	216
1994	4,100	1,600	4,000	100	–	240
1995	4,100	1,400	3,700	50	–	–
1996	4,100	1,200	3,600	40	–	–
1997	4,000	1,000	3,000	40	–	246
1998	3,400	1,200	3,000	50	–	–
1999	3,500	1,300	3,000	40	–	283
2000	4,300	1,400	3,200	40	–	363
2001	4,300	1,200	3,500	100	–	–
2002	3,400	1,200	2,700	40	–	288
2003	3,000	1,200	2,700	70	–	318
2004	4,200	1,000	2,600	60	–	497
2005	3,000	1,200	2,400	70	–	313

Non-referenced values were extracted from Emissions of Pollutants in Russia (1966–2006).
Other data sources:
[a] Ministry of the Environment of Finland (1991b).

in Nadvoitsy during 2 days of measurements in 2005 was 0.021 and 0.230 mg/m^3 (i.e., 4 and 46 times higher than the sanitary limit of 0.005 mg/m^3).

A doubled level of pollutant deposition (relative to the regional background) during the period 1986–1991 was recorded for 40 km^2 around Nadvoitsy (Prokacheva et al. 1992). At the same time, the footprint of the plant had not been detected in the course of mapping of the entire territory of Karelia for sulphur and metal concentrations in green mosses and forest litter (Fedorets et al. 1998).

Habitat Transformation due to Human Activity

Although we have not observed any dead stands near the smelter in 2004–2007, alarming newspaper reports claimed that 'emissions have turned a 100-m belt of the surrounding taiga into a lifeless zone of dead trees' (Baiduzhy 1994). Slight signs of forest deterioration (please see color plate 41 in Appendix II) are visible only within 1–2 km from the smelter, but local heating systems and recreation obviously share responsibility for the lower vitality of forests around Nadvoitsy.

Brief History of Environmental Research

We failed to locate any scientific publication analysing the impact of the aluminium smelter on terrestrial ecosystems. The data from environmental and health surveys conducted by scientists from the Petrozavodsk State University in the early 2000s have not been published and will remain unpublished according to the agreement with the Nadvoitsy aluminium plant that financed the research (T. Karaperyan, personal communication, 2008). On the basis of this study, the smelter administration reported an absence of ecological problems (Lukin & Gavrilova 2004), but it did not disclose the actual results of the research.

Perception of the Environmental Situation

Since the late 1980s–early 1990s, the Nadvoitsy aluminium plant has become the centre of the Greens' attention. An inspection (results of which we were unable to trace) produced a shocking report about an extreme excess of fluorine in the drinking water. A furor erupted, with a number of alarming papers published in newspapers and on the web (Baiduzhy 1994; Efron 1994; Kozlovich 2004). These papers claimed that there was a high degree of occupational illness, as well as outstanding development of fluorosis among the younger generation of Nadvoitsy (Efron 1994). However, according to another newspaper's report (Zlobin 2002), the significant worsening of the ecological situation around the works, which took place in the 1970s, is at present fully overcome. In 2001, emissions of substances detrimental to human health were reported to be within allowable limits. Thus, the 'ecological catastrophe' neither exists, nor is it anticipated (Zlobin 2002). However, it seems that these results have not convinced environmental activists (Kozlovich 2006), and the situation in Nadvoitsy was recently publicised in Northern Europe (Lorentzen 2005).

2.2.2.11 Nikel, Russia: Nickel-Copper Smelter, and Zapolyarnyy, Russia: Ore-Roasting Plant (Color Plates 43–47 in Appendix II)

Location and Environment

Nikel (the Finnish name is Kolosjoki), the administrative centre of the Pechengsky district, is an urban-type settlement (population 15,900; data from 2005) in the Murmansk Region. It is located on the shores of Kuetsjärvi (Kuets Lake) 195 km northwest of Murmansk and 7 km from the Norwegian border. Zapolyarnyy is a small (population 18,200; data from 2007) town located 25 km east of Nikel, half-way from Nikel to Pechenga (Fig. 2.12).

Both Nikel and Zapolyarnyy are situated in the north boreal and low-alpine vegetation regions close to the Arctic tree line. Pre-industrial vegetation of the landscape surrounding the towns, with hills of up to 600 m above sea level, was patchy, with sparse mixed forests of Scots pine and mountain birch to the south of Nikel (please see color plate 45 in Appendix II), and birch woodlands and tundras (please see color plate 44 in Appendix II) to the northeast.

Pre-industrial History

The ancestors of the indigenous population of Lapland, Saami, appeared in the Pechenga area in approximately 6000 BC. From the beginning of the sixteenth century, Russians (Pomors) built their temporary settlements here for fishing and trade. The Pechenga area (the Finnish name is Petsamo) was part of Russia between 1533 and 1920 and then became part of Finland. Discovery of the nickel ore deposits in the Pechenga Mountains (Petsamo-tunturi) in 1934 made this region very important in a strategic sense because nickel was in demand by military industry. The settlement of Kolosjoki was built 40 km southwest of the Pechenga Monastery, on approximately the same site where Pazrestkij Pogost (the Finnish name is Vanhatalvikylä) existed from the sixteenth century. The area became, again, a part of Russia (the Soviet Union at that time) in 1944.

Industrial History

In 1934, the Canadian company INCO obtained concessions for the entire ore-bearing area of 135 km^2. Preparatory works started in 1935, and mining was launched in 1940. In 1941, the area was occupied by Germany, and smelting (at the factory built by INCO of Canada) began in 1942. The smelter was very modern; it had the biggest convertors and the tallest (165 m) smokestack in Europe at that time. The mine and the smelter worked at full capacity until 1944, producing in total 16,000 t of nickel, 90% of which was exported to Germany. In 1944, there were 1,570 workers, including 360 war prisoners. The smelter was partially destroyed by retreating German troops in 1944, but it was reconstructed soon after (thanks to documentation bought by the Soviet government from the former Finnish owners for USD 20 million) and recommenced operations in 1946. The settlement of Zapolyarnyy was established in 1955 in association with the mining and processing of copper-nickel ores; it was granted town status in 1963.

Now Pechenganikel is a subsidiary of Norilsk Nikel, with four open pits, an enrichment plant, a roasting shop, and smelting and sulphuric acid production shops; it employs 10,000 workers. Its principal products are matte (processed on a tolling basis at the Severonikel smelter in Monchegorsk) and sulphuric acid. However, since smelting facilities at Nikel require extensive reparation, it is likely that the smelter will soon be (partially) closed, with simultaneous extension of the smelting facilities at Monchegorsk (BarentsObserver 2007).

Emissions Data

The smelter in Nikel contributed 98.0–99.9%, and the ore roasting factory in Zapolyarnyy contributed 90.6–99.7%, of aerial emissions reported for these towns during the period 1989–1999 (Tables 2.14, 2.15; Emissions of pollutants in Russia 1966–2006).

Pollution Loads and the Extent of the Contaminated Territory

Ambient concentrations of pollutants in both Nikel and Zapolyarnyy have monitored since 1967. The peak reported values for Nikel were 5.00 mg/m^3 of dust in 1978, 7.14 mg/m^3 of SO_2 in 1971, and 1.18 mg/m^3 of NO_x in 1978. The peak reported values for Zapolyarnyy were 3.25 mg/m^3 of SO_2 in 1979, and 1.87 mg/m^3 of NO_x in 1967 (Emissions of pollutants in Russia 1966–2006).

In spite of a substantial emissions decline (Table 2.14), high levels of ambient sulphur dioxide still occur at Nikel; for example, concentrations of 2.6 mg/m^3 (exceeding the sanitary limit of 0.05 mg/m^3 by a factor of 52) were officially reported on 4 and 15 July 2007 (www.kolgimet.ru). Importantly, the levels of ambient sulphur dioxide measured by an international research project from 1992 to 2005 were around three times higher than those officially reported by the Murmansk Hydrometeorological service, and they are considered unacceptably high (Stebel et al. 2007).

The long-term emissions impact resulted in slight soil acidification and severe contamination by metals. The peak reported metal concentrations in soils were 3,489 and 1,020 mg/kg of copper, and 2,990 and 2,230 mg/kg of nickel (for Nikel and Zapolyarnyy, respectively) (Niskavaara et al. 1996; Reimann et al. 1998; Kozlov & Zvereva 2007a). During the period 1987–1989, annual depositions of sulphur and nitrogen in Nikel were 3,300 and 200 kg/km^2, respectively (Vasilenko et al. 1991).

Table 2.14 Aerial emissions (t) from industrial enterprises at Nikel, Russia

Year	Dust	SO_2	CO	NO_x	Ni	Cu	Co
1943	–	55,000[a]	–	–	–	–	–
1971	–	131,800	–	–	–	–	–
1972	7,300	255,500	–	–	–	–	–
1973	7,300	255,500	–	14,600	–	–	–
1974	7,300	335,800	–	14,600	–	–	–
1975	7,300	335,800	–	14,600	–	–	–
1978	5,200	339,000	–	14,600	–	–	–
1979	5,200	339,000	–	14,600	–	–	–
1980	5,000	328,200	100	200	–	–	–
1981	4,800	290,100	100	–	–	–	–
1982	4,600	292,800	100	200	–	–	–
1983	4,700	293,200	200	400	240	170	10
1984	4,700	290,500	100	–	220	140	10
1985	4,400	274,900	200	100	210	130	10
1986	4,300	261,100	200	100	220	150	10
1987	4,000	257,900	200	100	190	130	6
1988	4,100	211,400	200	100	160	140	20
1989	3,900	199,600	200 800[b]	200	170	110	10
1990	3,900	190,100	200 4,200[b]	200 600[b]	136	92	5
1991	3,900	189,800	200 3,800[b]	200 600[b]	131	88	5
1993	3,500	160,600	300	200	130	87	5
1994	3,700	129,200	300	200	136	82	5
1995	3,900	175,400	340	160	137	85	5
1996	3,300	183,700	300	100	136	92	5
1997	3,700	183,500	300	100	155	96	5
1998	2,400	125,500	200	100	143	93	5
1999	2,500	90,100	700	200	155	100	5
2000	2,600	85,600	500	100	171	108	6
2001	2,600	85,600	500	100	171	108	–
2002	2,800	62,300	200	200	150	83	–
2003	2,400	60,600	100	200	149	83	5
2004	2,300	56,400	100	200	154	86	5
2005	2,200	55,500	70	10	157	88	5

Non-referenced values were extracted from Emissions of Pollutants in Russia (1966–2006).
Other data sources:
[a] Honkasalo (1989).
[b] Kryuchkov (1993a).

Satellite data showed that the total area affected by air pollution around Nikel increased from 400 km² in 1973 to more than 3,900 km² in 1988, and the area remained this size during the early 1990s (Høgda et al. 1995; Tømmervik et al. 1995). According to another estimate, between 1986 and 1991 a doubling of pollutant deposition (relative to the regional background) was recorded for 1,200 km² (within the Russian Federation only) around Nikel, Zapolyarnyy and Pechenga (Prokacheva et al. 1992).

Table 2.15 Aerial emissions (t) from industrial enterprises at Zapolyarnyy, Russia

Year	Dust	SO_2	CO	NO_x	Ni	Cu	Co
1971	9,900	52,900	–	–	–	–	–
1972	9,900	58,000	–	–	–	–	–
1973	10,950	60,200	–	3,700	–	–	–
1974	10,950	63,900	–	3,700	–	–	–
1975	10,950	63,900	–	3,700	–	–	–
1978	5,300	56,700	–	–	–	–	–
1979	5,300	60,000	–	8,600	–	–	–
1980	5,500	70,600	300	300	–	–	–
1981	6,300	71,300	300	–	–	–	–
1982	5,400	65,100	300	200	–	–	–
1983	5,500	75,500	300	300	190	90	–
1984	5,200	74,900	300	200	190	90	8
1985	4,600	79,300	300	200	170	90	10
1986	4,100	81,500	200	200	160		
1987	4,100	79,700	200	200	200	100	5
1988	4,100	78,600	300	200	200	100	5
1989	4,100	77,800	400	300	130	90	5
	4,500[a]		3,400[a]	1,200[a]			
1990	4,100	63,700	400	300	165	88	6
	4,500[a]		5,200[a]	1,500[a]			
1991	4,100	67,600	400	300	148	–	–
	4,500[a]		4,700[a]	1,400[a]			
1993	3,900	66,600	500	300	152	74	5
1994	6,400	69,200	500	300	161	81	5
1995	6,500	70,200	400	300	161	93	6
1996	5,900	52,500	400	200	162	85	5
1997	6,100	69,900	400	300	166	87	6
1998	5,800	65,200	400	200	180	95	6
1999	5,600	62,500	1,000	200	175	93	6
2000	5,500	65,600	900	200	183	98	6
2001	5,500	65,600	900	300	183	98	–
2002	5,500	61,600	70	300	183	86	–
2003	4,600	63,700	60	300	180	84	–
2004	4,300	56,000	100	300	175	82	–
2005	3,900	51,400	200	300	171	84	–

Non-referenced values were extracted from Emissions of Pollutants in Russia (1966–2006).
Other data sources:
[a] Kryuchkov (1993b).

Habitat Transformation due to Human Activity

Since the original vegetation was patchy, with a predominance of birch woodlands, the development of industrial barrens did not modify the landscape as much as in forested regions, such as Monchegorsk or Sudbury. The appearance of barrens is dated to the early 1960s for Nikel and the late 1960s for Zapolyarnyy. The overall extent of barren and semi-barren landscapes (please see color plates 45–47 in Appendix II) around these two towns with overlapping impact zones, was 340 km^2 in 1973 and 687 km^2 in 1999 (Tømmervik et al. 2003; Kozlov & Zvereva 2007a).

Brief History of Environmental Research

The effects of aerial emissions arising from Nikel and Zapolyarnyy on terrestrial biota were not properly documented until the late 1980s, in particular due to restricted access to this territory. However, the situation changed in the early 1990s, especially due to international concern about transboundary pollution effects. The studies mostly reported the effects on vegetation, including mosses, lichens, and vascular plants (Aamlid 1992; Tømmervik et al. 1998, 2003; Aamlid et al. 2000; Aamlid & Skogheim 2001; Chernenkova 2002; Bjerke et al. 2006). An extensive overview of recent studies is given by the AMAP (2006), with the most recent data summarised by Stebel et al. (2007).

Perception of the Environmental Situation

The industrial complex of Nikel and Zapolyarnyy poses its main threat to aquatic and terrestrial environments in the joint Norwegian, Finnish and Russian border area. The awareness of this situation exists both nationally and internationally (Stebel et al. 2007), although the Finnish people have not been concerned as much recently about the emissions from Nikel as they were in the 1980s–1990s. In 2007, pollution at Nikel became the focal point of several alarming publications (Popova 2007), and a criminal investigation was conducted in association with the excess of sulphur dioxide above sanitary limits and subsequent forest damage (nn.gazetazp.ru). The situation was widely debated, especially in Norway, and the Norwegian authorities even recommended evacuating the city (www.Euroarctic.com). People on the Norwegian side were indeed worried by this situation, but there was no panic (H. Tømmervik, personal communication, 2008).

2.2.2.12 Norilsk, Russia: Nickel-Copper Smelters (Color Plates 48–50 in Appendix II)

Location and Environment

Norilsk (population 209,300; data from 2007) is a heavily industrialised, major city in the Krasnoyarsk Region, located on the Taimyr Peninsula nearly 2,000 km to the north of Krasnoyarsk, and 300 km north of the Arctic Circle (Fig. 2.13). The city is situated on one of the largest nickel deposits on Earth, at the foot of the 1,700 m high Putoran Mountains. It is the northernmost city with a population over 100,000 on the planet, and one of two large cities in the continuous permafrost zone (the second one is Yakutsk, Russia).

Norilsk is situated approximately at the border between deciduous shrub tundra and sparse, 200–300 years old sub-tundra forests (please see color plate 48 in Appendix II). A detailed description of plant communities around Norilsk in 1962–1963 (Moskalenko 1965) to a large extent reflects the pre-industrial situation.

At about that time, the forests formed by Siberian larch or Gmelinii larch (*Larix gmelinii*) with white birch and Siberian spruce, covered 43% of the region (Kovalev & Filipchuk 1990).

Pre-industrial History

Archaeological evidence suggests that the copper ore deposits were first discovered and used in small amounts during the Bronze Age (Kunilov 1994; cited after Blais et al. 1999). For centuries, this area was only sparsely populated by natives who lived from hunting, fishing, and reindeer herding. The first known mining and smelting operations, resulting in production of 3 t of copper, are dated 1868. However, systematic exploration began only in the 1920s, and the first building in this unpopulated area was erected in 1921. The settlement of Norilsk was founded by the end of the 1920s; however, the official date is traditionally set to 1935, when Norilsk was expanded as a settlement for the mining-metallurgic complex. Town status was granted in 1953.

Industrial History

Initial exploration drilling began in the Norilsk area in 1923. The government of the USSR created in 1935 the Norilsk Combine, which used forced labour until the mid-1950s. The first mine to remove sulphide ores was operational in 1938, and the first copper-nickel matte was produced in 1939. The second line (the recent nickel smelter) was launched in 1942 using equipment evacuated from Monchegorsk; the copper smelter and the coke production plant began operations between 1947 and 1949. The third smelter, Nadezhda, was launched in 1979; it produces copper, fainstein, selenium, tellur, and concentrates of precious metals including platinum, palladium, Ro, Ru, Ir, Os, silver and gold. This smelter also has sulphur-processing facilities. By the late 1990s, these three smelters produced approximately 50% of Russia's nickel, 40% of its copper, 70% of its cobalt and 90% of its platinum group metals. A joint-stock company, Norilsk Nikel (created in 1993), now employs 56,000 inhabitants of Norilsk.

Emissions Data

Non-ferrous smelting is responsible for nearly 100% of emissions of sulphur dioxide, 77.5% of dust, and 67.6% of emissions of carbon oxide and nitrogen oxides at Norilsk (Savchenko 1998). The peak annual emissions of sulphur dioxide exceed 2.5 million tons (Table 2.16), contributing nearly 4% of global emissions (estimate after Lefohn et al. 1999).

Concentrations of nickel and copper in dust (according to Menshchikov & Ivshin 2006) suggest that metal emissions during the period 1990–1997 were much

Table 2.16 Aerial emissions (t) from industrial enterprises at Norilsk, Russia

Year	Dust	SO_2	CO	NO_x	Ni	Cu
1965	–	400,000[a]	–	–	–	–
1970	93,100	1,803,100	–	–	–	–
		750,000[a]				
1971	93,100	1,825,000	–	–	–	–
1972	143,100	1,825,000	–	–	–	–
1973	142,350	1,825,000	–	–	–	–
1974	142,350	1,825,000	–	–	–	–
1975	142,350	1,825,000	–	–	–	–
		1,225,000[a]				
1976	–	1,380,000[b]	–	–	–	–
1978	44,000	1,580,600	–	1,000	–	–
		1,490,000[b]				
1979	44,000	1,749,000	23,100	15,000	–	–
	59,750[c]	1,540,000[b]				
1980	41,000	1,622,700	27,800	31,800	–	–
	69,850[c]	1,742,000[b]				
1981	41,000	1,994,900[d]	29,400	29,400	–	–
	66,650[c]	1,880,000[b]				
1982	65,100	2,190,000	44,000	29,400	–	–
	62,900[c]	2,402,000[d]				
1983	73,300	2,520,300	58,800	29,000	–	–
	64,000[c]					
1984	52,600	2,195,000[b]	31,200	29,100	–	–
	71,450[c]	2,447,200				
		2,647,700[d]				
1985	47,400	1,890,000[b]	20,400	28,000	–	–
	59,750[c]	2,418,900[e]		29,600[e]		
		2,724,300[d]				
1986	35,400	1,850,000[b]	20,400	18,700	–	–
	54,950[c]	2,327,300		20,500[e]		
1987	33,800	1,760,000[b]	18,800	19,200	–	–
	42,150[c]	2,325,200[e]		20,800[e]		
1988	32,800	1,740,000[b]	18,600	16,900	–	–
	40,000[c]	2,244,300		18,600[e]		
		2,350,000[a]				
1989	30,900	1,710,000[b]	12,900	18,400	–	–
		2,207,500				
		2,300,000[a]				
1990	31,700	1,690,000[b]	15,000	20,200	–	–
		2,201,700				
1991	31,700	1,568,000[b]	7,790[a]	17,990[a]	1,300	3,021
		2,201,700	15,000	20,200		
		2,397,000[a]				
1992	29,710[a]	2,118,000[a]	10,040[a]	19,280[a]	–	–
1993	27,200	1,862,800	9,900	17,300	1,089	2,469
1994	26,600	1,860,300	10,400	14,700	1,284	2,382
1995	27,500	1,961,300	8,900	16,000	1,449	2,415
1996	24,100	2,014,400	37,400	12,200	1,198	2,151
1997	26,000	2,104,800	16,700	11,300	1,432	1,958
1998	20,100	2,064,900	15,400	11,600	916	1,750

(continued)

Table 2.16 (continued)

Year	Dust	SO_2	CO	NO_x	Ni	Cu
1999	20,300	2,104,000	12,400	10,700	3,000[f]	8,250[f]
					793	1,752
2000	16,800	2,078,800	13,200	10,500	596	1,069
2001	15,000	2,050,100	11,500	10,900	595	1,069
2002	14,200	1,965,000	12,400	9,300	629	–
2003	13,900	1,846,600	14,500	8,300	558	562
2004	–	2,011,200	–	8,200	556	661
2005	12,400	1,955,300	10,200	9,000	539	626

Non-referenced values extracted from Emissions of Pollutants in Russia (1966–2006).
Other data sources:
[a] Savchenko (1998).
[b] Kharuk (2000).
[c] Kharuk et al. (1996).
[d] Kovalev and Filipchuk (1990).
[e] Bezuglaya et al. (1991).
[f] Koutsenogii et al. (2002).

higher than officially reported (Table 2.16), comprising 3,900 t of copper, 3,100 t of nickel, and 84 t of cobalt annually. Other data sources reported annual emissions of 1,800 t of copper, 4,000 t of nickel, and 65 t of vanadium in the early 1990s (Nilsson et al. 1998; cited after Kharuk 2000) and of 3,000 t of copper and 8,250 t of nickel in 1999 (Koutsenogii et al. 2002). Ore chemistry and annual production data suggest that in 1994 smelters at Norilsk also emitted 150 t of lead, 29 t of zink, and less that 1 t of arsenic and antimony (Boyd et al. 2009). It was even claimed that 'heavy metal pollution in the area is so severe that the soil itself has platinum and palladium content which is feasible to mine' (Kramer 2007). All this information casts doubt on the reliability of official emissions data.

The strategic goal announced by Norilsk Nikel in 2005 is a fivefold decrease of sulphur dioxide emissions (i.e., to 426,000 t or less) by 2015 (Ershov et al. 2005). By 2010, Norilsk Nikel plans to implement large investment projects aimed at reconstruction of production facilities to reduce the environmental impact. Actions toward the gradual decrease of emissions resulted in 2006 in a decline of SO_2 emissions by 0.84%, and of dust emissions by 12.9%, relative to 2005 (Norilsk Nikel 2007).

Pollution Loads and the Extent of the Contaminated Territory

Ambient concentrations of pollutants in Norilsk have been monitored since 1968; the peak reported values were 8.50 mg/m^3 of dust in 1978, 68.4 mg/m^3 (sic!) of SO_2 in 1981, and 0.43 mg/m^3 of NO_x in 1978. Even in 1990, the average annual concentration of SO_2 exceeded the sanitary limit (0.05 mg/m^3) by a factor of 2–3, with a peak value of 19 mg/m^3 (Bezuglaya 1991).

In 1987–1989, annual depositions of sulphur and nitrogen in Norilsk were 12,200 and 1,200 kg/km^2, respectively (Vasilenko et al. 1991). The concentration of copper in the upper soil horizon 600 m from the copper smelter was 20,600 mg/kg (Gorshkov 1997), and soils within approximately 1,350 and 1,000 km^2 contained copper and nickel, respectively, in concentrations exceeding the regional background by a factor of 10 or more (Igamberdiev et al. 1994). Dust deposition exceeded the background values by a factor of 100 or more for 70 km^2, with peak values (1,000–1,100 t/km^2) 750 times higher than the regional background (Chekovich et al. 1993).

In spite of extreme emissions of sulphur dioxide, soil, snow and water of local lakes were slightly alkaline due to high concentrations of base cations (Blais et al. 1999; Menshchikov & Ivshin 2006). Still, the lowest pH values of soil in the impact zone were 4.4, compared with 5.3–7.0 in unpolluted areas (Kharuk 2000).

Norilsk is the only city whose individual footprint is clearly visible on a map showing the annual deposition of sulphur for the entire territory of the former USSR (Vasilenko et al. 1991). According to this map, the deposition exceeded the regional background (less than 250 kg/km^2) over 246,000 km^2. According to another estimate, the overall extent of the contaminated zone over a 'relatively clean' background in 1989 was approximately 34,200 km^2 (including both land and lakes); the strongly contaminated zone covered 2,300 km^2 (Klein & Vlasova 1992). At the same time, another report indicated a doubling of pollutant deposition (relative to the regional background) during the period 1986–1991 for 7,000 km^2 (Prokacheva et al. 1992).

Although the external border of the metal-contaminated area to the north of the smelter extended less than 100 km (Allen-Gil et al. 2002), aerosols emitted by Norilsk smelters were reported to reach both Severnaya Zemlya and Franz Josef Land, thus substantially contributing to regional contamination of the Arctic (Shevchenko et al. 2003).

Habitat Transformation due to Human Activity

We date the development of the industrial barrens around Norilsk to the mid-1950s. Different data sources suggest that the extent of barren and semi-barren sites around Norilsk in 1990s was around 4,000 km^2 (Kozlov & Zvereva 2007a). The initial deforestation, leading to the rapid development of industrial barrens, was created here for a very specific reason; since prisoners were used extensively for construction of the Norilsk plant, a buffer of 3–4 km was cut around each of the prison camps (Kharuk 2000).

The first report of forest decline near Norilsk is dated 1968; by this time, 50 km^2 of forest was dead (Kharuk et al. 1995). The extent of forest damage increased from 3,384 km^2 in 1976 (including 446 km^2 of dead forests) to 5,452 km^2 in 1986 (Kovalev & Filipchuk 1990). Other data sources suggest that the effects of emissions on vegetation in the late 1980s–early 1990s were detectable over 35,500 km^2 (Menshchikov 1992), or even 70,000 km^2 (Chekovich et al. 1993). The external border of declining forests (please see color plate 49 in Appendix II) in 1987 was located 60 km to the east and 120 km to the southeast. Visual signs of larch damage, such as needle discoloration, were detected up to 200 km from the polluter (Kovalev & Filipchuk 1990; Kharuk et al. 1995). In 2002, we observed necroses on white birch leaves at our most distant plot, located 96 km east of Norilsk. Still, some 'optimistic' publications report that '60–70 km from Norilsk [one can see] virgin sub-tundra forests in their original beauty' (Grachev 2004, p. 102).

Damage to terricolous lichens was detected at distances of 200–250 km (Kharuk 2000), or even 300 km from the smelters (Otnyukova 1997). The lichen desert area in 1989 was estimated to cover 3,000 km^2, and lichens over 6,000 km^2 exhibited retarded growth and visual signs of damage (Klein & Vlasova 1992).

Brief History of Environmental Research

Although scientific exploration of the pollution impact on terrestrial ecosystems around Norilsk started no later than the 1970s, almost no information was published before the early 1990s. The majority of publications about the Norilsk environment are too general and only rarely provide solid conclusions supported by both a description of methodology and statistical analysis. The book by Menshchikov and Ivshin (2006) is especially disappointing due to incomplete presentation of unbalanced data sets (or presentation of relative values only), a messy structure (i.e., distance of some study plots from the polluter remains unknown; data on plant growth and contamination originate from different subsets of study plots), and an absence of straightforward conclusions.

The reported biotic effects of pollution are limited to soil microbiota (Raguotis 1989; Kirtsideli et al. 1995), vitality of conifers (measured using an arbitrary index), decline in the radial increment of larch and spruce (albeit without statistically sound comparison between polluted and unpolluted sites; Simachev et al. 1992; Demyanov et al. 1996; Schweingruber & Voronin 1996; Menshchikov & Ivshin 2006), decline in forest biomass (Polyakov et al. 2005), changes in vegetation structure (Varaksin & Kuznetsova 2008), and remote sensing data outlining the extent of the pollution damage (Kharuk et al. 1995, 1996, Tutubalina & Rees 2001; Zubareva et al. 2003). For an overview of the environmental situation around Norilsk, consult Kharuk (2000). The acidity status of soils is discussed in the AMAP (2006) report.

Perception of the Environmental Situation

The Norilsk phenomenon can be seen as the largest ecological catastrophe on a global scale (Kharuk et al. 1996). The situation in Norilsk has long been attracting public attention on both national and international scales, and many alarming reports have been published in past years. The broad awareness of this particular case is emphasised by description of the pollution problem at Norilsk in Wikipedia (en.wikipedia.org). The Blacksmith Institute (2007) included Norilsk in its list of the ten most polluted places on Earth. Although there is no doubt that the problem is rather acute, some statements that have appeared in mass media, such as, 'Within 30 mi (48 km) of the nickel smelter there's not a single living tree' (Walsh 2007), are incorrect.

According to a recent BBC News (2007) report, Norilsk Nikel accepted responsibility for what has happened to the forest, but the company stressed that it was taking action to cut pollution. Voting arranged on a local website (www.norilskinfo.ru) for 2 weeks in January 2008 did not attract much attention; only two of 70 visitors identified local pollution problems as life-threatening.

2.2.2.13 Revda, Russia: Copper Smelter (Color Plates 51–54 in Appendix II)

Location and Environment

Revda (population 61,800; data from 2007) is an industrial town in the Sverdlovsk Region, located on the western slopes of the Middle Ural Mountains, 45 km west of Ekaterinburg (Fig. 2.14). Over 60% of the surrounding territory is covered by southern taiga, now represented mostly by secondary stands co-dominated by Siberian fir and Siberian spruce (please see color plate 52 in Appendix II).

Pre-industrial History

The first settlements on the Revda River, to the south of the recent position of the town, already existed 7–8,000 years ago. However, the territory remained sparsely populated until the beginning of the eighteenth century.

Industrial History

Development of Revda started in 1734 with the launching of an iron foundry. This factory, owned by Akinfij Demidov, produced 3,300–4,900 t of iron annually by the late 1700s. By 1907, production had increased to 16,000 t. In 1897, Revda had 8,000 inhabitants.

The decision by the Soviet government to build a copper smelter between Revda and Pervouralsk was made in 1931. Industrial development started from the building of roads and the railway and constructing a brick factory and wood processing facilities. Town status was acquired in 1935. The concentration mill was launched in 1937 and expanded in 1939. The building of the Middle Ural Copper Smelter started in 1938, and the first copper was produced in 1940. Development of the smelter continued after World War II; it has produced sulphuric acid since 1963 and super phosphate since 1972. Renovation of the smelting facilities began in 1984, and renovation of the shop producing sulphuric acid began in 2004. In 2006, the smelter employed 4,527 workers and produced 95,233 t of copper, 465,015 t of sulphuric acid, and substantial amounts of other chemical products (www.sumz.umn.ru). Revda also hosts iron and non-ferrous processing factories, a brick plant, a mechanical enterprise and several small-scale industries.

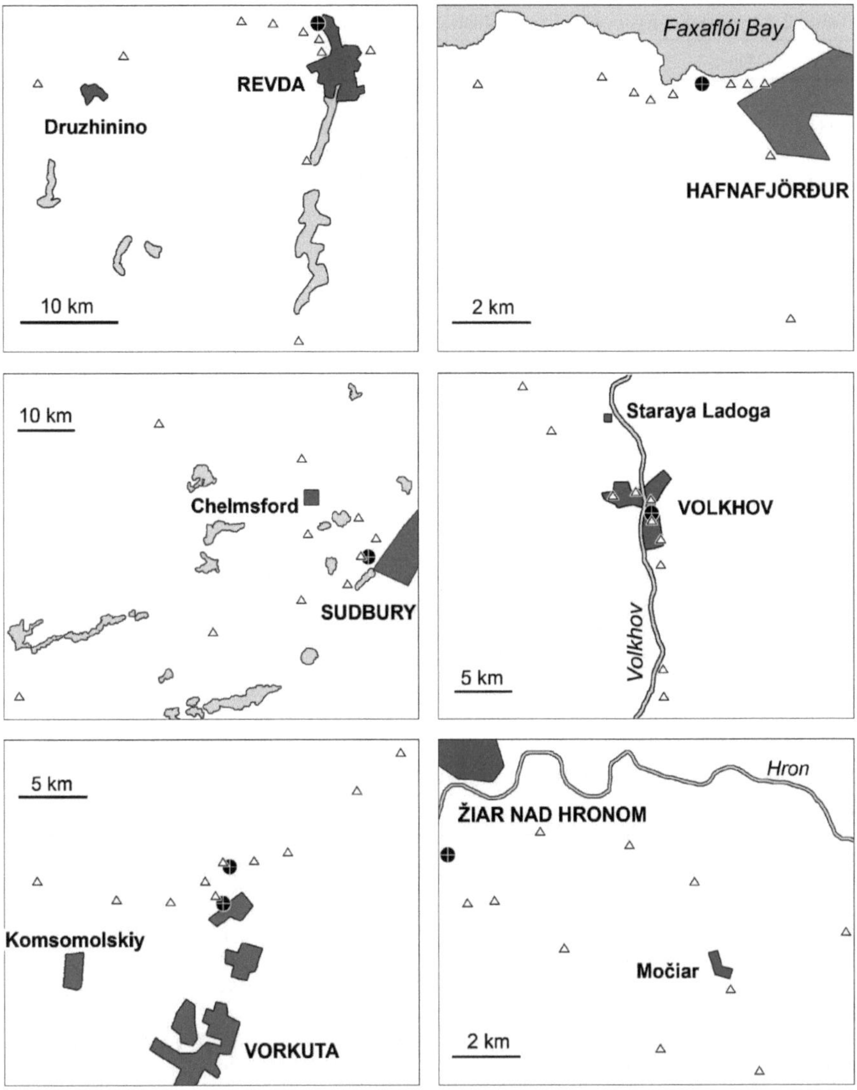

Figs. 2.14–2.19 Distribution of study sites (triangles) around investigated polluters (black circles with white cross inside). Names of larger settlements (usually towns) are capitalised. **2.14**, copper smelter in Revda, Russia; **2.15**, aluminium smelter in Straumsvík, Iceland; **2.16**, nickel-copper smelter in Copper Cliff near Sudbury, Canada; **2.17**, aluminium smelter in Volkhov, Russia; **2.18**, power plant and cement factory near Vorkuta, Russia; **2.19**, aluminium smelter in Žiar nad Hronom, Slovakia

Emissions Data

Metallurgy is responsible for 91–99% of all industrial emissions in Revda (Emissions of pollutants in Russia 1966–2006). The contribution of the copper

Table 2.17 Aerial emissions (t) from industrial enterprises at Revda, Russia

Year	Dust	SO_2	CO	NO_x	HF	As	Pb	Cu	Zn
1980	25,100	201,700	21,100	3,100	1,950	–	–	–	–
1981	28,900	143,800	20,200	3,200	1,600	–	–	–	–
1982	26,700	137,400	20,400	1,200	1,400	–	–	–	–
1983	26,799	136,200	20,300	1,200	1,458	–	–	–	–
1984	25,700	142,100	20,500	1,300	1,426	–	913	–	–
1985	25,300	141,100	20,000	1,100	1,340	–	890	–	–
1986	24,500	140,800	19,300	1,200	1,300	900	750	–	–
1987	23,600	139,500	19,300	1,200	1,240	639	565	–	–
1988	23,900	137,800	20,100	1,200	1,202	639	565	–	–
1989	23,500	134,400	20,100	1,200	1,016	639	564	–	–
1990	25,400	131,700	2,600	1,800	–	–	–	–	–
1993	14,600	91,100	2,400	1,100	191	310	401	–	–
1994	13,400	86,000	2,200	1,000	78	287	358	1,163	955
1995	10,400	64,600	2,400	1,100	42	205	292	848	804
1996	10,100	88,400	2,200	1,100	47	172	344	800	766
1997	11,500	86,200	2,000	900	84	216	354	761	903
1998	8,900	64,800	1,700	900	45	121	353	633	812
1999	7,700	58,200	2,200	200	32	116	353	405	828
2000	7,300	56,300	1,600	1,100	32	100	324	337	769
2001	7,300	52,800	1,400	1,000	27	60	302	207	561
2004	3,600	25,300	1,300	900	19	14	160	37	265
2005	3,000	24,300	900	600	20	18	146	49	255

All values were extracted from Emissions of Pollutants in Russia (1966–2006).

Table 2.18 Aerial emissions (t) from the Middle Ural Copper smelter at Revda, Russia

Year	Dust	SO_2	NO_x	HF	As	Pb	Cu	Zn
1980	–	–	–	–	943[a]	1,077[a]	4,400[a]	–
1986	20,910[b]	140,625[b]	470[b]	1,295[b]	908[b]	754[b]	–	–
1987	16,081[b]	139,325[b]	485[b]	1,242[b]	637[b]	565[b]	2,617[b]	1,779[b]
1988	16,110[b]	137,645[b]	485[b]	1,016[b]	639[b]	565[b]	2,918[b]	1,769[b]
1989	16,086[b]	134,089[b]	479[b]	1,011[b]	640[b]	564[b]	2,610[b]	1,753[b]
1990	15,967[b]	130,827[b]	477[b]	1,014[b]	620[b]	546[b]	2,532[b]	1,701[b]
1991	–	95,370[b]	375[b]	935[b]	455[b]	450[b]	1,480[b]	1,505[b]
1992	–	95,369[b]	361[b]	375[b]	353[b]	449[b]	1,479[b]	1,173[b]
1995	–	63,628[c]	–	–	–	292[c]	842[c]	791[c]
2003	–	29,600[c]	–	–	–	217[c]	88[c]	326[c]

All the values were extracted from publications that refer to the copper smelter as to the sole polluter.
Data sources:
[a]Zyrin et al. (1986).
[b]Yusupov et al. (1999).
[c]E. Vorobeichik (personal communication, 2006).

smelter is difficult to evaluate due to a shortage of reliable information. Therefore, we have grouped all the data into two tables, one reporting emissions from all industrial enterprises of Revda (Table 2.17) and the other referring to the copper smelter only (Table 2.18).

Pollution Loads and the Extent of the Contaminated Territory

Ambient concentrations of pollutants in Revda have been monitored since 1972; the peak reported values were 8.70 mg/m^3 of dust in 1975, 10.2 mg/m^3 of SO_2 in 1973, 0.65 mg/m^3 of NO_x in 1978, and 0.47 mg/m^3 of HF in 1975 (Emissions of pollutants in Russia 1966–2006). In the mid-1970s, an average multiyear ambient concentration of SO_2 within 2 km from the smelter was 2.04 mg/m^3 (Fimushin 1979).

From 1987 to 1989, annual depositions of sulphur and nitrogen in Revda were 11,700 and 1,200 kg/km^2, respectively (Vasilenko et al. 1991). Maximum annual depositions of metals (presumably in the early 1990s) were 1,044 kg/km^2 of copper, 443 kg/km^2 of zinc, 244 kg/km^2 of lead and 13 kg/km^2 of cadmium (Kaigorodova & Vorobeichik 1996). In the late 1990s, SO_2 concentrations 2 km from the smelter reached 15 mg/m^3 (Koroleva & Shavnin 2000). The pollution impact caused soil acidification (the lowest recorded pH = 2.9). Peak reported metal concentrations in soils were 14,000 mg/kg of copper (Sataeva 1992), 4,194 mg/kg of zinc, 2,348 mg/kg of lead, and 80 mg/kg of cadmium (Vorobeichik 2003a, b; Kozlov & Zvereva 2007a).

The first detailed survey of environmental contamination and ecosystem damage around Revda, with maps showing concentrations of metals in soil and plants, was performed in the early 1980s (Zyrin et al. 1986). The overall extent of the 'damaged' area was between 21 km^2 (soil 'damage') and 57 km^2 (visible vegetation damage) (Figs. 3.5 and 3.11 in Zyrin et al. 1986). Soil contamination by copper exceeded the regional background over 686 km^2 (Zyrin et al. 1986). Data from 1986–1991 provide a similar estimate, suggesting a doubling of pollutant deposition (relative to the regional background) over 700 km^2 (Prokacheva et al. 1992). Measurements from 1995–1998 revealed that soil contamination by metals expanded over 1,800 km^2 (Vorobeichik 2003b).

Habitat Transformation due to Human Activity

Development of the industrial barrens around Revda (please see color plates 53 and 54 in Appendix II) is tentatively dated to the early 1950s (Kozlov & Zvereva 2007a); their extent in the early 1960s was outlined by Tarchevskij (1964). Importantly, a part of this barren zone (to 2 km east and between 0.5 and 1.5 km southeast) had (at least until 1945) been used for agricultural purposes. Thus, the industrial barrens in this area differ from the industrial barrens located south and north to northeast of this polluter, which have developed directly from declining coniferous forests (Tarchevskij 1964). In the early 1980s, the extent of the industrial barrens was between 3.7 km^2 (as estimated from vegetation damage) and 5 km^2 (as estimated from soil erosion) (Zyrin et al. 1986, Figs. 3.5 and 3.11). In the 2000s, the industrial barrens occupied less than 3 km^2 (Kozlov & Zvereva 2007a).

In the mid-1970s, visible forest damage by pollution occurred over approximately 415 km^2 (Fimushin 1979). We are not aware of any publications reporting the overall recent extent of forest damage. However, severe forest damage in the late 1990s was recorded over 90 km^2 (Fomin & Shavnin 2001).

Brief History of Environmental Research

Exploration of the environmental effects caused by emissions of the copper smelter at Revda started in the early 1960s (Tarchevskij 1964). Researchers from the Institute of Plant and Animal Ecology (Ekaterinburg) contributed the most to the exploration of this impact zone. The history of environmental research at this institution has recently been outlined by Vorobeichik (2005), who also listed the most important publications related not only to Revda but also to other polluters in the Ural region. Several monographs are entirely (Vorobeichik et al. 1994; Yusupov et al. 1999; Bezel et al. 2001) or partly (Bezel 1987, 2006; Bezel et al. 1994, 2001; Shebalova & Babushkina 1999; Vasfilov 2002) based on environmental data collected around Revda. As a result, the impact zone of the copper smelter at Revda is one of the best explored polluted regions in the world; most functional and taxonomic groups of terrestrial biota have been studied there. The most detailed information has been accumulated on pollution-induced changes in soil quality and microbiological activity (Kovalenko et al. 1997; Shebalova & Babushkina 1999; Vorobeichik 1995, 2002, 2007), soil-dwelling invertebrates (Vorobeichik 1998; Ermakov 2004), mycorrhizal fungi (Veselkin 2004a, b) and xylotrophic basidiomycetes (Bryndina 2000), epiphytic lichens (Mikhailova & Vorobeichik 1999; Scheidegger & Mikhailova 2000; Mikhailova 2007), structure and productivity of plant communities (Khantemirova 1996; Goldberg 1997; Shavnin et al. 1999; Trubina 2002), and the diversity and breeding success of birds (Belskii & Lyakhov 2003; Eeva et al. 2006a) and small mammals (Lukyanova 1990; Mukhacheva 1996, 2007). Importantly, a substantial number of data were analysed in terms of dose–effect relationships, and this analysis demonstrated non-linearity in ecosystem responses, i.e., the existence of threshold pollution load at which the ecosystems undergo a rapid transition between two alternative, relatively stable, states (Vorobeichik et al. 1994; Vorobeichik & Khantemirova 1994; Vorobeichik 2003b). Unfortunately, the vast majority of the results are published in Russian, largely in low-circulation books.

Perception of the Environmental Situation

Although the Russian version of Wikipedia (ru.wikipedia.org) and several popular publications (Kočí 2006) claim the existence of a local ecological catastrophe, we did not find signs of specific public anxiety.

2.2.2.14 Straumsvík, Iceland: Aluminium Smelter (Color Plates 55–57 in Appendix II)

Location and Environment

The smelter is located on the northwestern coast of the Reykjanes Peninsula, 12 km southwest of Reykjavik, and 2 km outside of Hafnarfjörður (Hafnarfjord), the third largest settlement in Iceland (population 25,000; data from 2007) and one of the nation's largest fishing centres (Fig. 2.15). Although the area was presumably covered by birch woodlands when Iceland was first settled 1,100 years ago, currently the smelter is surrounded by a flat, open, treeless landscape (please see color plates 56 and 57 in Appendix II). Specific vegetation coverings of the lava fields near the smelter are classified as grasslands, grassy heaths, and heathlands (Magnússon & Thomas 2007).

Pre-industrial History

Cove Straumsvík, located between the smelter and the former farm Straumur, south of the town Hafnarfjörður, was a harbour and a trading post frequented by German traders during the late Middle Ages. Hafnarfjörður was first named in the medieval 'Book of Settlements', and the earliest reports of voyages to Hafnarfjörður date from the end of the fourteenth century. The settlement attained official municipal status in 1908.

Industrial History

In 1966, the Icelandic Parliament permitted the construction of the aluminium smelter. Construction started in 1967, and the smelter was launched in 1969. Soon the smelter at Straumsvík became a major polluter. For about a decade it was the largest industrial firm in Iceland without any pollution control, until dry cleaning was set up in the late 1970s. HF cleaning efficiency remained about 93% until the late 1980s, when it increased to 99.3% (G. M. Gíslason 2008, personal communication).

The initial production was 33,000 t of aluminium annually; since then, the smelter has been expanded four times, and its capacity increased to 162,000 t in the late 1990s (US Geological Survey 1999) and 179,000 t in 2006. Recently, the smelter staff included 500 people. A small harbour was constructed for the import of raw materials and the export of aluminium. Electricity is supplied by the hydroelectric power station at Burfell. From the beginning of operations to the end of 2006, the plant has produced 3.7 million tons of aluminium. The area of Straumsvík also houses a steel smelter, an asphalt factory, and several small-scale industries (Magnússon 2002, Magnússon & Thomas 2007).

Emissions Data

Data from 1980–2006 are summarised in Table 2.19. Tómasson and Thormar (1998) also reported aerial emissions of CF_4, which decreased from 62 t in 1988 to 3 t in 1996.

Pollution Loads and the Extent of the Contaminated Territory

Ambient concentrations of fluorides within 1 km of the smelter from 1977 to1980 exceeded 0.001 mg/m^3 (re-calculated for F) in 12 samples out of 16, with a peak

Table 2.19 Aerial emissions (t) from the aluminium smelter at Straumsvík, Iceland

Year	SO_2	Dust	Fluorine (total)
1980	1,444	2,408	1,137
1981	1,447	2,408	1,044
1982	1,577	1,401	681
1983	1,436	330	270
1984	1,489	552	536
1985	1,311	514	499
1986	1,473	588	477
1987	1,587	719	575
1988	1,485	428	576
1989	1,441	625	568
1990	1,480	290	433
1991	1,546	401	–
1992	1,558	70	95
1993	1,610	49	122
1994	1,680	50	139
1995	1,543	48	160
1996	–	68	122
1998	2,435[a]	162[a]	146[a]
1999	2,696[a]	100[a]	106[a]
2000	2,722[a]	129[a]	97[a]
2001	2,524[a]	177[a]	108[a]
2002	2,638[a]	160[a]	127[a]
2003	2,496[a]	167[a]	113[a]
2004	2,391[a]	173[a]	98[a]
2005	2,477[a]	151[a]	111[a]
2006	2,395[a]	141[a]	106[a]

Non-referenced data after Tómasson and Thormar (1998, Table 2.1; reprinted, in a modified form, with permission from Natturfræðingurinn).
[a]Environment and Food Agency of Iceland (personal communication, 2008)

value of 0.094 mg/m^3. Concentrations of sulphur dioxide exceeded 0.050 mg/m^3 in five samples out of nine, with a peak value of 0.090 mg/m^3 (Thormar & Jóhannesson 1981; cited after Gíslason & Helgason 1989). In 1994–1995 at Hvaleyrarholt (2.5 km northeast of Straumsvík), the average concentration of HF was 0.00005 mg/m^3 (peak value of 0.00049 mg/m^3), and that of SO_2 was 0.00088 mg/m^3 (peak value of 0.0107 mg/m^3) (Tómasson & Thormar 1998). Current measurements at this location are available at www.vista.is.

Fluorine analyses in vascular plants and mosses suggested that in 1971, the contaminated zone extended 10–11 km from the smelter; the peak concentration of 558 mg/kg was recorded at a distance of 900 m in the moss *Hylocomium splendens* (Comission on Fluorine Tolerance Limits 1971, cited after Gíslason & Helgason 1989). Concentrations of fluorine in grass have steadily declined since the late 1980s, following a decrease in emissions (Tómasson & Thormar 1998).

From 2000–2005, the extent of the metal-contaminated territory was less than 3 km from the smelter. The increased levels of arsenic (three times the regional background), nickel (five times the regional background) and sulphur (10% over the regional background) in mosses were attributed to the smelter, while 1.5–5 fold increases in cadmium, copper, lead and zinc most likely resulted from other industrial activities (Magnússon 2002). By 2005, the concentrations of nickel and sulphur near the smelter decreased relative to 2000, although they remained higher than in more distant localities (Magnússon & Thomas 2007).

Habitat Transformation due to Human Activity

Substantial changes in vegetation structure, including the replacement of declining mosses and lichens by crowberry (*Empetrum nigrum*), a decrease in plant species richness, and an increase in the area of exposed rock surfaces, were detected close to the smelter (300 m). Sites 2 km away were unaffected in terms of vegetation structure (Kristinsson 1998).

Brief History of Environmental Research

Very few environmental studies have been conducted in the vicinity of the smelter, and most of them concern concentrations of pollutants in ambient air and plants (Gíslason & Helgason 1989; Magnússon 2002, Magnússon & Thomas 2007). To our knowledge, a survey of plant communities by Kristinsson (1998) is the only publication reporting changes in terrestrial biota near the aluminium smelter at Straumsvík.

Perception of the Environmental Situation

Discussion of environmental issues has increased considerably in Iceland during 2000s, and Alcoa (the owner of the Straumsvík smelter) is listed as 'The Nature Killer' (www.savingiceland.org). Plans to increase production of aluminium at Straumsvík to 400,000 t were blocked by local people with a referendum (88 deciding votes). Inhabitants of Hafnarfjörđur voted against the expansion of the smelter in particular to prevent further hydroelectric development because a valuable area of Thjórsárver in the Central Highlands of Iceland would be partially inundated by a hydroelectric reservoir to supply sufficient electricity for the expanded Straumsvík smelter. Moreover, expansion of the smelter would reduce the area available for housing development in Hafnarfjörđur (G. M. Gíslason 2008, personal communication). Even after a referendum, the people of Hafnarfjörđur continued to protest, including by means of public actions such as blocking the gates of the smelter (Krater 2006).

2.2.2.15 Sudbury, Canada: Nickel-Copper Smelters (Color Plates 58–60 in Appendix II)

Location and Environment

Greater Sudbury (population 157,857; data from 2006) is a city in Northern Ontario, about 400 km North of Toronto (Fig. 2.16). It was created in 2001 by amalgamating the cities and towns of the former Regional Municipality of Sudbury.

Sudbury lies in the Great Lakes–St. Lawrence forest region. One part of this region was originally covered by stands of red and white pine (*Pinus resinosa* and *P. strobus*). In another part, white pine was mixed with hardwoods such as sugar maple (*Acer saccharum*) and yellow birch (*Betula alleghaniensis*). Although the pre-industrial state of vegetation within the severely degraded area was not described, stumps and vestiges hint at a mosaic of pine forests on the slopes and white cedar (*Thuja occidentalis*) swamps in many of the depressions.

Pre-industrial History

The area around the current location of Sudbury has been populated for at least 7,000 years. In 1824, the Hudson Bay Company established a fur-trading post near what was later to become Sudbury. Originally named Sainte-Anne-des-Pins, the community developed from a small lumber camp in McKim Township. Lumbering started in 1872 and until 1927 remained the dominant business, despite the emergence of the mining industry in the 1880s.

Industrial History

The first roast yard and smelter were set up in Copper Cliff in 1888, and open-bed roasting continued until 1929. The development of smelting facilities at Copper Cliff started in 1903, and the present smelter has been operating since 1930. An additional smelter was built in Coniston in 1913 (closed in 1972), and the third smelter was opened at Falconbridge in 1928. In 1972, the famous Super Stack, 381 m in height, was built at the Copper Cliff smelter. For more historical details, consult Gunn (1995) and Sudbury Area Risk Assessment (2008); statistics on annual metal production at the INCO and Falconbridge smelters and refineries are summarised in the Canadian Minerals Yearbooks (www.nrcan.gc.ca).

Emissions Data

By 1949, more than 900,000 t of sulphur dioxide had been discharged into the atmosphere (Linzon 1958; cited after Costescu & Hutchinson 1972). In the early 1960s, Sudbury's copper-nickel complex with three smelters (listed above) was one of the largest point sources of industrial pollution, contributing approximately 4% of global sulphur dioxide emissions. The peak emission of sulphur dioxide (2,560,000 t) was recorded in 1960, and since then it has steadily declined to approximately 700,000 t in the 1980s and 200,000 t in the 2000s, with 183,000 t in 2006. Annual data on SO_2 emission for the years 1960–1994 (total for all Sudbury smelters and separately for the Falconbridge smelter) were published by Gunn (1995, Figs. 4.6 and A4.2). Data for the years 1970–2006 are available on the INCO website (www.inco-cc-smelter.com). Average emissions from four different sources at the Copper Cliff smelter and the Falconbridge smelter from 1973–1981, along with annual data for SO_2 and particulate emissions through the superstack, were reported by Chan and Lusis (1985); separate data for the Copper Cliff and Falconbridge smelters were published by Pollution Probe (2003).

Data on particulate emissions are less detailed. It was estimated that they totalled approximately 35,000 t annually in the 1960s, approximately 10,000 t in 1976–1977, and approximately 3,500 t annually in the mid-1980s (Freedman & Hutchinson 1980a; Allum & Dreisinger 1987). Dust emissions during the period from 1973–1981 contained 1,800 t of iron, 700 t of copper, 500 t of nickel, 200 t of lead and 125 t of arsenic (Chan & Lusis 1985).

Recently, emissions of sulphur dioxide, total particulate matter, and some metals from smelting at Copper Cliff for the period 1930–2003 were summarized by Bouillon (2003; cited after Sudbury Area Risk Assessment 2008). Since the available information greatly differs in both spatial (different emission sources included) and temporal resolution (data are shown for different time periods), we do not provide a summary table for Sudbury emissions. This work remains to be done by someone who has access to the archives of the Sudbury smelters.

Pollution Loads and the Extent of the Contaminated Territory

The long-term emission impact resulted in soil acidification (the lowest reported pH was 2.0) and severe contamination by metals. The peak reported concentrations in soils were 9,700 mg/kg of copper, 12,300 mg/kg of nickel, 336 mg/kg of zinc, and 92 mg/kg of lead (Hutchinson & Whitby 1974; Hazlett et al. 1983; Dudka et al. 1995).

By the 1960s, sulphur dioxide adversely affected 5,300 km^2 (Dreisinger & McGovern 1970); increased concentrations of nickel and copper were reported up to 50–70 km from the nearest smelter (Hutchinson & Whitby 1974; Freedman & Hutchinson 1980a). The territory with average mean ambient concentrations of sulphur dioxide over 0.01 ppm (equivalent to approximately 0.013 mg/m^3), as estimated from lichen surveys, decreased from 480 km^2 in 1968 (LeBlanc et al. 1972) to 230 km^2 in 1978 (Beckett 1984) due to an emissions decline and construction of the superstack.

Habitat Transformation due to Human Activity

Development of industrial barrens on hilltops close to the roast yards started prior to 1920 (Allum & Dreisinger 1987). A combination of lumbering, forest fires, smelter emissions and soil erosion created a barren landscape that, prior to the beginning of reclamation, occupied approximately 100 km^2. This 'moonscape' was surrounded by 360 km^2 of open woodland dominated by stunted and coppiced trees of canoe birch (*Betula papyrifera*), red maple (*Acer rubrum*) and Northern red oak (*Quercus rubra*) (Winterhalder et al. 2001). In total, approximately 1,000 km^2 was classified as smelter-damaged lands, of which barren lands (please see color plate 59 in Appendix II) occupied 61.6 km^2 in 1970 and 35.7 km^2 in 1989. This difference can be attributed to both emissions reduction (that allowed vegetation to recover naturally) and implementation of the land reclamation program (McCall et al. 1995).

Brief History of Environmental Research

Environmental deterioration in the Sudbury area is perfectly documented (Courtin 1994; Gunn 1995; Munton 2002). Historically, the studies focussed on freshwater ecosystems (Gunn 1995; Nriagu et al. 1998; Yan et al. 2003), while terrestrial biota received less attention. On the other hand, Sudbury is one of only a few contaminated sites where development of metal-tolerant races of grasses (Cox & Hutchinson 1980) and mosses (Beckett 1986) has been properly documented.

Studies of forest damage had already started in the 1940s, but their first results were not widely available. Pioneering works by Gorham and Gordon (1960a, b), followed by those by Freedman and Hutchinson (1980a, b), describing both contamination and changes in vegetation structure are among the most cited data sources in the domain of pollution ecology. However, responses of terrestrial biota to pollution are properly documented only for soil microbiota (Anand et al. 2003) and vegetation (mosses: Gignac 1987; lichens: LeBlanc et al. 1972; Beckett 1984; Cox & Beckett 1993; vascular plants: Gorham & Gordon 1960a; Amiro & Courtin 1981; Freedman & Hutchinson 1980b). Although the studies were initiated to evaluate the pollution impact on local forests, no quantitative (i.e., suitable for meta-analysis) plant growth data have been published.

Perception of the Environmental Situation

Home of one of the world's largest metal smelting complexes, Sudbury for many years was widely known as a wasteland. However, this reputation has changed recently, with air pollution control efforts and implementation of the largest and most successful municipal land restoration program (Gunn 1995).

In the late 1970s, combined private, public, and commercial interests initiated an unprecedented 'regreening' effort (Gunn 1995). In 1992, Sudbury was one of 12 world cities given the Local Government Honours Award at the United Nations Earth Summit to recognise the city's community-based environmental reclamation strategies (en.wikipedia.org). The reclamation effort is described on several web sites (e.g., www.cyberbeach.net, www.inco.com). The emissions problem, although not as acute as it was, is still of importance to the local people, as can be seen from discussions on a local web site (www.northernlife.ca).

2.2.2.16 Volkhov, Russia: Aluminium Smelter (Color Plates 61–63 in Appendix II)

Location and Environment

Volkhov (Volkhovstroy from 1933 to1940) is a small (population 45,800; data from 2007) industrial town in the Leningrad Region, situated on the Volkhov River, 120 km east of St. Petersburg (Fig. 2.17). The region belongs to the mid-taiga zone, with mixed forests that were earlier dominated by Norway spruce.

Pre-industrial History

Volkhov is located only 12 km south of Staraya Ladoga (Ladoga before the eighteenth century), the oldest Russian settlement in this part of the country (known from the eighth century). Coniferous forests were repeatedly cut and burned to clear the area for agricultural use. They have now generally been replaced with deciduous forests formed by white and silver birches, black alder, and European aspen, as well as by meadows and agricultural landscapes (please see color plates 62 and 63 in Appendix II).

Industrial History

The building of the Volkhov hydroelectric plant was completed in 1926, and in 1932, the first Soviet aluminium plant (constructed with the assistance of the French company Ale Forge Comarg) was launched nearby to take advantage of local bauxite from the Tikhvin deposit. Volkhov acquired town status in 1933. In 1941, soon after the German attack on the USSR, equipment from the smelter was dismantled and sent to the Ural region. Production at Volkhov was restored in 1946, and in the 1950s the plant started integrated processing of nepheline ores transported from the Kola Peninsula. In the 1960s, facilities for production of various chemicals (including soda, potassium carbonate, sulphur acid, and super phosphate) were built to utilise the by-products of aluminium production. Cement and alumina production were suspended in 1995, and super phosphate production was restructured during the same year.

Data on the annual production of 45,000 t of aluminium in the late 1990s (US Geological Survey 1999) seem overestimated; other data sources reported an output of 24,000 t in 1994, 21,000 t in 2001, and 23,500 t in 2006 (Boltramovich et al. 2003; www.sual.ru). Since 2004, the smelter has been owned by SUAL Ltd., which in 2007 became a part of RUSAL Ltd., the world's largest producer of aluminium. Employment decreased from 1,417 in 2001 (Boltramovich et al. 2003) to 514 in 2006 (www.sual.ru).

The town also hosts chemical and cement enterprises. Volkhov has a large railway junction, and mooring and cargo handling facilities on the river that provide access to the St. Petersburg and Murmansk sea ports.

Emissions Data

To our knowledge, emissions data for the Volkhov aluminium smelter have only rarely been estimated. However, the smelter is responsible for a substantial part of the total emissions of sulphur dioxide and nitrogen oxides at Volkhov, and for the entire emission of fluorine (Table 2.20). Along with these data, emissions of 105 t of insoluble fluorides, 227 t of sulphuric acid, and 54 t of SO_3 were reported in 1990 (Ministry of the Environment of Finland 1991a).

Pollution Loads and the Extent of the Contaminated Territory

Ambient concentrations of pollutants in Volkhov have been monitored since 1970; the peak reported values were 6.30 mg/m^3 of dust in 1975, 4.00 mg/m^3 of SO_2 in 1974, and 2.66 mg/m^3 of HF in 1974. Importantly, high concentrations of HF (0.19–0.21 mg/m^3) were recorded up to 2 km, and concentrations of soluble fluorides ranging from 0.08 to 0.10 mg/m^3 were recorded up to 15 km from the smelter in 1971 (Emissions of pollutants in Russia 1966–2006). These concentrations are much lower than in the mid-1950s, when up to 30 mg/m^3 of SO_2 was sometimes observed (Kijamov 1959).

An analysis of fluorine in unwashed common birch leaves was conducted at 24 sites in 1989 and at 12 sites in 1994. The peak concentration from 1989 was around 1,000 mg/kg; in 1994, the concentrations of fluorine were only 10–40% of those recorded in 1989. The fluorine-contaminated area extended approximately 4 km north and 15 km south of the smelter (Kozlov & Zvereva 1997, and unpublished, 1994). The area contaminated by fluorine in the late 1980s covered 62 km^2, and that contaminated by sulphur covered 280 km^2 (Isachenko et al. 1990); however, the quality of the data behind this estimation remains questionable. On the other hand, the data from the period 1986–1991 suggest a doubling (relative to the regional background) of pollutant deposition over 195 km^2 (Prokacheva et al. 1992).

Habitat Transformation due to Human Activity

The severely contaminated area is densely populated, and plant communities are affected by urbanisation, recreation, and agriculture (please see color plate 63 in Appendix II). It therefore seems impossible to attribute any particular changes in vegetation structure to the effects of pollution.

Table 2.20 Aerial emissions (t) from industrial enterprises at Volkhov, Russia

Year	Dust	SO_2	CO	NO_x	HF
1971	–	8,750	–	–	730
1972	–	8,750	–	–	730
1973	–	10,950	–	–	730
1974	–	10,950	–	–	730
1975	–	10,950	–	–	730
1978	9,300	4,600	–	300	240
1979	9,300	4,600	2,300	800	330
1980	9,400	2,600	2,800	900	280
1981	8,900	2,500	2,500	800	222
1982	8,100	2,500	2,800	900	400
1983	9,400	2,900	3,200	1,100	369
1984	7,500	3,000	3,900	900	410
1985	6,700	2,900	2,400	800	320
1986	6,000	3,800	2,300	1,100	360
1987	5,000	3,600	2,100	1,200	250
1988	6,200	3,600	2,100	1,200	–
1989	4,900	3,000	2,000	1,100	378
1990	4,100	2,700	1,200	800	375
	1,816[a]	1,818[a]	1,318[a]	296[a]	162[a]
1991	2,800	2,900	1,900	900	238
1992	2,856[b]	2,388[b]	1,644[b]	–	253[b]
1993	2,300	2,600	1,500	600	133
1994	1,700	1,400	1,400	400	22
1995	1,400	2,900	1,100	300	7
1996	300	1,300	1,200	200	18
1997	200	1,300	900	100	22
1998	200	1,500	1,200	300	16
1999	200	1,600	1,300	300	16
2000	700	1,300	1,500	300	19
2001	719[b]	556[b]	1,600	200	16
	800	600			
2002	2,308[b]	938[b]	1,400	400	16
	2,400	1,500			
2003	2,400	1,500	1,400	400	15
2004	2,600	1,100	1,800	300	26
2005	2,000	1,000	3,100	300	24

Non-referenced values were extracted from Emissions of Pollutants in Russia (1966–2006).
Other data sources:
[a] Ministry of the Environment of Finland (1991a), data refer to the aluminium plant only.
[b] HELCOM (2004), data refer to the aluminium plant only.

Brief History of Environmental Research

Our team started collecting environmental data for bioindication purposes in 1986 and continued monitoring birch-feeding insect herbivores until 2005. However, the information collected in the course of this research remains largely unpublished, except for the data on needle longevity of the Norway spruce (Kozlov 1991), densities of birch-feeding leaf rollers (Kozlov 1991), and abundancies of several groups of flies (Zvereva 1994; Kozlov & Zvereva 1997). We are not aware of any other pollution-related research conducted in this region.

Perception of the Environmental Situation

Extensive health problems (respiratory diseases, fluorosis) had already been reported in the mid-1950s (Kijamov 1959), but their public perception remains unknown. Over the past decades, international concern about pollution in Volkhov (www.pollutedplaces.org) has been much larger than just a local concern. Volkhov had been identified as a 'hot spot' in terms of pollution (mostly discharge of wastewater; www.blacksmithinstitute.org) by the early 1990s, and measures taken by the smelter had positive effects on environmental quality by 2002 (HELCOM 2004). Voting arranged on a Volkhov website (63clan.ru) for 2 weeks in January 2008 did not attract much attention; only three of 82 visitors identified local pollution problems as life-threatening.

2.2.2.17 Vorkuta, Russia: Power Plant and Cement Factory (Color Plates 64–67 in Appendix II)

Location and Environment

The industrial city of Vorkuta (population 90,100; data from 2002) is located in the northwestern Russian tundra (Fig. 2.18), within the permanent permafrost zone, 80 km north of the treeline formed by Siberian spruce. The topography in the region is relatively flat, with elevation varying from 100 m above sea level in the deepest river valleys to 250 m atop the smooth hills. Shrub tundra dominated by dwarf birch (*Betula nana*) is the most common vegetation type, associated with better drained and slightly elevated sites. Depressions are occupied by willow-dominated (*Salix glauca, S. phylicifolia, S. lanata*), often paludified, vegetation (please see color plate 66 in Appendix II). A detailed description of the pre-industrial state of the plant communities is given by Rebristaya (1977).

Pre-industrial History

Before the 1930s, the territory was only sparsely populated by aboriginal people subsisting on reindeer herding and hunting.

Industrial History

The Vorkuta region is the largest industrial centre in the tundras of European Russia. It consists of the main city and more than ten subcentres located near coal mines and other industrial units. The city and the first coal mines were established in the 1930s.

The population increased from 30,000 in the early 1950s to over 180,000 in the 1960s, to 216,000 in 1991, and then declined to 127,500 by 2005. For a detailed account of the early industrial development and the use of forced labour, consult Negretov (1977).

The two main air pollutant sources are the Vorkuta cement factory (established in 1950) and the coal-fired power plant (TEZ-2, established in 1956); both are located about 15 km north of the town of Vorkuta. However, their emissions are included in the total emissions of Vorkuta in government reports (Emissions of pollutants in Russia 1966–2006).

Emissions Data

Atmospheric pollution in the area surrounding Vorkuta is mainly associated with dust from open coal mines, emissions from the power plant and cement factory, and burning of waste rock near the coal mines (Table 2.21). Contributions from two

Table 2.21 Aerial emissions (t) from industrial enterprises at Vorkuta, Russia

Year	Dust	SO_2	CO	NO_x
1975	175,200	51,100	34,700	–
1978	107,300	41,800	13,800	11,700
1979	103,500	54,000	26,700	12,000
1980	141,200	47,000	25,200	8,200
1981	135,700	34,000	18,900	7,200
1982	90,700	35,400	4,100	5,600
1983	91,500	43,900	4,000	5,500
1984	108,800	47,000	4,500	4,900
1985	102,700	41,500	4,500	3,400
1986	67,300	41,100	3,500	2,500
1987	62,800	41,400	4,200	4,600
1988	64,100	44,200	3,900	4,600
1989	118,500	46,300	16,600	5,500
1990	128,700	46,400	18,000	5,500
1991	131,800	47,500	17,700	5,800
1992	130,667[a]	50,442[a]	16,235[a]	6,572[a]
1993	111,600	47,500	13,600	6,800
1994	93,500	45,300	11,000	7,400
1995	85,500	46,300	14,100	7,800
1996	79,000	44,100	13,100	7,400
1997	72,100	36,900	10,000	7,300
1998	58,300	36,400	9,400	7,200
1999	58,200	39,900	9,300	7,200
2000	57,900	38,700	9,300	6,900
2001	53,200	33,600	8,600	6,900
2002	53,200	33,600	8,600	6,900
2003	35,000	34,200	7,300	5,900
2004	33,300	33,000	7,100	7,000
2005	26,800	32,300	5,500	6,900

Non-referenced values were extracted from Emissions of Pollutants in Russia (1966–2006).
Other data sources:
[a] Solovieva et al. (2002, and personal communication, 2008).

power plants and the cement factory to local aerial emissions in the late 1980s-early 1990s were 53–56% and 32–34%, respectively (Emissions of pollutants in Russia 1966–2006). Data from 1999–2000 (Tables 2.22 and 2.23) demonstrated that one of the power plants (TEZ-2) is responsible for two thirds of the sulphur dioxide emissions in the Vorkuta region (summarised in Table 2.21). Annual emissions of SO_2 from the two power plants ranged from 37–39,000 t in 1994–1996 (Getsen 2000). Peak emissions from the cement factory reached 60,000 t in the late 1980s (Regional Committee of Nature Protection at Vorkuta, personal communication 2001).

Pollution Loads and the Extent of the Contaminated Territory

Ambient concentrations of pollutants in and around Vorkuta have been monitored since 1974; the peak reported values were 2.70 mg/m^3 of dust in 1978, 1.70 mg/m^3 of SO_2 in 1975, and 0.36 mg/m^3 of NO_x in 1978 (Emissions of pollutants in Russia 1966–2006).

The peak concentrations of dust measured in snow in the mid-1970s suggests an annual deposition rate of approximately 1,000 t/km^2 (Kuliev & Lobanov 1978); values from the late 1990s are about 300 t/km^2 (Getsen 2000). Deposition of calcium-containing dust caused strong alkalisation near the polluters, with soil pH ranging from 6.7 to 8.9 (Getsen et al. 1994), while the soil pH outside of the impacted territory varied from approximately 5 to less than 4.5 (Virtanen et al. 2002). From 1987 to 1989, annual depositions of sulphur and nitrogen in Vorkuta were 1,400 and 700 kg/km^2, respectively (Vasilenko et al. 1991). Pollution caused a strong increase in the soil concentration of strontium (up to 25 times higher than in background regions), along with moderate (two to ten times above the background) increases in zinc, lead, iron, cadmium and chromium concentrations (Krasovskaya 1996).

In the 1970s, smoke from the local polluters was visually observed up to 50 km away. A characteristic smell was detected up to 30 km away, and the presence of cement dust was recorded up to 25 km from the factory (Kuliev & Lobanov 1978). During the period 1986–1991, a doubling of pollutant deposition (relative to the regional background) was recorded for 3,000 km^2 around Vorkuta and adjacent

Table 2.22 Aerial emissions (t) from the cement factory at Vorkuta, Russia

Year	Dust	SO_2	NO_x	HF
1999	18,887	393	300	6
2000	19,115	415	318	6

Unpublished data received from the Regional Committee of Nature Protection at Vorkuta.

Table 2.23 Aerial emissions (t) from the power plant 'TEZ-2' at Vorkuta, Russia

Year	Dust	SO_2	NO_x
1999	22,698	26,900	4,122
2000	21,832	25,072	3,932

Unpublished data received from the Regional Committee of Nature Protection at Vorkuta.

settlements (Prokacheva et al. 1992). At the end of the 1990s, the local gradient in deposition of alkaline ash extended to 25–40 km, with increased concentrations of Ca, Ba, Sr and other alkaline earth metals recorded within approximately 30 km from Vorkuta (Walker et al. 2003a, b). Satellite image analysis (Landsat TM, from 31 July 1988) suggests that the effects of pollution on vegetation are detectable for 600–900 km^2, of which 150–200 km^2 are strongly affected by pollution (Virtanen et al. 2002).

Habitat Transformation due to Human Activity

Substantial changes in terrestrial ecosystems around Vorkuta are caused by a combination of different disturbances, including pollution, coal mining, sand and gravel mining for building purposes, the use of track vehicles, and agriculture (mostly pasture) (Druzhinina 1985). The relative importance of other kinds of disturbances is expected to increase with the decline of pollution (Virtanen et al. 2002; Walker et al. 2006).

Brief History of Environmental Research

Studies on the pollution impact on terrestrial ecosystems around the industrial complex of Vorkuta are scarce, and they are mostly published in Russian (for the list, consult Getsen 2000). Studies from 1976–1978 (Kuliev 1977, 1979; Kuliev & Lobanov 1978) revealed the extent of the pollution impact by analysing the concentration of particles in snow and related these to changes in vegetation structure (in particular, a decline in lichens, accompanied by increases in moss and grass cover) and increased growth of several plants. Studies from the 1990s are summarised in the collection of scientific papers on bioindication (Getsen et al. 1996). A review is given by Virtanen et al. (2002).

Perception of the Environmental Situation

Although Walker et al. (2006) concluded that 'Vorkuta's inhabitants perceived air pollution as the primary environmental threat', we did not find signs of specific public anxiety in local webpages.

2.2.2.18 Žiar nad Hronom, Slovakia: Aluminium Smelter (Color Plates 68–70 in Appendix II)

Location and Environment

Žiar nad Hronom (Svätý Kríž nad Hronom until 1955) is a small (population 19,750; data from 2005) industrial town situated on the northern bank of the river Hron, 150 km northeast of Bratislava (Fig. 2.19). The entire mountainous region belongs to the western Carpathians, with altitudes ranging 300–700 m above sea level. The original vegetation surrounding the polluter (below 600 m above sea level) was oak-beech and beech forest (please see color plates 69 and 70 in Appendix II).

Pre-industrial History

Although the region was populated long ago, the first written reference to Svätý Kríž dates to 1237. During the Middle Ages, Svätý Kríž was a toll-collecting town on an important trade route. In the early 1950s, it was a village of 1,400 people.

Industrial History

Construction of the aluminium plant began in 1951, and production of primary aluminium from bauxite ore by electrolysis was launched in 1953; production of anode matter started in 1954. The coal-fired power plant (a source of both sulphur dioxide and heavy metals) was built in 1956, and a second aluminium electrolysis plant was built in 1958. Additional facilities were built in 1967, and a new heating plant has operated since 1986. In 1985, the company running the aluminium plant started a capital expenditure program to invent modern smelting technology. SLOVALCO, founded in 1993, took over construction of the partly built new smelter and ancillary facilities. Production of aluminium reached 60,000 t in 1966, about 110,000 t annually in the late 1990s, and 176,000 t in 2006 (Maňkovská & Steinnes 1995; US Geological Survey 1999; Maňkovská & Kohút 2002). In the 2000s, the smelter employed between 600 and 700 people (www.slovalco.sk). Recently, metallurgy has formed the basis of local economy.

Emissions Data

Before effective filters were installed, production resulted in the emission of fly ash with high contents of heavy metals and fluorides (gaseous hydrogen fluoride and fluoride minerals such as cryolite), particles of aluminium oxide, sulphur dioxide, and many other substances. Accurate emission records are available starting from 1990 (Table 2.24). The smelter also emitted fluorine-containing dust (from 18 t in 1997 to 0.2 t in 2007: www.slovalco.sk). In the 1990s, the production process was gradually updated, and the closing of the old smelter in 1996 resulted in a reduction of dust emissions. The company also implemented the Environmental Remediation Program (www.slovalco.sk).

Pollution Loads and the Extent of the Contaminated Territory

In the early 1970s, annual deposition of fluorine exceeded 1,000 kg/km^2 up to 2.2 km south of the polluter (Hajdúk 1974), and soils within 16 km^2 contained over 200 mg/kg

Table 2.24 Aerial emissions (t) from aluminium smelter and associated power plant at Žiar nad Hronom, Slovakia

Year	Dust	SO_2	CO	NO_x	HF
1958–1959	–	–	–	–	1,022[a]
1960–1966	–	–	–	–	1,096[a]
1966	–	4,723[a]	–	–	–
1967–1973	–	–	–	–	657[a]
1973	–	11,315[a]	–	–	–
1974–1975	–	–	–	–	563[a]
1975	–	9,606[a]	–	–	–
1990	1,580	6,556	840	810	848
1991	1,718	5,557	932	788	809
1992	1,859	3,879	706	616	557
1993	718	3,555	892	596	364
1994	303	3,168	975	588	319
1995	317	1,943	951	360	326
1996	253	2,373	11,160	378	114
1997	195	2,595	10,603	389	60
	98[b]	1,009[b]	10,499[b]	39[b]	42[b]
1998	177	2,267	10,589	321	58
	97[b]	1,009[b]	10,499[b]	41[b]	40[b]
1999	193	2,651	8,503	486	34
	119[b]	1,511[b]	8,439[b]	230[b]	32[b]
2000	186	2,477	7,960	584	30
	115[b]	1,178[b]	7,887[b]	293[b]	28[b]
2001	190	2,431	7,937	577	30
	118[b]	1,176[b]	7,865[b]	295[b]	28[b]
2002	88[b]	1,293[b]	10,220[b]	403[b]	22[b]
2003	95[b]	1,334[b]	11,618[b]	474[b]	23[b]
2004	104[b]	1,107[b]	13,010[b]	541[b]	24[b]
2005	146[b]	1,310[b]	12,994[b]	689[b]	46[b]
2006	104[b]	1,324[b]	12,956[b]	565[b]	10[b]
2007	98[b]	1,326[b]	12,942[b]	559[b]	9[b]

Non-referenced values after ŽSNP, a.s. (personal communication).
Other data sources:
[a] Sobocky (1977), the annual emissions of HF are averaged for several years.
[b] www.slovalco.sk, data refer to aluminium smelter only.

of fluorine (Kontrišová 1980, Fig. 2.11). The contaminated area approached 500 km^2 (Maňkovská & Steinnes 1995; Krištín & Žilinec 1997). The polluted region was subdivided into zones with different levels of environmental contamination by Maňkovská (1979) and Kontrišová (1980). By the early 2000s, a statistically significant increase in fluorine concentrations over the regional background was detected only within the most polluted zone (61 km^2); however, other zones were still distinguished on the basis of sulphur concentrations (Maňkovská & Kohút 2002). According to another data source, the content of water soluble fluorine in soils in the early 1990s exceeded the sanitary limit (10 mg/kg) for 37 km^2 (Andersen 2000), i.e., within approximately 3.5 km from the smelter.

Habitat Transformation due to Human Activity

The pollution impact did not cause deforestation; however, the signs of pollution damage are easily recognisable in forests located close to the smelter (please see color plate 70 in Appendix II). The proportion of dying trees reached 32% near the polluter, compared to 4.4% in a distant (background) site (Cicák & Mihál 1996). In 1990, heavily damaged stands were observed up to 3 km from the smelter (Bucha & Maňkovská 2002).

Brief History of Environmental Research

The surroundings of Žiar nad Hronom are among the most contaminated areas of the Slovak Republic (Kellerová 2005). The adverse effects of air pollution on forests had already been detected in the late 1950s (Štefančík 1995), and since then, the accumulation of pollutants (inorganic: Maňkovská & Steinnes 1995; Wilcke & Kaupenjohann 1998; Maňkovská & Kohút 2002; polycyclic aromatic hydrocarbons: Wilcke et al. 1996), soil acidification (Löffler 1983; Wilcke & Kaupenjohann 1998) and biotic responses to pollution have been described. Most studies were forestry-oriented, documenting forest vitality (Cicák & Mihál 1996), the occurrence of mycorrhizatl and parasitic fungi (Cicák & Mihál 1996; Mihál & Bučinová 2005), and damage by herbivorous insects (Šušlík & Kulfan 1993; Kulfan et al. 2002). Plant diversity (Hajdúk 1974) as well as abundance, species richness and the breeding success of birds declined near the polluter (Feriancová-Masárová & Kalivodová 1965a, b; Newman 1977; Krištín & Žilinec 1997, and references therein). Data on mammals are restricted to reports on fluorosis in roe deer (*Capreolus capreolus*) (Hell et al. 1995).

A recent decrease in emissions has already resulted in a lower accumulation of pollutants in the foliage of forest trees (Maňkovská 2001, 2004) and in a steady improvement of forest vitality (Bucha & Maňkovská 2002).

Perception of the Environmental Situation

Soon after the launch of the smelter, the pollution problem became extremely acute for the inhabitants of the small village of Horné Opatovice, located less than 1 km from the smelter. The emissions first killed bees, then caused the death of cattle and trees, and worsened health conditions of the inhabitants. Although scientific data on this environmental disaster have not been disclosed, the government of the Czechoslovak Socialist Republic decided to abolish the village in 1960, and this name disappeared from official maps in 1969. Inhabitants (1,380 people living in 228 houses) were mostly relocated to Žiar nad Hronom. However, after reconstruction of the smelter in 1995 and the subsequent emissions decline, pollution is not perceived as a life-threatening problem by local people (J. Kulfan 2008, personal communication).

2.3 Study Sites and General Sampling Design

We have adopted a uniform sampling design for all investigated impact zones. Around each polluter, we selected ten study sites, grouped in two blocks (transects) of five, located in two different directions from the polluter (Figs. 2.2–2.19). The sites are coded by a transect number followed by the site number, with site 1 being the closest to and site 5 the farthest from the polluter.

Deviations from this sampling scheme occurred near Monchegorsk and Žiar nad Hronom. In the impact zone of the Monchegorsk smelter, which was the focus of our research for decades, we chose 17 sampling sites (Fig. 2.10) due to the practical impossibility of collecting all the data (mostly related to plant vigour) from the same set of ten sites; however, each individual analysis is based on ten sites only. Moreover, in this impact zone, we were forced to select both controls to the south of the polluter due to an overlap between the northern part of the impact zone of Monchegorsk smelter with the southern part of the impact zone of the iron ore processing factory in Olenegorsk. In the impact zone of the aluminium plant situated at Žiar nad Hronom, attribution of the study sites to two transects is somewhat arbitrary (Fig. 2.19), since the mountain relief, forestry and agriculture substantially restricted the extent of areas suitable for establishment of study sites.

Two impact zones include more than one point polluter. The smelter at Nikel and ore-roasting plant at Zapolyarnyy are located 25 km apart, and their impact zones overlap substantially (Tømmervik et al. 2003; Kozlov & Zvereva 2007a). Similarly, pollution at Norilsk is imposed by three plants located 5–10 km apart. Therefore, two transects selected in these impact zones start from different polluters (Figs. 2.12 and 2.13). We have chosen to consider the Copper Cliff smelter at Sudbury, Canada, as the sole polluter, since the Coniston smelter located nearby was closed in 1972, and the Falconbridge plant emitted into the ambient air about ten times less SO_2 and about 10–50 times less metal than the Copper Cliff smelter (Pollution Probe 2003).

We started selection of study sites from surveys of all available ecological and environmental information. In particular, at this stage we made a decision on the approximate location of our most distant study sites (5 to 96 km from the polluter), taking into account both the extent of the impact zone under study and its overlap with the impact zones of other polluters. Our intention was to establish the most distant study sites at localities representing the regional background in terms of pollution load. At this stage, we also made a preliminary decision on the type of plant community to be explored.

In the course of the reconnaissance work (1–2 days, depending on the size of the impact zone and road network), we usually visited 15–30 localities potentially suitable for establishment of our study sites. Whenever possible, we explored study sites used by other researchers who had been working in these impact zones earlier. In the course of this reconnaissance work, we made pilot observations on the types of plant communities, stand age and composition (for forested habitats), the occurrence of plant species to be sampled for vitality indices, approximate levels of the pollution impact (evaluated, e.g., by needle longevity in conifers), and disturbances other than the pollution impact (e.g., fellings, fires, recreation). This information was summarised in the form of a table and used to make an optimal selection of study sites. Some pictures taken from the sites not selected for detailed survey, included in the colour section of this book, are labelled as taken at additional sites.

The positions of study sites (Tables 2.25–2.42) were located using GPS. As a rule, all data were collected within 50 m from the point located using GPS; under specific circumstances, such as rarity of one of the selected plant species, some samples were collected up to 100 m from the central point of the study site. At the time of the first visit, we photographed each study site from its approximate central point toward the main compass directions.

Whenever possible, we collected the data over several sampling sessions (Table 2.43); collection of some data over 2 or more years allowed us to account for the repeatability of measurements (discussed in Chapters 3–7). However, due to financial limitations and logistic constrains, two of our impact zones (around Norilsk and Bratsk) were visited only once, and therefore the amount of information collected from these areas is lower than from other (repeatedly visited) study areas. We were unable to measure chlorophyll fluorescence around Sudbury and Vorkuta and soil respiration around Vorkuta.

Data collection was performed by the same team. M. Kozlov collected the data in all impact zones except for Vorkuta; V. Zverev collected the data in all impact zones except for Vorkuta and Sudbury; E. Zvereva collected the data around Harjavalta, Kandalaksha, Monchegorsk, Nikel, Straumsvík, Volkhov and Vorkuta.

Table 2.25 Location of study plots and concentrations of pollutants in the impact zone of power plant at Apatity, Russia

	Plot position				Pollutant concentrations, mg/kg (mean ± SE)[a]	
Plot code	Latitude (N)	Longitude (E)	Altitude, m a.s.l.	Distance from polluter, km	Fe	Sr
Polluter	67°35′55″	33°25′26″	160	0	–	–
1–1	67°35′52″	33°24′54″	160	0.4	45.0 ± 3.2	30.0 ± 3.9
1–2	67°36′36″	33°24′00″	160	1.6	47.8 ± 5.5	20.0 ± 2.3
1–3	67°37′06″	33°22′23″	170	3.1	32.5 ± 6.0	33.2 ± 6.5
1–4	67°37′47″	33°19′46″	160	5.3	18.8 ± 1.2	86.2 ± 6.6
1–5	67°38′22″	33°17′11″	150	7.4	16.0 ± 1.6	37.8 ± 3.2
2–1	67°36′02″	33°26′35″	150	0.9	32.5 ± 1.2	22.8 ± 4.2
2–2	67°35′26″	33°27′44″	160	1.9	40.2 ± 4.3	17.2 ± 2.4
2–3	67°35′08″	33°28′57″	190	2.9	31.8 ± 6.0	10.8 ± 2.3
2–4	67°34′59″	33°30′19″	190	3.9	17.8 ± 1.2	14.2 ± 2.1
2–5	67°34′57″	33°33′02″	190	5.7	11.7 ± 1.7	49.3 ± 12.4

[a] In leaves of speckled alder, *Alnus incana* (four samples per site; collected in 1992); meta-analyses are based on correlations with iron.

Table 2.26 Location of study plots and concentrations of fluorine in the impact zone of aluminium smelter at Bratsk, Russia

Plot code	Plot position: Latitude (N)	Longitude (E)	Altitude, m a.s.l.	Distance from polluter, km	Fluorine concentrations, mg/kg (mean ± SE)[a]
Polluter	56°07′44″	101°27′10″	380	0	–
1–1	56°09′00″	101°26′19″	340	1.6	97.2 ± 7.2
1–2	56°08′17″	101°30′31″	400	3.6	52.1 ± 21.5
1–3	56°12′13″	101°37′23″	520	13.0	21.7 ± 4.8
1–4	56°20′15″	101°39′51″	460	27.0	13.4 ± 6.6
1–5	56°35′22″	101°31′35″	400	56.0	2.3 ± 0.1
2–1	56°07′40″	101°29′56″	420	1.8	283.2 ± 68.8
2–2	56°06′00″	101°26′09″	480	2.6	35.3 ± 5.2
2–3	56°05′13″	101°12′00″	400	16.0	11.8 ± 1.9
2–4	55°54′00″	101°05′41″	520	30.0	4.5 ± 0.4
2–5	55°34′50″	101°06′49″	460	65.0	3.6 ± 0.8

[a] In leaves of white birch, *Betula pubescens* (two samples per site; collected in 2002).

Table 2.27 Location of study plots and concentrations of pollutants in the impact zone of nickel-copper smelter at Harjavalta, Finland

Plot code	Plot position: Latitude (N)	Longitude (E)	Altitude, m a.s.l.	Distance from polluter, km	Pollutant concentrations, mg/kg (mean ± SE)[a]: Ni	Cu
Polluter	61°19′11″	22°07′16″	30	0	–	–
1–1	61°19′40″	22°07′05″	30	0.93	315.8 ± 38.2	27.2 ± 5.7
1–2	61°19′31″	22°06′28″	30	1.0	903.8 ± 184.7	38.8 ± 6.7
1–3	61°20′13″	22°05′23″	25	2.6	58.2 ± 6.0	6.0 ± 1.9
1–4	61°20′44″	22°03′53″	20	4.2	70.4 ± 8.8	4.8 ± 1.2
1–5	61°23′47″	21°52′21″	25	15.8	20.4 ± 11.7	3.2 ± 0.9
2–1	61°19′14″	22°07′57″	30	0.6	196.4 ± 18.9	39.0 ± 3.5
2–2	61°19′05″	22°08′11″	30	0.8	242.2 ± 11.1	42.8 ± 3.3
2–3	61°18′44″	22°08′57″	30	1.7	154.6 ± 22.0	21.6 ± 4.4
2–4	61°17′42″	22°11′26″	50	4.6	47.8 ± 3.2	7.8 ± 1.6
2–5	61°15′15″	22°18′36″	35	12.5	22.4 ± 2.8	6.6 ± 1.2

[a] In leaves of goat willow, *Salix caprea* (five samples per site; collected in 1999); meta-analyses are based on correlations with nickel.

Table 2.28 Location of study plots and concentrations of pollutants in the impact zone of fertilizing factory at Jonava, Lithuania

Plot code	Plot position			Distance from polluter, km	Pollutant concentrations, mg/kg (mean ± SE)[a]		
	Latitude (N)	Longitude (E)	Altitude, m a.s.l.		Sr	Zn	Ni
Polluter	55°05′02″	24°19′36″	35	0	–	–	–
1–1	55°04′25″	24°19′21′″	39	1.2	28.0 ± 1.5	95.6 ± 12.5	1.51 ± 0.17
1–2	55°03′46″	24°18′60″	72	2.4	13.0 ± 2.7	176.9 ± 37.0	0.87 ± 0.10
1–3	55°02′38″	24°13′43″	69	7.7	18.4 ± 1.5	96.1 ± 1.2	1.82 ± 0.28
1–4	55°01′40″	24°10′38″	55	11.4	13.1 ± 3.2	175.9 ± 47.8	1.53 ± 0.25
1–5	54°57′40″	24°01′47″	80	24.3	21.7 ± 1.7	183.6 ± 4.8	2.51 ± 0.28
2–1	55°05′00″	24°18′51″	33	0.8	13.4 ± 1.4	153.4 ± 39.7	1.36 ± 0.05
2–2	55°05′56″	24°21′55″	69	3.0	35.5 ± 10.4	64.7 ± 21.9	1.27 ± 0.33
2–3	55°06′32″	24°25′31″	88	6.9	14.3 ± 0.2	88.3 ± 10.0	0.86 ± 0.04
2–4	55°08′59″	24°27′47″	69	11.4	37.9 ± 7.4	287.1 ± 32.5	2.44 ± 0.33
2–5	55°10′58″	24°34′04″	48	18.9	11.4 ± 1.1	157.0 ± 28.7	2.73 ± 0.32

[a] In leaves of common birch, *Betula pendula* (three samples per site; collected in 2005); meta-analyses are based on correlations with strontium.

Table 2.29 Location of study plots and concentrations of fluorine in the impact zone of aluminium smelter at Kandalaksha, Russia

	Plot position			Distance	Fluorine concentrations, mg/kg (mean ± SE)	
Plot code	Latitude (N)	Longitude (E)	Altitude, m a.s.l.	from polluter, km	1998[a]	2002[b]
Polluter	67°11′43″	32°25′51″	80	0	–	–
1–1	67°12′20″	32°26′25″	90	1.2	26.9 ± 9.5	99.4 ± 22.0
1–2	67°12′57″	32°26′09″	110	2.3	15.1 ± 2.4	117.4 ± 20.0
1–3	67°13′47″	32°24′52″	150	3.9	15.0 ± 2.1	54.9 ± 1.6
1–4	67°14′57″	32°25′35″	140	6.0	15.6 ± 5.3	44.0 ± 8.6
1–5	67°16′53″	32°27′12″	130	9.7	8.0 ± 2.0	27.2 ± 3.6
2–1	67°11′27″	32°26′11″	70	0.6	21.1 ± 3.7	142 ± 20.5
2–2	67°10′40″	32°25′24″	80	2.0	19.2 ± 5.3	84.1 ± 9.2
2–3	67°09′39″	32°26′56″	60	3.9	–	36.6 ± 11.8
2–4	67°08′42″	32°26′01″	40	5.6	13.3 ± 2.7	20.9 ± 2.7
2–5	67°09′45″	32°07′44″	85	13.6	–	6.9 ± 1.2

[a] In needle of Scots pine, *Pinus sylvestris* (five samples per site; collected in 1998).
[b] In leaves of mountain birch, *Betula pubescens* ssp. *czerepanovii* (two samples per site; collected in 2002); meta-analyses are based on correlations with data of 2002.

Table 2.30 Location of study plots and concentrations of pollutants in the impact zone of copper smelter at Karabash, Russia

	Plot position			Distance	Pollutant concentrations, mg/kg (mean ± SE)[a]	
Plot code	Latitude (N)	Longitude (E)	Altitude, m a.s.l.	from polluter, km	Cu	Zn
Polluter	55°28′03″	60°12′07″	360	0	–	–
1–1	55°29′00″	60°13′17″	380	2.2	164.6 ± 5.2	1,107 ± 50
1–2	55°30′01″	60°15′35″	340	5.2	45.8 ± 3.3	327 ± 33
1–3	55°31′42″	60°20′08″	320	10.8	37.8 ±10.8	312 ± 82
1–4	55°36′44″	60°25′19″	280	21.3	13.9 ± 1.7	260 ± 92
1–5	55°42′47″	60°28′17″	260	32.2	14.3 ± 2.2	173 ± 67
2–1	55°27′13″	60°12′20″	330	1.6	238.2 ± 19.9	1,282 ± 200
2–2	55°26′18″	60°13′05″	330	3.4	129.2 ± 22.3	517 ± 54
2–3	55°24′39″	60°08′55″	330	7.1	51.9 ± 2.5	455 ± 162
2–4	55°13′19″	60°09′00″	320	27.5	15.9 ± 1.8	265 ± 52
2–5	55°06′47″	59°57′11″	420	42.5	11.3 ± 1.4	175 ± 43

[a] In leaves of *Betula pendula* (three samples per site; collected in 2003); meta-analyses are based on correlations with copper.

Table 2.31 Location of study plots and concentrations of pollutants in the impact zone of iron pellet plant at Kostomuksha, Russia

	Plot position				Pollutant concentrations, mg/kg (mean ± SE)[a]	
Plot code	Latitude (N)	Longitude (E)	Altitude, m a.s.l.	Distance from polluter, km	Cu	Fe
Polluter	64°38′52″	30°45′07″	200	0	–	–
1–1	64°39′04″	30°43′55″	200	1.0	4.40 ± 0.09	1,296 ± 19
1–2	64°41′42″	30°45′15″	200	5.3	4.01 ± 0.45	441 ± 14
1–3	64°42′31″	30°52′26″	170	9.0	3.88 ± 0.18	224 ± 22
1–4	64°45′14″	30°47′50″	190	12.0	4.24 ± 0.49	805 ± 190
1–5	64°49′46″	30°42′36″	150	20.4	4.75 ± 0.58	592 ± 158
2–1	64°38′52″	30°45′40″	190	0.5	5.08 ± 0.49	3,758 ± 160
2–2	64°37′53″	30°43′08″	200	2.4	8.06 ± 3.08	1,278 ± 249
2–3	64°37′06″	30°39′46″	190	5.4	4.10 ± 0.56	354 ± 57
2–4	64°34′26″	30°53′19″	170	10.5	6.27 ± 1.38	170 ± 20
2–5	64°29′49″	31°07′32″	150	24.6	2.80 ± 0.10	95 ± 7

[a] In leaves of white birch, *Betula pubescens* (three to five samples per site; collected in 2003); meta-analyses are based on correlations with iron.

Table 2.32 Location of study plots and concentrations of pollutants in the impact zone of copper smelter at Krompachy, Slovakia

	Plot position				Pollutant concentrations, mg/kg (mean ± SE)[a]	
Plot code	Latitude (N)	Longitude (E)	Altitude, m a.s.l.	Distance from polluter, km	Cu	Ni
Polluter	48°55′20″	20°52′56″	370	0	–	–
1–1	48°55′45″	20°52′52″	540	0.7	15.6 ± 1.98	4.30 ± 0.53
1–2	48°56′09″	20°50′43″	500	3.0	13.9 ± 1.93	2.30 ± 0.55
1–3	48°56′35″	20°49′33″	410	4.7	11.2 ± 0.91	2.53 ± 0.19
1–4	48°55′11″	20°46′49″	460	7.4	6.43 ± 0.61	0.79 ± 0.17
1–5	48°55′54″	20°43′10″	430	12.0	16.6 ± 7.73	2.62 ± 0.64
2–1	48°55′14″	20°54′04″	450	1.5	35.6 ± 17.8	2.24 ± 0.47
2–2	48°54′30″	20°53′16″	460	1.7	11.1 ± 1.46	1.88 ± 0.28
2–3	48°53′47″	20°53′16″	730	2.9	8.77 ± 0.12	3.11 ± 0.44
2–4	48°54′33″	20°56′32″	390	4.6	6.50 ± 1.13	2.97 ± 0.51
2–5	48°48′38″	21°00′25″	550	15.4	7.03 ± 0.24	1.75 ± 0.41

[a] In leaves of European beech, *Fagus sylvatica* (three samples per site; collected in 2002); meta-analyses are based on correlations with copper.

Table 2.33 Location of study plots and concentrations of pollutants in the impact zone of nickel-copper smelter at Monchegorsk, Russia

	Plot position				Pollutant concentrations, mg/kg (mean ± SE)[b]	
Plot code	Latitude (N)	Longitude (E)	Altitude, m a.s.l.	Distance from polluter,[a] km	Ni	Cu
Polluter	67°55′15″	32°50′18″	140	0	–	–
1–1	67°56′04″	32°49′08″	180	1.6	237.1 ± 25.2	86.5 ± 13.2
1–2	67°58′07″	32°52′29″	140	5.0	145.8 ± 24.0	67.5 ± 7.3
1–3	67°59′37″	32°54′57″	160	8.1	(72.2 ± 5.5)	(20.2 ± 1.9)
1–4	68°00′59″	32°57′03″	180	11.1	44.8 ± 5.4	16.9 ± 1.5
1–5	68°02′20″	33°00′53″	180	14.6	(32.8 ± 4.5)	(10.4 ± 0.7)
2–1	67°54′49″	32°48′30″	190	1.1	(119.2 ± 8.6)	(56.2 ± 5.7)
2–2	67°52′59″	32°46′40″	210	4.3	318.0 ± 29.4	134.3 ± 14.2
2–3	67°51′58″	32°47′50″	260	5.7	586.8 ± 92.7	323.9 ± 60.5
2–4	67°51′01″	32°48′10″	240	7.5	315.8 ± 39.1	147.2 ± 32.8
2–5	67°48′03″	32°46′56″	140	13.0	163.3 ± 6.4	67.4 ± 4.78
2–6	67°46′36″	32°47′45″	200	15.6	131.6 ± 5.2	80.3 ± 3.8
2–7	67°45′31″	32°48′29″	210	17.5	(88.8 ± 5.2)	(38.0 ± 4.9)
2–8	67°40′39″	32°49′27″	220	26.7	52.3 ± 3.3	28.0 ± 1.6
2–9	67°38′21″	32°45′00″	170	31.1	23.8 ± 2.1	13.2 ± 1.0
2–10	67°34′38″	32°32′54″	140	39.7	19.3 ± 0.4	10.0 ± 0.4
2–11	67°34′46″	33°35′22″	220	49.5	(12.0 ± 2.3)	(9.6 ± 0.4)
2–12	67°32′16″	33°57′52″	240	64.0	5.9 ± 0.3	5.7 ± 0.3

[a] Measured from the nearest smokestack.

[b] In leaves of mountain birch, *Betula pubescens* ssp. *czerepanovii* (three samples per site; collected in 2003, except for values in parentheses that are each based on five samples collected in 1993); consult Kozlov (2005a) for comparability of contamination data; meta-analyses are based on correlations with nickel.

Table 2.34 Location of study plots and concentrations of pollutants in the impact zone of aluminium smelter at Nadvoitsy, Russia

	Plot position				
Plot code	Latitude (N)	Longitude (E)	Altitude, m a.s.l.	Distance from polluter, km	Fluorine concentrations, mg/kg (mean ± SE)[a]
Polluter	63°52′50″	34°15′50″	100	0	–
1–1	63°53′15″	34°15′53″	100	0.8	60.7 ± 2.0
1–2	63°53′44″	34°14′04″	110	2.2	59.4 ± 10.8
1–3	63°54′31″	34°12′55″	110	3.9	25.8 ± 2.5
1–4	63°55′44″	34°07′47″	100	8.5	14.8 ± 4.7
1–5	64°01′45″	34°04′12″	110	19.1	11.2 ± 4.2
2–1	63°53′08″	34°16′34″	100	0.8	85.8 ± 20.7
2–2	63°52′31″	34°17′39″	100	1.6	65.4 ± 3.8
2–3	63°52′47″	34°20′40″	100	4.0	15.8 ± 0.2
2–4	63°52′28″	34°24′12″	90	6.9	12.4 ± 3.1
2–5	63°51′30″	34°29′58″	80	11.8	8.6 ± 0.8

[a] In leaves of white birch, *Betula pubescens* (two samples per site; collected in 2004).

Table 2.35 Location of study plots and concentrations of pollutants in the impact zone of nickel-copper smelter at Nikel (Polluter 2) and ore-roasting plant at Zapolyarnyy (Polluter 1), Russia

	Plot position				Pollutant concentrations, mg/kg (mean ± SE)[a]	
Plot code	Latitude (N)	Longitude (E)	Altitude, m a.s.l.	Distance from the nearest polluter, km	Ni	Cu
Polluter 1	69°24′25″	30°47′49″	160	0	–	–
1–1	69°24′40″	30°47′51″	140	0.5	366.3 ± 53.7	175.3 ± 23.8
1–2	69°25′30″	30°52′45″	70	3.8	67.9 ± 3.9	23.5 ± 1.3
1–3	69°26′50″	31°02′10″	90	10.4	36.0 ± 3.7	7.8 ± 0.4
1–4	69°27′46″	31°29′44″	170	28.1	14.8 ± 2.3	6.4 ± 0.2
1–5	69°26′34″	31°56′42″	160	45.3	7.5 ± 0.5	5.2 ± 0.6
Polluter 2	69°24′46″	30°14′30″	100	0	–	–
2–1	69°24′42″	30°16′32″	230	1.4	273.3 ± 24.0	139.0 ± 13.2
2–2	69°23′32″	30°10′45″	70	3.4	96.6 ± 9.4	48.9 ± 5.7
2–3	69°21′15″	30°03′11″	90	9.9	47.7 ± 7.0	18.6 ± 2.0
2–4	69°16′14″	30°04′56″	230	17.1	29.2 ± 2.6	16.7 ± 1.8
2–5	69°04′28″	30°12′15″	200	37.8	33.1 ± 2.1	9.4 ± 0.1

[a] In leaves of mountain birch, *Betula pubescens* ssp. *czerepanovii* (five samples per site; collected in 2001); meta-analyses are based on correlations with nickel.

Table 2.36 Location of study plots and concentrations of pollutants in the impact zone of nickel-copper smelters at Norilsk, Russia

	Plot position				Pollutant concentrations, mg/kg (mean ± SE)[a]	
Plot code	Latitude (N)	Longitude (E)	Altitude, m a.s.l.	Distance from the nearest polluter, km	Ni	Cu
Polluter 1	69°19′25″	87°58′20″	220	0	–	–
Polluter 2	69°21′45″	88°08′10″	90	0	–	–
Polluter 3	69°19′00″	88°12′10″	100	0	–	–
1–1	69°20′35″	88°01′20″	120	2.5	133.9 ± 10.6	459.6 ± 37.7
1–2	69°21′55″	87°37′20″	140	14.5	21.5 ± 0.3	31.3 ± 5.5
1–3	69°22′40″	87°20′10″	120	25.9	12.8 ± 1.1	17.8 ± 2.0
1–4	69°23′25″	86°46′50″	70	47.5	6.7 ± 0.6	15.6 ± 1.0
1–5	69°25′00″	86°23′05″	60	64.0	7.6 ± 0.5	10.1 ± 0.9
2–1	69°19′35″	88°18′35″	50	4.0	82.7 ± 14.0	62.1 ± 4.2
2–2	69°21′30″	88°23′20″	40	9.0	53.4 ± 9.8	33.9 ± 5.1
2–3	69°32′40″	88°19′30″	80	21.0	31.0 ± 3.2	22.7 ± 0.8
2–4	69°20′40″	89°00′55″	50	31.8	12.9 ± 1.1	16.1 ± 1.4
2–5	69°28′00″	90°36′00″	50	96.0	2.8 ± 0.4	7.0 ± 1.1

[a] In leaves of woolly willow, *Salix lanata* (three samples per site; collected in 2002); meta-analyses are based on correlations with nickel.

Table 2.37 Location of study plots and concentrations of pollutants in the impact zone of copper smelter at Revda, Russia

	Plot position				Pollutant concentrations, mg/kg (mean ± SE)[a]	
Plot code	Latitude (N)	Longitude (E)	Altitude, m a.s.l.	Distance from polluter, km	Cu	Zn
Polluter	56°51′11″	59°54′06″	360	0	–	–
1–1	56°50′40″	59°52′41″	360	1.7	58.2 ± 1.4	888 ± 183
1–2	56°51′05″	59°49′33″	380	4.6	13.8 ± 1.5	640 ± 156
1–3	56°51′15″	59°46′23″	420	7.8	11.3 ± 0.5	649 ± 110
1–4	56°49′19″	59°34′13″	330	20.5	6.5 ± 0.9	170 ± 16
1–5	56°47′51″	59°25′36″	380	29.7	5.2 ± 0.5	326 ± 56
2–1	56°50′15″	59°54′17″	360	1.7	50.3 ± 8.8	576 ± 104
2–2	56°49′37″	59°54′33″	350	3.0	41.1 ± 5.8	841 ± 130
2–3	56°49′41″	59°59′30″	340	6.2	33.2 ± 10.6	575 ± 40
2–4	56°43′47″	59°53′00″	350	13.8	11.3 ± 2.0	309 ± 77
2–5	56°34′10″	59°52′06″	360	31.7	7.6 ± 0.1	338 ± 33

[a] In leaves of white birch, *Betula pubescens* (three samples per site; collected in 2003); meta-analyses are based on correlations with copper.

Table 2.38 Location of study plots and concentrations of fluorine in the impact zone of aluminium smelter at Straumsvík, Iceland

	Plot position				
Plot code	Latitude (N)	Longitude (W)	Altitude, m a.s.l.	Distance from polluter, km	Fluorine concentrations, mg/kg (mean ± SE)[a]
Polluter	64°02′44″	22°01′39″	15	0	–
1–1	64°02′26″	22°02′17″	15	0.77	24.4 ± 5.0
1–2	64°02′17″	22°02′48″	10	1.26	0.26 ± 0.24
1–3	64°02′10″	22°03′12″	20	1.71	0.81 ± 0.79
1–4	64°02′15″	22°04′19″	20	2.35	1.61 ± 1.51
1–5	64°01′30″	22°07′22″	30	5.19	0.04 ± 0.00
2–1	64°02′54″	22°00′48″	15	0.70	6.0 ± 0.8
2–2	64°02′58″	22°00′28″	15	1.06	4.4 ± 2.4
2–3	64°03′05″	21°59′59″	20	1.51	1.0 ± 0.5
2–4	64°02′20″	21°58′58″	25	2.31	0.9 ± 0.3
2–5	64°00′40″	21°56′35″	85	5.64	0.05 ± 0.02

[a] In leaves of dwarf birch, *Betula nana* (two samples per site; collected in 2002).

Table 2.39 Location of study plots and concentrations of pollutants in the impact zone of nickel-copper smelter at Copper Cliff, Sudbury, Canada

	Plot position				Pollutant concentrations, mg/kg (mean ± SE)[a]	
Plot code	Latitude (N)	Longitude (W)	Altitude, m a.s.l.	Distance from polluter, km	Ni	Cu
Polluter	46°28′36″	81°03′33″	300	0	–	–
1–1	46°30′18″	81°02′37″	315	3.4	192.0 ± 7.4	79.4 ± 6.8
1–2	46°32′02″	81°04′47″	320	6.5	96.2 ± 17.7	35.9 ± 2.6
1–3	46°30′37″	81°12′00″	315	11.4	38.6 ± 5.5	14.4 ± 1.4
1–4	46°37′30″	81°12′43″	300	20.2	20.9 ± 4.2	9.0 ± 0.6
1–5	46°40′37″	81°32′38″	450	43.3	7.8 ± 0.8	21.9 ± 4.3
2–1	46°28′28″	81°04′39″	285	1.4	225.3 ± 9.3	47.7 ± 5.5
2–2	46°26′00″	81°06′28″	280	6.1	180.8 ± 19.2	50.9 ± 1.1
2–3	46°24′34″	81°12′59″	270	14.2	27.0 ± 1.1	10.7 ± 0.6
2–4	46°21′48″	81°24′56″	260	30.1	13.2 ± 1.7	6.1 ± 0.9
2–5	46°15′46″	81°51′24″	210	65.8	4.3 ± 0.3	5.4 ± 0.2

[a] In leaves of canoe birch, *Betula papyrifera* (three samples per site; collected in 2007); meta-analyses are based on correlations with nickel.

Table 2.40 Location of study plots and concentrations of fluorine in the impact zone of aluminium smelter at Volkhov, Russia

	Plot position				
Plot code	Latitude (N)	Longitude (E)	Altitude, m a.s.l.	Distance from polluter, km	Fluorine concentrations, mg/kg (mean ± SE)[a]
Polluter	59°54′38″	32° 21′22″	30	0	–
1–1	59°55′12″	32° 20′46″	30	1.2	38.1 ± 0.9
1–2	59°55′39″	32°19′50″	30	2.0	17.8 ± 2.3
1–3	59°55′40″	32°18′28″	40	3.3	13.0 ± 1.6
1–4	59°56′52″	32°13′37″	40	8.3	5.5 ± 0.2
1–5	59°59′35″	32°10′32″	40	13.7	5.8 ± 0.6
2–1	59°54′31″	32°21′02″	30	0.3	146.1 ± 29.4
2–2	59°53′30″	32°21′48″	30	2.1	54.6 ± 8.4
2–3	59°52′19″	32°21′36″	30	4.4	16.9 ± 0.3
2–4	59°47′32″	32°21′44″	20	13.2	7.2 ± 0.7
2–5	59°46′20″	32°21′38″	20	15.4	6.6 ± 0.5

[a] In leaves of white birch, *Betula pubescens* (two samples per site; collected in 2002).

Table 2.41 Location of study plots and concentrations of pollutants in the impact zone of power plant at Vorkuta, Russia

	Plot position				Pollutant concentrations[a] (Mean ± SE), mg/kg	
Plot code	Latitude (N)	Longitude (E)	Altitude, m a.s.l.	Distance from polluter, km	Cu	Fe
Polluter	67°37′20″	64°05′30″	160	0	–	–
1–1	67°36′30″	64°04′00″	160	0.8	5.44 ± 1.32	71.0 ± 10.4
1–2	67°37′00″	64°02′40″	160	2.0	8.05 ± 0.55	302.3 ± 35.6
1–3	67°36′30″	63°59′35″	160	4.0	7.37 ± 0.41	163.7 ± 7.9
1–4	67°36′45″	63°53′30″	160	8.0	6.73 ± 0.74	74.1 ± 4.3
1–5	67°36′10″	63°45′20″	180	14.0	5.44 ± 1.32	71.0 ± 10.4
2–1	67°37′40″	64°04′30″	140	0.3	8.49 ± 0.26	414.1 ± 46.3
2–2	67°37′00″	64°08′40″	180	2.1	7.56 ± 0.40	248.5 ± 32.9
2–3	67°38′30″	64°12′00″	180	4.6	10.33 ± 0.11	268.3 ± 30.2
2–4	67°40′20″	64°21′30″	200	12.0	7.32 ± 0.41	83.3 ± 7.4
2–5	67°42′25″	64°26′40″	160	17.5	5.72 ± 0.41	79.5 ± 6.3

[a] In leaves of woolly willow, *Salix lanata* (three samples per site; collected in 2001); meta-analyses are based on correlations with iron.

Table 2.42 Location of study plots and concentrations of fluorine in the impact zone of aluminum smelter at Žiar nad Hronom, Slovakia

	Plot position				
Plot code	Latitude (N)	Longitude (E)	Altitude, m a.s.l.	Distance from polluter, km	Fluorine concentrations, mg/kg (mean ± SE)[a]
Polluter	48°34′00″	18°50′50″	250	0	–
1–1	48°33′15″	18°51′20″	350	1.5	79.0 ± 4.7
1–2	48°33′19″	18°51′53″	350	1.7	47.5 ± 3.4
1–3	48°32′40″	18°53′30″	580	4.1	10.0 ± 0.7
1–4	48°31′05″	18°52′40″	680	8.4	5.9 ± 0.4
1–5	48°31′00″	18°57′40″	600	10.0	3.8 ± 0.4
2–1	48°34′17″	18°52′54″	300	2.6	8.3 ± 0.4
2–2	48°34′06″	18°54′53″	440	4.9	4.0 ± 0.1
2–3	48°33′35″	18°56′19″	460	6.7	2.5 ± 0.5
2–4	48°32′05″	18°57′05″	560	8.5	3.1 ± 2.1
2–5	48°32′52″	18°59′38″	390	11.0	1.8 ± 0.2

[a] In leaves of European beech, *Fagus sylvatica* (two samples per site; collected in 2002).

Table 2.43 Data collection timetable

Site	Dates of sampling sessions by study year									
	Prior 1999	1999	2000	2001	2002	2003	2004	2005	2006	2007
Apatity	3–7.7.1992; 9.8.1997	–	–	–	–	–	–	–	**30.7;** 19.8	–
Bratsk	–	–	–	–	**2–4.8**	–	20–25.1	–	–	–
Harjavalta	28.8.1997; 26.8.1998	31.8	–	26.9	**25.8**	–	–	–	7.10	7.11
Jonava	–	–	–	–	–	–	–	**3–5.9**	–	12.10
Kandalaksha	10.6.1998	–	–	15.9	**26.6, 16–17.7**	–	5.8	25.7	12.7	11.7
Karabash	–	–	–	–	–	**23–25.7**	–	–	–	31.8–1.9
Kostomuksha	–	–	–	11–12.9	20.3, **18.7**	–	–	–	21.8	–
Krompachy	–	–	–	–	**2–5.9**	–	6–8.10	–	11.11	–
Monchegorsk[a]	12–14.7.1993; 17.6–6.7.1997	10.7–11.8; 10–13.9	22–27.6	16.6–5.8	**14–20.8**	17.6–2.8; 12–19.10	21–27.8	11.7–20.8	10.7–13.8	–
Nadvoitsy	–	–	–	–	–	–	31.5; **25–27.7;** 25–26.8	23.8	18.6	16.8
Nikel	–	–	–	**17–18.7**	–	11–12.6; 29.6–1.7; 12–13.10	18–19.8	5–7.8	24–26.6; 8–9.8	5–6.7
Norilsk	–	–	–	–	**23–30.7**	–	–	–	–	–
Revda	–	–	–	–	–	**19–21.7**	–	–	–	24–26.8

Straumsvík	–	–	–	–	**11–13.7**	–	–	–	–	30.10–1.11
Sudbury	–	–	–	–	18–19.5	–	–	–	–	**19–21.10**
Volkhov	18.9.1994, 9.9.1998	5.9.	–	–	9.6; **8–9.8;** 11.10	–	–	–	17.6	17.8
Vorkuta	–	–	–	**8–10.7**	–	–	–	–	8–9.7	–
Žiar nad Hronom	–	–	–	–	**29.8–1.9**	–	3–5.10	–	9.11	–

The dates of the principal sampling session (when the majority of the data have been collected) are boldfaced.
[a] In Monchegorsk we also estimated longevity of Scots pine on 12.6.2008.

Although assisting personnel changed with sampling year and impact zone, this was not expected to cause variation in the outcomes of our surveys because all measurements that require specific training were performed by the authors, who conducted training sessions to assure the uniformity of methods across the entire study.

2.4 Environmental Contamination at Study Sites

2.4.1 Sampling and Preservation

As the basic estimate of the pollution load at our study sites we used total concentrations of selected pollutants in unwashed samples of plant leaves. Thus, our samples reflect both root and canopy uptake of pollutants (Kozlov et al. 2000).

Woody plants with large, preferably pubescent, leaves were selected for analyses of pollutants. We sampled white birch (eight impact zones), common birch (2), European beech (2), woolly willow, *Salix lanata* (2), goat willow, *Salix caprea* (1), dwarf birch (1), canoe birch (1), speckled alder (1) and Scots pine (1). Samples were taken from five mature plants randomly selected at each site (the same individuals from which vigour indices were measured; consult Chapter 4), within approximately 25 m from the marked centre of the study plot. One branch with 30–50 shoots was cut at a height of 1.2–1.4 m (except for dwarf birch in Iceland, where the uppermost twig was collected), packed in a plastic bag and transported to the laboratory. The next day after sampling, the leaves were cut by scissors in such a way that the basal parts of petioles and buds were not included in the sample; care was taken to avoid cross-contamination. In birches, only short-shoot leaves (Fig. 5.1) were sampled. Unwashed leaf samples were packed in paper bags, dried for 12–24 h at 80°C, and preserved for analysis. The parts of samples that were not used for our analyses were deposited in the Paljakka Environmental Bank of the Finnish Forest Research Institute.

2.4.2 Analyses

2.4.2.1 Metals

Concentrations of metals in samples from 1992–1993 (from Apatity and Monchegorsk) and from 1999 (from Harjavalta) were determined by X-ray fluorescence (Spectrace 5000 spectrometer, Tractor X-Ray, USA) at St. Petersburg State University, Russia. Samples from 2001–2005 were analysed by ICP (Analist 800, Perkin Elmer, USA) at the Institute of North Industrial Ecology, Apatity, Russia. Samples from 2007 (from Sudbury) were analysed by ICP-OES (IRIS Intrepid II XSP, Thermo Electron Corporation, USA) at the University of Joensuu, Finland.

The quality of the analytical data was checked by replicate analyses of the same samples, both by blind tests with the same analytical procedure in the original laboratory and by comparing the results obtained in different laboratories. For more details on sample preparation, analytical procedures and intercalibrations, consult Kozlov et al. (1995, 2000) and Kozlov (1996, 2005a).

2.4.2.2 Fluorine

Total concentrations of fluorine were determined by Enviroservis, s.r.o., Žiar nad Hronom, Slovakia. Foliar samples were pulverised and dried at 75°C to a constant weight. Each sample was subdivided into two parts that were processed independently, according to internal procedure PP OZP 006 cl. 5.7; the results were averaged for a sample-specific value.

Fluoride concentrations were measured by potenciometry with a fluoride ion selective electrode (Orion model 96-09, Thermo Electron Corporation, USA) in the automatic system SINTALYZER (SINTEF ADR 7034, Trondheim, Norway) and compared to the blank test conducted with a certified reference material GBW07604 (poplar leaves) with a fluoride concentration of 22 ± 4 mg/kg.

2.5 Statistical Approaches

2.5.1 Traditional Analyses

2.5.1.1 Data Inspection and Transformation

Raw data were first inspected for obvious errors. Then distributions of the data were analysed by SAS UNIVARIATE procedure (SAS Institute 2007). Following removal of the identified outliers (usually less than 0.5% of all measurements), the data were averaged by observational units (sensu Kozlov & Hurlbert 2006) and log-transformed whenever necessary.

2.5.1.2 Variation Between Sites

Distributions of individual measurements, or measurements averaged by observational units (sensu Kozlov & Hurlbert 2006), were used to test the null hypothesis, that there were no differences between the ten study sites within the impact zone of a given polluter. This analysis was performed using either ANOVA for normally distributed variables (SAS GLM procedure) or the Kruskal-Wallis test based on X^2 for variables with highly skewed distributions (SAS NPAR1WAY procedure).

Since the later analyses were based on plot-specific values, we chose not to include standard errors or other measures of within-plot variation in the data tables.

2.5.1.3 Test for Non-linearity of Responses

We checked whether second-degree polynomial regression fit the data better than linear regression. These tests were only performed using log-transformed distances as explanatory variables, and only for biotic response variables (i.e., concentrations of pollutants and soil quality indices, except for soil respiration, have been excluded from this analysis). For the data sets where the quadratic model had a significant second-degree component, we compared residual variations for linear and second-degree regression models using F-statistics (Motulsky & Christopoulos 2004; web calculator at http://graphpad.com/quickcalcs/AIC1.cfm), with one degree of freedom subtracted from the second-degree regression model to account for the additional parameter.

2.5.1.4 Correlation Analysis

Relationships between response variables and both distance from the polluter and concentration of pollutants (indicated in Tables 2.25–2.42) were explored by correlation analyses (SAS CORR procedure; SAS Institute 2007); correlations with both explanatory variables are reported in all data tables in Chapters 3–7. Logarithmic transformation was applied to the distance data to make them roughly proportional to the deposition of pollutants. As a rule, we calculated the Pearson linear correlation coefficient; however, if the distribution of response variables was highly skewed (e.g., herbivore density data), the Spearman rank correlation coefficient was calculated instead.

2.5.2 Meta-analyses

2.5.2.1 General Approaches

Meta-analysis is a quantitative synthetic research method that statistically integrates results from individual studies to find common trends and differences (Gurevitch & Hedges 2001). This method has been demonstrated to be a much better tool than traditional narrative review for synthesising results from multiple studies with diverse outcomes. Although meta-analysis is becoming increasingly popular in basic ecology, it is relatively rarely applied to integrate field collected environmental data (but see Ruotsalainen & Kozlov 2006; Menzel et al. 2006; Kozlov & Zvereva 2007b; Zvereva et al. 2008; Zvereva & Kozlov 2009).

In this book, we use the meta-analytic approach to summarise collected information, search for general patterns, and identify the sources of variation in changes of structural and functional parameters of terrestrial ecosystems. In general, our approach follows the methodology described by Zvereva et al. (2008). However, in this book we also explore whether correlations with distance and correlations with pollutant concentrations generally yield the same results, and whether contrast between the two most polluted and two control sites is sufficiently informative in comparison with the gradient approach involving ten study sites.

2.5.2.2 Calculation of Effect Sizes

To calculate effect sizes (ESs) based on the gradient approach involving ten study sites, individual correlation coefficients were *z*-transformed and weighed by their sample size. Note that the sign of the correlation coefficient between response variables and distance was changed by multiplying by (−1) to make these data consistent with correlations calculated for concentrations of pollutants; in our database, positive ESs denote an increase in the response variable with an increase in pollution.

Hedge's d (another measure of the ES) was calculated as the difference between the means based on the two most polluted and two control plots divided by the pooled standard deviation.

It should be stressed that the expressions 'positive effect of pollution' and 'negative effect of pollution' used throughout the text should not be perceived as 'beneficial' and 'adverse' effects. These expressions only denote that the parameter under study in polluted site(s) attained higher and lower values than in control site(s), respectively.

2.5.2.3 Overall Effect and Sources of Variation

If not stated otherwise, our analyses are based on ESs calculated from correlations between the measured parameter and the concentration of the selected pollutant (indicated in Tables 2.25–2.42). The mean ESs were calculated and compared using a random effect model with 95% confidence intervals (CIs). If the number of ESs in an individual analysis was seven or less, then we used a bootstrap estimate of the CI.

Separate meta-analyses were conducted for each class of response variables. Pollution was considered to have a statistically significant effect if the 95% CI of the mean effect size did not overlap zero (Gurevitch & Hedges 2001).

Variation in ES among classes of categorical variables was explored for each response variable by calculating between-class heterogeneity (Q_B) and testing it against the X^2 distribution (Gurevitch & Hedges 2001).

All calculations described above were performed using the MetaWin program (version 2.1.4; Rosenberg et al. 2000).

2.5.2.4 Repeatability of Results

When measurements of some index were performed more than once for two or more study areas, we used a meta-analysis to check for repeatability of the observed patterns.

For this analysis, we calculated all pairwise Pearson linear correlation coefficients between site-specific values obtained from the same study area and then calculated ES from these coefficients as described above (Section 2.5.2.2). The mean ESs were calculated and compared using a random effect model. A pattern was considered repeatable if the 95% CI of the mean effect size did not overlap zero.

2.5.2.5 Regression Analysis of Effect Sizes

We applied stepwise regression analysis (SAS REG procedure; SAS Institute 2007) to better understand the sources of variation in the responses of selected structural and functional characteristics of terrestrial ecosystems to pollution. In particular, it was used to estimate the contribution of different geographical (latitude of the polluter's location), climatic (mean July and January temperatures and annual precipitation in the locality) and polluter (emissions of principal pollutants and duration of the impact) characteristics to the variation of the effect size. The distribution of variables was checked, and appropriate transformation (usually logarithmic) was applied whenever necessary.

2.5.3 Significance of Effects

Through the entire book, the presence of an effect means that the effect in question was significant at $P = 0.05$; the absence of an effect means that, for the given test, $P > 0.05$. Sometimes we use expressions such as 'tended to be' for effects whose significance approached $P = 0.05$; an actual probability level is always reported in these situations.

2.6 Summary

The selected 18 polluters differ in the type and amount of emissions, as well as in pollution history, climate and type of primary vegetation within the impact zone. Considered together, they are sufficiently representative to reveal general patterns in the effects of industrial polluters on terrestrial ecosystems of temperate to northern regions of the Northern hemisphere. Uniformity in sampling design and proper replication at all hierarchical levels allowed us to decrease the impact of methodology-related variation (which is typical in meta-analyses of published data) on the outcome of the comparative analysis. To study possible causal links between pollution impacts and biotic effects, as well as general patterns and sources of variation, we used a combination of traditional statistical procedures (analysis of variance, correlation and regression analyses) and meta-analyses based on correlations with distances, correlations with pollution loads, and relative differences between the most and least polluted sites.

Chapter 3
Soil Quality

3.1 Introduction

Soil, like air and water, is an integral component of our environment. The value of soil is difficult to overestimate. One definition of soil especially stresses that this naturally occurring, unconsolidated or loose covering of broken rock particles and decaying organic matter on the surface of the Earth is *capable of supporting life* (Voroney 2006).

Soil, together with water, constitutes the most important natural resource, and therefore soil quality has been discussed widely within soil science and agronomy. Soil quality was recently defined as 'the fitness… to function within its capacity and within natural and managed ecosystem boundaries, to sustain plant and animal productivity, maintain air and water quality and support human health and habitation' (Karlen et al. 1997). Doran (2002) defines soil health similarly, but speaks of the capacity of a living soil to function.

Many studies have been conducted concerning soil quality, but there is still no well-defined universal methodology to characterise soil quality by means of a set of clear indicators (Bouma 2002). Different researchers have listed six to 12 physical, chemical and biological indicators to characterise soil quality in agro-ecosystems (Doran & Jones 1996; Gomez et al. 1996; Arshad & Martin 2002). Several of these indicators, such as pH, concentrations of pollutants, total and extractable nitrogen, phosphorous, potassium, calcium and magnesium, are frequently measured in studies addressing the impact of industrial polluters on soils. Some others, such as topsoil depth, infiltration, and electrical conductivity, are reported only occasionally, probably because ecologists generally consider soil as an environment for plant roots and animals, and not as a study object (Kaigorodova & Vorobeichik 1996; Pankhurst et al. 1997).

Little is known about changes of soil morphology under pollution impacts (Kryuchkov 1993a, c; Kaigorodova & Vorobeichik 1996), and an integrated approach to soil quality assessment, to our knowledge, has not been used in studies of industrially contaminated terrestrial ecosystems. Moreover, changes in soil quality are only rarely accounted for by researchers exploring pollution effects on biota. For example, we were unable to locate data on changes in soil pH around 12 of 60

M.V. Kozlov et al. *Impacts of Point Polluters on Terrestrial Biota*,
DOI 10.1007/978-90-481-2467-1_3, © Springer Science+Business Media B.V. 2009

polluters involved in a meta-analysis of pollution impacts on the diversity of vascular plants (Zvereva et al. 2008), and around 23 of 74 polluters in a meta-analysis of changes in diversity, abundance and fitness of terrestrial arthropods (Zvereva & Kozlov 2009).

In our study, we recorded five indices of soil quality: topsoil depth, stoniness, pH, electrical conductivity, and soil respiration. These indices, which are relatively easy to measure, provided basic information on changes in soil, which is necessary for understanding ecosystem-level effects of industrial pollution.

3.2 Materials and Methods

3.2.1 Topsoil Depth and Soil Stoniness

Topsoil, sometimes referred to as the soil organic layer or the A horizon (although labelling of soil horizons varies among different classificatory systems), is the upper, outermost layer of soil. It has the highest concentration of organic matter and microorganisms, and most of the Earth's biological soil activity occurs in this layer. The depth of the organic layer is the outcome of litter production and decomposition rates. Both of these parameters depend on site condition, including productivity and climate. Plants generally concentrate their roots in and obtain most of their nutrients from the soil organic layer.

The depth of the organic layer (including litter) was measured (by a ruler, to the nearest 2 mm) in soil cores (30 mm diameter; 10–25 per site, depending on intrasite variation) collected at 2 m intervals along a line crossing the study site. If we discovered a surface rock fragment at the selected sampling point, we then considered topsoil depth to be zero (in cases where this fragment was large enough, so there was no topsoil under the stone), or we measured topsoil depth several centimetres away. The boundary between the topsoil and the subsoil was drawn where soil colour and texture changed abruptly, or in the middle of the intermediate layer if soil properties changed gradually. We are aware of the fact that by using this protocol, we may have measured different soil horizons in different regions. However, since we do not compare absolute values obtained around different polluters, but only patterns of changes along pollution gradients, the regional differences in soil morphology are unlikely to influence our conclusions.

In some heavily contaminated areas, a high level of stoniness is the main evidence of extensive erosion (please see color plates 34 and 54 in Appendix II). Different researchers define and measure stoniness in an inconsistent way, by percentage of stone cover, using an arbitrarily defined scale with four to six ranks, or by weighing stones after sieving the soil sample (the minimum size of rock fragments varied from 2 to 20 mm). In the present study, we combined two measures of stoniness, the percentage of surface stone cover and average stone depth (i.e., average thickness of the soil layer that does not contain stones) measured using the rod penetration method (Eriksson & Holmgren 1996). We visually estimated the percentage of stone cover in ten 1 × 1 m plots; this was done simultaneously with an evaluation of vegetation cover (consult Section 6.2.2 for details). The depth of the first stone was measured using a pointed stainless steel probe (7 mm diameter, 500 mm length). We forced this probe into the soil (perpendicular to the soil surface) until it was hampered by a stone. These measurements were performed simultaneously with measurements of topsoil thickness (see above), 10 or 25 measurements per site, depending on the variability of the results.

Due to a relatively large number of zero values in measurements of stone cover, variation of this characteristic between study sites was explored using a non-parametric Kruskal-Wallis test. In calculation of site-specific stone depths, we estimated summary statistics of these data using the robust method described by Helsel (1990).

We excluded five impact zones (Harjavalta, Jonava, Kostomuksha, Revda, and Vorkuta) from the meta-analysis of stoniness because the topsoil in these areas contained almost no stones, and therefore soil erosion could not be explored by measuring stoniness. For example, soil probing detected a stone in 7 of 100 measurement points around Harjavalta, and stone cover equalled zero in all 100 plots around Jonava.

3.2.2 *Soil pH and Electrical Conductivity*

Samples for measurements of pH and electrical conductivity were collected from the upper part of the subsoil, usually at a depth of 5–15 cm (measuring from the upper surface of litter). We have chosen to measure these parameters in subsoil because topsoil was completely washed out from many barren sites close to large polluters (please see color plates 20, 29, 34, 46 and 54 in Appendix II).

A small shovel was used to remove topsoil before collecting approximately 100 g of subsoil. Five samples were taken from each study site, approximately 10 m apart, at least 1 m from the nearest large woody plant. Samples were sieved through a plastic screen with 2.5 mm openings. All samples collected around one polluter were processed and analysed simultaneously. If immediate processing of samples was impossible, they were stored in a refrigerator for a period not exceeding 5 days.

Samples (15 cm^3 of compacted soil) were extracted by 1 min shaking in distilled water (1:2.5 vol:vol) in plastic bottles and then left for 24 h at room temperature. Then the solid and liquid phases were mixed, and measurements were performed by digital pH- and EC-meters (Nieuwkoop Aalsmeer, The Netherlands). The pH-meter (PH-93), with an epoxy gel electrode, had an accuracy of 0.1 pH within the range from 0 to 14 pH; the EC-meter (EC-93), with an automatic compensation for temperature, had an accuracy of 0.1 μS/cm within the range from 0 to 2,000 μS/cm. Both meters were calibrated prior to each series of measurements. The electrode was kept in the extract for 60 s prior to recording pH and EC values.

3.2.3 Soil Respiration

The soil respiration rate (g $CO_2/m^2 \cdot h$) was measured *in situ*, by using a soil CO_2 flux system (PP Systems International, UK). This system employed the method described by Parkinson (1981); a chamber of known volume (SRC-1, 100 mm diameter) was placed on the soil, and the rate of increase in CO_2 within the chamber was monitored. In this system, the air was continuously sampled using a closed circuit through the infrared gas analyser (EGM-2) which calculated, displayed and recorded the soil respiration rate. This system ensured that air within the chamber was carefully mixed without generating pressure differences that would affect the evolution of CO_2 from the soil surface.

Five measurements were performed at each study site during a sampling session. Representative microsites were selected after surveying of both polluted and control sites in each study area, and then measurements at each study site were conducted in the first five discovered microsites of the selected type that were at least 10 m apart. For example, in the impact zones of the Monchegorsk, Nikel, and Kandalaksha smelters, we measured soil respiration in open microsites covered by crowberry; at sites where pollution killed this species (please see color plate 38 in Appendix II), we conducted measurements in spots that were covered by dead stems of crowberry. This allowed us to minimise variation between individual measurements performed at the same study site, that was outside the scope of the present study. The point selected for measurement was carefully cleaned of green parts of vegetation; however, care was taken not to disturb litter. All measurements were conducted in the daytime, at least 4 h after any heavy rain. Care was taken to randomise the sampling pattern in such a way that distance to the polluter was not confounded by the daily rhythms of illumination and temperature.

3.3 Results

3.3.1 Interdependence of Soil Quality Indices

The site-specific values of five indices of soil quality varied independently from each other. At a table-wide probability level of $P = 0.05$, we discovered only a weak positive correlation between soil pH and electrical conductivity (r = 0.32). This result suggests that separate analyses of the five measured indices of soil quality were not redundant.

3.3.2 Topsoil Depth and Soil Stoniness

Variation in topsoil depth between study sites was significant in all 11 data sets and in stoniness in 16 of 17 data sets. However, only four of 22 individual correlations (with both distance and pollution) were significant for topsoil depth, and four of 34 were significant for stoniness (Tables 3.1–3.16).

Pollution had no overall effect on topsoil depth (Fig. 3.1); this result did not depend on the method used to calculate ES ($Q_B = 1.38$, df = 2, $P = 0.79$). The effect

Table 3.1 Soil quality indices in the impact zone of the power plant at Apatity, Russia

Site	Topsoil depth, mm	pH	Electrical conductivity, μS/cm	Soil respiration, g $CO_2/m^2 \cdot h$
1–1	23.8	7.24	47.2	0.33
1–2	23.0	7.09	49.6	0.52
1–3	20.0	7.41	31.2	0.23
1–4	34.2	5.81	23.8	0.86
1–5	19.0	7.06	34.2	0.66
2–1	16.5	7.67	40.5	0.92
2–2	39.8	6.02	38.8	0.32
2–3	70.2	6.66	45.8	2.36
2–4	76.0	5.66	37.6	0.54
2–5	40.2	6.04	53.8	0.76
ANOVA: *F/P*	42.0/<0.0001	22.0/<0.0001	4.16/0.0008	39.4/<0.0001
Dist.: *r/P*	0.29/0.42	−0.52/0.12	−0.36/0.30	0.17/0.64
Poll.: *r/P*	−0.29/0.42	0.50/0.14	0.28/0.43	−0.15/0.68

Measurements were conducted on 30.7.2006. Topsoil depth is based on 20 measurements per site. ANOVA: test for significance of variation between study sites. Dist.: Pearson linear correlation between site-specific means and log-transformed distance from polluter. Poll.: Pearson linear correlation between site-specific means and concentration of the selected pollutant.

Table 3.2 Soil quality indices in the impact zones of the aluminium smelter at Bratsk, Russia and nickel-copper smelters at Norilsk, Russia

Site	Bratsk		Norilsk	
	Stone cover, %	Soil respiration, g $CO_2/m^2 \cdot h$	Stone cover, %	Soil respiration, g $CO_2/m^2 \cdot h$
1–1	0	1.50	0	0.17
1–2	0	1.11	1.3	0.85
1–3	3.0	0.88	0	0.61
1–4	0	1.74	0	0.70
1–5	0	1.90	0	0.58
2–1	4.3	1.18	1.0	0.48
2–2	0	1.23	3.3	0.48
2–3	0	1.26	1.7	0.85
2–4	0	1.53	0	0.64
2–5	0	2.16	0.3	0.91
ANOVA: *F* (X^2)/*P*	28.7/0.0007	3.95/0.0012	21.5/0.01	1.94/0.07
Dist.: *r/P*	−0.37/0.29	0.65/0.04	−0.37/0.29	0.73/0.02
Poll.: *r/P*	0.47/0.17	−0.34/0.34	0.38/0.28	−0.82/0.004

Stone cover was analysed by non-parametric methods: Kruskal-Wallis test for between-site variation, and Spearman rank correlation coefficients for relationships with both distance and pollution load. Measurements were conducted on 2–4.8.2002 (Bratsk) and 23–28.7.2002 (Norilsk). For other explanations, consult Table 3.1.

Table 3.3 Soil quality indices in the impact zone of the nickel-copper smelter at Harjavalta, Finland

Site	Topsoil depth, mm	pH	Electrical conductivity, μS/cm	Soil respiration, g $CO_2/m^2 \cdot h$
1–1	16.9	4.60	58.0	0.22
1–2	41.5	4.48	62.2	0.62
1–3	18.7	4.74	101.6	0.38
1–4	17.3	5.29	44.4	0.47
1–5	54.5	4.91	54.2	0.74
2–1	16.0	5.63	52.8	0.71
2–2	51.5	5.92	53.4	0.70
2–3	30.2	5.14	70.8	0.35
2–4	52.5	4.49	58.4	0.98
2–5	32.0	5.77	81.3	1.16
ANOVA: *F*/*P*	13.8/<0.0001	13.0/<0.0001	3.00/0.0085	3.69/0.0019
Dist.: *r*/*P*	0.32/0.36	−0.02/0.95	0.17/0.63	0.51/0.14
Poll.: *r*/*P*	0.07/0.86	−0.33/0.35	−0.15/0.69	−0.22/0.55

Measurements were conducted on 25.8.2002 (respiration), 7.10.2006 (pH and Electrical conductivity), and 7.11.2007 (topsoil depth). Topsoil depth is based on ten measurements per site. For other explanations, consult Table 3.1.

Table 3.4 Soil quality indices in the impact zone of the fertiliser factory at Jonava, Lithuania

Site	Topsoil depth, mm	pH	Electrical conductivity, μS/cm	Soil respiration, g $CO_2/m^2 \cdot h$
1–1	28.5	5.79	72.0	0.74
1–2	38.3	3.97	47.8	0.61
1–3	29.3	4.31	27.6	0.52
1–4	29.7	3.74	48.0	1.37
1–5	34.0	3.80	48.4	0.77
2–1	34.0	3.66	67.0	0.64
2–2	37.2	3.35	83.6	0.71
2–3	19.7	4.55	40.8	0.61
2–4	25.6	3.68	29.0	0.94
2–5	27.5	4.08	38.4	0.49
ANOVA: *F*/*P*	2.04/0.04	41.3/<0.0001	10.3/<0.0001	2.63/0.02
Dist.: *r*/*P*	−0.34/0.34	−0.28/0.43	−0.67/0.03	0.21/0.56
Poll.: *r*/*P*	0.05/0.90	−0.04/0.91	0.26/0.47	0.16/0.66

Measurements were conducted on 3–5.9.2005 (respiration) and 12.11.2007 (all other indices). Topsoil depth is based on ten measurements per site. For other explanations, consult Table 3.1.

of pollution on topsoil depth did not depend on either the polluter's impact on soil pH (Fig. 3.1; $Q_B = 1.33$, df = 2, $P = 0.52$) or its geographical position (Fig. 3.1; $Q_B = 0.69$, df = 1, $P = 0.41$). The only source of variation was the type of polluter (Fig. 3.1; $Q_B = 4.57$, df = 1, $P = 0.03$); significant decreases in topsoil depth were detected near non-ferrous smelters, while the effects of aluminium plants were not significant.

Table 3.5 Soil quality indices in the impact zone of the aluminium smelter at Kandalaksha, Russia

Site	Topsoil depth, mm	Stone depth, mm	Stone cover, %	pH	Electrical conductivity, µS/cm	Soil respiration, g $CO_2/m^2 \cdot h$	Soil respiration, g $CO_2/m^2 \cdot h$
1–1	38.5	150	0	5.69	31.6	0.51	0.30
1–2	33.2	70	0	5.14	42.2	1.42	0.73
1–3	38.0	67	0.3	4.70	51.6	1.29	0.88
1–4	30.5	93	0	5.90	23.4	1.04	0.93
1–5	36.2	28	0.3	5.91	27.2	0.76	0.60
2–1	45.0	96	0.7	6.23	106.8	0.35	0.54
2–2	28.2	88	0.3	4.92	61.2	0.39	1.70
2–3	34.0	88	1.7	4.56	49.8	0.84	0.28
2–4	33.8	98	2.0	4.55	76.4	0.87	0.88
2–5	31.8	270	0.0	4.58	44.8	0.51	0.67
ANOVA: F (X^2)/P	4.24/<0.0001	80.9/<0.0001	11.5/0.24	19.23/<0.0001	1.73/0.11	3.60/0.0023	6.20/<0.0001
Dist.: r/P	−0.56/0.10	0.22/0.55	−0.13/0.73	−0.42/0.23	−0.54/0.11	0.27/0.45	0.05/0.88
Poll.: r/P	0.49/0.15	−0.25/0.49	−0.16/0.64	0.50/0.13	0.40/0.25	−0.11/0.77	−0.01/0.98

Measurements were conducted on 15.9.2001 (respiration, first measurement), 5.8.2004 (topsoil and stone depth), 16–17.7.2005 (respiration, second measurement; stone cover), and 12.7.2006 (pH and Electrical Conductivity). Topsoil and stone depths are based on 20 and 50 measurements per site, respectively. Stone cover was analysed by non-parametric methods: Kruskal-Wallis test for between-site variation, and Spearman rank correlation coefficients for relationships with both distance and pollution load. For other explanations, consult Table 3.1.

Table 3.6 Soil quality indices in the impact zone of the copper smelter at Karabash, Russia

Site	Topsoil depth, mm	Stone depth, mm	Stone cover, %	pH	Electrical conductivity, µS/cm	Soil respiration, g $CO_2/m^2 \cdot h$
1–1	3.3	77	25.0	4.48	170.0	0.35
1–2	13.7	350	0	6.08	69.6	0.92
1–3	34.3	149	0.7	6.02	101.0	1.46
1–4	11.1	343	0	6.19	115.0	1.58
1–5	24.7	236	0	6.58	111.8	1.80
2–1	0	23	21.7	4.56	138.2	0.26
2–2	8.7	62	0	6.02	89.0	0.52
2–3	31.2	178	0	5.93	66.2	1.10
2–4	29.5	107	0	6.09	91.2	1.20
2–5	36.9	229	0	6.12	95.8	1.46
ANOVA: F (X^2)/P	29.6/<0.0001	12.2/<0.0001	26.4/0.0018	49.0/<0.0001	3.39/0.0035	10.9/<0.0001
Dist.: r/P	0.76/0.01	0.24/0.51	−0.62/0.06	0.78/0.0072	−0.36/0.31	0.92/0.0002
Poll.: r/P	−0.77/0.01	−0.45/0.19	0.62/0.06	−0.88/0.0007	0.58/0.08	−0.91/0.0002

Measurements were conducted on 23–25.7.2003 (respiration) and 31.8–1.9.2007 (all other indices). Topsoil depth is based on ten measurements per site. Stone cover was analysed by non-parametric methods: Kruskal-Wallis test for between-site variation, and Spearman rank correlation coefficients for relationships with both distance and pollution load. For other explanations, consult Table 3.1.

Table 3.7 Soil quality indices in the impact zone of the iron pellet plant at Kostomuksha, Russia

Site	pH	Electrical conductivity, μS/cm	Soil respiration, g $CO_2/m^2·h$	Soil respiration, g $CO_2/m^2·h$
1–1	4.98	46.6	0.83	2.05
1–2	4.60	47.2	1.51	2.03
1–3	4.60	50.0	1.45	1.84
1–4	4.27	77.2	2.10	2.11
1–5	4.46	37.0	–	2.60
2–1	6.63	66.6	1.38	2.05
2–2	4.57	60.6	0.95	1.81
2–3	5.66	13.6	1.58	1.87
2–4	4.47	32.8	–	1.36
2–5	4.52	35.0	2.39	1.92
ANOVA: *F*/*P*	31.3/<0.0001	6.52/<0.0001	2.15/0.08	1.23/0.30
Dist.: *r*/*P*	−0.73/0.02	−0.34/0.33	0.78/0.02	0.06/0.87
Poll.: *r*/*P*	0.78/0.008	0.54/0.11	−0.37/0.36	0.20/0.59

Measurements were conducted on 11–12.9.2001 and 18.7.2002 (respiration, first and second measurements, respectively) and 21.8.2006 (pH and electrical conductivity). For other explanations, consult Table 3.1.

Table 3.8 Soil quality indices in the impact zone of the copper smelter at Krompachy, Slovakia

Site	pH	Electrical conductivity, μS/cm	Soil respiration, g $CO_2/m^2·h$	Soil respiration, g $CO_2/m^2·h$
1–1	5.30	59.8	0.55	0.36
1–2	4.87	51.2	0.53	0.63
1–3	4.71	70.6	0.80	0.61
1–4	8.15	208.0	0.80	0.93
1–5	7.30	110.6	0.99	0.59
2–1	4.95	49.2	0.44	0.49
2–2	4.98	96.8	0.59	0.80
2–3	5.24	91.8	0.83	0.65
2–4	4.94	79.8	1.69	0.65
2–5	4.60	44.4	0.54	0.63
ANOVA: *F*/*P*	29.1/<0.0001	9.37/<0.0001	2.62/0.02	2.88/0.0071
Dist.: *r*/*P*	0.36/0.30	0.30/0.41	0.32/0.37	0.45/0.19
Poll.: *r*/*P*	−0.13/0.73	−0.37/0.30	−0.43/0.22	−0.54/0.11

Measurements were conducted on 2–4.9.2002 and 6–8.10.2004 (respiration, first and second measurements, respectively) and 11.11.2006 (pH and electrical conductivity). For other explanations, consult Table 3.1.

Since stone depth decreases and stone cover increases with an increase in stoniness, we changed the sign of the effect sizes (ESs) for stone depth prior to further analysis. These two measures of stoniness showed a consistent response to pollution (correlation between site-specific means: $r_S = 0.47$, $N = 59$ sites, $P = 0.0002$; comparison between ESs for six impact zones: $Q_B = 1.94$, df = 1, $P = 0.16$). Therefore, we combined these two measures and calculated an overall effect on

Table 3.9 Soil quality indices in the impact zone of the nickel-copper smelter at Monchegorsk, Russia

Site	Topsoil depth, mm	Stone depth, mm	Stone cover, %	pH	Electrical conductivity, μS/cm	Soil respiration, g $CO_2/m^2 \cdot h$	Soil respiration, g $CO_2/m^2 \cdot h$	Soil respiration, g $CO_2/m^2 \cdot h$
1–1	27.4	209	1.0	4.70	43.8	0.16	0.21	0.18
1–2	17.1	472	1.3	4.80	87.0	0.26	0.34	0.35
1–4	26.5	200	0	4.93	45.8	0.27	0.31	0.65
1–5	24.2	259	0.7	5.27	69.4	0.37	0.32	0.47
2–3	6.4	170	4.3	4.50	45.8	0.02	0.13	0.08
2–4	11.2	74	9.3	4.68	30.0	0.13	0.22	0.33
2–5	30.4	344	0.3	4.56	49.0	0.31	0.75	0.67
2–8	36.0	238	0	4.62	40.0	0.64	0.73	0.88
2–10	25.6	180	0.3	4.70	45.5	0.49	1.10	1.12
2–12	31.4	154	0	4.61	39.4	0.53	0.73	0.83
ANOVA: F (X^2)/P	22.7/<0.0001	32.2/<0.0001	21.2/0.01	6.92/<0.0001	3.91/0.0012	9.67/<0.0001	13.1/<0.0001	10.3/<0.0001
Dist.: r/P	0.47/0.17	−0.20/0.58	−0.60/0.07	−0.02/0.95	−0.22/0.55	0.80/0.005	0.78/0.008	0.86/0.001
Poll.: r/P	−0.79/0.006	0.20/0.57	0.62/0.06	−0.46/0.18	−0.17/0.64	−0.82/0.003	−0.61/0.06	−0.78/0.008

Measurements were conducted on 16–18.6, 7.7, and 12.7.2001 (respiration; first, second, and third measurements, respectively), 18–31.7.2003 (topsoil depth and stoniness) 8–10.7.2006 (pH and electrical conductivity). Topsoil and stone depths are based on 25 and 50 measurements per site, respectively. Stone cover was analysed by non-parametric methods: Kruskal-Wallis test for between-site variation, and Spearman rank correlation coefficients for relationships with both distance and pollution load. For other explanations, consult Table 3.1.

Table 3.10 Soil quality indices in the impact zone of the aluminium smelter at Nadvoitsy, Russia

Site	Topsoil depth, mm	Stone depth, mm	pH	Electrical conductivity, μS/cm	Soil respiration, g CO_2/m²·h
1–1	25.6	359	4.83	48.6	0.49
1–2	57.0	>500	4.54	96.6	2.31
1–3	19.1	206	4.67	35.0	1.15
1–4	15.4	269	4.50	37.4	1.35
1–5	32.5	227	4.53	61.2	2.57
2–1	23.1	305	5.22	38.2	1.29
2–2	17.0	151	5.54	67.0	1.62
2–3	23.8	313	4.29	54.8	1.16
2–4	19.2	440	4.27	48.6	2.18
2–5	38.7	333	4.88	31.8	2.26
ANOVA: *F/P*	9.86/<0.0001	4.42/<0.0001	11.6/<0.0001	6.60/<0.0001	9.76/<0.0001
Dist.: *r/P*	0.07/0.86	−0.22/0.57	−0.54/0.10	−0.17/0.64	0.65/0.04
Poll.: *r/P*	0.09/0.80	0.02/0.97	0.70/0.03	0.31/0.39	−0.37/0.30

Measurements were conducted on 25–27.7.2004 (respiration), 18.6.2006 (pH and electrical conductivity), and 16.8.2007 (topsoil and stone depths, each based on ten measurements per site). For other explanations, consult Table 3.1.

stoniness, which increased with an increase in stone cover and with a decrease in stone depth.

We excluded Žiar nad Hronom from this analysis, because significant increases in stoniness with distance from the polluter (Table 3.16) resulted from the geology of the study area, not from environmental effects of emissions; the impact zone is mountainous, with the smelter located at the bottom of a river valley (Fig. 2.19, Table 2.42).

Stoniness slightly increased with pollution (Fig. 3.2); this result did not depend on the method used to calculate ES ($Q_B = 0.46$, df = 2, $P = 0.79$). Pollution effects did not depend on the type of polluter (Fig. 3.2; $Q_B = 0.05$, df = 1, $P = 0.81$), the impact on soil pH (Fig. 3.2; $Q_B = 0.68$, df = 2, $P = 0.71$) or the geographical position of the polluter (Fig. 3.2; $Q_B = 0.87$, df = 1, $P = 0.35$).

3.3.3 Soil pH and Electrical Conductivity

Variation in pH between study sites was significant in all 16 data sets and in electrical conductivity (EC) in 15 of 16 data sets. However, only ten of 32 individual correlations (with both distance and pollution) were significant for pH, and eight of 32 were significant for EC (Tables 3.1–3.14).

We divided our polluters into three groups according to their effects on soil acidity: acidifying, alkalysing and neutral (Table 2.1). This classification is based on the results of meta-analysis; we attributed our polluters to one of these classes in such

Table 3.11 Soil quality indices in the impact zone of the of the nickel-copper smelter at Nikel and ore-roasting plant at Zapolyarnyy, Russia

Site	Topsoil depth, mm	Stone depth, mm	Stone cover, %	pH	Electrical conductivity, μS/cm	Soil respiration, g $CO_2/m^2 \cdot h$	Soil respiration, g $CO_2/m^2 \cdot h$	Soil respiration, g $CO_2/m^2 \cdot h$
1–1	25.2	224	0.3	3.73	402.6	0.22	0.13	0.27
1–2	31.8	>500	0	4.65	100.6	0.50	0.22	0.27
1–3	24.0	424	0	4.88	62.0	0.65	0.25	1.30
1–4	86.8	364	0	5.18	63.8	0.68	0.44	0.94
1–5	87.2	176	2.3	5.06	54.4	0.56	0.26	0.73
2–1	61.4	330	2.7	4.36	140.8	0.42	0.16	0.17
2–2	62.1	291	2.7	5.25	73.6	0.13	0.25	0.21
2–3	37.2	366	0	4.72	114.6	0.57	0.50	0.46
2–4	44.8	184	0.3	4.83	55.2	0.52	0.29	0.51
2–5	29.5	609	1.7	4.70	31.4	0.38	0.39	0.90
ANOVA: F (X^2)/P	12.8/<0.0001	26.6/<0.0001	17.7/0.04	7.46/<0.0001	34.1/<0.0001	13.6/< 0.0001	3.03/0.0077	5.94/<0.0001
Dist.: r/P	0.36/0.30	0.27/0.48	−0.13/0.72	0.72/0.02	−0.81/0.005	0.60/0.06	0.68/0.03	0.68/0.03
Poll.: r/P	−0.25/0.49	−0.24/0.54	0.21/0.56	−0.85/0.002	0.88/0.001	−0.59/0.07	−0.67/0.03	−0.59/0.08

Measurements were conducted on 17–18.7.2001, 29.6–1.7.2003, and 5–7.8.2005 (respiration; first, second, and third measurements, respectively), 17–18.7.2001 (stone cover), 12–13.10.2003 (topsoil and stone depths) and 24–26.6.2006 (pH and electrical conductivity). Topsoil and stone depths are based on 20 and 50 measurements per site, respectively. Stone cover was analysed by non-parametric methods: Kruskal-Wallis test for between-site variation, and Spearman rank correlation coefficients for relationships with both distance and pollution load. For other explanations, consult Table 3.1.

Table 3.12 Soil quality indices in the impact zone of the copper smelter at Revda, Russia

Site	Topsoil depth, mm	Stone depth, mm	pH	Electrical conductivity, μS/cm	Soil respiration, g $CO_2/m^2 \cdot h$	Soil respiration, g $CO_2/m^2 \cdot h$
1–1	26.0	395	5.18	50.2	0.70	1.08
1–2	33.1	384	5.01	81.8	1.18	1.04
1–3	32.9	362	5.48	71.2	0.84	1.48
1–4	26.6	>500	5.66	86.6	0.93	1.31
1–5	20.1	>500	5.63	69.6	1.36	1.33
2–1	28.8	387	5.10	84.8	0.72	0.91
2–2	40.9	260	5.19	85.4	0.49	1.52
2–3	43.9	282	5.29	122.4	0.88	2.05
2–4	40.7	359	5.38	53.6	1.00	1.18
2–5	58.5	392	5.34	71.0	0.69	1.74
ANOVA: *F/P*	8.86/<0.0001	3.34/0.0014	4.80/0.0002	5.65/<0.0001	3.40/0.0035	2.62/0.02
Dist.: *r/P*	0.22/0.53	0.50/0.14	0.77/0.009	−0.07/0.85	0.49/0.15	0.38/0.27
Poll.: *r/P*	−0.14/0.71	−0.46/0.19	−0.64/0.05	0.07/0.86	−0.61/0.06	−0.21/0.56

Measurements were conducted on 19–21.7.2003 (respiration, first measurement) and 24–26.8.2007 (respiration, second measurement; all other indices). Topsoil and stone depths are each based on ten measurements per site. For other explanations, consult Table 3.1.

Table 3.13 Soil quality indices in the impact zone of the aluminium smelter at Straumsvík, Iceland

Site	Topsoil depth, mm	Stone depth, mm	Stone cover, %	pH	Electrical conductivity, μS/cm	Soil respiration, g $CO_2/m^2 \cdot h$
1–1	1.8	5	66.5	5.99	24.2	0.31
1–2	4.3	170	3.8	6.01	45.0	0.11
1–3	5.1	77	4.8	6.10	42.8	0.11
1–4	3.1	20	32.2	6.01	58.0	0.62
1–5	4.5	98	6.6	6.32	54.8	0.36
2–1	12.4	19	19.1	5.75	52.6	0.09
2–2	4.0	61	2.2	6.16	54.4	0.15
2–3	2.6	54	4.1	6.09	48.4	0.17
2–4	4.0	92	3.0	5.88	44.4	0.25
2–5	4.0	42	1.7	5.82	31.8	0.22
ANOVA: F (X^2)/P	3.90/0.0003	2.54/0.01	36.6/<0.0001	10.5/<0.0001	7.30/<0.0001	0.96/0.49
Dist.: r/P	–0.30/0.40	0.20/0.57	–0.31/0.38	0.23/0.53	0.06/0.86	0.39/0.26
Poll.: r/P	–0.14/0.69	–0.53/0.11	0.47/0.17	–0.13/0.72	–0.59/0.07	0.08/0.83

Measurements were conducted on 11–13.7.2002 (respiration) and 31.10–1.11.2007 (all other indices). Topsoil and stone depths are each based on ten measurements per site. Stone cover was analysed by non-parametric methods: Kruskal-Wallis test for between-site variation, and Spearman rank correlation coefficients for relationships with both distance and pollution load. For other explanations, consult Table 3.1.

Table 3.14 Soil quality indices in the impact zone of the nickel-copper smelter at Sudbury, Canada

Site	Topsoil depth, mm	Stone depth, mm	Stone cover, %	pH	Electrical conductivity, μS/cm	Soil respiration, g $CO_2/m^2 \cdot h$
1–1	11.5	95	33.3	5.57	57.0	0.11
1–2	32.0	298	11.7	3.90	53.6	0.35
1–3	44.5	148	6.0	3.57	88.6	0.99
1–4	28.5	115	0.7	5.03	73.8	0.65
1–5	40.5	338	0.3	4.16	73.8	1.86
2–1	6.2	38	63.3	4.33	36.0	0.12
2–2	26.7	208	0.7	4.53	31.6	0.39
2–3	41.0	301	0	4.68	52.4	0.69
2–4	30.0	227	0.3	4.17	55.2	0.58
2–5	24.5	355	0	4.46	38.4	0.81
ANOVA: F (X^2)/P	8.19/<0.0001	5.21/<0.0001	25.1/0.03	15.1/<0.0001	8.68/<0.0001	11.1/<0.0001
Dist.: r/P	0.61/0.06	0.57/0.09	−0.84/0.002	−0.12/0.74	0.01/0.97	0.72/0.02
Poll.: r/P	−0.76/0.01	−0.51/0.14	0.84/0.002	0.31/0.38	−0.51/0.13	−0.70/0.02

Measurements were conducted on 19–21.10.2007. Topsoil and stone depths are each based on ten measurements per site. Stone cover was analysed by non-parametric methods: Kruskal-Wallis test for between-site variation, and Spearman rank correlation coefficients for relationships with both distance and pollution load. For other explanations, consult Table 3.1.

Table 3.15 Soil quality indices in the impact zones of the aluminium smelter at Volkhov, Russia and power plant at Vorkuta, Russia

	Volkhov			Vorkuta	
Site	pH	Electrical conductivity, µS/cm	Soil respiration, g $CO_2/m^2 \cdot h$	pH	Electrical conductivity, µS/cm
1–1	7.12	177.0	1.05	8.31	176.6
1–2	6.90	160.8	2.05	8.05	223.0
1–3	6.76	155.0	1.68	7.87	195.6
1–4	4.57	24.2	1.81	8.20	211.2
1–5	5.98	63.6	1.24	5.93	45.2
2–1	7.21	196.8	–	8.49	236.0
2–2	6.91	209.8	1.76	–	–
2–3	6.53	120.6	1.20	–	–
2–4	5.49	66.6	1.48	–	–
2–5	5.14	96.4	1.19	–	–
ANOVA: *F/P*	20.1/<0.0001	13.5/<0.0001	2.47/0.03	71.8/<0.0001	9.48/<0.0001
Dist.: *r/P*	−0.81/0.005	−0.83/0.003	−0.46/0.18	−0.71/0.12	−0.62/0.19
Poll.: *r/P*	0.57/0.08	0.65/0.04	0.44/0.21	0.43/0.39	0.59/0.22

Measurements around Volkhov were conducted on 8–9.8.2002 (respiration) and 17.6.2006 (pH and electrical conductivity), around Vorkuta – on 8–9.7.2006. For other explanations, consult Table 3.1.

Table 3.16 Soil quality indices in the impact zone of the aluminium smelter at Žiar nad Hronom, Slovakia

Site	Stone cover, %	pH	Electrical conductivity, μS/cm	Soil respiration, g $CO_2/m^2{\cdot}h$
1–1	0	5.90	128.8	0.46
1–2	0	6.38	125.6	0.83
1–3	0.2	6.15	76.8	0.65
1–4	0.2	5.20	81.6	1.27
1–5	0.5	5.86	65.6	1.01
2–1	0	6.60	106.0	0.71
2–2	0	6.10	94.4	1.66
2–3	1.3	6.15	84.0	1.78
2–4	0.7	6.25	126.6	1.13
2–5	4.3	5.75	67.0	1.26
ANOVA: F $(X^2)/P$	22.8/0.0066	3.81/0.0015	4.76/0.0002	5.31/<0.0001
Dist.: r/P	0.83/0.003	−0.49/0.15	−0.70/0.03	0.62/0.06
Poll.: r/P	−0.85/0.002	0.08/0.83	0.66/0.04	−0.62/0.06

Measurements were conducted on 29.8–1.9.2002 (respiration and stone cover) and 9.11.2006 (pH and electrical conductivity). Stone cover was analysed by non-parametric methods: Kruskal-Wallis test for between-site variation, and Spearman rank correlation coefficients for relationships with both distance and pollution load. For other explanations, consult Table 3.1.

a way that the classes differed significantly from each other ($Q_B = 3.08 \ldots 40.3$, $P = 0.01 \ldots 0.0001$), no matter how the ESs were computed. Importantly, there was no overlap between confidence intervals for these three classes of polluters (Fig. 3.3). An overall effect of the studied polluters on soil pH did not differ from zero ($d = -0.01$, CI = −0.39…0.38, $N = 16$) and did not depend on the method used to calculate ES ($Q_B = 0.29$, df = 2, $P = 0.86$).

The effect of pollution on soil pH depended on the type of polluter (Fig. 3.3; $Q_B = 13.0$, df = 2, $P = 0.002$). Non-ferrous smelters caused an overall decrease in soil pH (i.e., acidification), although two of them (at Harjavalta and Sudbury) were classified as neutral (Table 2.1). Alkalysing effects were observed around aluminium smelters and power plants (Fig. 3.3). Effects of northern and southern polluters on soil pH (Fig. 3.3) did not differ for either non-ferrous smelters ($Q_B = 0.64$, df = 1, $P = 0.42$) or aluminium plants ($Q_B = 0.02$, df = 1, $P = 0.88$).

Although the overall effect of pollution on EC did not depend on the method used to calculate ES ($Q_B = 1.06$, df = 2, $P = 0.50$), the effect was significant for correlations with distance from polluter (d = 0.39, CI = 0.14 … 0.63, $N = 16$), but not significant for correlations with pollution load (Fig. 3.4). The effect of pollution on EC did not depend on either the type of polluter (Fig. 3.4; $Q_B = 0.53$, df = 2, $P = 0.77$) or its geographical position (Fig. 3.4; $Q_B = 0.06$, df = 1, $P = 0.81$). The only source of variation was the polluter's type of impact on soil pH (Fig. 3.4; $Q_B = 7.73$, df = 2, $P = 0.02$); a significant increase in EC was detected only near alkalysing polluters.

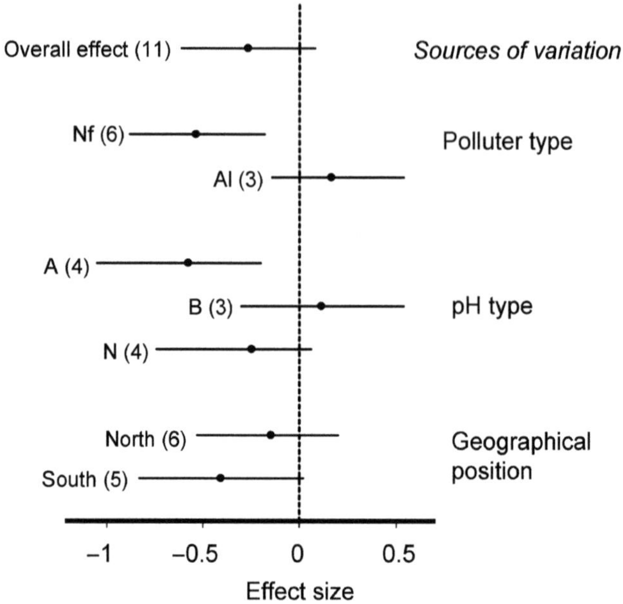

Fig. 3.1 Effects of point polluters on topsoil depth. Horizontal lines denote 95% confidence intervals; sample sizes are shown in brackets; an asterisk denotes significant ($P < 0.05$) between-class heterogeneity. For classifications of polluters and abbreviations consult Table 2.1

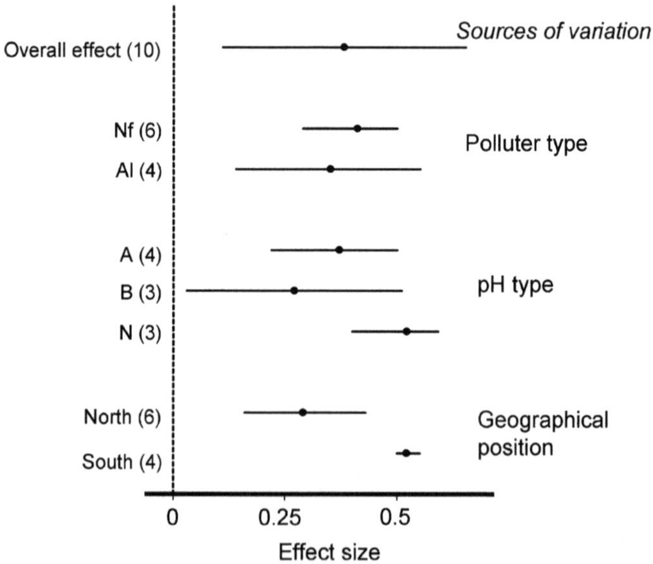

Fig. 3.2 Effects of point polluters on soil stoniness. For explanations, consult Fig. 3.1

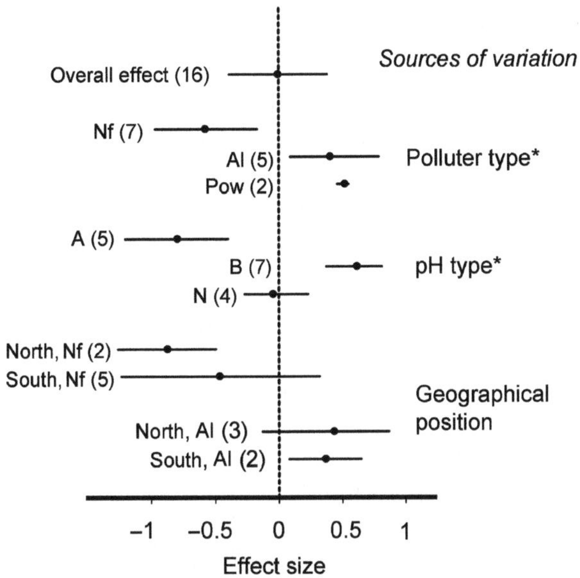

Fig. 3.3 Effects of point polluters on soil pH. For explanations, consult Fig. 3.1

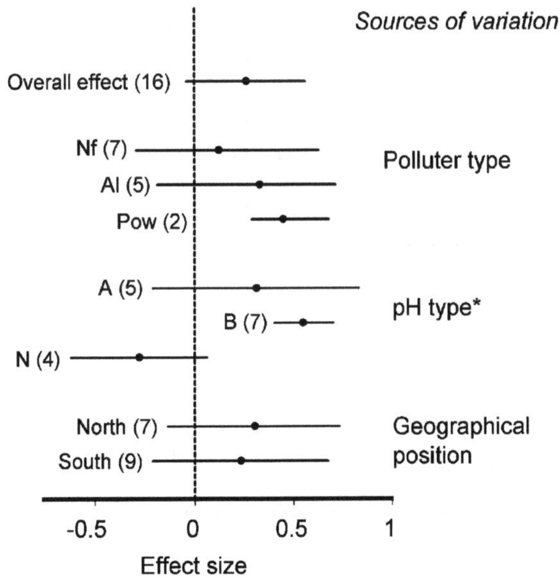

Fig. 3.4 Effects of point polluters on soil electrical conductivity. For explanations, consult Fig. 3.1

3.3.4 Soil Respiration

Variation in soil respiration between study sites was significant in 21 of 25 data sets. However, only 17 of 50 individual correlations (with both distance and pollution) were significant (Tables 3.1–3.16).

We measured soil respiration twice around Kandalaksha, Kostomuksha, Krompachy and Revda and three times around Monchegorsk and Nikel. The site-specific values from different measurements generally correlated with each other (for methods of calculation, consult Section 2.5.2.4; $z_r = 0.60$, CI = 0.16...1.04, $N = 10$), demonstrating sufficient repeatability of spatial patterns in soil respiration.

Soil respiration decreased with pollution (Fig. 3.5); this result did not depend on the method used to calculate ES ($Q_B = 2.17$, df = 2, $P = 0.34$). The effects of non-ferrous smelters were much stronger than the effects of aluminium smelters (Fig. 3.5; $Q_B = 8.80$, df = 1, $P = 0.003$). Consistently, only acidifying polluters caused a significant decrease in soil respiration (Fig. 3.5). Northern and southern polluters had similar impacts on soil respiration (Fig. 3.5; $Q_B = 0.01$, df = 1, $P = 0.91$).

We detected three non-linear response patterns in 25 data sets. Soil respiration peaked in moderately contaminated sites around the Kandalaksha smelter and reached its lowest values in moderately contaminated sites around the Bratsk and Kostomuksha smelters.

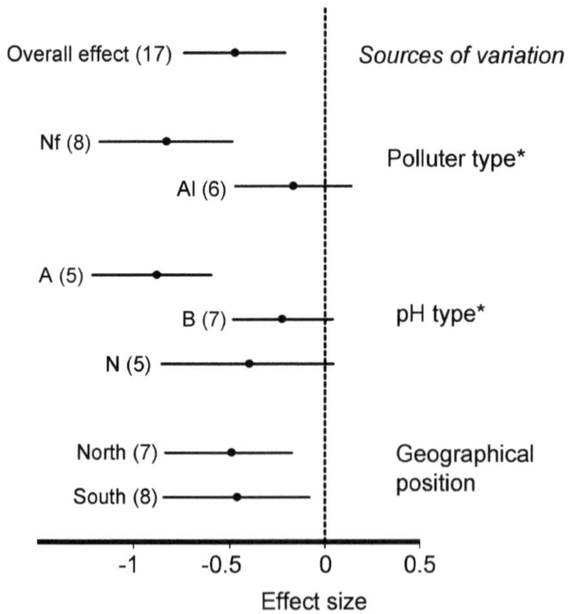

Fig. 3.5 Effects of point polluters on soil respiration. For explanations, consult Fig. 3.1

3.4 Discussion

In terms of our study, this chapter has an auxiliary function. Keeping in mind the huge number of both case studies and review papers describing the effects of pollution on soil chemical properties, we did not aim at broad generalization on the basis of our data. Instead, we use these data to (1) justify the selection of polluters and impact zones for exploration of biotic responses to environmental pollution and (2) evaluate the explanatory value of the selected indices of soil quality for understanding and predicting the impacts of point polluters on terrestrial biota.

3.4.1 Soil Erosion

Topsoil depth estimates the rooting volume for plants and is therefore considered one of the soil quality indicators in agricultural systems (Arshad & Martin 2002). Loss of topsoil due to erosion has long been known to result in proportional losses of crop yield (Stallings 1964). In forest ecosystems, degradation of the topsoil reduces the regrowth capacity of the forest cover by hampering establishment of seedlings and decreasing growth of instant trees (FAO 1992; Williamson & Neilsen 2003).

We detected a significant decrease in the depth of the soil organic layer near non-ferrous smelters (Fig. 3.1), which generally impose the most severe adverse effects on biota (Kozlov & Zvereva 2007a). On average, topsoil depth in the most polluted sites was reduced to 40% of the control value. To our knowledge, this is the first quantitative estimate of topsoil loss in heavily polluted areas. It can be used in several models, in particular to evaluate direct economic losses due to a decrease of soil fertility (US$16/t of eroded soil; Uri & Lewis 1998) and to estimate effects on forest regeneration (Hein & van Ierland 2006). However, more research efforts are needed to properly document the spatial pattern of soil erosion around big polluters, in particular accounting for relief of the impacted areas.

Significant increases in stoniness with an increase in pollution (Fig. 3.2) are consistent with loss of topsoil (Fig. 3.1) and indicative of erosion triggered by pollution. Stoniness appeared a more sensitive indicator of erosion than topsoil depth, probably due to both lower subjectivity and a generally higher number of measurements of stone depth made using the rod penetration method. In terms of soil quality indicators (Arshad & Martin 2002), increases in stoniness can be translated to increases in soil bulk density, i.e., the ratio of mass of dry solids to bulk volume of soil. Increases in stoniness have long been known to adversely affect forest production through changes in soil hydrology and soil temperature (reviewed by Eriksson & Holmgren 1996).

Importantly, models of forest ecosystems, which supply wood and secure erosion control, assume linear decreases in forest regeneration with decreases of topsoil below a depth of 30 mm, with a ceasing of regrowth following a complete loss of topsoil (Hein & van Ierland 2006). This gives new insight into the problem of ecosystem recovery following a decline in emissions. Earlier, we assumed that metal toxicity is a principal factor hampering natural regeneration of heavily contaminated barren areas (Kozlov 2004; Kozlov & Zvereva 2007a; Zverev et al. 2008; Zverev 2009). Data on topsoil loss suggest that further studies should be conducted in industrial barrens to partition the effects of soil toxicity and the reduction of the soil organic layer. This is crucial to elucidate the role of erosion in promoting environmental deterioration and slowing vegetation recovery around big polluters. These studies should, in particular, identify critical limits beyond which restoration of topsoil (e.g., by application of a mulch cover, as described by Helmisaari et al. 2007) is the only way to assure vegetation recovery within reasonable time limits.

Topsoil recovery depends upon sedimentation of soil particles by water or wind and upon deposition of organic material from standing biomass. For modelling purposes, Hein & van Ierland (2006) assumed zero recovery in the absence of forests to annual accumulations of 1 mm of topsoil for totally forested areas. Since vegetation cover in industrial barrens is below 5%, we can hardly expect that topsoil recovery will be fast. We are even uncertain whether topsoil recovery has recently occurred in the most degraded barren sites around Karabash, Monchegorsk and Nikel. In any case, relationships between soil erosion and topsoil recovery should be accounted for in estimations of time lags between pollution decline and the beginning of vegetation recovery.

3.4.2 Changes in Soil pH

Industrial pollution may cause both increases and decreases in soil pH, which lead in particular to changes in the bioavailability and toxicity of metal pollutants for plants and animals (Rooney et al. 2006; Spurgeon et al. 2006). Both extreme acidification and extreme alkalinisation have drastic consequences for natural ecosystems, including the development of industrial barrens (Kozlov & Zvereva 2007a). However, acidification is studied more frequently, possibly because acidifying pollutants are transferred long distances, resulting in regional increases in acidic depositions (AMAP 2006). Another reason is that the term 'acid rain' long ago became permanently included in the lexicon of politicians. High general interest in acidification problems is suggested by the fact that 123 of 1,810 publications with 'acid rain' in the title, listed by the ISI Web of Science database, appeared in the two most prestigious journals, '*Nature*' and '*Science*'.

Several meta-analyses of published data revealed a consistent research bias towards the exploration of environmental effects of acidifying polluters (Table 3.17). Although alkalysing polluters are selected for studies of biotic effects with about the same frequency as acidifying polluters ($X^2 = 0.33$, df = 1, $P = 0.56$), the overall amount of information collected around acidifying polluters is slightly higher ($X^2 = 2.08$, df = 1, $P = 0.15$), comprising 56% of all ESs calculated in four meta-analyses (Table 3.17). Therefore, conclusions based on published data may well be biased, with acidifying polluters contributing more to the estimates of overall pollution effects on biota.

Under these circumstances, we give specific value to a somewhat trivial result: zero overall effect on soil pH in a sample of 18 polluters selected for the present

Table 3.17 Numbers of polluters and effect sizes considered by four meta-analyses of published data on the pollution impact on terrestrial biota

Study group	Response variables	Numbers of polluters		Numbers of effect sizes		Data sources
		Acidifying	Alkalysing	Acidifying	Alkalying	
Soil microfungi	Diversity and abundance	36	25	45	28	Ruotsalainen & Kozlov 2006
Vascular plants	Diversity	35	21	127	71	Zvereva et al. 2008
Vascular plants	Growth and reproduction	41	64	270	348	Roitto et al. 2009
Terrestrial arthropods	Diversity, abundance and fitness	33	39	239	197	Zvereva & Kozlov 2009

study. This result suggests that our sample is well balanced in terms of pollution impacts on soil pH, allowing us to distinguish biotic effects mediated by acidification and alkalinisation of soils from other effects of industrial pollution.

3.4.3 Soil Electrical Conductivity

Electrical conductivity (EC) is one of the key indicators for soil quality assessment (Arshad & Martin 2002). EC, determined by the total amount of dissolved ions in a soil solution, has been used as a rough estimate of nutrient availability (Hartsock et al. 2000). This parameter is becoming increasingly popular in mapping soil quality for agricultural purposes (Corwin & Lesch 2003). Several case studies have indicated that EC can be used to explain spatial patterns of plant distribution (Guretzky et al. 2004) and to predict, in combination with some other factors, crop yield (Kravchenko et al. 2003).

EC has only rarely been measured in impact zones of point polluters (Wilcke et al. 2000; Gago et al. 2002; Seguin et al. 2004; Shukurov et al. 2006), and we failed to find any study that linked changes in EC with the biotic effects of pollution. However, in metal-contaminated soils, higher EC values were indicative of higher levels of toxic elements (Nagamori et al. 2007).

Our results on EC changes near point polluters are therefore difficult to interpret, since EC is positively related to concentrations of both macronutrients and water-soluble pollutants. Pollution of non-ferrous smelters is known to decrease concentrations of macronutrients (due to displacement of base cations by heavy metals) and to increase concentrations of metals in affected soils (Freedman & Hutchinson 1980a; Barcan & Kovnatsky 1998; Derome & Lindroos 1998; Lukina & Nikonov 1999). Since we did not detect any changes in EC around non-ferrous smelters (Fig. 3.4), we conclude that these two effects, namely, loss of nutritional quality and increase in toxicity with increase in pollution load, counterbalanced each other. We therefore concluded that our attempt to use EC for quantifying changes in soil quality in impact zones of industrial polluters did not yield useful information, and we disregarded this index in further analyses and discussion.

3.4.4 Changes in Soil Respiration

Soil respiration is a key component of terrestrial ecosystem carbon processes. It can strongly influence net carbon uptake from the atmosphere and net ecosystem production - the balance between photosynthesis and ecosystem respiration (Ryan & Law 2005). On average, 80% of photosynthetic carbon uptake is respired back to the atmosphere (Law et al. 2002). Soil in temperate forests contributes around two thirds of overall ecosystem respiration (Goulden et al. 1996; Law et al. 1999; Janssens et al. 2001). Measurements of soil respiration are highly valued in global change research (Rustad et al. 2001), but they are only rarely performed in impact zones of point polluters (Vanhala & Ahtiainen 1994; Fritze et al. 1996).

Pollution caused an overall decrease in soil respiration measured *in situ* (Fig. 3.5). On average, respiration in most polluted sites was reduced to 56% of controls. This result is not surprising: the adverse effects of pollution on the activity of soil processes have been recognised for decades (Fritze 1987; Bååth 1989). However, in terms of our study, we interpret this result as proof of a significant overall adverse impact of investigated polluters on terrestrial ecosystems.

In spite of a straightforward net effect, interpretation of detected decreases in soil respiration is indeed difficult. First, our data reflect root and associated mycorrhizal respiration plus decomposition of recently produced root and leaf litter. In intact ecosystems, these processes have roughly equal contributions to net respiration (Ryan & Law 2005), while data on polluted ecosystems are insufficient for generalization. Second, we have almost no data on the pollution impact on relative contributions of different functional groups of soil microbiota to the cumulative release of carbon dioxide. Moreover, these contributions are likely to change with pollution, since the adverse effects of pollution on mycorrhizal fungi were larger than on saprotrophic fungi (Ruotsalainen & Kozlov 2006).

On the other hand, an overall decrease in soil respiration measured *in situ* can be higher or lower than the decrease in soil microbial activity, because soil respiration is regulated by multiple factors. These include, for example, temperature, moisture, soil pH, topsoil depth and several others (Howard & Howard 1993; Singh & Gupta 1977; Vanhala 2002; Vanhala et al. 2005); many of these are modified in the course of pollution-induced environmental deterioration (Kozlov & Zvereva 2007a, and references therein).

Non-ferrous smelters included in our study had a stronger effect on soil respiration than aluminium smelters, reducing it to 39% and 63% of controls, respectively. This result is consistent with a pattern discovered for soil microfungi: non-ferrous smelters caused the most pronounced adverse effects (Ruotsalainen & Kozlov 2006). Although the importance of direct toxicity of metal pollutants in depression of soil respiration seems obvious, our data still do not allow partitioning these direct effects from indirect effects of pollution, mediated by environmental deterioration and changes in microclimate (consult Section 9.2.3 for discussion).

However, in spite of obvious limitations imposed both by the method used to measure soil respiration and a limited number of measurements, our data provide the very first estimate of regional changes in the carbon budget of terrestrial ecosystems due to pollution impacts.

3.4.5 *Sources of Variation*

All explored indicators of soil quality demonstrated pronounced variation among study sites around most of the polluters. However, only a minor part of this variation was related to the pollution impact, emphasising that proper replication of study sites is mandatory to distinguish pollution effects from confounding environmental variation.

Pollution-induced changes in soil pH depended on the polluter type (Fig. 3.1), a pattern that is easily explained by the composition of emissions. Non-ferrous smelters emit mostly sulphur dioxide that causes acidification, while particulate emissions of metal oxides are relatively minor and cannot compensate for SO_2-mediated changes in soil pH. Other polluters (aluminium smelters, iron pellet plants and power plants) emit large amount of alkaline dusts, which compensates for or even overbalances the effects of acidifying gaseous pollutants.

In contrast to soil chemistry, changes in topsoil depth and stoniness showed no variation in respect to polluter type or its impact on soil pH (Figs. 3.3 and 3.4). This result suggests that changes in soil morphology and soil physical properties are mostly mediated by pollution effects on vegetation, which gradually lose control over erosion and produce less litter, slowing recovery of topsoil.

Patterns of variation in soil respiration (Fig. 3.5), the only measured biotic index of soil quality, showed a large affinity for changes in soil chemical properties. Surprisingly, we did not detect geographical variation in pollution effects on soil respiration, although it was anticipated due to higher bioavailability and toxicity of pollutants in a warmer climate (Zvereva et al. 2008, and references therein).

3.5 Summary

Soil quality generally decreased with pollution. Changes in soil chemical properties (measured by pH and electrical conductivity) were industry-specific, whereas changes in morphology or physical properties (measured by topsoil depth and soil stoniness) were similar across the studied polluters. Both losses of topsoil and overall decreases in soil respiration indicate adverse effects of industrial polluters on ecosystem services. Our results, as early and rough estimates, can be used to account for pollution effects on regional carbon fluxes and on soil erosion. Further studies of mechanisms governing soil erosion in polluted areas are badly needed to predict ecosystem development under different emission and climatic scenarios and to develop optimal remediation measures.

Chapter 4
Plant Growth and Vitality

4.1 Introduction

4.1.1 The Vitality of Plants

Many different ways to define plant vitality or vigour have been suggested. Although definitions differ in details, they generally refer to the capacities to live or grow, as well as to resist stress (reviewed by Dobbertin 2005). Importantly, the hypothetical 'optimal' plant vitality remains a theoretical concept: it can neither be measured directly nor predicted on the basis of other measurements. However, it is generally accepted that plants experiencing environmental stress differ in some characteristics from plants growing under optimal conditions, and these characteristics can therefore be considered as indices of vitality.

Plants' responses to environmental changes, including pollution, have been explored from the molecular to the community levels (Kozlowski 1980; Treshow & Anderson 1989; Sandermann 2004; Dobbertin 2005; DalCorso et al. 2008). Consequently, a number of vitality indices have been suggested (Waring 1987; Stolte et al. 1992; Schulz & Härtling 2003; Dobbertin 2005; Polak et al. 2006). Although molecular indicators may appear most suitable to detect plant responses to experimental manipulations (DalCorso et al. 2008; Nesatyy & Suter 2008), they are difficult to use for predicting responses of plant organisms to chronic pollution impacts. Moreover, the use of molecular and biochemical methods in field assessment programs is limited by a shortage of qualified workers and generally high costs (Dobbertin 2005). Therefore, we restricted our study to several cost-effective methods that allow for evaluation of processes reflecting the accumulation of plant biomass, i.e., plant growth.

Changes in primary productivity are seen as one of the very basic responses of ecosystems to various disturbances (Odum 1985; Rapport et al. 1985; Sigal & Suter 1987). Since growth and biomass accumulation critically depend on photosynthesis, we have chosen the efficiency of photosynthetic system II as the first index of plant vitality. Further on, we measured the size of the photosynthetic organs (leaves in deciduous plants and needle in conifers) and plant growth in terms of shoot length and radial increment. Finally, we assessed needle longevity in conifers, since premature

M.V. Kozlov et al. *Impacts of Point Polluters on Terrestrial Biota*,
DOI 10.1007/ 978-90-481-2467-1_4, © Springer Science+Business Media B.V. 2009

shedding of foliage may have adverse affects on both forest products and forest services (Smith 1992). The selected characteristics reflect different aspects of biomass accumulation, and their combination is indicative of plant productivity (Sigal & Suter 1987).

4.1.2 Indices of Plant Vitality Used in Our Study

4.1.2.1 Chlorophyll Fluorescence

Although the processes resulting in chlorophyll fluorescence are complicated, the principle underlying fluorescence analysis is relatively straightforward. Light energy absorbed by chlorophyll molecules can be used for photosynthesis, dissipated as heat, or re-emitted as light; the latter process is called chlorophyll fluorescence. Since fluorescence reflects the primary processes of photosynthesis, it can be used to obtain information on changes in the efficiency of the photochemical reaction in photosystem II (Krause & Weis 1988; Maxwell & Johnson 2000).

Many environmental stressors directly or indirectly affect the function of photosystem II (Öquist 1987). Therefore, chlorophyll fluorescence is a frequently used index in the assessment of plant stress (Daley 1995; Nesterenko et al. 2007), including stress imposed by pollution (Adams et al. 1989; Snel et al. 1991; Saarinen 1993; Kitao et al. 1997; Odasz-Albrigtsen et al. 2000; Zvereva & Kozlov 2005). However, chlorophyll fluorescence is influenced by numerous factors in a complex manner, and therefore, an exact interpretation of the observed phenomena is often difficult (Krause & Weis 1988; Snel et al. 1991; Maxwell & Johnson 2000).

4.1.2.2 Leaf/Needle Size and Shoot Length

Retarded growth and decreased leaf area are commonly considered as general and well-known plant responses to industrial emissions, including sulphur dioxide, fluorine and heavy metals (National Research Council of Canada 1939; Scurfield 1960a, b; Odum 1985; Treshow & Anderson 1989; Armentano & Bennett 1992; Dobbertin 2005). Although 'positive' effects of these pollutants have also been documented (Bennett et al. 1974; Lechowicz 1987; Zvereva et al. 1997a; Zvereva & Kozlov 2001; Kozlov & Zvereva 2007b), a meta-analysis of published data demonstrated significant decreases in leaf/needle size and weight, shoot length, root growth and radial increment with pollution, while leaf number and shoot weight were not affected (Roitto & Kozlov 2007; Roitto et al. 2009). At the same time, another data set on herbaceous plants revealed a decrease in plant size near industrial polluters, but detected no effects on aboveground biomass due to an increase in the number of leaves and flowers/inflorescences (Kozlov & Zvereva 2007b). The diversity of responses, as well as discrepancies between outcomes of the meta-analyses of published and original data, give special importance to further investigation of plant growth responses to industrial pollution, and especially to identification of the factors affecting the direction and magnitude of the effects.

4.1.2.3 Radial Increment

Adverse effects of pollution on the radial growth of trees were documented long ago (National Research Council of Canada 1939), and in the middle of the twentieth century, measurements of tree rings were routinely used to estimate economic losses of foresters due to pollution (Treshow 1984). Therefore, we did not intend to measure radial increments of woody plants when designing our project. However, evaluation of the published data demonstrated that the results of many dendrochronological studies are not suitable for meta-analysis. Although the published evidence, such as the abrupt decline in the width of annual rings during the first years of a polluter's operation (Bunce 1979; Fox et al. 1986; Nöjd & Reams 1996; Kobayashi et al. 1997; Long & Davis 1999), is quite impressive, the data generally do not allow for calculation of ESs in the same manner as for other variables used in our analyses. This is mostly due to reporting of plot-specific means only (usually in a graphical form) and frequent use of temporal control, i.e., growth of the same stand(s) prior to the beginning of pollution impact, instead of spatial control, i.e., growth of another stand(s) outside the polluted area. As the result, by the beginning of 2007, we had identified only 28 published data sets that were suitable for meta-analysis (Roitto & Kozlov 2007). This unexpected shortage of information forced us to measure the radial increment of Scots pine around some of our polluters.

4.1.2.4 Needle Longevity in Conifers

Most conifers are typical evergreens with long-living foliage. Every year about one class of the oldest needles shed at the end of the growing season is replaced by new needles in the spring of the following year. Longevity of needles is generally defined by the number of age classes simultaneously occurring on a plant, but sometimes it is corrected for the proportion of needle survival in different age classes (Lamppu 2002). The accumulated needle mass contains a considerable reserve of mobile nutrients, and prolonged needle longevity in conditions of low nutrient availability maximises nutrient use efficiency (Lamppu & Huttunen 2003, and references therein). Thus, needle longevity is often considered as an important ecophysiological trait related to both carbon and nutrient balances (Aerts 1995).

The effects of pollution on needle longevity have been known for decades (Treshow 1984; Kryuchkov & Makarova 1989), and a decrease in the number of needle age classes has often been suggested as one of the vitality indices for bioindication of pollution impact on forests (Dässler 1976; Schubert 1985). However, since reductions in needle longevity can be caused by other stressors (including diseases, climate, and soil nutritional quality), this measure can serve an indicator of air pollution stress only when other factors leading to accelerated needle abscission are taken into account. The feasibility of the needle age structure as an objective and reliable vitality indicator for Scots pine was recently confirmed by Lamppu and Huttunen (2001). However, national and international monitoring programs addressing the health conditions of forest ecosystems in Europe and North America do not use this index, but instead visually estimate crown defoliation and discoloration (UN-ECE 2006).

4.1.3 Carbon Allocation and Allometric Relationships

Pollution may not only reduce carbon assimilation, but also alter carbon allocation within a plant (Waring 1987; Kozlowski & Pallardy 1996). Thus, the outcome of vitality analysis may change with the measured characteristic, and obtaining an adequate estimate of plant response to pollution requires simultaneous investigation of different vitality indices.

Waring (1987) ranked growth processes in the order of their decreasing importance for a tree as follows: foliage, root, bud, storage tissue, stem, defensive compounds, and reproductive growth. Although comparing responses of different growth processes to pollution seems a relatively easy task, to our knowledge this has not yet been performed except for our meta-analysis of published data (Roitto et al. 2009). Therefore, we specifically aimed to address resource partitioning effects by comparing pollution-induced changes in different vitality indices measured from the same individuals of woody plants.

4.2 Materials and Methods

4.2.1 Selection of Study Objects and Plant Individuals

To avoid object selection bias, we have chosen plant species for our measurements without *a priori* knowledge on their responses to pollution. The first criterion was plant abundance in the study area: we selected the most common species, including forest-forming trees and shrubs dominating field layer vegetation. The second criterion was the balance between plant taxa and life forms: whenever possible, we preferred to measure one coniferous and one deciduous tree species, instead of measuring two conifers or two deciduous plants. Similarly, we attempted to keep a balance between top-canopy plants (trees and large shrubs) and field layer vegetation (dwarf shrubs), and to measure both evergreen and deciduous species. Third, to allow for comparisons among polluters, we preferred species that were studied around other polluters. As a result, our samples included four species of Gymnosperms and 39 species of Angiosperms, among which were 17 species of deciduous trees and large shrubs, six species of dwarf shrubs, and 16 species of herbs. The largest numbers of measurements were obtained from Scots pine, cowberry (*Vaccinium vitis-idaea*) and white/mountain birch (around 12, 11 and 10 of 18 polluters, respectively).

The five sampled trees or shrubs were the first mature individuals with accessible foliage found at the study site. In practice, we chose trees that were closest to the centres of the selected study sites. For small (dwarf) shrubs,[1] we selected ramets growing at least 5 m apart (usually 10–15 m apart) to minimise the probability that two or more of the sampled ramets belonged to the same plant individual. For abundant herbs, we sampled ten individuals that were growing nearest to points located 2 m apart along a line crossing the study site. For all plants, we disregarded individuals bearing signs of severe damage not attributable to pollution impact (broken main stem, intensive browsing, etc.).

From plant individuals selected for chlorophyll fluorescence measurements, we always collected information on growth (both leaf/needle size and shoot length) and needle longevity (from Scots pine only), as well as samples for measurements of fluctuating asymmetry (see below, Chapter 5). All measurements performed in woody plants refer to vegetative shoots. Radial increment was explored in other individuals of Scots pines than were used for measurements of chlorophyll fluorescence, needle size and shoot growth.

4.2.2 Chlorophyll Fluorescence

In each impact zone, we measured chlorophyll fluorescence in two or three species of woody plants. All measurements were conducted during the second half of the growth season, when growth of shoots and leaves had already terminated. As a rule, all measurements around the polluter were performed on the same day; the order of plots was randomised whenever possible.

Prior to the beginning of the project, we demonstrated that measurements conducted with freshly detached leaves (not later than 20 min after sampling) yield the same results as measurements conducted on intact leaves. Therefore, all measurements were conducted by using detached leaves/needles, three leaves (or

[1]Dwarf shrubs, or chamaephytes in the Raunkiaer's classification of life forms, are woody plants with perennating buds borne close to the ground, no more than 25 cm above soil surface. Within this book, dwarf shrubs are restricted to *Vaccinium* and *Empetrum* species, which form substantial part of field layer vegetation in boreal forests.

groups of needle fascicles) per individual. For trees and large shrubs, we took leaves/needles from different sides of the crown at a height of 1–3 m. In small shrubs, we sampled leaves from shoots located at approximately one half of the ramet's height.

In birches, which possess two distinct types of vegetative shoots (Fig. 5.1), we sampled short-shoot leaves; in other deciduous trees shoot length varied continuously, and therefore leaves were selected from a random sample of shoots irrespective of shoot length. We always collected the largest leaf from the selected shoot. In Scots pine, we collected current-year needles from the terminal shoot of the first-order branches (Fig. 5.3).

A lightweight leaf cuvette assuring dark adaptation was placed on the collected leaves/needles at the time of sampling, and samples were placed into a plastic box to minimise desiccation. Within 15–20 min after sampling (an amount of time sufficient for dark adaptation), chlorophyll fluorescence was measured using a portable fluorometer (Biomonitor S.C.I. AB, Umeå, Sweden) with a light level of 200 μmol photons/m^2×s. The indices measured were the ratio of variable to maximum fluorescence yielded under the artificial light treatment (F_v/F_m) and the time needed for the leaf to reach half of its F_m ($T_{1/2}$). In total, we obtained approximately 5,700 measurements of each of two indices from 13 plant species in impact zones of 16 polluters.

4.2.3 *Leaf/Needle Size and Shoot Length*

Sampling of needles and leaves of woody plants followed the same protocol as described for measurements of chlorophyll fluorescence (Section 4.2.1). Leaf and needle size of woody plants were measured in the laboratory from dried and mounted samples prepared for fluctuating asymmetry analysis (Section 5.2.1). We measured (with a ruler, to the nearest 1 mm) the length of the leaf lamina (i.e., excluding petiole) of large-leaved plants and the length of the longest needle in each pair of Scots pine needles. In small-leaved plants (leaf length generally less than 10 mm), the length of the leaf lamina was measured using a dissecting microscope with an ocular scale (to the nearest 0.1 mm). Most samples were measured twice to minimise the probability of occasional errors. As a rule, we measured ten leaves and 20 needles from one individual of woody plants.

Shoot length of Norway spruce was also measured in the laboratory from dried and mounted samples of annual whorls prepared for fluctuating asymmetry analysis (Section 5.2.1). In these samples, we measured (with a ruler, to the nearest 1 mm) the length of an apical shoot (Fig. 5.2a), i.e., the annual increment of the first-order branch. Shoot length of other trees and dwarf shrubs (long shoots in birches, Fig. 5.1) was measured either in the field or in the laboratory from field-collected branches. As a rule, we measured ten shoots from one individual of woody plants.

In contrast to woody plants, herbaceous plants were sampled only from the two most and two least polluted sites. The samples were transported to the laboratory, where we measured (with a ruler) the length of the longest leaf (except for grasses) and plant height, which was presumed to be equivalent to the length of the annual shoot in woody plants. Accuracy of measurements was 1 mm for values not exceeding 100 mm and 5 mm for larger values.

4.2.4 Radial Increment

Radial increment was measured at the two most and two least polluted sites, in contrast to all other measurements of woody plants that were performed at ten study sites. At each site we selected five Scots pine trees of the dominant size class, avoiding trees that were too close to their neighbours (less than half of the stand-specific value), and cored these at a height of 1.3 m by using a standard increment borer. An approximate tree age was estimated immediately after coring. This information allowed us to modify the sampling scheme whenever necessary in order to sample trees of about the same age from both polluted and unpolluted sites.

In the laboratory, annual rings were counted under a dissecting microscope, and the total width of rings formed during the past 10 years (excluding the year of sampling) was measured to the nearest 0.2 mm. Analysis was always based on three trees per study site, selected on the basis of their age in such a way that between-site variation in age was kept to a minimum. ANCOVA with tree age as a covariate was used to distinguish between-site variation from age effects. However, none of the analyses detected a significant effect of tree age (data not shown), and therefore ANOVA was used in the final analyses. Hedge's d was calculated on the basis of site-specific means.

4.2.5 Needle Longevity in Conifers

Foliage longevity in many conifer species can be measured by counting nodes on branches back from the branch tip to the oldest whorl, with each node separating an annual whorl of needles or needle fascicles corresponding to 1 year of growth. Several methods have been developed, accounting in particular for needle loss in each age class; needle losses were either reported by age class (Choi et al. 2006) or combined into a composite index called mean longevity (Lamppu 2002). However, accurate estimation of mean longevity is laborious and may appear somewhat subjective due to visual estimation of the proportion of needle loss. Therefore, we used the maximum longevity, i.e., the age (in years) of the oldest green needle recorded in a sampling branch.

Estimations of the maximum needle age within each pollution gradient were performed by the same observer (either V.E.Z. or M.V.K.). They were conducted in mature (aged 20 years or more) trees, on two first-order branches selected from opposite sides of the crown of each tree at a height of 0.5–2.5 m. We surveyed ten trees per site; tree-specific values (averaged from measurements of two branches) were used to explore between-plot variation, while correlation analysis was based on plot-specific means. Lower needle longevity was considered a sign of decreased vitality.

4.2.6 Identification of Traits Associated with Sensitivity

We compared individual ESs between species, and compared species-specific mean ESs between Raunkiaer life forms (classification follows Hill et al. 2004). We also correlated species-specific mean ESs with axis scores for the 'competitor', 'stress tolerator' and 'ruderal' components for each species according to Grime's CSR strategy (Grime 1979; data extracted from the Modular Analysis of Vegetation Information System 'MAVIS' package http://www.ceh.ac.uk/products/software/CEHSoftware-MAVIS.htm) and with Ellenberg's scores for habitat requirements (light, temperature, continentality, humidity, pH and nitrogen; data extracted from Ellenberg et al. 1992).

4.3 Results

4.3.1 Chlorophyll Fluorescence

Two indices reflecting chlorophyll fluorescence, F_v/F_m and $T_{1/2}$, responded differently to pollution (correlation between ESs calculated for these indices: $r = 0.09$, $N = 39$ data sets, $P = 0.61$), confirming that their separate analysis was not redundant.

Variation in both indices between study sites was generally significant (32 of 39 tests for F_v/F_m and 31 of 39 tests for $T_{1/2}$); however, only 22 of 156 correlation coefficients (with both distance and pollution load) were significant (Tables 4.1–4.17).

To explore the repeatability of the results, we conducted measurements on the same set of individuals of mountain birch in the impact zone of the Monchegorsk smelter over a period of 4 years (Table 4.1). Repeated measures ANOVA demonstrated a large ($P < 0.0001$) interaction for both F_v/F_m and $T_{1/2}$ between study sites and study years, indicating that the relationship between chlorophyll fluorescence parameters and pollution varied between study years. The latter result was also evident from correlation analysis: even the signs of correlations between chlorophyll fluorescence parameters and distance from the smelter (or pollution load) changed with the study year (Table 4.1). Consequently, the average correlation

Table 4.1 Repeated chlorophyll fluorescence measurements of the same set of mountain birch (*Betula pubescens*) individuals in the impact zone of the nickel-copper smelter at Monchegorsk, Russia

	F_v/F_m				$T_{1/2}$, ms			
Site	2002	2005	2006	2007	2002	2005	2006	2007
1–1	0.410	0.702	0.706	0.745	262	383	506	475
1–2	0.504	0.640	0.486	0.732	351	574	382	418
1–4	0.520	0.713	0.450	0.722	431	499	468	418
1–5	0.523	0.712	0.604	0.590	270	449	412	541
2–2	0.480	0.697	0.455	0.687	285	446	451	495
2–4	0.533	0.673	0.583	0.741	338	551	492	483
2–6	0.572	0.606	0.748	0.691	399	460	500	489
2–8	0.576	0.758	0.773	0.682	356	466	377	420
2–9	0.657	0.767	0.672	0.644	307	316	369	418
2–10	0.629	0.714	0.742	0.690	366	419	339	383
ANOVA: *F*/*P*	4.46/0.0005	5.27/<0.0001	16.1/<0.0001	2.75/0.01	2.54/0.02	4.58/0.0003	2.86/0.01	3.93/0.0012
Dist.: *r*/*P*	0.94/<0.0001	0.38/0.28	0.44/0.21	−0.55/0.10	0.41/0.23	−0.24/0.50	−0.64/0.04	−0.41/0.24
Poll.: *r*/*P*	−0.63/0.05	−0.42/0.23	−0.34/0.34	0.54/0.10	−0.34/0.34	0.31/0.38	0.63/0.05	0.45/0.19

Decreases in F_v/F_m and increases in $T_{1/2}$ indicate lower vitality of plants. ANOVA: test for significance of variation between study sites. Dist.: Pearson linear correlation between site-specific means and log-transformed distance from polluter. Poll.: Pearson linear correlation between site-specific means and concentration of the selected pollutant. Long dash indicates: for woody plants - species not found in study site; for herbaceous plants - site not included in the sampling scheme.

Table 4.2 Chlorophyll fluorescence measurements in the impact zone of the power plant at Apatity, Russia

	Alnus incana		*Betula pubescens*	
Site	F_v/F_m	$T_{1/2}$, ms	F_v/F_m	$T_{1/2}$, ms
1–1	0.661	317	0.687	345
1–2	0.581	345	0.658	290
1–3	0.676	347	0.637	315
1–4	0.634	310	0.602	308
1–5	0.674	323	0.550	311
2–1	0.678	303	0.642	249
2–2	0.641	324	0.647	287
2–3	0.654	363	0.686	231
2–4	0.644	274	0.657	309
2–5	0.713	278	0.611	402
ANOVA: *F*/*P*	1.95/0.07	1.27/0.28	1.77/0.11	2.43/0.03
Dist.: *r*/*P*	0.19/0.60	−0.16/0.67	−0.70/0.02	0.20/0.58
Poll.: *r*/*P*	−0.55/0.10	0.58/0.07	0.67/0.03	−0.37/0.29

Measurements conducted on 30.7.2006. For other explanations, consult Table 4.1.

Table 4.3 Chlorophyll fluorescence measurements in the impact zone of aluminium smelter at Bratsk, Russia

	Betula pubescens		*Larix sibirica*		*Pinus sylvestris*	
Site	F_v/F_m	$T_{1/2}$, ms	F_v/F_m	$T_{1/2}$, ms	F_v/F_m	$T_{1/2}$, ms
1–1	0.742	306	0.754	118	0.673	181
1–2	0.742	286	0.757	101	0.727	180
1–3	0.698	345	0.736	99	0.676	155
1–4	0.679	245	0.655	111	0.492	133
1–5	0.609	246	0.649	98	0.472	163
2–1	0.671	337	0.642	91	0.612	181
2–2	0.700	272	0.698	94	0.545	129
2–3	0.654	288	0.687	90	0.589	137
2–4	0.701	250	0.730	90	0.642	137
2–5	0.553	240	0.730	221	0.580	145
ANOVA: *F*/*P*	4.47/0.0004	2.16/0.05	11.6/<0.0001	1.13/0.37	25.7/<0.0001	9.97/<0.0001
Dist.: *r*/*P*	−0.72/0.02	−0.66/0.04	−0.17/0.63	0.40/0.25	−0.49/0.15	−0.51/0.13
Poll.: *r*/*P*	0.21/0.55	0.63/0.05	−0.30/0.40	−0.21/0.56	0.23/0.52	0.62/0.05

Measurements conducted on 2–4.8.2002. For other explanations, consult Table 4.1.

between plot-specific values obtained in different years did not differ from zero for either F_v/F_m (for methods of calculation, consult Section 2.5.2.4; $z_r = 0.10$, CI = −0.19…0.39, $N = 6$) or $T_{1/2}$ ($z_r = 0.13$, CI = −0.21…0.42, $N = 6$).

Table 4.4 Chlorophyll fluorescence measurements in the impact zone of the nickel-copper smelter at Harjavalta, Finland

	Betula pubescens		*Pinus sylvestris*		*Salix caprea*	
Site	F_v/F_m	$T_{1/2}$, ms	F_v/F_m	$T_{1/2}$, ms	F_v/F_m	$T_{1/2}$, ms
1–1	0.710	342	0.479	142	0.742	224
1–2	0.636	484	0.489	157	0.647	255
1–3	0.666	326	–	–	0.674	317
1–4	0.643	192	0.527	253	0.647	254
1–5	0.704	289	0.524	245	0.748	217
2–1	0.756	213	0.594	123	0.774	186
2–2	0.701	339	0.537	108	0.674	173
2–3	0.782	236	0.518	95	0.694	211
2–4	0.742	241	0.609	129	0.663	265
2–5	0.684	210	0.516	143	0.694	227
ANOVA: *F*/*P*	5.60/<0.0001	5.75/<0.0001	2.56/0.03	5.07/0.0004	1.79/0.10	7.15/<0.0001
Dist.: *r*/*P*	−0.17/0.65	−0.41/0.24	0.03/0.94	0.57/0.11	−0.08/0.84	0.28/0.43
Poll.: *r*/*P*	−0.35/0.32	0.83/0.0029	−0.45/0.23	−0.18/0.65	−0.23/0.52	−0.02/0.95

Measurements conducted on 25.8.2002. Missing values resulted from absence of the selected plant species in some of study sites. For other explanations, consult Table 4.1.

Table 4.5 Chlorophyll fluorescence measurements in the impact zone of the fertiliser factory at Jonava, Lithuania

	Betula pendula		*Pinus sylvestris*	
Site	F_v/F_m	$T_{1/2}$, ms	F_v/F_m	$T_{1/2}$, ms
1–1	0.620	334	0.558	262
1–2	0.739	348	0.706	179
1–3	0.686	295	0.660	178
1–4	0.761	261	0.571	164
1–5	0.712	306	0.630	205
2–1	0.710	369	0.647	169
2–2	0.723	330	0.635	182
2–3	0.771	240	0.688	168
2–4	0.715	328	0.737	193
2–5	0.777	274	0.740	195
ANOVA: *F*/*P*	4.45/0.0004	3.18/0.0055	7.73/<0.0001	6.11/<0.0001
Dist.: *r*/*P*	0.50/0.14	−0.68/0.03	0.32/0.37	−0.18/0.63
Poll.: *r*/*P*	−0.47/0.17	0.34/0.33	−0.06/0.87	0.41/0.24

Measurements conducted on 3–5.9.2008. For other explanations, consult Table 4.1.

Pollution had no overall effect on F_v/F_m (Fig. 4.1) but caused a small increase in $T_{1/2}$ which is indicative of decreased efficiency of the photosynthetic system (Fig. 4.4). This result did not depend on the method used to calculate ES (F_v/F_m: $Q_B = 0.23$,

Table 4.6 Chlorophyll fluorescence measurements in the impact zone of the aluminium smelter at Kandalaksha, Russia

	Betula pubescens		*Pinus sylvestris*		*Salix caprea*	
Site	F_v/F_m	$T_{1/2}$, ms	F_v/F_m	$T_{1/2}$, ms	F_v/F_m	$T_{1/2}$, ms
1–1	0.775	302	0.658	153	0.757	287
1–2	0.717	255	0.567	128	0.597	286
1–3	0.768	321	0.739	269	0.739	373
1–4	0.727	267	0.633	149	0.632	287
1–5	0.663	306	0.674	168	0.764	362
2–1	0.734	354	0.643	176	0.658	374
2–2	0.722	400	0.626	157	0.711	400
2–3	0.742	271	0.672	141	0.785	275
2–4	0.803	262	0.739	161	0.735	409
2–5	0.784	273	0.746	169	0.748	331
ANOVA: *F*/*P*	3.61/0.0023	2.63/0.02	11.3/<0.0001	2.37/0.03	8.54/<0.0001	7.21/<0.0001
Dist.: *r*/*P*	0.00/0.99	−0.50/0.14	0.52/0.13	0.06/0.86	0.33/0.35	−0.01/0.98
Poll.: *r*/*P*	−0.20/0.59	0.43/0.22	−0.70/0.03	−0.12/0.75	−0.57/0.09	−0.08/0.83

Measurements conducted on 16–17.7.2002. For other explanations, consult Table 4.1.

Table 4.7 Chlorophyll fluorescence measurements in the impact zone of the copper smelter at Karabash, Russia

	Betula pendula		*Pinus sylvestris*	
Site	F_v/F_m	$T_{1/2}$, ms	F_v/F_m	$T_{1/2}$, ms
1–1	0.580	382	–	–
1–2	0.628	378	0.697	187
1–3	0.705	303	0.720	193
1–4	0.742	231	0.669	179
1–5	0.624	292	0.432	315
2–1	0.695	373	–	–
2–2	0.610	458	0.616	229
2–3	0.651	377	0.708	188
2–4	0.737	372	0.714	184
2–5	0.784	237	0.786	194
ANOVA: *F*/*P*	6.83/<0.0001	4.54/0.0004	21.7/<0.0001	2.12/0.07
Dist.: *r*/*P*	0.61/0.06	−0.72/0.02	−0.04/0.92	0.18/0.67
Poll.: *r*/*P*	−0.42/0.23	0.57/0.08	−0.09/0.84	0.04/0.92

Measurements conducted on 23–25.7.2003. Missing values resulted from absence of the selected plant species in some of study sites. For other explanations, consult Table 4.1.

df = 2, $P = 0.89$; $T_{1/2}$: $Q_B = 0.58$, df = 2, $P = 0.75$). Individual polluters did not differ in their effects on $T_{1/2}$ (Fig. 4.5; $Q_B = 16.7$, df = 15, $P = 0.43$). However, variation in F_v/F_m response was significant (Fig. 4.2; $Q_B = 33.4$, df = 15, $P = 0.04$). The latter

Table 4.8 Chlorophyll fluorescence measurements in the impact zone of the iron pellet plant at Kostomuksha, Russia

	Betula pubescens		*Pinus sylvestris*		*Salix caprea*	
Site	F_v/F_m	$T_{1/2}$, ms	F_v/F_m	$T_{1/2}$, ms	F_v/F_m	$T_{1/2}$, ms
1–1	0.649	184	0.583	115	0.718	259
1–2	0.725	265	0.726	175	0.721	291
1–3	0.743	285	0.648	150	0.683	328
1–4	0.749	313	0.687	190	0.719	304
1–5	0.764	291	0.743	173	0.726	345
2–1	0.751	284	0.745	163	0.713	307
2–2	0.698	254	0.485	130	0.560	241
2–3	0.723	244	0.500	138	0.586	346
2–4	0.647	265	0.548	127	0.634	252
2–5	0.656	334	0.528	124	0.651	295
ANOVA: *F*/*P*	3.77/0.0017	4.09/0.0009	8.94/<0.0001	7.34/<0.0001	2.43/0.03	2.06/0.06
Dist.: *r*/*P*	0.02/0.95	0.61/0.06	−0.07/0.85	0.18/0.63	−0.04/0.91	0.32/0.36
Poll.: *r*/*P*	0.27/0.44	−0.13/0.72	0.37/0.29	0.15/0.68	0.22/0.53	−0.09/0.80

Measurements conducted on 18.7.2002. For other explanations, consult Table 4.1.

Table 4.9 Chlorophyll fluorescence measurements in the impact zone of the copper smelter at Krompachy, Slovakia

	Fagus sylvatica		*Pinus sylvestris*	
Site	F_v/F_m	$T_{1/2}$, ms	F_v/F_m	$T_{1/2}$, ms
1–1	0.682	306	0.668	229
1–2	0.718	273	0.530	194
1–3	0.784	226	0.597	169
1–4	0.749	272	0.556	171
1–5	0.797	191	0.716	206
2–1	0.683	388	0.500	228
2–2	0.771	265	0.567	175
2–3	0.778	276	0.628	191
2–4	0.751	254	0.742	174
2–5	0.720	185	0.512	150
ANOVA: *F*/*P*	3.31/0.0042	4.50/0.0004	13.4/<0.0001	4.65/0.0003
Dist.: *r*/*P*	0.50/0.14	−0.78/0.0080	0.03/0.94	−0.66/0.04
Poll.: *r*/*P*	−0.49/0.15	0.69/0.03	−0.29/0.42	0.77/0.01

Measurements conducted on 2–4.9.2002. For other explanations, consult Table 4.1.

index decreased near six polluters, increased near five polluters, and showed no change around five polluters (Fig. 4.2).

The effect of pollution on chlorophyll fluorescence did not depend on either the type of the polluter (F_v/F_m: Fig. 4.1; $Q_B = 2.99$, df = 4, $P = 0.56$; $T_{1/2}$: Fig. 4.4;

Table 4.10 Chlorophyll fluorescence measurements in the impact zone of the nickel-copper smelter at Monchegorsk, Russia

	Pinus sylvestris		*Salix caprea*	
Site	F_v/F_m	$T_{1/2}$, ms	F_v/F_m	$T_{1/2}$, ms
1–1	0.484	163	0.574	435
1–2	0.476	182	0.522	441
1–4	0.529	177	0.521	388
1–5	0.591	179	0.577	356
2–2	0.474	182	0.511	431
2–4	0.532	185	0.588	431
2–6	0.597	180	0.579	375
2–8	0.663	245	0.575	333
2–9	0.588	158	0.521	423
2–10	0.499	202	0.498	452
ANOVA: *F*/*P*	5.56/<0.0001	3.55/0.0027	1.29/0.27	1.94/0.07
Dist.: *r*/*P*	0.63/0.05	0.42/0.23	−0.19/0.60	−0.35/0.32
Poll.: *r*/*P*	−0.51/0.13	−0.21/0.56	0.23/0.51	0.42/0.23

Measurements conducted on 14.8.2002. For data on *Betula pubescens* consult Table 4.1. For other explanations, consult Table 4.1.

Table 4.11 Chlorophyll fluorescence measurements in the impact zone of the aluminium smelter at Nadvoitsy, Russia

	Betula pubescens		*Pinus sylvestris*	
Site	F_v/F_m	$T_{1/2}$, ms	F_v/F_m	$T_{1/2}$, ms
1–1	0.716	293	0.641	158
1–2	0.738	290	0.735	184
1–3	0.690	368	0.632	177
1–4	0.786	207	0.726	178
1–5	0.773	357	0.721	163
2–1	0.675	368	0.572	161
2–2	0.739	207	0.532	184
2–3	0.694	322	0.523	152
2–4	0.716	285	0.594	167
2–5	0.748	280	0.638	164
ANOVA: *F*/*P*	2.69/0.02	7.78/<0.0001	24.3/<0.0001	0.66/0.74
Dist.: *r*/*P*	0.61/0.06	−0.04/0.92	0.40/0.25	−0.01/0.97
Poll.: *r*/*P*	−0.42/0.23	0.05/0.92	−0.23/0.53	0.18/0.62

Measurements conducted on 25–27.7.2004. For other explanations, consult Table 4.1.

$Q_B = 1.08$, df = 4, $P = 0.90$) or it's impact on soil pH (F_v/F_m: Fig. 4.1; $Q_B = 2.70$, df = 2, $P = 0.26$; $T_{1/2}$: Fig. 4.4; $Q_B = 1.91$, df = 2, $P = 0.38$). The only source of variation identified in this database was the geographical position of polluters that affected the $T_{1/2}$ response to pollution (Fig. 4.4; $Q_B = 5.37$, df = 1, $P = 0.02$):

Table 4.12 Chlorophyll fluorescence measurements in the impact zone of the nickel-copper smelter at Nikel and ore-roasting plant at Zapolyarnyy, Russia

	Betula pubescens		*Salix glauca*	
Site	F_v/F_m	$T_{1/2}$, ms	F_v/F_m	$T_{1/2}$, ms
1–1	0.694	579	0.654	593
1–2	0.766	499	0.609	632
1–3	0.666	510	0.638	558
1–4	0.647	619	0.634	559
1–5	0.597	581	0.602	603
2–1	0.733	503	0.657	602
2–2	0.685	455	–	–
2–3	0.758	606	0.681	686
2–4	0.778	566	0.736	570
2–5	0.655	775	0.619	542
ANOVA: *F*/*P*	5.92/<0.0001	8.24/<0.0001	2.50/0.03	2.33/0.04
Dist.: *r*/*P*	−0.42/0.23	0.53/0.12	−0.15/0.71	−0.29/0.46
Poll.: *r*/*P*	0.19/0.59	−0.24/0.51	0.10/0.79	0.10/0.79

Measurements conducted on 17–18.7.2001. Missing values resulted from absence of the selected plant species in some of study sites. For other explanations, consult Table 4.1.

Table 4.13 Chlorophyll fluorescence measurements in the impact zone of the nickel-copper smelters at Norilsk, Russia

	Larix sibirica		*Betula nana*		*Vaccinium uliginosum*	
Site	F_v/F_m	$T_{1/2}$, ms	F_v/F_m	$T_{1/2}$, ms	F_v/F_m	$T_{1/2}$, ms
1–1	0.582	160	0.635	275	0.646	363
1–2	0.536	122	0.607	349	0.600	405
1–3	0.509	130	0.493	335	0.620	398
1–4	0.551	128	0.659	299	0.599	443
1–5	0.631	172	0.688	312	0.634	411
2–1	0.355	46	0.568	243	0.596	235
2–2	0.687	139	0.687	176	0.597	315
2–3	0.573	73	0.635	230	0.623	283
2–4	0.588	154	0.581	300	0.556	339
2–5	0.713	189	0.695	272	0.673	265
ANOVA: *F*/*P*	16.5/<0.0001	8.36/<0.0001	6.60/<0.0001	7.06/<0.0001	1.09/0.39	10.4/<0.0001
Dist.: *r*/*P*	0.48/0.16	0.47/0.17	0.27/0.45	0.35/0.32	0.17/0.64	0.25/0.48
Poll.: *r*/*P*	−0.30/0.41	−0.25/0.49	−0.06/0.87	−0.39/0.27	0.10/0.78	−0.29/0.42

Measurements conducted on 23–28.8.2002. For other explanations, consult Table 4.1.

adverse effects (increase in the time needed for the leaf to reach half of its F_m) were significant only near the southern polluters. The pattern of F_v/F_m changes did not differ between the northern and southern polluters (Fig. 4.1; $Q_B = 2.76$, df = 1, $P = 0.10$).

Table 4.14 Chlorophyll fluorescence measurements in the impact zone of the copper smelter at Revda, Russia

	Betula pubescens		*Pinus sylvestris*	
Site	F_v/F_m	$T_{1/2}$, ms	F_v/F_m	$T_{1/2}$, ms
1–1	0.726	321	0.732	237
1–2	0.713	395	0.708	225
1–3	0.668	387	0.634	183
1–4	0.561	377	0.481	171
1–5	0.724	403	0.624	184
2–1	0.692	376	0.705	235
2–2	0.766	319	0.760	277
2–3	0.728	285	0.657	197
2–4	0.689	266	0.691	222
2–5	0.583	309	0.590	271
ANOVA: *F*/*P*	5.53/<0.0001	2.90/0.01	15.0/<0.0001	11.6/<0.0001
Dist.: *r*/*P*	−0.59/0.07	0.03/0.94	−0.75/0.01	−0.34/0.34
Poll.: *r*/*P*	0.54/0.11	−0.24/0.51	0.67/0.04	0.43/0.21

Measurements conducted on 19–21.7.2003. For other explanations, consult Table 4.1.

Table 4.15 Chlorophyll fluorescence measurements in the impact zone of the aluminium smelter at Straumsvík, Iceland

	Salix herbacea		*Vaccinium uliginosum*	
Site	F_v/F_m	$T_{1/2}$, ms	F_v/F_m	$T_{1/2}$, ms
1–1	0.660	296	0.639	423
1–2	0.651	319	0.620	379
1–3	0.612	352	0.577	430
1–4	0.584	329	0.587	373
1–5	0.664	281	0.666	320
2–1	0.663	312	0.607	398
2–2	0.620	383	0.646	391
2–3	0.626	323	0.653	354
2–4	0.638	350	0.599	337
2–5	0.670	338	0.710	282
ANOVA: *F*/*P*	1.40/0.22	1.07/0.40	3.71/0.0019	3.49/0.0029
Dist.: *r*/*P*	0.09/0.81	−0.09/0.81	0.45/0.19	−0.83/0.0030
Poll.: *r*/*P*	0.25/0.49	−0.31/0.38	0.01/0.98	0.52/0.13

Measurements conducted on 11–13.7.2002. For other explanations, consult Table 4.1.

We have not discovered any variation in the responses of photosystem II among plant species (F_v/F_m: Fig. 4.3; $Q_B = 3.79$, df = 6, $P = 0.71$; $T_{1/2}$: Fig. 4.6; $Q_B = 2.99$, df = 6, $P = 0.81$). Similarly, we found no differences between Gymnosperms and Angiosperms (F_v/F_m: $Q_B = 0.03$, df = 1, $P = 0.87$; $T_{1/2}$: $Q_B = 1.79$, df = 1, $P = 0.21$),

Table 4.16 Chlorophyll fluorescence measurements in the impact zone of the aluminium smelter at Volkhov, Russia

	Betula pubescens		*Salix caprea*	
Site	F_v/F_m	$T_{1/2}$, ms	F_v/F_m	$T_{1/2}$, ms
1–1	0.738	310	0.636	384
1–2	0.652	278	0.518	476
1–3	0.684	255	0.624	314
1–4	0.667	260	0.613	366
1–5	0.662	302	0.573	326
2–1	0.739	286	0.588	413
2–2	0.751	337	0.620	417
2–3	0.695	303	0.637	357
2–4	0.662	246	0.673	313
2–5	0.680	300	0.592	368
ANOVA: *F*/*P*	1.80/0.10	1.32/0.26	2.18/0.04	3.23/0.0050
Dist.: *r*/*P*	−0.70/0.02	−0.24/0.51	0.18/0.61	−0.59/0.07
Poll.: *r*/*P*	0.71/0.02	0.24/0.50	−0.10/0.77	0.42/0.22

Measurements conducted on 8–9.8.2002. For other explanations, consult Table 4.1.

Table 4.17 Chlorophyll fluorescence measurements in the impact zone of the aluminium smelter at Žiar nad Hronom, Slovakia

	Fagus sylvatica		*Carpinus betulus*		*Quercus petraea*	
Site	F_v/F_m	$T_{1/2}$, ms	F_v/F_m	$T_{1/2}$, ms	F_v/F_m	$T_{1/2}$, ms
1–1	0.831	133	0.821	179	0.717	370
1–2	0.827	114	0.835	133	0.783	209
1–3	0.791	170	0.796	143	–	–
1–4	0.801	152	0.830	151	0.807	151
1–5	0.819	121	0.806	197	–	–
2–1	0.804	127	0.810	155	0.679	341
2–2	0.800	165	0.830	136	0.655	379
2–3	0.826	130	0.818	119	–	–
2–4	0.765	150	0.799	131	–	–
2–5	0.798	169	0.794	189	0.767	261
ANOVA: *F*/*P*	4.60/0.0003	7.07/<0.0001	5.81/<0.0001	3.42/0.0033	2.73/0.05	6.81/0.0007
Dist.: *r*/*P*	−0.48/0.16	0.42/0.22	−0.45/0.19	0.16/0.67	0.28/0.59	−0.38/0.46
Poll.: *r*/*P*	0.57/0.08	−0.41/0.24	0.40/0.25	0.16/0.66	0.06/0.91	0.21/0.69

Measurements conducted on 29.8–1.9.2003. Missing values resulted from absence of the selected plant species in some of study sites. For other explanations, consult Table 4.1.

as well as between deciduous and evergreen species (F_v/F_m: $Q_B = 0.05$, df = 1, $P = 0.80$; $T_{1/2}$: $Q_B = 0.13$, df = 1, $P = 0.73$).

We have detected significant non-linear responses in five of 78 data sets (four U-shaped and one dome-shaped, all for F_v/F_m).

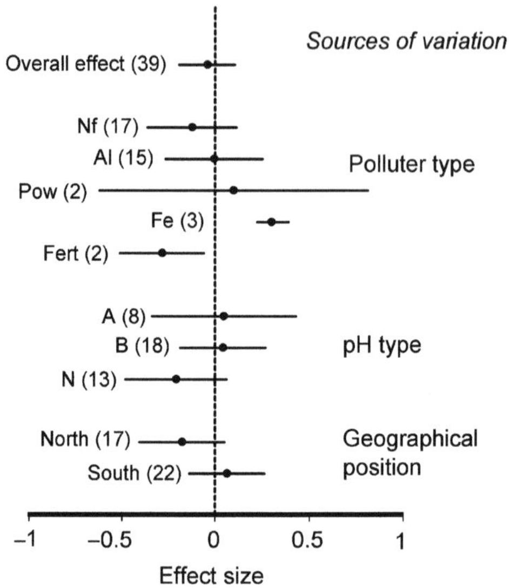

Fig. 4.1 Overall effect and sources of variation in the responses of the ratio of variable to maximum fluorescence yielded under the artificial light treatment (F_v/F_m). Decreases in F_v/F_m indicate lower plant vitality. Horizontal lines denote 95% confidence intervals; sample sizes are shown in brackets; an asterisk denotes significant ($P < 0.05$) between-class heterogeneity. For classifications of polluters and abbreviations, consult Table 2.1

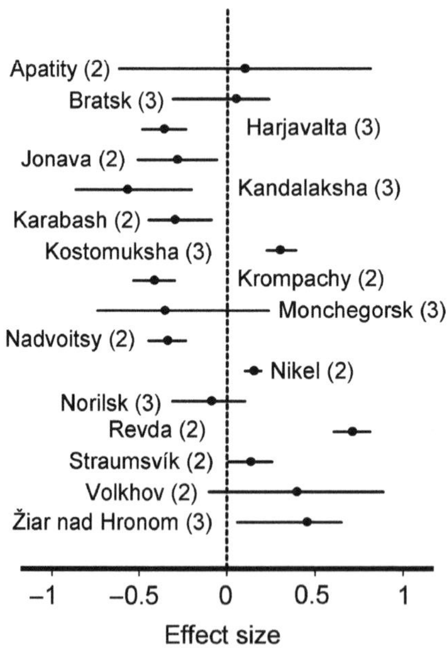

Fig. 4.2 Effects of individual polluters on the ratio of variable to maximum fluorescence yielded under the artificial light treatment (F_v/F_m). For explanations, consult Fig. 4.1

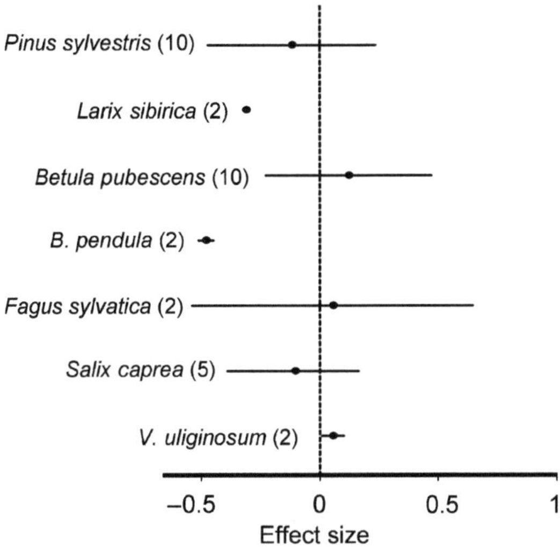

Fig. 4.3 Effects of point polluters on the ratio of variable to maximum fluorescence yielded under the artificial light treatment (F_v/F_m) in woody plant species. For explanations, consult Fig. 4.1

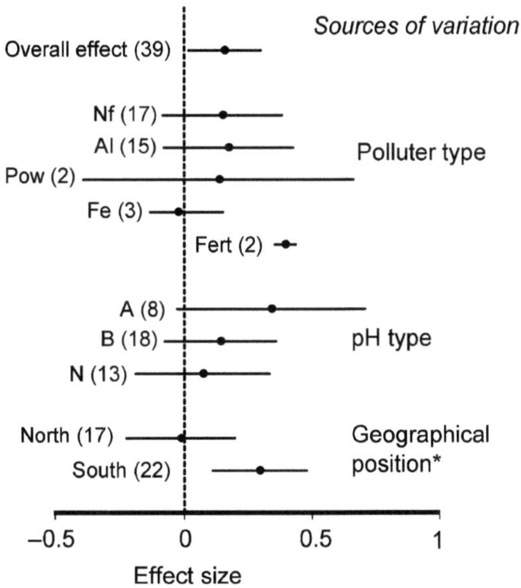

Fig. 4.4 Overall effect and sources of variation in the responses of the time needed for the leaf to reach half of its F_m ($T_{1/2}$). Increases in $T_{1/2}$ indicate lower plant vitality. For explanations, consult Fig. 4.1

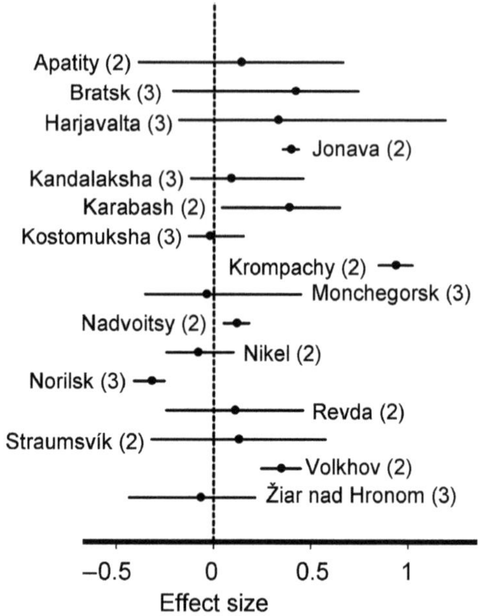

Fig. 4.5 Effects of individual polluters on the time needed for the leaf to reach half of its F_m ($T_{1/2}$). For explanations, consult Fig. 4.1

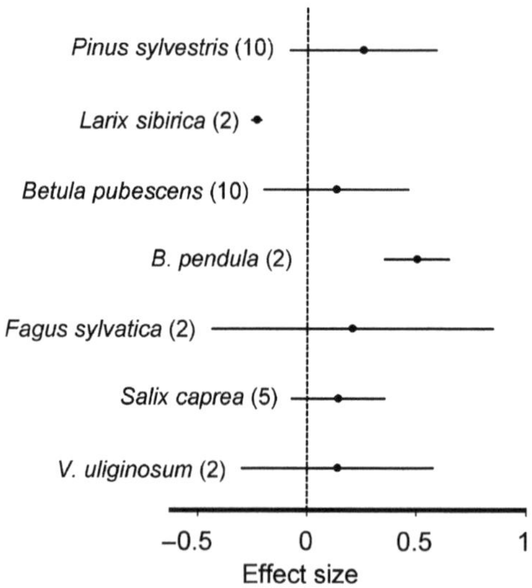

Fig. 4.6 Effects of point polluters on the time needed for the leaf of woody plant species to reach half of its F_m ($T_{1/2}$). For explanations, consult Fig. 4.1

4.3.2 Leaf/Needle Size

Variation in leaf/needle size between study sites was generally significant (60 of 90 data sets); however, only 23 of 146 correlation coefficients (with both distance and pollution load; calculated only for woody plants) were significant (Tables 4.18–4.35).

The magnitude of the pollution effect on the leaf size of woody plants did not depend on the method used to calculate ES ($Q_B = 0.50$, df = 2, $P = 0.78$). Therefore, the analyses of herbaceous plants, as well as of pooled data, are based on Hedge's d values because data on herbs were collected only from the most and least polluted study sites. However, in the analysis of woody plants we employed ESs based on correlations with pollution, in line with all other characteristics analysed in this book.

An overall pollution effect on the leaf/needle size did not differ from zero (Fig. 4.7). However, leaf size in Angiosperms significantly decreased with pollution (Fig. 4.7), although the difference between Angiosperms and Gymnosperms was not significant ($Q_B = 1.87$, df = 1, $P = 0.17$). Trees, dwarf shrubs, and herbs ($Q_B = 1.48$, df = 2, $P = 0.48$) as well as evergreen and deciduous woody plants ($Q_B = 0.50$, df = 1, $P = 0.48$) responded similarly to pollution (Fig. 4.7). We found no differences among plants belonging to four Raunkiaer life forms (phanerophytes, chamaephytes, non-bulbous geophytes and hemicriptophytes) ($Q_B = 2.79$, df = 3, $P = 0.26$).

Effect size (averaged by plant species) positively correlated with the Ellenberg's indicator value for light ($r_S = 0.36$, $N = 28$ species, $P = 0.06$), but did not correlate with indicator values for temperature, continentality, humidity, pH and nitrogen ($r_S = 0.09...0.33$, $N = 15–24$ species, $P = 0.11...0.74$). Similarly, we found no correlation with axis scores for Grime's CSR strategy ($r_S = -0.28...0.19$, $N = 13$ species, $P = 0.35...0.54$).

Table 4.18 Leaf size (L, mm) and shoot length (S, mm) of woody plants in the impact zone of the power plant at Apatity, Russia

	Alnus incana		*Betula pubescens*		*Vaccinium vitis-idaea*	
Site	L (1997)	S (2006)	L (2006)	S (2006)	L (2006)	S (2006)
1–1	62.4	105.1	54.7	99.1	–	–
1–2	64.6	87.6	47.8	84.0	26.0	58.4
1–3	56.5	74.3	45.0	102.3	22.5	43.4
1–4	55.7	86.6	42.0	68.9	23.3	55.8
1–5	58.1	93.7	43.9	70.1	24.5	62.4
2–1	60.0	76.3	52.7	92.9	20.6	48.6
2–2	50.2	118.1	50.6	113.5	25.0	58.2
2–3	61.5	162.3	48.6	106.5	24.8	49.9
2–4	57.2	74.5	41.9	87.6	23.3	53.1
2–5	56.9	78.1	47.8	52.1	24.7	44.1
ANOVA: *F/P*	1.75/0.11	0.98/0.47	2.55/0.02	2.30/0.03	3.87/0.0022	6.55/<0.0001
Dist.: *r/P*	−0.41/0.24	−0.12/0.73	−0.85/0.0018	−0.56/0.09	0.30/0.43	0.10/0.79
Poll.: *r/P*	0.36/0.31	0.30/0.39	0.66/0.04	0.71/0.02	0.18/0.64	0.15/0.69

Sampling year shown in parentheses. For other explanations, consult Table 4.1.

Table 4.19 Leaf/needle size (L, mm) and shoot length (S, mm) of woody plants in the impact zone of the aluminium smelter at Bratsk, Russia

	Betula pubescens		*Larix sibirica*	*Picea abies*	*Pinus sylvestris*		*Vaccinium vitis-idaea*	
Site	L (2002)	S (2002)	S (2002)	S (2004)	L (2002)	S (2002)	L (2002)	S (2002)
1–1	52.4	99.6	50.8	46.2	48.9	36.4	17.3	35.8
1–2	57.2	121.0	68.5	30.7	40.5	23.9	17.7	38.4
1–3	53.4	113.6	56.5	51.8	40.7	29.2	17.4	32.8
1–4	51.3	110.0	64.5	58.5	54.5	26.8	18.4	30.4
1–5	53.6	100.8	67.9	47.9	52.2	38.2	17.9	40.3
2–1	45.1	91.2	47.1	53.0	55.8	58.2	16.5	27.7
2–2	53.4	93.4	53.7	42.1	40.4	23.1	18.3	34.7
2–3	46.0	94.2	60.4	53.2	43.0	36.0	18.2	29.9
2–4	50.4	99.5	54.8	38.0	46.5	38.6	19.5	40.6
2–5	46.4	68.6	53.6	83.8	49.4	27.6	19.5	29.6
ANOVA: *F*/*P*	1.88/0.08	1.01/0.45	1.84/0.09	3.11/0.0079	3.99/0.0013	4.72/0.0003	0.75/0.66	2.17/0.05
Dist.: *r*/*P*	−0.16/0.65	−0.24/0.51	0.43/0.22	0.49/0.15	0.18/0.61	−0.21/0.56	0.70/0.03	0.10/0.78
Poll.: *r*/*P*	−0.36/0.31	−0.09/0.81	−0.57/0.09	−0.07/0.84	0.43/0.22	0.76/0.01	−0.72/0.02	−0.40/0.25

Sampling year shown in parentheses. For other explanations, consult Table 4.1.

Table 4.20 Leaf/needle size (L, mm), shoot length (S, mm) of woody plants and height (H, mm) of herbaceous plants in the impact zone of the nickel-copper smelter at Harjavalta, Finland

	Achillea millefolium		*Betula pubescens*		*Picea abies*	*Pinus sylvestris*		*Salix caprea*				*Vaccinium vitis-idaea*	
Site	L (2007)	H (2007)	L (2002)	S (2002)	S (2001)	L (2002)	S (2002)	S (1997)	S (1998)	L (2002)	S (2002)	L (2002)	S (2002)
1–1	169	526	44.9	77.7	–	40.8	50.7	144.7	71.9	84.3	86.9	–	–
1–2	237	638	40.7	134.4	–	51.4	65.4	191.7	79.6	77.8	94.6	19.3	30.2
1–3	–	–	40.7	105.1	78.2	–	–	184.5	118.7	73.8	97.9	19.3	32.5
1–4	–	–	46.3	96.8	70.8	61.2	220.2	196.1	81.1	86.1	68.0	21.8	46.8
1–5	125	499	41.4	102.6	55.2	42.7	113.4	72.7	49.3	71.0	61.8	22.8	55.8
2–1	–	–	40.2	110.7	–	49.4	98.6	279.2	125.0	80.3	56.3	–	–
2–2	–	–	36.2	90.4	63.4	42.0	86.1	144.8	126.8	74.3	75.5	–	–
2–3	–	–	43.0	126.2	65.5	52.0	67.8	142.0	76.2	68.6	69.6	20.2	32.6
2–4	102	461	39.9	142.1	70.4	42.1	42.3	146.9	60.1	74.1	48.3	18.9	47.3
2–5	–	–	–	123.4	67.4	34.3	96.9	84.9	58.2	72.3	85.6	–	–
ANOVA: *F/P*	6.41/0.0014	3.22/0.03	2.00/0.07	1.20/0.32	0.85/0.54	3.43/0.0061	6.84/<0.0001	1.70/0.13	2.24/0.04	1.74/0.11	2.11/0.05	2.37/0.07	8.33/0.0001
Dist.: *r/P*	–	–	0.23/0.55	0.21/0.57	−0.29/0.53	−0.28/0.46	0.31/0.42	−0.70/0.02	−0.69/0.03	−0.38/0.28	−0.17/0.63	0.70/0.12	0.94/0.0053
Poll.: *r/P*	–	–	−0.11/0.78	0.17/0.64	−0.12/0.80	0.23/0.56	−0.31/0.41	0.31/0.39	0.11/0.77	0.24/0.50	0.45/0.20	−0.37/0.47	−0.58/0.23

Sampling year shown in parentheses. For other explanations, consult Table 4.1.

Table 4.21 Leaf/needle size (L, mm) and shoot length (S, mm) of plants in the impact zone of the fertiliser factory at Jonava, Lithuania

	Betula pendula		*Fragaria vesca*	*Frangula alnus*		*Picea abies*	*Pinus sylvestris*		*Quercus robur*	
Site	L (2005)	S (2005)	L (2007)	L (2007)	S (2007)	S (2005)	L (2005)	S (2005)	L (2007)	S (2007)
1–1	49.3	95.0	33.9	–	–	–	55.9	53.1	112.1	25.1
1–2	64.4	121.6	–	74.3	69.1	79.1	61.3	86.1	119.1	36.2
1–3	50.7	93.2	–	72.9	65.9	71.8	59.5	113.5	114.2	28.2
1–4	60.7	115.1	–	71.6	62.4	78.7	62.1	46.1	128.9	28.7
1–5	61.3	138.9	44.0	75.7	62.3	58.2	58.7	42.1	105.6	33.6
2–1	55.4	134.7	46.3	81.2	127.8	–	58.2	139.5	121.7	25.5
2–2	47.1	99.7	–	87.6	81.6	75.1	63.3	47.6	102.8	34.4
2–3	52.5	116.5	–	78.2	38.5	74.1	52.4	97.0	118.9	33.0
2–4	64.6	166.9	–	68.0	97.8	68.1	50.9	58.7	110.6	61.3
2–5	54.8	123.1	39.5	74.8	47.6	51.0	48.2	213.9	110.4	24.7
ANOVA: *F/P*	3.68/0.002	2.94/0.009	2.86/0.05	1.53/0.19	2.60/0.03	3.48/0.0069	2.03/0.06	6.55/<0.0001	1.18/0.33	1.44/0.21
Dist.: *r/P*	0.35/0.32	0.31/0.38	–	–0.54/0.13	–0.65/0.06	–0.76/0.03	–0.33/0.36	0.05/0.90	–0.20/0.57	0.26/0.46
Poll.: *r/P*	–0.14/0.70	0.16/0.66	–	0.10/0.80	0.33/0.38	0.06/0.90	0.02/0.97	–0.57/0.09	–0.65/0.04	0.63/0.05

Sampling year shown in parentheses. For other explanations, consult Table 4.1.

Table 4.22 Leaf/needle size (L, mm), shoot length (S, mm) of woody plants and height (H, mm) of herbaceous plants in the impact zone of the aluminium smelter at Kandalaksha, Russia

	Betula pubescens		*Empetrum nigrum*	*Epilobium angustifolium*		*Linnaea borealis*		*Picea abies*	*Pinus sylvestris*
Site	L (2002)	S (2002)	S (2002)	L (2006)	H (2006)	L (2006)	H (2006)	S (2001)	L (1998)
1–1	40.8	86.5	25.4	110	590	12.2	91	25.6	28.4
1–2	46.0	91.5	26.6	–	–	–	–	29.0	28.7
1–3	35.2	74.1	25.2	–	–	–	–	28.6	29.0
1–4	53.8	132.3	27.8	–	–	–	–	34.3	37.6
1–5	37.3	95.1	27.6	105	548	12.2	99	24.1	27.2
2–1	33.2	38.7	23.7	106	720	11.6	86	34.1	32.0
2–2	45.1	102.6	20.1	–	–	–	–	31.7	27.6
2–3	36.6	116.6	22.5	–	–	–	–	30.9	30.3
2–4	44.1	98.0	25.2	–	–	–	–	34.4	31.3
2–5	33.1	97.2	13.1	88	485	12.4	100	30.9	29.1
ANOVA: *F/P*	7.33/<0.0001	4.01/0.0011	2.09/0.05	1.88/0.15	4.98/0.0054	0.42/0.73	2.67/0.06	0.95/0.50	1.81/0.10
Dist.: *r/P*	0.04/0.91	0.61/0.06	–0.22/0.53	–	–	–	–	–0.11/0.75	0.03/0.94
Poll.: *r/P*	0.04/0.92	–0.61/0.06	0.27/0.45	–	–	–	–	0.01/0.96	–0.06/0.88

Sampling year shown in parentheses. For other explanations, consult Table 4.1.

Table 4.22 (continued)

Site	*Salix caprea*		*Trientalis europaea*		*Vaccinium myrtillus*		*Vaccinium uliginosum*		*Vaccinium vitis-idaea*	
	L (2002)	S (2002)	L (2006)	H (2006)	L (2005)	S (2005)	L (2005)	S (2005)	L (2001)	S (2001)
1-1	67.3	34.4	44.2	138	24.8	14.0	28.5	19.9	21.3	32.0
1–2	65.0	56.7	–	–	21.7	15.4	25.5	22.5	15.9	29.5
1–3	67.3	39.3	–	–	25.2	14.9	23.5	20.2	18.7	37.3
1–4	74.9	64.8	–	–	23.9	17.0	21.6	20.1	20.0	36.9
1–5	66.4	30.6	48.5	112	29.5	15.7	30.8	22.4	15.2	31.2
2–1	71.8	110.7	38.6	113	24.8	12.4	33.9	23.4	19.3	34.9
2–2	67.2	29.0	–	–	35.1	16.9	35.0	22.5	19.4	32.5
2–3	63.8	34.0	–	–	27.4	15.5	28.3	22.3	18.3	35.2
2–4	68.0	38.4	–	–	35.5	18.3	36.4	24.5	20.1	37.3
2–5	71.9	34.2	49.5	92	24.2	13.4	28.8	19.8	20.9	37.2
ANOVA: *F/P*	1.28/0.28	6.63/<0.0001	4.48/0.009	3.84/0.02	2.01/0.06	6.81/<0.0001	1.53/0.17	1.41/0.22	3.83/0.0015	1.07/0.41
Dist.: *r/P*	0.09/0.80	−0.57/0.08	–	–	0.12/0.74	0.37/0.29	−0.23/0.53	−0.24/0.51	−0.13/0.71	0.34/0.34
Poll.: *r/P*	−0.06/0.87	0.63/0.05	–	–	−0.31/0.38	−0.42/0.22	0.07/0.84	0.18/0.62	−0.08/0.82	−0.52/0.12

Sampling year shown in parentheses. For other explanations, consult Table 4.1.

Table 4.23 Leaf/needle size (L, mm), shoot length (S, mm) of woody plants and height (H, mm) of herbaceous plants in the impact zone of the copper smelter at Karabash, Russia

	Alnus incana		*Betula pendula*		*Fragaria vesca*	*Orthilia secunda*		*Pinus sylvestris*		*Vaccinium vitis-idaea*	
Site	L (2007)	S (2007)	L (2003)	S (2003)	L (2007)	L (2007)	H (2007)	L (2003)	S (2003)	L (2003)	S (2003)
1–1	–	–	30.2	58.2	–	–	–	–	–	–	–
1–2	78.4	51.4	39.1	45.8	34.4	20.5	124	47.7	45.4	22.3	33.0
1–3	86.1	95.5	46.3	46.3	–	–	–	45.8	42.5	21.5	36.8
1–4	77.8	55.1	43.9	73.8	–	–	–	45.4	42.0	22.2	43.6
1–5	92.7	144.7	48.6	75.9	40.7	32.6	138	53.6	53.1	23.0	39.4
2–1	77.4	158.6	36.3	46.6	–	–	–	–	–	–	–
2–2	60.1	29.3	47.9	60.2	32.1	24.4	105	36.4	19.3	19.4	30.1
2–3	87.8	65.8	41.1	67.1	–	–	–	39.1	37.5	20.5	41.9
2–4	76.8	105.0	43.7	58.7	–	–	–	52.2	42.5	22.6	39.3
2–5	86.7	76.3	48.1	78.3	40.2	34.9	163	43.5	26.2	20.5	35.8
ANOVA: *F*/*P*	2.39/0.04	3.98/0.0018	6.34/<0.0001	1.92/0.08	3.14/0.04	25.3/<0.0001	25.9/<0.0001	1.70/0.014	3.24/0.01	1.36/0.26	1.09/0.39
Dist.: *r*/*P*	0.52/0.15	0.02/0.96	0.71/0.02	−0.51/0.13	–	–	–	0.62/0.10	0.32/0.44	0.49/0.21	0.02/0.96
Poll.: *r*/*P*	−0.47/0.20	0.30/0.44	−0.65/0.04	−0.10/0.79	–	–	–	−0.73/0.04	−0.63/0.09	−0.72/0.05	0.53/0.17

Sampling year shown in parentheses. For other explanations, consult Table 4.1.

Table 4.24 Leaf/needle size (L, mm), shoot length (S, mm) of woody plants and height (H, mm) of herbaceous plants in the impact zone of the iron pellet plant at Kostomuksha, Russia

	Betula pubescens		*Melampyrum sylvaticum*	*Picea abies*	*Pinus sylvestris*		*Salix caprea*		*Vaccinium vitis-idaea*	
Site	L (2002)	H (2002)	H (2006)	S (2001)	L (2002)	S (2002)	L (2002)	S (2002)	L (2001)	S (2001)
1–1	44.0	83.4	173	24.6	34.9	60.0	64.2	51.0	20.4	38.7
1–2	40.8	98.8	–	36.1	34.8	93.9	80.9	81.0	19.8	36.0
1–3	36.5	136.5	–	28.5	–	81.6	74.6	52.3	21.2	37.5
1–4	42.9	127.8	–	26.4	36.1	63.0	85.2	60.7	19.9	32.8
1–5	37.7	83.2	202	31.4	37.4	64.1	73.2	59.3	20.9	52.4
2–1	39.6	93.9	205	31.2	40.1	53.7	66.8	55.2	21.0	45.5
2–2	42.1	80.6	–	35.1	30.5	41.4	58.8	53.6	21.3	40.3
2–3	42.6	109.6	–	34.2	30.6	54.7	59.2	59.7	18.0	35.9
2–4	32.6	71.3	–	34.4	34.0	43.4	71.0	66.4	16.6	45.8
2–5	41.1	79.2	236	38.5	36.2	–	74.3	52.9	19.6	43.6
ANOVA: *F*/*P*	3.40/0.0035	3.99/0.0011	4.94/0.0061	3.49/0.0030	3.20/0.0076	6.95/<0.0001	4.06/0.0009	1.82/0.09	4.41/0.0005	4.84/0.0002
Dist.: *r*/*P*	−0.33/0.35	0.10/0.78	–	0.32/0.37	−0.10/0.81	0.24/0.53	0.53/0.12	0.20/0.58	−0.30/0.40	0.11/0.75
Poll.: *r*/*P*	0.20/0.59	−0.10/0.79	–	−0.25/0.48	0.52/0.16	−0.28/0.46	−0.29/0.42	−0.26/0.47	0.43/0.21	0.19/0.61

Sampling year shown in parentheses. For other explanations, consult Table 4.1.

Table 4.25 Leaf/needle size (L, mm) and shoot length (S, mm) of woody plants in the impact zone of the copper smelter at Krompachy, Slovakia

	Betula pendula		*Fagus sylvatica*			*Picea abies*	*Pinus sylvestris*	
Site	L (2004)	S (2004)	L (2002)	S (2002)	S (2004)	S (2002)	L (2002)	S (2002)
1–1	63.9	185.4	80.5	77.8	20.6	38.4	47.6	61.4
1–2	71.7	223.7	72.5	75.1	67.8	60.9	41.1	44.9
1–3	48.9	147.8	73.9	35.4	26.6	52.7	50.2	35.2
1–4	91.3	240.2	71.8	71.4	62.5	41.9	47.1	46.0
1–5	71.6	241.6	77.9	56.5	4.5	56.7	56.7	55.6
2–1	67.2	181.9	70.7	77.0	54.6	58.6	48.3	108.8
2–2	82.1	217.1	71.2	75.3	30.2	42.7	45.5	25.9
2–3	66.4	197.5	77.7	71.1	12.8	39.9	47.8	53.0
2–4	62.8	218.1	83.2	83.7	22.7	51.0	43.7	55.3
2–5	72.9	230.5	79.1	72.1	62.1	65.0	50.4	58.3
ANOVA: *F/P*	9.45/<0.0001	1.29/0.27	2.11/0.05	1.33/0.25	5.37/<0.0001	3.75/0.0017	0.79/0.62	3.13/0.0060
Dist.: *r/P*	0.18/0.61	0.52/0.12	0.16/0.64	–0.35/0.32	0.13/0.72	0.51/0.13	0.45/0.19	–0.21/0.57
Poll.: *r/P*	–0.17/0.64	–0.34/0.34	–0.40/0.26	0.07/0.85	0.10/0.78	0.28/0.44	0.16/0.66	0.79/0.0062

Sampling year shown in parentheses. For other explanations, consult Table 4.1.

Table 4.26 Leaf/needle size (L, mm), shoot length (S, mm) of woody plants and height (H, mm) of herbaceous plants in the impact zone of the nickel-copper smelter at Monchegorsk, Russia

	Betula nana		*Betula pubescens*			*Cornus suecica*		*Dactylorhiza maculata*		*Empetrum nigrum*
Site	L (2001)	S (2001)	L (2002)	S (2002)	S (2005)	L (2006)	H (2006)	L (2006)	H (2006)	S (2002)
1–1	11.3	55.6	39.0	75.9	57.8	–	–	–	–	10.8
1–2	11.9	52.0	34.1	62.7	53.1	–	–	–	–	12.0
1–4	12.9	51.4	34.0	64.5	45.5	–	–	–	–	6.3
1–5	12.0	45.2	41.3	86.9	67.4	–	–	–	–	8.2
2–2	10.9	52.5	39.5	92.6	55.5	–	–	53.6	237	9.7
2–3	–	–	–	–	–	26.7	99	45.7	240	–
2–4	12.1	52.0	44.1	67.6	42.4	22.8	91	–	–	7.7
2–5	12.4	–	37.6	91.1	61.5	–	–	–	–	12.0
2–6	–	56.9	–	–	–	–	–	–	–	–
2–8	11.3	40.0	42.4	53.5	35.5	29.5	154	–	–	15.6
2–9	11.7	64.5	45.3	70.0	44.8	32.1	230	68.2	333	21.2
2–10	12.7	62.8	41.8	62.4	57.5	–	–	60.8	294	–
2–12	–	–	–	–	–	–	–	–	–	35.0
ANOVA: *F*/*P*	0.81/0.61	1.09/0.39	2.53/0.02	1.15/0.35	1.64/0.14	14.6/<0.0001	45.1/<0.0001	11.1/<0.0001	7.39/0.0004	16.2/<0.0001
Dist.: *r*/*P*	0.26/0.50	0.13/0.71	0.47/0.20	−0.34/0.34	−0.24/0.51	–	–	–	–	0.70/0.03
Poll.: *r*/*P*	−0.08/0.83	−0.05/0.88	−0.02/0.95	0.40/0.25	0.03/0.93	–	–	–	–	−0.50/0.14

	Epilobium angustifolium		*Equisetum sylvaticum*	*Picea abies*	*Pinus sylvestris*		*Populus tremula*	*Rubus chamaemorus*	*Salix borealis*	
Site	L (2006)	H (2006)	H (2006)	S (2002)	L (2002)	S (2002)	L (2002)	L (2006)	L (1997)	S (1997)
1–1	99.7	620	331	23.4	31.3	26.9	51.3	–	32.4	32.0
1–2	–	–	–.	24.2	32.5	36.6	–	–	39.2	55.4
1–3	–	–	–	–	–	–	39.8	–	–	–
1–4	–	–	–	20.1	23.1	28.3	–	–	35.9	35.4
1–5	–	–	–	28.1	34.5	15.0	–	–	33.8	30.8
2–1	–	–	–	–	–	–	–	–	34.0	28.0
2–2	–	–	–	23.7	34.1	33.4	–	37.9	–	–
2–3	98.7	566	–	–	–	–	38.2	–	–	–
2–4	–	–	235	20.6	31.8	30.5	–	34.8	–	–
2–5	–	–	–	30.0	32.0	47.6	40.4	–	41.6	50.8
2–6	–	–	–	–	–	–	41.5	–	–	–
2–7	–	–	–	–	–	–	–	–	41.7	49.9
2–8	–	–	–	24.3	34.2	20.6	42.4	–	41.6	33.5
2–9	92.1	551	381	26.2	31.6	32.0	43.5	–	36.1	30.5
2–10	85.4	473	311	–	31.4	36.3	44.4	49.6	37.4	41.7
2–11	–	–	–	–	–	–	46.3	–	–	–
2–12	–	–	–	28.7	–	–	47.1	42.2	–	–
ANOVA: *F/P*	0.92/0.44	1.53/0.22	11.0/<0.0001	6.22/<0.0001	1.40/0.22	3.86/0.0014	6.61/<0.0001	11.2/<0.0001	4.45/0.0004	1.85/0.09
Dist.: *r/P*	–	–	–	0.48/0.16	0.09/0.81	−0.02/0.95	−0.16/0.65	–	0.54/0.13	0.18/0.61
Poll.: *r/P*	–	–	–	−0.41/0.23	−0.28/0.47	0.26/0.48	−0.09/0.80	–	−0.29/0.45	0.25/0.48

(continued)

Table 4.26 (continued)

Site	*Salix caprea*			*Solidago virgaurea*		*Vaccinium myrtillus*	*Vaccinium uliginosum*		*Vaccinium vitis-idaea*	
	S (1997)	L (2002)	S (2002)	L (2006)	H (2006)	S (1999)	L (2005)	S (2005)	L (2001)	S (1999)
1–1	45.0	59.6	28.7	–	–	25.3	15.1	20.3	16.3	18.9
1–2	–	56.1	33.5	–	–	26.7	17.5	14.5	17.3	15.6
1–4	34.7	49.2	30.5	–	–	36.0	17.7	23.6	14.2	26.2
1–5	–	56.7	44.7	–	–	36.7	16.5	28.7	16.4	24.7
2–1	48.6	–	–	–	–	–	–	–	–	–
2–2	–	55.8	34.8	84.0	335	51.1	20.3	32.9	19.1	16.0
2–4	–	52.3	34.7	89.0	280	29.7	19.7	28.3	13.5	21.4
2–5	44.8	60.2	44.6	–	–	26.7	21.0	21.6	14.5	22.6
2–8	30.5	59.0	31.3	–	–	38.4	19.9	38.6	18.8	15.2
2–9	–	57.8	59.8	128.2	467	36.6	25.2	29.6	15.9	19.5
2–10	57.9	61.9	50.4	133.1	478	30.0	17.9	46.6	17.6	18.0
ANOVA: *F*/*P*	1.44/0.25	2.39/0.03	1.66/0.13	1.71/0.20	3.27/0.05	1.57/0.16	9.41/<0.0001	5.50/<0.0001	4.62/0.0003	4.00/0.0010
Dist.: *r*/*P*	–0.10/0.85	0.18/0.65	0.68/0.03	–	–	0.13/0.72	0.48/0.19	0.65/0.04	0.01/0.98	0.09/0.80
Poll.: *r*/*P*	0.08/0.88	–0.51/0.16	–0.49/0.15	–	–	0.06/0.86	–0.46/0.21	–0.30/0.39	0.41/0.31	–0.22/0.54

Sampling year shown in parentheses. For other explanations, consult Table 4.1.

Table 4.27 Leaf/needle size (L, mm), shoot length (S, mm) of woody plants and height (H, mm) of herbaceous plants in the impact zone of the aluminium smelter at Nadvoitsy, Russia

	Betula pubescens		*Epilobium angustifolium*		*Empetrum nigrum*	*Picea abies*	*Pinus sylvestris*	
Site	L (2004)	S (2004)	L (2007)	H (2007)	S (2004)	S (2004)	L (2004)	S (2004)
1–1	45.0	91.9	149.3	759	46.7	42.2	30.8	24.2
1–2	44.1	105.7	–	–	34.2	40.1	25.2	19.0
1–3	46.5	95.1	–	–	48.1	45.7	30.0	21.4
1–4	45.5	83.8	–	–	36.7	46.0	30.3	26.2
1–5	49.1	66.5	119.4	538	37.4	49.1	28.6	25.5
2–1	46.7	81.5	117.3	673	39.4	48.0	35.8	26.8
2–2	45.3	93.5	–	–	50.4	45.5	38.6	28.2
2–3	47.0	57.0	–	–	42.6	37.0	31.4	25.7
2–4	48.0	69.5	–	–	45.3	35.4	32.6	19.5
2–5	49.1	69.2	152.7	869	42.9	42.6	28.2	19.7
ANOVA: *F*/*P*	0.70/0.71	1.30/0.27	7.42/0.0005	5.94/0.0021	2.54/0.02	1.09/0.40	4.83/0.0002	1.16/0.38
Dist.: *r*/*P*	0.69/0.03	–0.55/0.10	–	–	–0.28/0.43	0.00/0.99	–0.45/0.18	–0.27/0.45
Poll.: *r*/*P*	–0.63/0.05	0.62/0.05	–	–	0.07/0.84	0.26/0.47	0.44/0.20	0.32/0.36

Table 4.27 (continued)

Site	*Vaccinium myrtillus*		*Vaccinium uliginosum*		*Vaccinium vitis-idaea*	
	L (2005)	S (2005)	L (2005)	S (2005)	L (2004)	S (2004)
1–1	16.3	45.0	22.8	39.4	20.8	45.7
1–2	16.4	38.4	23.7	42.7	27.1	51.8
1–3	15.2	28.1	23.6	23.7	21.5	45.2
1–4	14.4	31.4	19.5	19.3	25.9	71.4
1–5	16.8	33.8	25.6	43.1	23.4	63.2
2–1	15.6	35.3	21.5	32.4	24.4	54.5
2–2	15.8	36.3	20.9	49.7	25.7	46.3
2–3	17.0	38.9	22.1	39.0	24.2	62.8
2–4	16.2	37.8	24.7	38.7	21.0	58.7
2–5	17.4	36.9	26.7	39.0	25.2	63.3
ANOVA: *F*/*P*	1.39/0.33	0.84/0.59	2.46/0.03	1.84/0.09	4.05/0.001	5.32/<0.0001
Dist.: *r*/*P*	0.21/0.55	−0.44/0.21	0.49/0.15	−0.13/0.71	0.06/0.87	0.71/0.02
Poll.: *r*/*P*	−0.22/0.55	0.31/0.39	−0.43/0.21	0.24/0.51	0.17/0.64	−0.65/0.04

Sampling year shown in parentheses. For other explanations, consult Table 4.1.

In woody plants, the effect depended on the polluter type ($Q_B = 10.4$, df = 4, $P = 0.04$): effects of power plants were positive, while effects caused by other types of polluters did not differ from zero (Fig. 4.8). Effects of polluters also differed in relation to their impact on soil pH ($Q_B = 6.64$, df = 2, $P = 0.04$): only acidifying polluters caused a significant negative effect (Fig. 4.8). The geographical position of polluters did not influence their effect on leaf/needle size (Fig. 4.8; $Q_B = 0.55$, df = 1, $P = 0.46$). Individual polluters differed in their effects on leaf/needle length ($Q_B = 63.5$, df = 17, $P < 0.0001$); two non-ferrous smelters (Karabash, Revda) and one aluminium smelter (Žiar nad Hronom) caused significant negative effects, while we detected significant increase in leaf/needle size in the vicinity of four polluters (Apatity, Sudbury, Volkhov, Vorkuta) (Fig. 4.9). Woody plant species responded similarly to pollution ($Q_B = 7.53$, df = 11, $P = 0.76$); leaf size decreased with pollution only in silver birch, European aspen, and European beech (Fig. 4.10).

In herbaceous plants (Fig. 4.11), the effect did not depend on the polluter type ($Q_B = 0.44$, df = 1, $P = 0.51$) or changes in soil pH ($Q_B = 2.64$, df = 2, $P = 0.27$), or geographical position of polluters ($Q_B = 0.03$, df = 1, $P = 0.86$). Individual polluters did not differ in their effects on leaf length of herbaceous plants ($Q_B = 4.58$, df = 4, $P = 0.33$); significant negative effects were recorded only around Karabash and Volkhov (Fig. 4.12).

We have detected significant non-linear responses in six of 73 data sets on woody plants (two dome-shaped and four U-shaped), which is nearly twice as high as the number of dome-shaped patterns that may be expected to occur by chance.

4.3.3 Shoot Growth

Variation in shoot length between study sites was generally significant (77 of 120 data sets); however, only 24 of 200 correlation coefficients (with both distance and pollution load; calculated only for woody plants) were significant (Tables 4.18–4.35).

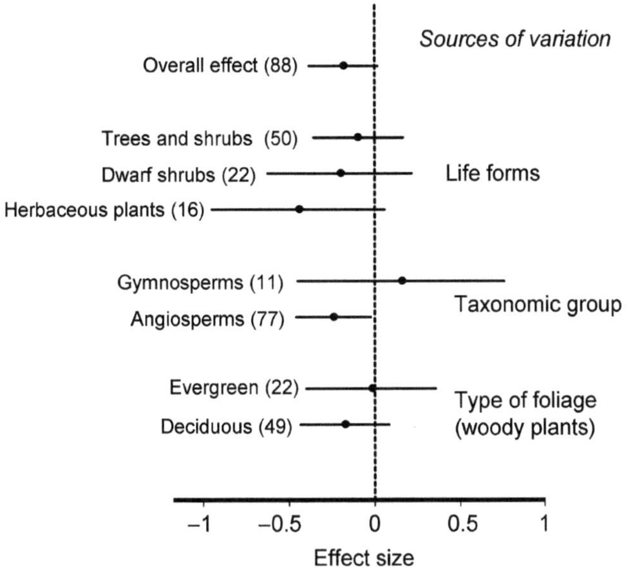

Fig. 4.7 Overall effect and sources of variation in the responses of leaf/needle length of vascular plants. Effect sizes are Hedge's d based on comparison of two most polluted and two control sites. Needle length of Scots pine (*Pinus sylvestris*) measured near aluminium smelter at Volkhov (Table 4.33) is excluded from this figure. For explanations, consult Fig. 4.1

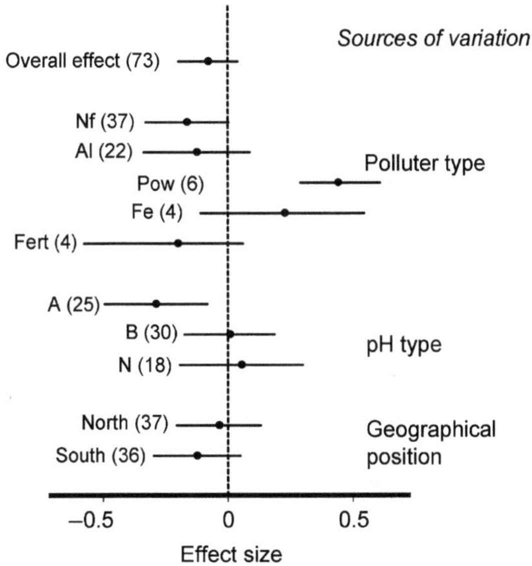

Fig. 4.8 Overall effect and sources of variation in the responses of leaf/needle length of woody plants. For explanations, consult Fig. 4.1

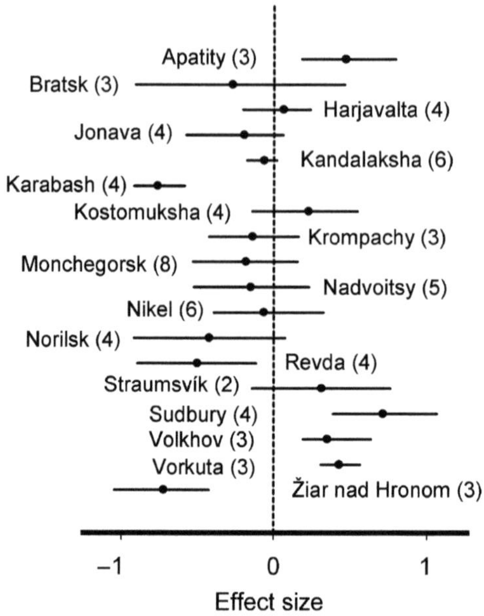

Fig. 4.9 Effects of individual polluters on the leaf/needle length of woody plants. For explanations, consult Fig. 4.1

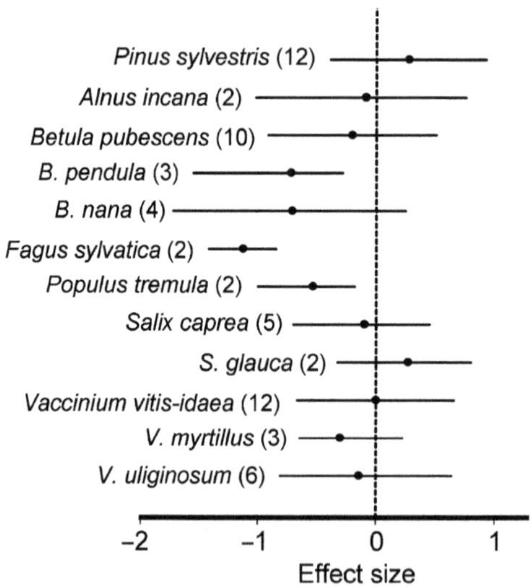

Fig. 4.10 Effects of point polluters on leaf/needle length of woody plant species. For explanations, consult Fig. 4.1

Table 4.28 Leaf size (L, mm), shoot length (S, mm) of woody plants and height (H, mm) of herbaceous plants in the impact zone of the nickel-copper smelter at Nikel and ore-roasting plant at Zapolyarnyy, Russia

	Betula nana		*Betula pubescens*		*Empetrum nigrum*	*Epilobium angustifolium*		*Equisetum sylvaticum*	*Eriophorum vaginatum*
Site	L (2001)	S (2001)	L (2001)	S (2001)	S (2003)	L (2006)	H (2006)	H (2006)	H (2006)
1–1	8.4	30.0	33.6	53.3	–	–	–	–	202
1–2	11.6	39.8	32.0	68.3	10.0	–	–	–	–
1–3	11.2	40.7	31.8	64.4	10.6	–	–	–	–
1–4	9.1	32.8	30.9	75.4	22.9	–	–	–	–
1–5	10.0	37.5	28.8	61.6	13.6	–	–	–	–
2–1	10.3	43.4	35.0	60.6	12.3	90.5	431	177	324
2–2	11.6	44.0	39.8	145.0	16.3	111.6	472	197	382
2–3	12.2	43.5	41.2	94.7	8 .7	–	–	–	–
2–4	10.8	49.9	36.9	90.8	20.1	124.7	726	225	–
2–5	10.7	39.7	33.3	53.5	11.6	113.0	550	377	464
ANOVA: *F/P*	2.85/0.01	0.79/0.63	5.88/<0.0001	4.01/0.0010	2.97/0.0084	5.76/0.0025	11.3/<0.0001	23.9/<0.0001	37.0/<0.0001
Dist.: *r/P*	0.18/0.61	0.17/0.63	−0.28/0.43	−0.03/0.93	0.29/0.45	–	–	–	–
Poll.: *r/P*	−0.49/0.15	−0.34/0.34	0.11/0.76	−0.23/0.52	−0.21/0.58	–	–	–	–

Sampling year shown in parentheses. For other explanations, consult Table 4.1.

(continued)

Table 4.28 (continued)

Site	*Rubus chamaemorus*	*Salix glauca*		*Vaccinium myrtillus*		*Vaccinium uliginosum*		*Vaccinium vitis-idaea*	
	L (2006)	L (2001)	S (2001)	L (2005)	S (2005)	L (2005)	S (2005)	L (2001)	S (2001)
1–1	–	35.1	29.7	–	–	–	–	–	–
1–2	–	34.5	24.1	11.7	19.3	14.8	28.5	12.7	16.0
1–3	–	35.4	22.3	13.6	18.4	18.7	22.9	13.2	33.4
1–4	–	35.5	28.8	12.3	27.1	17.0	35.0	11.9	21.2
1–5	–	34.0	32.3	11.1	26.8	17.8	29.1	12.6	20.0
2–1	38.4	36.0	35.2	13.0	28.3	14.6	28.2	–	33.5
2–2	45.0	–	–	12.0	17.0	15.0	26.9	15.2	27.2
2–3	–	35.5	32.7	11.3	15.5	16.2	19.7	14.6	31.5
2–4	37.6	38.9	32.7	12.3	24.5	14.6	22.9	11.4	17.6
2–5	32.3	38.4	34.5	15.1	24.5	18.0	22.9	13.1	25.8
ANOVA: *F/P*	3.54/0.02	0.75/0.65	0.98/0.46	3.20/0.0076	1.78/0.11	3.46/0.0046	0.85/0.57	1.87/0.11	4.56/0.0007
Dist.: *r/P*	–	0.26/0.51	0.11/0.78	0.12/0.76	0.27/0.48	0.69/0.04	0.01/0.98	–0.54/0.17	–0.38/0.31
Poll.: *r/P*	–	–0.16/0.69	0.16/0.67	0.11/0.78	0.21/0.59	–0.55/0.13	0.09/0.82	0.71/0.05	0.49/0.18

Sampling year shown in parentheses. For other explanations, consult Table 4.1.

Table 4.29 Leaf size (L, mm), shoot length (S, mm) of woody plants and height (H, mm) of herbaceous plants in the impact zone of the nickel-copper smelters at Norilsk, Russia

	Betula nana		*Empetrum nigrum*	*Larix sibirica*	*Salix lanata*		*Vaccinium uliginosum*		*Vaccinium vitis-idaea*	
Site	L (2002)	S (2002)	S (2002)	S (2002)	L (2002)	S (2002)	L (2002)	S (2002)	L (2002)	S (2002)
1–1	9.3	13.4	6.2	51.4	–	23.1	11.7	9.2	12.8	14.2
1–2	12.1	34.1	10.5	57.3	50.2	34.0	11.7	9.8	12.8	11.4
1–3	13.2	43.7	12.5	41.4	54.6	30.9	10.9	10.3	12.4	14.1
1–4	13.6	42.2	18.3	78.9	55.9	37.2	12.2	13.2	12.9	15.5
1–5	14.2	41.9	14.6	75.6	51.2	48.0	12.7	19.4	14.2	17.8
2–1	8.6	12.0	13.0	35.3	45.9	19.9	13.3	5.6	9.7	11.3
2–2	11.9	20.0	20.8	51.1	53.3	27.7	14.6	8.7	12.8	28.8
2–3	15.8	33.8	24.8	90.3	60.5	50.1	12.9	11.1	16.2	23.5
2–4	13.0	40.3	13.5	61.3	45.1	26.6	10.4	14.8	14.5	14.2
2–5	13.6	49.8	18.4	63.2	52.9	33.9	12.3	17.3	16.1	21.9
ANOVA: *F/P*	7.30/<0.0001	7.51/<0.0001	6.93/<0.0001	9.47/<0.0001	3.35/0.0059	3.91/0.0012	3.15/0.0058	4.54/0.0004	3.63/0.0023	5.01/0.0002
Dist.: *r/P*	0.81/0.0044	0.96/<0.0001	0.44/0.20	0.57/0.09	0.28/0.46	0.62/0.05	−0.22/0.54	0.85/0.0017	0.64/0.05	0.21/0.55
Poll.: *r/P*	−0.80/0.006	−0.91/0.0002	−0.46/0.18	−0.45/0.19	−0.28/0.47	−0.57/0.08	0.22/0.54	−0.65/0.04	−0.48/0.16	−0.14/0.71

Sampling year shown in parentheses. For other explanations, consult Table 4.1.

Table 4.30 Leaf/needle size (L, mm), shoot length (S, mm) of woody plants and height (H, mm) of herbaceous plants in the impact zone of the copper smelter at Revda, Russia

	Abies sibirica	*Betula pubescens*		*Equisetum silvaticum*	*Linaria vulgaris*	*Picea abies*	*Pinus sylvestris*		*Populus tremula*		*Vaccinium vitis-idaea*	
Site	S (2007)	L (2003)	S (2003)	H (2007)	H (2007)	S (2003)	L (2003)	S (2003)	L (2007)	S (2007)	L (2003)	S (2003)
1–1	36.9	43.3	59.5	373	444	39.7	34.2	36.5	38.9	35.8	19.7	26.5
1–2	58.3	48.4	40.4	–	–	42.7	39.8	40.2	48.4	59.3	–	–
1–3	55.7	54.2	72.6	–	–	39.1	40.0	79.9	48.4	5.9	–	–
1–4	60.8	53.3	90.2	–	–	60.5	40.6	37.4	44.6	45.1	19.5	41.9
1–5	52.7	47.3	75.2	447	959	49.2	37.5	65.8	43.9	41.5	19.3	50.3
2–1	64.9	42.4	79.8	445	497	45.2	42.9	44.5	43.9	8.7	22.0	39.4
2–2	63.2	47.2	81.1	–	–	50.0	42.0	37.4	47.0	55.2	22.7	49.5
2–3	53.5	50.4	53.6	–	–	42.3	40.9	63.0	43.3	8.8	20.5	30.8
2–4	62.9	52.0	83.4	–	–	39.6	42.4	31.4	43.6	22.9	22.7	52.6
2–5	58.1	53.6	82.6	598	911	41.9	45.0	42.2	43.9	92.4	23.2	36.7
ANOVA:												
F/*P*	2.66/0.02	2.51/0.02	1.68/0.13	15.6/<0.0001	29.9/<0.0001	1.77/0.10	1.37/0.23	4.37/0.0005	1.23/0.30	3.98/0.0011	2.67/0.03	5.53/0.0003
Dist.: *r*/*P*	0.19/0.59	0.73/0.02	0.40/0.25	–	–	0.27/0.44	0.28/0.44	0.19/0.61	0.12/0.75	0.39/0.26	−0.05/0.90	0.40/0.33
Poll.: *r*/*P*	−0.33/0.35	−0.80/0.0059	−0.22/0.54	–	–	−0.21/0.57	−0.28/0.43	−0.24/0.50	−0.46/0.18	−0.31/0.39	0.01/0.98	−0.52/0.18

Sampling year shown in parentheses. For other explanations, consult Table 4.1.

Table 4.31 Leaf size (L, mm), shoot length (S, mm) of woody plants and height (H, mm) of herbaceous plants in the impact zone of the aluminium smelter at Straumsvík, Iceland

	Empetrum nigrum		*Juncus trifidus*	*Luzula spicata*	*Salix herbacea*			*Vaccinium uliginosum*	
Site	S (2002)	S (2007)	H (2007)	H (2007)	L (2002)	S (2002)	S (2007)	L (2002)	S (2002)
1–1	4.6	73.2	153	292	12.7	11.4	6.9	14.8	30.0
1–2	6.3	77.4	–	–	12.8	9.2	8.1	13.1	19.0
1–3	5.1	86.0	–	–	15.0	12.4	10.4	13.3	15.8
1–4	4.8	127.4	–	–	13.0	9.4	6.4	12.3	18.6
1–5	5.2	118.2	183	487	11.9	11.7	18.8	14.1	20.8
2–1	5.3	53.0	158	320	14.1	12.5	4.4	14.3	24.1
2–2	5.4	66.4	–	–	13.6	8.6	5.2	12.3	18.3
2–3	4.4	46.4	–	–	14.1	10.6	4.6	12.1	14.9
2–4	5.4	69.0	–	–	13.6	9.3	–	12.2	22.0
2–5	5.0	91.6	255	375	12.6	9.3	14.1	12.5	15.4
ANOVA: *F/P*	1.41/0.22	3.22/0.005	10.9/<0.0001	8.73/0.0002	0.88/0.55	0.32/0.96	4.17/0.0013	1.03/0.43	2.34/0.03
Dist.: *r/P*	–0.07/0.85	0.65/0.04	–	–	–0.47/0.18	–0.21/0.56	0.84/0.005	–0.32/0.36	–0.48/0.16
Poll.: *r/P*	–0.35/0.32	–0.22/0.54	–	–	–0.14/0.71	0.27/0.45	–0.31/0.42	0.64/0.05	0.83/0.0032

Sampling year shown in parentheses. For other explanations, consult Table 4.1.

Table 4.32 Leaf/needle size (L, mm) and shoot length (S, mm) in the impact zone of the nickel-copper smelter at Sudbury, Canada

	Betula papyrifera		*Pinus resinosa*		*Populus tremuloides*		*Vaccinium angustifolium*	
Site	L (2007)	S (2007)	L (2002)	S (2002)	L (2007)	S (2007)	L (2007)	S (2007)
1–1	72.2	207.8	–	–	40.1	41.4	–	–
1–2	48.9	148.7	126.2	126.7	44.7	163.8	22.2	42.5
1–3	78.3	151.5	121.2	43.9	53.8	79.6	16.1	17.2
1–4	49.4	168.9	97.0	58.8	52.7	117.2	20.8	39.7
1–5	77.5	168.1	102.7	27.8	54.9	82.7	21.3	19.3
2–1	78.7	185.6	–	–	57.8	33.7	–	–
2–2	67.4	132.4	118.7	94.6	53.3	16.5	32.9	38.7
2–3	60.7	133.8	110.5	139.7	43.2	64.1	19.0	30.6
2–4	50.7	87.5	96.0	58.6	34.5	24.1	19.1	16.5
2–5	75.5	130.1	79.7	74.5	49.2	67.2	21.0	20.6
ANOVA: *F/P*	9.79/<0.0001	4.33/0.0005	6.67/<0.0001	6.92/<0.0001	4.57/0.0006	2.85/0.01	6.17/0.0002	11.0/<0.0001
Dist.: *r/P*	–0.14/0.70	–0.55/0.10	–0.91/0.0018	–0.54/0.17	–0.16/0.66	0.18/0.63	–0.41/0.32	–0.66/0.07
Poll.: *r/P*	0.27/0.45	0.53/0.11	0.65/0.08	0.41/0.31	0.18/0.61	–0.36/0.30	0.84/0.0086	0.62/0.10

Sampling year shown in parentheses. For other explanations, consult Table 4.1.

Table 4.33 Leaf/needle size (L, mm), shoot length (S, mm) of woody plants and height (H, mm) of herbaceous plants in the impact zone of the aluminium smelter at Volkhov, Russia

	Achillea millefolium		*Betula pubescens*		*Picea abies*	*Pinus sylvestris*	*Salix caprea*		*Tanacetum vulgare*	
Site	L (2007)	H (2007)	L (2002)	S (2002)	S (2002)	L (1998)	L (2002)	S (2002)	L (2007)	H (2007)
1–1	72.1	168.1	44.5	134.3	34.5	–	63.2	63.4	202.4	1021
1–2	–	–	49.4	109.6	51.0	–	69.2	57.0	–	–
1–3	–	–	56.6	142.4	–	–	79.5	98.4	–	–
1–4	–	–	52.0	128.0	45.2	53.7	71.6	101.0	–	–
1–5	58.9	187.6	48.6	123.7	50.6	42.0	61.8	74.2	188.8	1172
2–1	72.7	172.1	51.2	82.6	59.2	58.3	76.9	69.9	252.3	1024
2–2	–	–	51.4	140.5	41.7	–	75.2	65.7	–	–
2–3	–	–	47.9	142.6	60.2	56.8	80.0	65.0	–	–
2–4	–	–	49.3	149.8	38.9	50.6	66.0	62.7	–	–
2–5	57.7	186.3	32.8	90.7	41.8	51.6	64.0	50.2	205.3	977
ANOVA: *F*/*P*	3.37/0.03	0.21/0.89	6.99/<0.0001	2.24/0.04	1.76/0.13	4.59/0.04	2.16/0.05	2.94/0.0093	6.67/0.0011	2.89/0.05
Dist.: *r*/*P*	–	–	–0.33/0.35	0.31/0.38	–0.30/0.44	–0.68/0.14	–0.44/0.20	0.04/0.92	–	–
Poll.: *r*/*P*	–	–	0.19/0.59	–0.51/0.13	0.38/0.32	0.56/0.25	0.36/0.31	–0.10/0.78	–	–

Sampling year shown in parentheses. For other explanations, consult Table 4.1.

Table 4.34 Leaf size (L, mm) and shoot length (S, mm) in the impact zone of the power plant at Vorkuta, Russia

	Betula nana		*Salix glauca*		*Vaccinium vitis-idaea*	
Site	L (2001)	S (2001)	L (2001)	S (2001)	L (2001)	S (2001)
1–1	11.7	28.9	32.1	34.8	9.0	13.0
1–2	9.6	16.8	29.6	24.6	10.5	13.3
1–3	10.0	12.7	25.6	31.2	9.1	10.9
1–4	9.6	18.1	31.4	37.2	10.0	15.2
1–5	11.5	23.0	28.9	28.0	10.5	22.8
2–1	12.1	27.0	27.9	19.0	11.4	14.9
2–2	10.2	17.5	33.1	28.2	9.3	12.9
2–3	10.0	18.1	28.1	33.4	9.0	16.4
2–4	8.8	15.5	31.9	30.8	9.6	17.3
2–5	8.9	16.8	31.3	33.0	8.5	13.3
ANOVA: *F*/*P*	4.51/0.0004	3.64/0.0021	2.26/0.04	2.18/0.04	4.78/0.0002	8.63/<0.0001
Dist.: *r*/*P*	−0.67/0.03	−0.60/0.07	0.13/0.72	0.54/0.11	−0.41/0.24	0.41/0.24
Poll.: *r*/*P*	0.30/0.40	0.12/0.75	−0.38/0.28	0.76/0.01	0.51/0.13	−0.24/0.50

Sampling year shown in parentheses. For other explanations, consult Table 4.1.

Site-specific values of shoot length of the same species measured during 2 different years correlated with each other ($z_r = 0.34$, CI = 0.02…0.66, $N = 8$), indicating repeatability of our results.

The magnitude of the pollution effect on the shoot length of woody plants did not depend on the method used to calculate ES ($Q_B = 1.95$, df = 2, $P = 0.38$). Therefore, the analyses of herbaceous plants, as well as of pooled data, are based on Hedge's d values because data on herbs were collected only from the most and least polluted study sites. However, in the analysis of woody plants we employed ESs based on correlations with pollution, in line with all other characteristics analysed in this book.

In general, shoot length (including height of herbaceous plants) decreased with pollution (Fig. 4.13). This effect was pronounced in Angiosperms, while Gymnosperms showed no response to pollution (Fig. 4.13; $Q_B = 0.22$, df = 1, $P = 0.64$). Changes in shoot length did not depend on plant life form ($Q_B = 1.65$, df = 2, $P = 0.44$); evergreen and deciduous woody plants responded similarly to pollution ($Q_B = 0.38$, df = 1, $P = 0.54$) (Fig. 4.13). We found no differences among plants belonging to four Raunkiaer life forms (listed in Section 4.3.2) ($Q_B = 3.17$, df = 3, $P = 0.17$).

Effect sizes (averaged by plant species) did not correlate with any of the six Ellenberg's indicator values listed in Section 4.3.2 ($r_S = 0.10…0.31$, N = 21–36 species, $P = 0.14…0.55$). Similarly, we found no correlation with axis scores for Grime's CSR strategy ($r_S = −0.34…0.44$, $N = 14$ species, $P = 0.11…0.97$).

In woody plants (Fig. 4.14), the effect did not depend on polluter type ($Q_B = 2.23$, df = 4, $P = 0.69$), pollution effects on soil pH ($Q_B = 0.42$, df = 2, $P = 0.81$), or geographical position of polluters ($Q_B = 0.001$, df = 1, $P = 0.95$). Individual polluters differed in their effects on shoot length ($Q_B = 45.1$, df = 17, $P = 0.0002$).

Table 4.35 Leaf size (L, mm), shoot length (S, mm) of woody plants in the impact zone of the aluminium smelter at Žiar nad Hronom, Slovakia

	Carpinus betulus			*Fagus sylvatica*				*Quercus petraea*	
Site	L (2002)	S (2002)	S (2004)	L (2002)	S (2002)	S (2004)	S (2006)	L (2002)	S (2002)
1–1	42.1	11.6	13.8	65.1	40.0	12.6	43.2	82.1	22.0
1–2	49.1	10.4	34.3	70.9	49.1	42.0	7.8	92.1	23.7
1–3	57.9	13.8	49.4	77.4	61.1	41.0	22.3	–	–
1–4	58.9	26.2	–	75.7	44.1	73.1	40.6	89.8	16.5
1–5	54.8	26.4	50.3	81.0	60.3	40.3	30.8	–	–
2–1	67.1	23.7	33.1	82.4	47.2	54.8	2.2	83.7	42.7
2–2	56.5	12.4	18.0	69.2	40.6	39.2	26.7	95.9	32.1
2–3	52.0	14.7	41.0	72.9	41.4	65.6	27.6	–	–
2–4	56.3	14.8	26.5	70.3	20.8	34.2	8.9	–	–
2–5	60.6	18.3	84.1	77.4	50.1	28.5	26.2	87.0	47.1
ANOVA: *F/P*	2.98/0.0085	2.06/0.06	4.27/0.0011	3.50/0.0028	1.90/0.08	3.14/0.0059	0.86/0.57	1.03/0.42	1.49/0.24
Dist.: *r/P*	0.44/0.20	0.51/0.13	0.59/0.09	0.39/0.26	–0.03/0.93	0.30/0.40	0.19/0.60	0.33/0.53	0.31/0.55
Poll.: *r/P*	–0.78/0.0082	–0.49/0.15	–0.47/0.20	–0.61/0.06	–0.05/0.89	–0.55/0.10	0.22/0.53	–0.40/0.43	–0.50/0.31

Sampling year shown in parentheses. For other explanations, consult Table 4.1.

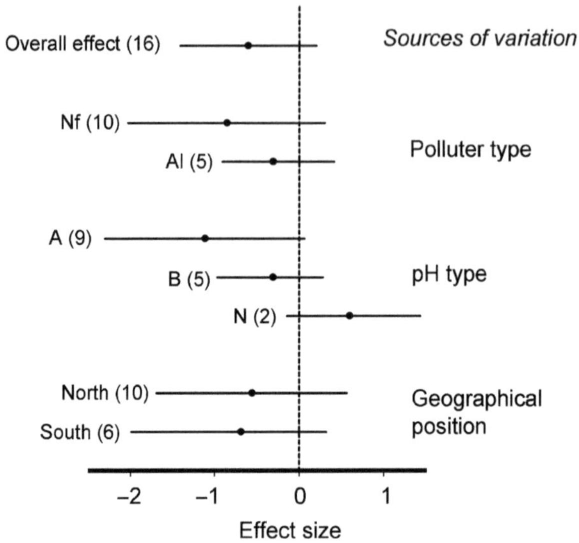

Fig. 4.11 Overall effect and sources of variation in the responses of leaf length of herbaceous plants (including height of herbaceous plants) to pollution. Effect sizes are Hedge's d based on comparison of two most polluted and two control sites. For explanations, consult Fig. 4.1

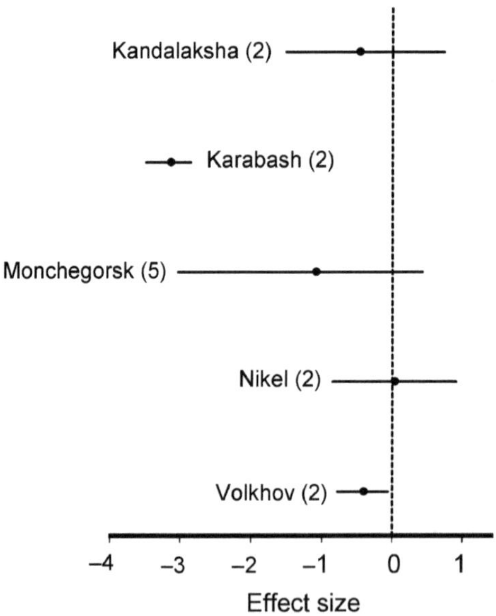

Fig. 4.12 Effects of individual polluters on the leaf length of herbaceous plants. Effect sizes are Hedge's d based on comparison of two most polluted and two control sites. For explanations, consult Fig. 4.1

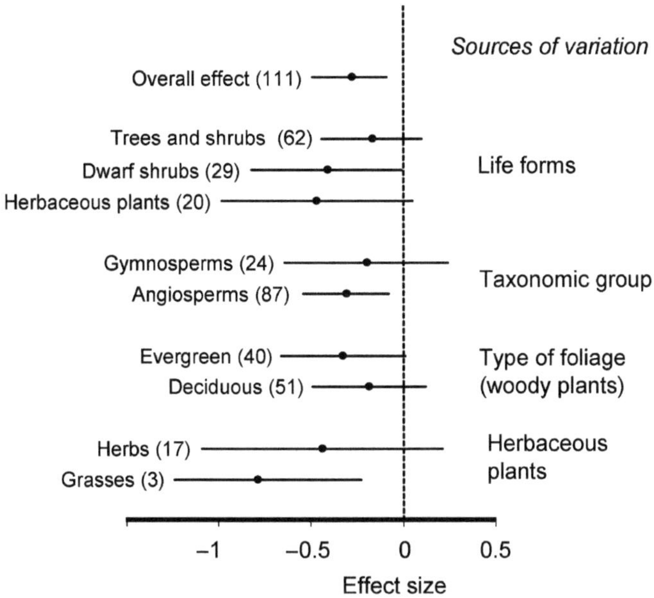

Fig. 4.13 Overall effect and sources of variation in the responses of shoot length of vascular plants (including height of herbaceous plants). Effect sizes are Hedge's d based on comparison of the two most polluted and two control sites. For explanations, consult Fig. 4.1

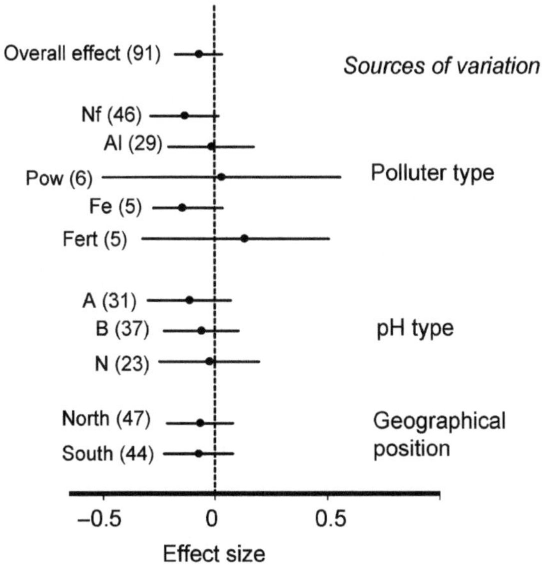

Fig. 4.14 Overall effect and sources of variation in the responses of shoot length of woody plant. For explanations, consult Fig. 4.1

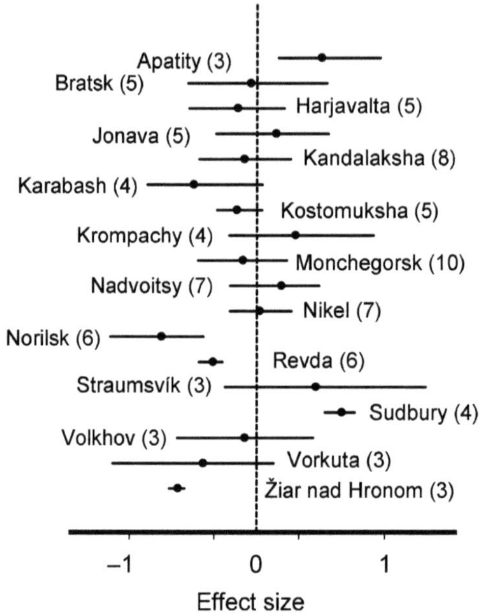

Fig. 4.15 Effects of individual polluters on the shoot length of woody plants. For explanations, consult Fig. 4.1

Three polluters caused significant negative effects (Norilsk, Revda, and Žiar nad Hronom), and two polluters caused positive effects (Apatity and Sudbury); the effects of other polluters were not significant (Fig. 4.15). Individual species of woody plants responded similarly to pollution ($Q_B = 12.8$, df = 13, $P = 0.46$); a significant decrease in shoot length was detected only in Siberian larch, *Larix sibirica* (Fig. 4.16).

Only non-ferrous smelters caused a decrease in the height of herbaceous plants (Fig. 4.17), although the difference between non-ferrous and aluminium smelters was not significant ($Q_B = 2.66$, df = 1, $P = 0.10$). Correspondingly, the effects depended on the pollution impact on soil pH ($Q_B = 5.61$, df = 2, $P = 0.06$): herbs were smaller only around acidifying polluters (Fig. 4.17). Pollution's effects on growth of herbaceous plants were independent of the geographical position of the polluters (Fig. 4.17; $Q_B = 0.09$, df = 1, $P = 0.77$). Among individual polluters ($Q_B = 6.66$, df = 5, $P = 0.24$), effects of three non-ferrous smelters were negative, while aluminium smelters caused both negative (Straumsvík) and positive (Kandalaksha and Volkhov) effects (Fig. 4.18).

Pollution effects on shoot length and on leaf/needle size did not differ ($Q_B = 0.14$, df = 1, $P = 0.71$) when calculated for the same data sets (species by polluter) and positively correlated to each other ($r = 0.44$, $N = 70$, $P = 0.0001$). Individual polluters imposed similar effects on these two vitality indices ($r = 0.66$, $N = 18$, $P = 0.0027$).

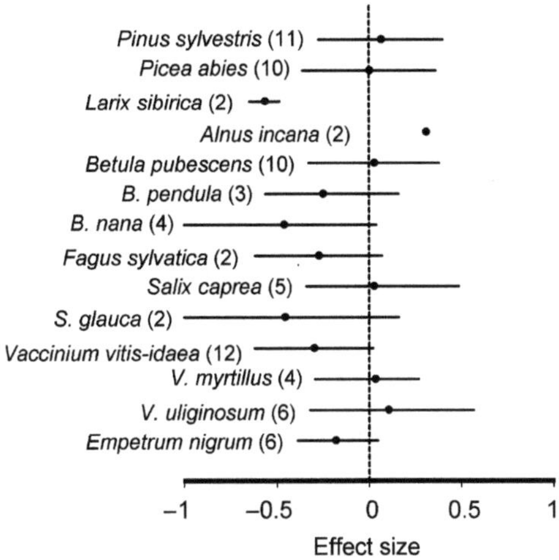

Fig. 4.16 Effects of point polluters on the shoot length of individual species of woody plants. For explanations, consult Fig. 4.1

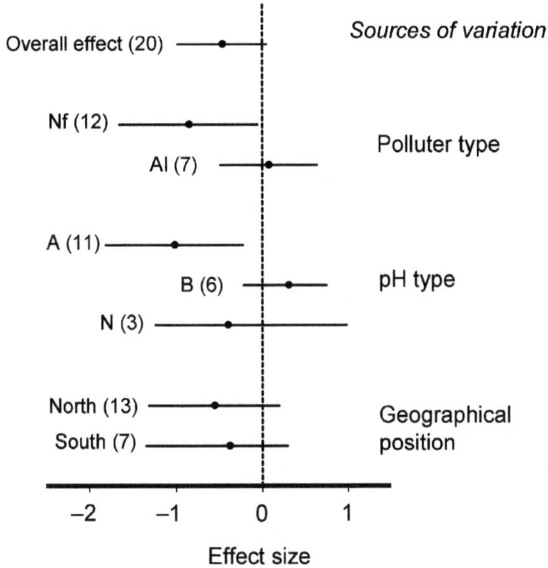

Fig. 4.17 Overall effect and sources of variation in the responses of height (equivalent to shoot length) of herbaceous plants. Effect sizes are Hedge's d based on comparison of two most polluted and two control sites. For explanations, consult Fig. 4.1

We have detected significant non-linear responses in ten of 100 data sets on woody plants (three dome-shaped and seven U-shaped), which is twice as high as the number of dome-shaped patterns that may be expected to occur by chance.

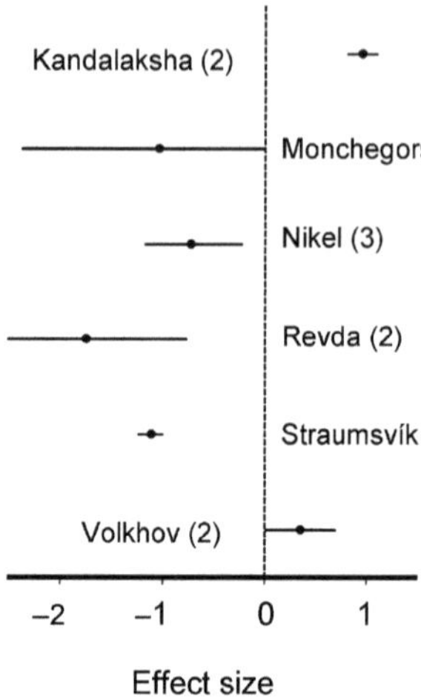

Fig. 4.18 Effects of individual polluters on height (equivalent to shoot length) of herbaceous plants. For explanations, consult Fig. 4.1

4.3.4 *Radial Growth*

Variation in tree ring width between study sites, in spite of small sample sizes, was significant (or nearly significant: $P = 0.06$) in five of ten data sets (Table 4.36). Radial increment in polluted sites tended to be lower than in clean sites (Fig. 4.19: $d = -0.77$, CI = −1.55…0.02, $N = 10$), although the effect did not reach significance. The effect did not depend on the polluter type ($Q_B = 2.50$, df = 1, $P = 0.11$) or its effects on soil pH ($Q_B = 2.51$, df = 2, $P = 0.29$), although significant decreases in radial increment were observed only around acidifying polluters (Fig. 4.19). Similarly, although the differences between Northern and Southern polluters were not significant ($Q_B = 0.49$, df = 1, $P = 0.50$), adverse effects were observed only around the Northern polluters (Fig. 4.19).

Changes in radial growth of Scots pine ($d = -0.72$, CI = −1.56…0.13, $N = 9$) tended to be larger than changes in shoot length around the same polluters ($d = -0.24$, CI = −1.03…0.55, $N = 9$), but the difference was not significant ($Q_B = 0.91$, df = 1, $P = 0.34$). We found no correlation between polluter-specific effects on radial growth and shoot length ($r = -0.03$, $N = 9$, $P = 0.94$).

4.3.5 *Needle Longevity in Conifers*

Variation in needle longevity between study sites was significant in all 29 data sets; however, only 28 of 58 correlation coefficients (with both distance and pollution load) appeared significant (Tables 4.37 and 4.38).

Table 4.36 Width (mm) of annual rings of Scots pine (*Pinus sylvestris*), averaged for the past 10 years prior sampling

Sites	Apatity[a]	Harjavalta[b]	Jonava	Kandalaksha	Karabash[c]	Krompachy	Monchegorsk[d]	Nadvoitsy	Nikel[e]	Revda
	2006	2007	2005	2007	2007	2002	2005	2005	2007	2007
1–1	1.63 (60)	0.61 (43)	1.83 (31)	0.59 (49)	0.64 (130)	3.50 (33)	0.42 (47)	1.63 (49)	0.72 (29)	1.47 (59)
1–5	1.83 (49)	2.09 (42)	2.17 (35)	1.57 (52)	0.55 (134)	4.01 (28)	1.06 (38)	1.79 (60)	0.50 (35)	1.44 (65)
2–1	1.22 (59)	1.19 (45)	1.95 (39)	0.68 (54)	1.29 (59)	2.32 (38)	0.62 (32)	1.72 (44)	0.22 (48)	0.98 (63)
2–5	1.88 (56)	1.38 (52)	1.42 (51)	1.75 (48)	1.08 (61)	1.94 (51)	0.90 (44)	1.86 (59)	1.29 (41)	1.55 (63)
ANOVA: *F*/*P*	2.25/0.16	6.52/0.02	2.47/0.14	10.3/0.0041	3.63/0.06	4.36/0.04	2.17/0.17	0.23/0.87	4.87/0.04	0.62/0.62

Plot-specific means are each based on three tree-specific values; mean age of sampled trees is given in parentheses; sampling year shown under the locality name.

[a] Sampled sites: 1–2, 1–5, 2–1, 2–4.

[b] Sampled sites: 1–1, 1–5, 1–2, 2–4.

[c] Sampled sites: 1–2, 1–5, 2–2, 2–5.

[d] Sampled sites: 1–2, 2–9, 2–4, 2–10.

[e] Sampled sites: 2–1, 2–4, 2–2, 2–5.

For other explanations, consult Table 4.1.

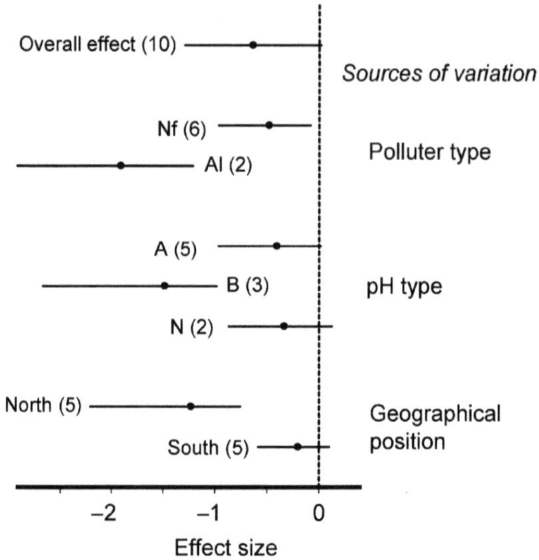

Fig. 4.19 Overall effect and sources of variation in the responses of radial increment of Scots pine (*Pinus sylvestris*). Effect sizes are Hedge's d based on comparison of two most polluted and two control sites. For explanations, consult Fig. 4.1

To explore the repeatability of the results, we estimated the needle longevity of Siberian/Norway spruce needles twice around Harjavalta, Kandalaksha and Monchegorsk, and three times around Krompachy (Table 4.38). Needle longevity of Scots pine was estimated twice around Harjavalta and Monchegorsk (Table 4.37). The measurements conducted in different years strongly correlated with each other ($z_r = 0.95$, CI = 0.51…1.39, $N = 7$), demonstrating high repeatability of needle longevity estimates. Repeatability was equally high in both species ($Q_B = 0.09$, df = 1, $P = 0.92$).

Pollution generally caused a decrease in needle longevity (Fig. 4.20); this result did not depend on the method used to calculate ES ($Q_B = 1.08$, df = 2, $P = 0.58$). Pollution effect on needle longevity did not differ between Siberian/Norway spruce and Scots pine ($Q_B = 0.56$, df = 1, $P = 0.45$), allowing us to combine all species in further analyses.

Individual polluters differ in their effects on needle longevity (Fig. 4.21; $Q_B = 18.3$, df = 8, $P = 0.02$) from significant decreases (around seven of nine polluters) to significant increases with pollution (near the fertilising factory at Jonava). This variation was not linked with either the type of the polluter (Fig. 4.20; $Q_B = 6.23$, df = 3, $P = 0.10$) or with the changes in soil pH (Fig. 4.20; $Q_B = 3.72$, df = 1, $P = 0.16$). However, the significant negative effects of acidifying and alkalysing polluters differed (Fig. 4.20; $Q_B = 3.85$, df = 1, $P = 0.05$) from non-significant effects of polluters whose impact did not change soil pH. The Northern polluters negatively

Table 4.37 Needle longevity (years) in pines[a]

	Bratsk	Harjavalta		Jonava	Kandalaksha	Karabash	Kostomuksha	Krompachy	Monchegorsk		Nadvoitsy	Nikel	Revda	Sudbury
Site[b]	2002	2001	2007	2004	2001	2003	2002	2002	2001	2008	2004	2003	2003	2002
1–1	5.45	2.40	3.10	3.75	5.80	–	5.25	3.45	3.05	3.55	4.85	–	4.61	–
1–2	6.35	2.65	2.85	3.40	6.40	4.70	5.29	4.00	3.15	3.60	5.65	–	4.50	3.40
1–3	6.15	2.85	2.75	3.55	6.80	5.15	5.00	3.95	4.10	5.80	5.65	–	3.90	4.62
1–4	6.10	3.00	2.60	3.35	7.65	4.10	5.72	3.50	4.75	6.75	5.90	–	4.55	5.45
1–5	6.80	2.85	3.35	3.25	6.85	5.15	4.75	3.15	2.25	3.80	5.70	–	3.95	3.95
2–1	4.05	2.83	3.75	2.67	5.25	–	4.77	4.00	2.40	3.25	5.10	3.85	4.20	–
2–2	6.00	3.65	3.20	3.58	5.60	4.50	5.05	3.30	3.85	5.50	5.05	4.80	4.30	3.40
2–3	6.05	3.95	3.55	3.70	6.95	5.25	5.65	3.92	4.80	6.85	5.20	5.85	3.75	4.05
2–4	6.15	2.65	2.65	3.30	6.50	5.00	4.85	3.30	5.45	6.35	5.65	5.30	5.10	3.90
2–5	6.05	2.90	3.00	3.40	6.70	4.55	4.45	2.90	5.10	6.61	4.20	6.25	4.50	4.10
ANOVA: *F/P*	6.39/ <0.0001	7.78/ <0.0001	3.42/ 0.0013	6.60/ <0.0001	4.47/ <0.0001	2.50/ 0.02	4.27/ <0.0001	3.57/ 0.0009	21.4/ <0.0001	35.8/ <0.0001	4.35/ <0.0001	12.6/ <0.0001	3.89/ 0.0003	5.20/ 0.0005
Dist.: *r/P*	0.62/ 0.05	–0.12/ 0.73	–0.30/ 0.40	0.17/ 0.63	0.78/ 0.0072	–0.01/ 0.99	–0.13/ 0.72	–0.47/ 0.17	0.81/ 0.0046	0.85/ 0.0019	0.22/ 0.53	0.92/ 0.02	0.06/ 0.87	0.29/ 0.49
Poll.: *r/P*	–0.93/ <0.0001	–0.19/ 0.60	–0.04/ 0.91	0.25/ 0.48	–0.71/ 0.02	–0.10/ 0.81	–0.12/ 0.74	0.42/ 0.23	–0.86/ 0.0012	–0.75/ 0.01	–0.17/ 0.64	–0.89/ 0.04	–0.07/ 0.84	–0.53/ 0.17

[a] Scots pine (*Pinus sylvestris*) in all localities except for red pine (*Pinus resinosa*) in Sudbury; sampling year shown under the locality name.
[b] For all polluters except for Monchegorsk, where the sites are ordered as follows: 1–1, 1–2, 1–4, 1–5, 2–3, 2–4, 2–5, 2–8, 2–9, 2–10.
For other explanations, consult Table 4.1.

Table 4.38 Needle longevity (years) in spruces[a]

	Bratsk	Harjavalta		Jonava	Kandalaksha		Kostomuksha	Krompachy			Monchegorsk		Nadvoitsy	Revda	Volkhov
Site[b]	2002	2001	2007	2004	2001	2007	2002	2002	2004	2006	2001	2005	2004	2003	2002
1–1	9.90	–	–	–	10.65	8.75	13.35	5.44	6.25	5.65	4.10	5.80	7.95	8.60	8.75
1–2	10.60	–	6.83	8.75	10.15	10.65	10.55	9.10	8.15	6.60	3.65	7.90	11.15	8.95	5.00
1–3	12.60	7.15	8.35	9.20	14.10	12.45	11.30	8.00	8.16	7.25	8.25	7.65	13.55	10.50	6.17
1–4	12.80	7.56	8.80	7.10	13.95	13.65	11.00	9.05	8.70	8.00	9.40	10.95	13.70	10.30	8.85
1–5	14.00	9.59	8.70	9.55	12.60	13.95	14.55	8.90	9.05	10.10	4.15	6.65	14.95	11.40	9.20
2–1	5.20	–	–	–	7.67	8.00	10.00	6.75	5.45	6.80	6.55	6.35	8.40	6.20	10.00
2–2	9.40	8.83	9.88	8.50	8.62	7.00	12.35	6.50	7.05	6.00	7.50	7.25	9.45	8.20	8.89
2–3	11.00	10.80	10.55	10.60	11.05	11.70	10.65	9.65	9.65	7.45	10.55	8.90	10.60	9.20	8.25
2–4	14.20	11.35	11.35	10.65	13.90	13.05	11.25	7.05	9.90	7.50	9.70	10.95	13.20	11.05	8.75
2–5	13.05	11.65	8.20	7.85	13.10	12.65	14.00	9.10	10.30	7.95	10.80	10.20	14.10	10.35	8.25
ANOVA:	18.2/	9.53/	5.43/	4.37/	8.17/	5.16/	8.38/	7.49/	13.6/	6.72/	27.8/	15.2/	15.2/	9.95/	2.36/
F/P	<0.0001	<0.0001	0.0001	<0.0001	<0.0001	<0.0001	<0.0001	<0.0001	<0.0001	<0.0001	<0.0001	<0.0001	<0.0001	<0.0001	0.02
Dist.:	0.84/	0.33/	−0.02/	0.01/	0.80/	0.84/	0.39/0.27	0.74/	0.79/	0.84/	0.91/	0.80/	0.95/	0.86/	−0.04/
r/P	0.0022	0.47	0.97	0.98	0.0054	0.0019		0.02	0.0060	0.0023	0.0002	0.0058	<0.0001	0.0013	0.91
Poll.:	−0.89/	−0.15/	−0.53/	0.42/	−0.81/	−0.78/	−0.32/0.37	−0.38/	−0.76/	−0.14/	−0.72/	−0.70/	−0.85/	−0.83/	0.43/
r/P	0.0004	0.74	0.18	0.31	0.0050	0.0071		0.28	0.01	0.69	0.02	0.02	0.0017	0.0032	0.22

[a] Siberian spruce (*Picea abies* ssp. *obovata*) in all localities except for Norway spruce (*Picea abies* ssp. *abies*) in Jonava and Krompachy; sampling year shown under the locality name.

[b] For all polluters except for Monchegorsk, where the sites are ordered as follows: 1–1, 1–2, 1–4, 1–5, 2–3, 2–4, 2–5, 2–8, 2–9, 2–10.

For other explanations, consult Table 4.1.

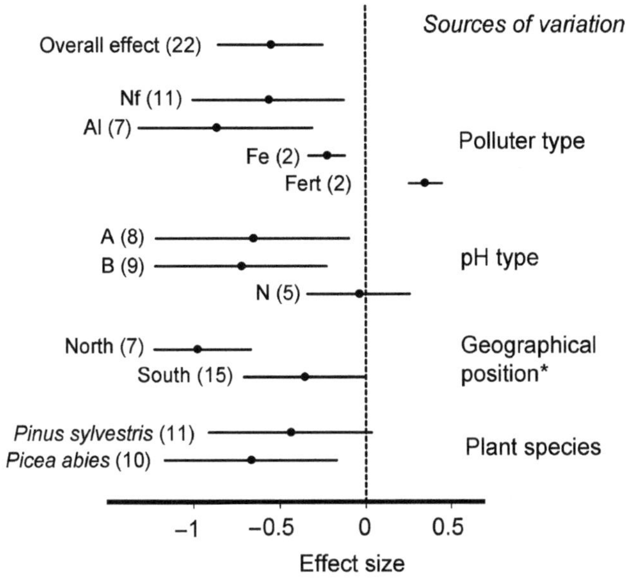

Fig. 4.20 Overall effect and sources of variation in the responses of needle longevity in coniferous plants. For explanations, consult Fig. 4.1

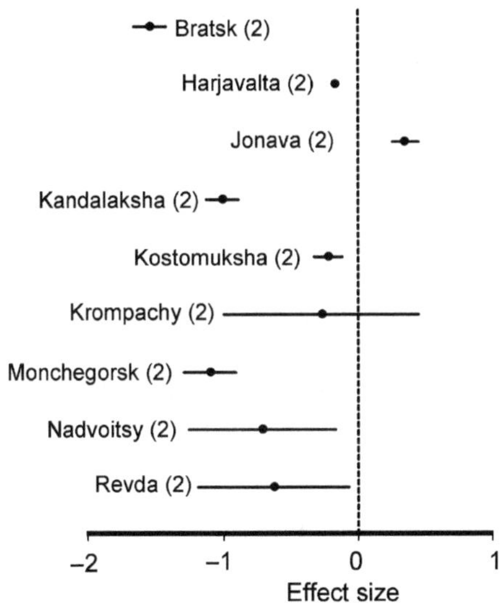

Fig. 4.21 Effects of individual polluters on needle longevity in coniferous plants. For explanations, consult Fig. 4.1

affected needle longevity, whereas southern polluters did not cause any effect (Fig. 4.20; $Q_B = 4.48$, df = 1, $P = 0.03$).

Pollution effects on needle longevity in pines and spruces were larger than the effects on needle size ($Q_B = 4.34$, df = 1, $P = 0.04$) and shoot length ($Q_B = 9.73$, df = 1, $P = 0.002$) when calculated for the same data sets (species by polluter). We found no correlation between polluter-specific effects on needle longevity and either shoot length ($r = -0.14$, $N = 21$, $P = 0.54$) or needle size ($r = -0.11$, $N = 11$, $P = 0.75$).

Only three of 27 data sets were better fitted by the second-order (dome-shaped) function than by the linear model.

4.4 Discussion

4.4.1 Overall Effects of Pollution on Vitality Indices

4.4.1.1 Chlorophyll Fluorescence

Negative effects of different pollutants, like fluorine, sulphur dioxide and heavy metals, on photosynthesis of several species have been detected both in experiments (Snel et al. 1991; Strand 1993; Cook et al. 1997; Łukaszek & Poskuta 1998; Ouzounidou et al. 2006) and in field studies using plants naturally growing in polluted areas (Saarinen 1993; Odasz-Albrigtsen et al. 2000; Andreucci et al. 2006). On the other hand, some researchers did not detect adverse effects of pollution on chlorophyll fluorescence either in experiments (Boese et al. 1995; Sahi et al. 2007) or in field conditions (Lepedus et al. 2005; Zvereva & Kozlov 2005; Divan et al. 2007). The factors contributing to discrepancies between studies have not, to our knowledge, been identified.

The absence of repeatability in multiyear measurements conducted on the same birch trees around Monchegorsk (Table 4.1) is indeed frustrating. This result demonstrated that environmental variables (some of which have not been controlled in the course of our study) can substantially modify responses of the photosynthetic system to pollution. In light of this information, the significant variation between both study sites (Tables 4.1–4.17) and individual polluters (Figs. 4.1 and 4.4) is difficult to interpret. While this variation may reflect more or less stabile differences between study sites (e.g., in soil contamination and nutritional quality, as well as in leaf area index), it may also result from short-term variation (e.g., in temperature, illumination, and soil moisture at the time of sampling). All of these factors are known to influence photosynthesis (Mohammed et al. 1995; Martinez-Carrasco et al. 2005; Qaderi et al. 2006; Kitao et al. 2007), making interpretation of observational data collected at multiple plots a difficult task. Thus, it is not surprising that we failed to confirm the negative effect of pollution on F_v/F_m in the vicinity of Nikel that was detected by Odasz-Albrigtsen et al. (2000). On the other hand, the absence of a pollution effect on F_v/F_m in Scots pine near Revda is in line with the results of Shavnin et al. (1997).

Meta-analysis revealed no pollution effect on F_v/F_m, suggesting that the significant decline of this index in plants growing in polluted habitats reported in earlier studies (Saarinen 1993; Saarinen & Liski 1993; Odasz-Albrigtsen et al. 2000) can be observed only under specific environmental conditions. The detected absence of F_v/F_m changes around industrial polluters is in line with the conclusion by Bussotti et al. (2008), who found that this index is less sensitive to ozone exposure than other parameters of chlorophyll fluorescence. More generally, F_v/F_m is considered most suitable for evaluation of plant responses to short-term impacts, primarily to temperature extremes (Lichtenthaler & Rinderle 1988; Sayed 2003), while its applicability for exploration of consequences of chronic stress is questioned (Venediktov et al. 1999).

In contrast to F_v/F_m, the time needed for the leaf to reach half of its F_m ($T_{1/2}$) in our data sets increased with an increase in pollution load (Fig. 4.4), indicating a slowing down of the photochemical reaction. Differential responses of two indices of chlorophyll fluorescence to environmental variation have previously been detected in several studies addressing both abiotic and biotic stress on mountain birch (Eränen & Kozlov 2006, 2008). Although in environmental studies $T_{1/2}$ has been used less frequently than F_v/F_m, our results suggest that this parameter may be more informative, or less influenced by factors other than pollution, and therefore deserves more attention from environmental scientists. Other indices reflecting different aspects of the induction and attenuation of chlorophyll fluorescence (reviewed by Van Kooten & Snel 1990; Nesterenko et al. 2007) may also be useful in exploring the consequences of chronic impacts of industrial pollutants.

4.4.1.2 Leaf/Needle Size, Shoot Growth and Radial Increment

Surprisingly, the overall effect of pollution on leaf/needle size appeared non-significant (Fig. 4.7; $d = -0.22$, CI = −0.46…0.02, $N = 88$). This result strongly contrasts with a meta-analysis of published data (Roitto et al. 2009), which demonstrated a substantial decrease in leaf/needle size with pollution ($d = -1.08$, CI = −1.35…−0.80, $N = 204$). Adverse effects on shoot length detected from our data sets (Fig. 4.13; $d = -0.29$, CI = −0.49…−0.08, $N = 111$) better fit both the general theory and meta-analysis of published data, although the magnitude of the effect was about one third of that calculated from published studies ($d = -1.06$, CI = −1.27…−0.84, $N = 164$). Similarly, ES based on published data on radial increment (d = −1.45, CI = −2.08 ... −0.81; N = 40) was twice as large as ES based on original data (Fig. 4.19: $d = -0.77$, CI = −1.55…0.02, $N = 10$).

Of course, the two meta-analyses are based on data of different structure. For example, 50% of the original data on shoot length were collected around non-ferrous polluters, compared with 25% of the published data. Since non-ferrous smelters generally impose stronger effects on biota than other polluters (Kozlov &

Zvereva 2007a; Zvereva et al. 2008), we expected to find a stronger effect relative to the published data. Therefore, as in several other situations (Sections 6.4.3 and 7.4.1), we suggest that the detected differences result from both research and publication biases. The smallest difference between the original and published data was observed for radial increment, further supporting this conclusion since assessment of the width of annual rings (in contrast to leaf size or shoot length) is less susceptible to influence by unintentional non-random selection of study sites.

Since a meta-analysis of published data (Roitto et al. 2009) demonstrated that pollution similarly affected the size and weight of plant metamers (leaves/needles and shoots), the decrease in shoot length with pollution can be interpreted as a decline in biomass production. However, although our results supported the somewhat trivial (Scurfield 1960a, b; Odum 1985; Treshow & Anderson 1989; Armentano & Bennett 1992; Dobbertin 2005; Roitto & Kozlov 2007; Roitto et al. 2009) conclusion on adverse effects of pollution on plant growth, we demonstrated that this effect is usually overestimated.

4.4.1.3 Needle Longevity in Conifers

Our result of a significant decrease in needle longevity near industrial polluters generally agrees with the published data. Substantial decreases in needle longevity were reported for Siberian spruce near the Monchegorsk nickel-copper smelter (Kryuchkov & Makarova 1989; Stjernquist et al. 1998) and Kandalaksha aluminium smelter (Kryuchkov & Makarova 1989); for Scots pine near the Monchegorsk nickel-copper smelter (Yarmishko 1993, 1997; Jalkanen 1996; Lamppu & Huttunen 2003), near the Kostomuksha iron pellet plant (Lamppu & Huttunen 2003), and in industrial regions of Eastern Germany (Schulz et al. 1998); and for both Korean pine (*Pinus koraiensis*) and Pitch pine (*P. rigida*) in the Ansan industrial region of Korea (Choi et al. 2006). An absence of effects was reported only exceptionally: pollution of the oil shale industry in northeast Estonia did not influence needle longevity in Scots pine (Pensa et al. 2000, 2004).

The decrease in needle longevity with pollution is opposite to changes observed along other environmental gradients. Plants growing in less favourable conditions, including lower temperatures during the growth season, tend to compensate for reduced photosynthesis by increased longevity of needles (Ewers & Schmid 1981; Schoettle 1990; Jalkanen 1995; Pensa et al. 2007). Needle longevity is generally higher on less fertile soils (Lamppu & Huttunen 2003; Pensa et al. 2007); fertilisation decreased needle longevity of both Douglas fir (*Pseudotsuga menziesii* var. *glauca*) and grand fir (*Abies grandis*) (Balster & Marshall 2000). Thus, premature shedding of foliage is a specific response to pollution rather than a general response to environmental stress; it may result from acceleration of aging processes due to pollution impact (Wulff et al. 1996). Importantly, shedding of older needle age classes does not necessarily reduce primary production: thinning of the tree crown may increase the levels of photosynthetically active radiation reaching the remaining (younger) needles (Beyschlag et al. 1994).

Although pollution research often focuses on Scots pine, we conclude that needle longevity in Siberian/Norway spruce is a more sensitive indicator of pollution impact. This is particularly due to the generally higher number of age classes (up to 17 in the northernmost regions) retained by spruces in unpolluted regions (Table 4.37), which makes the difference between polluted and control sites larger in absolute value.

Thus, our data support an earlier conclusion (Schubert 1985; Kryuchkov & Makarova 1989) that needle longevity may serve as a handy indicator of pollution impact on the vitality of conifers. However, this indicator is far from being universal: it is applicable only to polluters that change soil pH, and its power decreases from North to South.

4.4.2 Sources of Variation in Pollution Impacts on Vitality Indices

4.4.2.1 Variation Between Study Years

Weather conditions of both previous and current seasons greatly influence plant growth and vitality (Hustich 1978; Junttila & Heide 1981; Valkama & Kozlov 2001; Morison & Morecroft 2006; Jonas et al. 2008) and are likely to modify pollution effects on plant vitality (Armentano & Bennett 1992). However, except for dendrochronological studies, this source of variation remains almost unexplored and is therefore routinely neglected in pollution ecology.

The vitality indices measured in the course of our study demonstrated different levels of annual variation. Needle longevity showed the highest correspondence between measurements conducted in different years (Section 4.3.5). Shoot growth measurements also correlated between study years, although to a lesser extent than needle length (Section 4.3.3). Finally, annual variation in both indices of photosynthetic efficiency was so large that even the sign of the correlation with pollution load changed with study year. These results clearly demonstrate that annual variation in plant responses to pollution is substantial and should therefore be accounted for in environmental monitoring and assessment programs. Long-term monitoring in polluted regions is the only way to obtain the data for parameterisation of phenomenological models accounting for the combined effects of pollution and weather conditions on plant growth.

4.4.2.2 Variation Between Polluters

Individual polluters generally differ in their impacts on woody plants: polluter-specific changes of all vitality indices varied from significantly negative to significantly positive (Figs. 4.2, 4.5, 4.9, 4.15, 4.21). Importantly, vitality indices (except for leaf/needle size and shoot length) showed individualistic (uncoordinated) responses to the impacts of the investigated polluters.

The detected variation between individual polluters in general was not related to the type of the polluter. This result contrasts with the meta-analyses of published data, which consistently detected significant variation among the classes of polluters (Ruotsalainen & Kozlov 2006; Zvereva et al. 2008; Zvereva & Kozlov 2009, Roitto et al. 2009). The difference most likely resulted from the limited number of polluter types involved in our study. In particular, we did not survey chemical factories, which caused the largest effects on plant growth (Roitto et al. 2009) and significantly altered insect abundance (Zvereva & Kozlov 2009).

Only changes in the leaf/needle size of woody plants and in the height of herbaceous plants depended on pollution effect on soil pH. Adverse effects were stronger around acidifying polluters.

To conclude, our data suggest that only a minor part of the variation in plant vitality changes around industrial polluters can be explained by the type of the polluter or by its impact on soil pH. Thus, other sources of variation need to be explored in greater detail to allow building of phenomenological models.

4.4.2.3 Variation Between Plant Species

Investigated plants similarly responded to pollution: we did not detect differences between species, life forms, or between evergreen and deciduous plants, in any of the vitality indices considered in the present study. We also identified only one marginally significant relationship between the pollution-induced changes in leaf length and ecological habitat requirements, as shown by the Ellenberg's indicator values: species with higher light requirements tended to respond positively to pollution. We think that this regularity reflects plant responses to pollution-induced habitat deterioration, primarily forest decline leading to higher light availability, rather than direct effects of industrial pollutants. Although we did not find significant correlations between the ESs and scores of Grime's CSR strategy, this result should be viewed as tentative, since the scores were available for only 15 of 43 investigated species.

Similarly, Gymnosperms and Angiosperms did not differ in their responses to pollution. This result clearly contrasts with the repeatedly expressed opinion on higher sensitivity of conifers to industrial pollution relative to deciduous plants (Crowther & Steuart 1914; Bohne 1971; Freedman 1989; Vike 1999; Hijano et al. 2005; Ozolincius et al. 2005). Importantly, a meta-analysis of published data (Roitto et al. 2009) yielded stronger adverse effects on Gymnosperms than on Angiosperms in shoot size ($d = -1.59$ vs. -0.61) but similar changes in leaf/needle size ($d = -0.99$ vs. -1.12). While original data demonstrated that the decrease in both leaf size and shoot length was significant in Angiosperms, but did not differ from zero in Gymnosperms (Figs. 4.7 and 4.13). These patterns may indicate the existence of both research and publication biases. Damage to conifers (independent of its cause), due to their higher economical importance, obviously received more attention than damage to other groups of plants, and studies supporting the general paradigm were published more readily.

Last but not least, damage to conifers has sometimes been attributed to pollution erroneously. One of the most recent examples is forest damage observed in Finnish Lapland in the late 1980s. Forest dieback was originally believed to be the result of pollution transfer from the adjacent industrial areas of the Kola Peninsula. However, detailed investigations revealed that the amounts of industrial emissions reaching Finnish Lapland would hardly cause forest damage. The marked premature shedding of needles and other signs of forest damage are now explained by exceptional weather conditions in the previous autumn and winter, when fast freezing of the soil caused root damage, in combination with the extensive epidemy of the scleroderris cancer (Tikkanen & Niemelä 1995).

Thus, although we do not question different pollution sensitivity of plant species, our results suggest that between-species variation in plant responses to pollution does not depend on their taxonomic affinity (in terms of Gymnosperms vs. Angiosperms), growth form or life habit, and is only weakly (if at all) related to their ecological habitat requirements.

4.4.2.4 Geographical Variation

Responses of several vitality indices (such as needle longevity and the rate of photochemical reactions) differ between the northern and southern polluters, while variations in other indices were independent of the location of the polluters.

Significant decreases in both needle longevity and radial increment were observed only around the northern polluters. We suggest that the differential effects on needle longevity were to a certain extent due to well-known geographical variation in this index (Ewers & Schmid 1981; Schoettle 1990; Jalkanen 1995): there are no 'spare' needles in southern regions that can be shed under pollution impact.

Stronger adverse effects of the northern polluters on radial increment of Scots pine may be explained in two ways. First, stand density around the southern polluters is generally higher than around the northern polluters (Tables 6.15–6.26), indicating a greater importance of competition in the southern relative to the northern regions. If plant growth is limited by competition, then we may expect no effects of pollution, or even better growth of the survivors released from competition pressure due to decreased stand density in polluted regions. Second, additional stress from pollution may cause a greater reduction of plant growth in the less favourable northern environment.

In contrast to needle longevity and radial increment, adverse effects of pollution on photosynthesis (in terms of $T_{1/2}$) were significant only near the southern polluters (Fig. 4.4). Although the mechanisms behind this pattern cannot be revealed from our data, we hypothesise that pollution-induced forest deterioration results in more pronounced climatic differences between polluted and unpolluted sites in the harsh climate of the northern taiga and subtundra zone. On sunny days in particular, polluted (more open) sites are warmer than the surrounding forests (Hursh 1948; Wołk 1977; Kozlov & Haukioja 1997; Winterhalder 2002), and the positive effects of this temperature increase on photosynthesis (Sage & Kubien 2007) can mask or even counterbalance the adverse impacts of toxicants.

4.4.3 Carbon Allocation and Allometric Relationships

Pollution differentially affected growth of plant parts (Kozlov & Zvereva 2007b; Section 4.4.1.2), thus influencing the allometric relationships. However, this research field remains almost unexplored. A notable exception is the shoot/root ratio; however, it was studied almost exclusively in experimental conditions (Rennenberg et al. 1996). Field data exist for tree crown structure, changes of which were documented in several case studies (Sokov & Rozhkov 1975; Yarmishko 1993) but never generalized. We are aware of a single study explicitly addressing the impact of industrial pollution on plant allometry (Elkarmi & Eideh 2006). This acute shortage of information hampers understanding of pollution impact on carbon allocation within the plant.

Stress supposedly alters not only photosynthesis but also subsequent carbon allocation in a tree in such a way that the most important processes are last affected (Waring 1987). A ranking of foliage, shoot and trunk growth responses to pollution in woody plants based on our data (Figs. 4.7, 4.13, 4.19) only partially agrees with the order of their importance for a tree as suggested by Waring (1987). Leaf/needle size, which is considered most important, was not affected by pollution (correlation re-calculated from the ES: $r = -0.08$), and trunk increment, which is least important, showed the largest decrease with pollution ($r = -0.35$). On the other hand, we detected no effect of pollution on shoot growth ($r = -0.09$), which is considered less important for a tree than the growth of foliage (Waring 1987).

Effect sizes calculated for individual vitality indices still show either positive correlations to each other (for leaf/needle size and shoot length) or no correlation. Thus, we did not detect any trade-offs in responses of different growth processes to pollution. However, since these correlations were based on site-specific values, we can only conclude that resource allocation showed no consistent response to pollution, while individualistic responses may well exist. This research field obviously deserves further investigation.

4.5 Summary

Although studies of pollution impact on plant vitality in general and plant growth in particular started more than a century ago, the amount of reliable and comprehensive information remains insufficient to explain variation in plant responses to pollution.

These responses depend more on the individual polluter than on the affected plant species, and vitality indices measured from the same plants often show uncoordinated responses to pollution. In woody plants, we found no effects on the efficiency of photosynthesis (measured by F_v/F_m) or leaf/needle size. Slight adverse effects were detected on the rate of photochemical reaction (measured by $T_{1/2}$) and on shoot length, while radial increment strongly decreased with pollution. Responses of all vitality indices demonstrated annual variation, frequently resulting in inconsistency of results obtained in different years. Still our data confirm that pollution generally decreases plant vitality and productivity, although these effects are much smaller than could have been expected from published studies. Needle longevity is the best operational index of pollution impact on the vitality of conifers.

Chapter 5
Fluctuating Asymmetry of Woody Plants

5.1 Introduction

5.1.1 Definitions and Theoretical Background

Fluctuating asymmetry (FA) represents small, random deviations from symmetry of a bilaterally symmetrical trait (Ludwig 1932). This variation is non-directional, with a normal distribution of signed right minus left differences whose mean is zero. Some researchers, however, suggest that the distribution of these values is generally leptokurtic, i.e., with a more acute peak around the mean and fatter tails than in a normally distributed variable (Gangestad & Thornhill 1999). Two other kinds of asymmetry are directional asymmetry (DA), when one side of the body is consistently larger than the other, i.e., the mean value of signed right minus left differences differs from zero; and antisymmetry, when the mean value does not differ from zero but the distribution of signed right minus left differences is platykurtic, i.e., broad-peaked or even bimodal (Palmer & Strobeck 1986). It is generally accepted that only FA can serve as a measure of developmental instability, reflecting the inability of an individual to control development under genetic and environmental stress (Møller & Swaddle 1997; Leamy 1999; Palmer & Strobeck 1986, 2003).

The concept of developmental stability unites many research fields, including developmental biology, genetics, ecology and evolution (Polak 2003). However, within the present study we explore only one of the regularities that, at least until recent years, was believed to be general: developmental stability is disturbed by environmental stress, and the consequent increase in developmental instability can be detected by measuring FA.

5.1.2 Use of Fluctuating Asymmetry in Pollution Research

An increase in FA had been reported for many organisms collected from polluted habitats, including birds (Eeva et al. 2000), mammals (Pankakoski et al. 1992), fishes (Valentin et al. 1973), insects (Rabitsch 1997), and plants (Freeman et al.

M.V. Kozlov et al. *Impacts of Point Polluters on Terrestrial Biota,*
DOI 10.1007/978-90-481-2467-1_5, © Springer Science+Business Media B.V. 2009

1993; Kozlov et al. 1996c). On the other hand, studies reporting an absence of pollution effects on FA began to appear long ago (e.g., Owen & McBee 1990), and the steady accumulation of such results (Zvereva et al. 1997a; Valkama & Kozlov 2001; Dauwe et al. 2006; Ambo-Rappe et al. 2008; Talloen et al. 2008) demonstrates that the positive association between pollution and FA is far from being a general regularity (Palmer 1996; Bjorksten et al. 2000).

Bjorksten et al. (2000, p. 165) suggested 'to abandon the search for a general link between FA and environmental stress'. Furthermore, combined with an earlier statement on 'poor use of research money' for 'measuring different stresses and traits in different organisms' (Bjorksten et al. 2000, p. 165), this opinion clearly votes for blacklisting of FA studies. As the result, the number of studies of FA and developmental instability published in evolutionary journals steadily declined during the 1999–2005 period (Van Dongen 2006). However, in our opinion the inconsistency of results is intriguing rather than disappointing, because it suggests that the regularities remain to be discovered; in contrast, consistency of results would signal that further accumulation of evidence is redundant.

The general pattern of FA responses to stress can be identified, and sources of variation explored, by analysing reliable and diverse data sets. However, an extensive search of publications reporting FA changes in plants along pollution gradients revealed only a few studies (Table 5.1) that are suitable for meta-analysis according to the following criteria, coined by Kozlov and Zvereva (2003) and further developed by Zvereva et al. (2008):

1. The study was conducted near a point polluter.
2. The polluter was influencing surrounding habitats primarily via the ambient air.
3. The study involved natural ecosystems, not modified by experimental treatments.
4. The data have been collected from organisms naturally inhabiting the study area.
5. The study involved both impacted and non-impacted ecosystems, allowing their comparison.
6. The study provided numerical information, allowing calculation of the effect size (ES).

Many of the identified studies are 'quasireplications' as defined by Palmer (2000), i.e., the 'replication' of previous studies using different species or systems. Only one of 24 identified data sets (Table 5.1) was tested for statistical significance of FA relative to measurement error (ME). Five of these data sets involved only one polluted and one control plot, i.e., the analyses of pollution effects suffered from pseudoreplication (as defined by Hurlbert 1984). Therefore, we decided to invest the maximum possible effort into accumulation of primary data in order to collect representative information suitable for exploration of factors influencing relationships between pollution and FA of woody plants.

Table 5.1 Summary of observational studies reporting plant FA in impact zones of point polluters

Location of the polluter	Polluter type	Plant species	Measured character(s)	FA index[a]	Sample size[b]	Test for relationship with pollution	Effect size[c]	Reference
Chapaevsk, Russia	Chemical industries	*Betula pendula*	Five foliar traits combined	FA2	5/45/450	t-test	1.08	Kryazheva et al. 1996
Chapaevsk, Russia	Chemical industries	*Trifolium pratense*	Nine foliar traits combined	FA2	5/100/200	t-test (?)	0.54	Zakharov et al. 2001
Harjavalta, Finland	Nickel-copper smelter	*Pinus sylvestris*	Difference in needle length	FA2	4/156/3,120	Duncan	0.73	Kozlov et al. 2002
Harjavalta, Finland	Nickel-copper smelter	*Betula pendula*	Difference in width of leaf halves	FA2	5/30/300	Spearman correlation	1.10	Kozlov et al. 1996c
Harjavalta, Finland	Nickel-copper smelter	*Betula pubescens*	Difference in width of leaf halves	FA2	5/30/300	Spearman correlation	1.47	Kozlov et al. 1996c
Harjavalta, Finland	Nickel-copper smelter	*Salix caprea*	Difference in width of leaf halves	FA2	10/50/500	ANOVA	0.86	Zvereva & Kozlov 2001
Karabuma, Serbia	Petrochemical industry	*Plantago major*	Difference in width of leaf halves	FA4	2/70/70 (p)	ANOVA	–0.25	Veličković & Perisic 2006
Kostomuksha, Russia	Iron pellet plant	*Betula pubescens*	Five foliar traits combined	FA2	7/70/700	Not tested	0.79	Pimenov 2003
Monchegorsk, Russia	Nickel-copper smelter	*Betula pubescens*	Difference in width of leaf halves	FA2[+]	20/100/7,400	Pearson correlation	0.24	Valkama & Kozlov 2001
Monchegorsk, Russia	Nickel-copper smelter	*Pinus sylvestris*	Difference in needle length	FA2	10/50/1,000	Pearson correlation	0.93	Kozlov & Niemelä 1999
Monchegorsk, Russia	Nickel-copper smelter	*Betula pubescens*	Difference in width of leaf halves	FA2	6/120/600	Spearman correlation	1.19	Kozlov et al. 1996c
Monchegorsk, Russia	Nickel-copper smelter	*Picea abies*	Difference in branch length	FA2[+]	2/20/40 (p)	Kruskal–Wallis test	0.43	Kozlov et al. 2001
Monchegorsk, Russia	Nickel-copper smelter	*Salix borealis*	Difference in width of leaf halves	FA2	10/50/500	ANOVA	–0.14	Zvereva & Kozlov 2001
Monchegorsk, Russia	Nickel-copper smelter	*Salix caprea*	Difference in width of leaf halves	FA2	6/30/300	ANOVA	0.20	Zvereva & Kozlov 2001
Monchegorsk, Russia	Nickel-copper smelter	*Vaccinium myrtillus*	Difference in width of leaf halves	FA2	4/20/200	ANOVA	0.04	Zvereva & Kozlov 2005

(continued)

Table 5.1 (continued)

Location of the polluter	Polluter type	Plant species	Measured character(s)	FA index[a]	Sample size[b]	Test for relationship with pollution	Effect size[c]	Reference
Monchegorsk, Russia	Nickel-copper smelter	*Vaccinium vitis-idaea*	Difference in width of leaf halves	FA2	4/20/200	ANOVA	0.04	Zvereva & Kozlov 2005
Nizhny Novgorod, Russia	Car factory	*Tilia cordata*	Five foliar traits combined	FA2	6/60/600	t-test	1.22	Sherzhukova et al. 2002
Novgorod, Russia	Nitrogen fertilizer factory	*Acer platanoides*	Differences in angles between three major veins on leaf halves	FA1	2/11/62 (p)	ANOVA	0.18	Freeman et al. 1993
Novgorod, Russia	Nitrogen fertilizer factory	*Aegopodium podagraria*	Blade length	FA1	4/60/120	ANOVA	0.67	Freeman et al. 1993
Novgorod, Russia	Nitrogen fertilizer factory	*Epilobium angustifolium*	Difference in width of leaf halves	FA1	4/80/320	ANOVA	0.51	Freeman et al. 1993
Novi Bilyari, Ukraine	Ammonia production	*Convolvulus arvensis*	Leaf lobe width	FA1	3/?/120 (p)	Student–Newman–Keuls	0.34	Graham et al. 1993; Freeman et al. 1993
Novi Bilyari, Ukraine	Ammonia production	*Robinia pseudoacacia*	Phyllotaxy	FA1	3/?/875 (p)	Student–Newman–Keuls	0.31	Graham et al. 1993; Freeman et al. 1993
Syasstroy, Russia	Pulp and paper mill	*Betula pubescens*	Difference in width of leaf halves	FA2	6/30/300	Spearman correlation	1.02	Kozlov et al. 1996c
Volkhov, Russia	Aluminium plant	*Betula pubescens*	Difference in width of leaf halves	FA2	6/30/300	Spearman correlation	1.74	Kozlov et al. 1996c
Voronezh, Russia	Chemical factory	*Betula pendula*	Five foliar traits combined	FA2	2/?/200 (p)	Not tested	0.03	Sapelnikova 1997

[a] For explanations of different indices used to measure FA consult Palmer and Strobeck (2003). Plus sign denotes the only study where significance of FA was tested against measurement error.

[b] Number of study plots/total number of plant individuals/total number of measured objects (i.e. leaves, shoots forming annual whorl in Norway spruce, and needle pairs in Scots pine). Question mark denotes missing information; (p) indicates non-replicated sampling design (i.e., one polluted and one control site).

[c] Calculated by us as z_r, a z-transformed correlation coefficient with pollution load or as (−1) multiplied by the correlation (r) with the distance from the polluter. If the primary data only allowed calculation of Hedge's d (a normalised difference between polluted and unpolluted study sites), this index was converted into r by using the equation: $r^2 = d^2/(d^2 + 4)$ (Rosenberg et al. 2000).

5.1.3 *Misuse and Abuse*

An enthusiasm that emerged soon after the very first studies of FA response to environmental and genomic stressors led in particular to numerous suggestions to use FA as a handy indicator of stress experienced by organisms (Zakharov 1990; Clarke 1992; Parsons 1992; Freeman et al. 1993; Hume 2001). More specifically, FA was advertised for evaluation of environmental health (Zakharov & Clarke 1993; Zakharov et al. 2001) and identification of the extinction risk for threatened populations (Anciaes & Marini 2000).

These publications still remain influential among researchers who are not deeply involved in studies of developmental instability (and therefore not familiar with the recent methodological literature) but are seeking a 'simple stress index' to use in their own research. This is especially true for the Russian scientists who uncritically follow the methodology suggested by Zakharov et al. (2001), which was already both oversimplified and outdated at the time of publication (discussed by Kozlov 2001a).

Since the reliability of conclusions based on measurements of fluctuating asymmetry critically depends on methodology (Palmer & Strobeck 2003; Van Dongen 2006), in this chapter we describe the methods of data collection and processing in much more details than for other plant measurements. This is intended to allow other researchers to check for data quality and to assure that the utmost care was taken to obtain representative and unbiased data sets.

5.2 Materials and Methods

5.2.1 *Sampling and Preservation*

At each plot, samples for FA measurements were collected from vegetative shoots of five randomly selected individuals of each plant species – the same individuals from which we measured chlorophyll fluorescence and growth indices (Section 4.2). All materials were collected and preserved in a uniform manner.

In trees and large shrubs, we took leaves/shoots from different sides of the crown at a height of 1-3 m. In small shrubs, e.g., dwarf birch, and dwarf shrubs, we selected ramets growing at least 5 m apart and sampled leaves from shoots located approximately at one half of the ramet's height.

In birches, which possess two distinct types of vegetative shoots (Fig. 5.1), we sampled short-shoot leaves; in other trees the shoot length varied continuously, and therefore shoots were selected irrespectively of their length. We always collected the largest leaf from the selected shoot. If the largest leaf was damaged (e.g., part of the leaf lamina was removed by insect herbivores) in a manner that prevented accurate measurement, then the shoot was disregarded and another shoot was sampled.

Fig. 5.1 Branching pattern in white birch (*Betula pubescens*). Note the difference between short (S) and long (L) shoots

In spruces, we sampled apical annual whorls consisting of two opposite second-order lateral shoots (Figs. 5.2a, 5.4b) from first-order branches (i.e., branches arising from the tree stem). Apical whorls with only one lateral shoot (Fig. 5.2b) or with three or more lateral shoots (Fig. 5.2c) were disregarded.

In Scots pine and red pine, we also selected terminal shoots from the first-order branches (Fig. 5.3a). The current-year or previous-year needle fascicles (each consisting of two needles) were collected from approximately the middle of three annual shoots (Fig. 5.3b) per tree.

Sampled leaves, branches, or needle fascicles were packed into plastic bags and transported to the laboratory for mounting and drying. The leaves (ten per plant individual) were mounted on strong paper using tape (Fig. 5.4a). Annual whorls of spruces (ten per plant individual) were stapled to a sheet of paper (Fig. 5.4b). Two methods were used for mounting pine needle pairs (20 per plant individual). The tips of short and strong needle pairs were inserted into small holes in a bent sheet of paper, assuring fixation in a position best suited for measurements (Fig. 5.4c). If the needles were long and soft, they were placed over a sheet of paper and fixed by thread (Fig. 5.4d). Care was taken to keep the needles of each pair tightly pressed to each other to allow accurate measurement of the difference in needle length. All samples are now deposited in the Paljakka Environmental Bank of the Finnish Forest Research Institute.

Fig. 5.2 Branching pattern of Norway spruce (*Picea abies* ssp. *abies*): regular (**a**) and irregular (**b**, **c**) annual whorls. Epicormic shoots (not measured in this study) are marked with asterisks

5.2.2 Measurements

Measurement error (ME) seriously complicates tests for differences in FA among traits of different sizes. In particular, at the given accuracy of the measurements, larger individuals and larger traits may exhibit lower FA than smaller ones simply because ME contributes less to between-sides variation (Kozlov 2001a; Palmer & Strobeck 2003). To partially compensate for this effect, smaller plant organs were measured with higher accuracy: we used a dissecting microscope to measure differences

Fig. 5.3 Branching pattern of Scots pine (*Pinus sylvestris*): (**a**) collector holds first-order branch to be sampled; (**b**) the region of apical shoot from which needle fascicles were collected

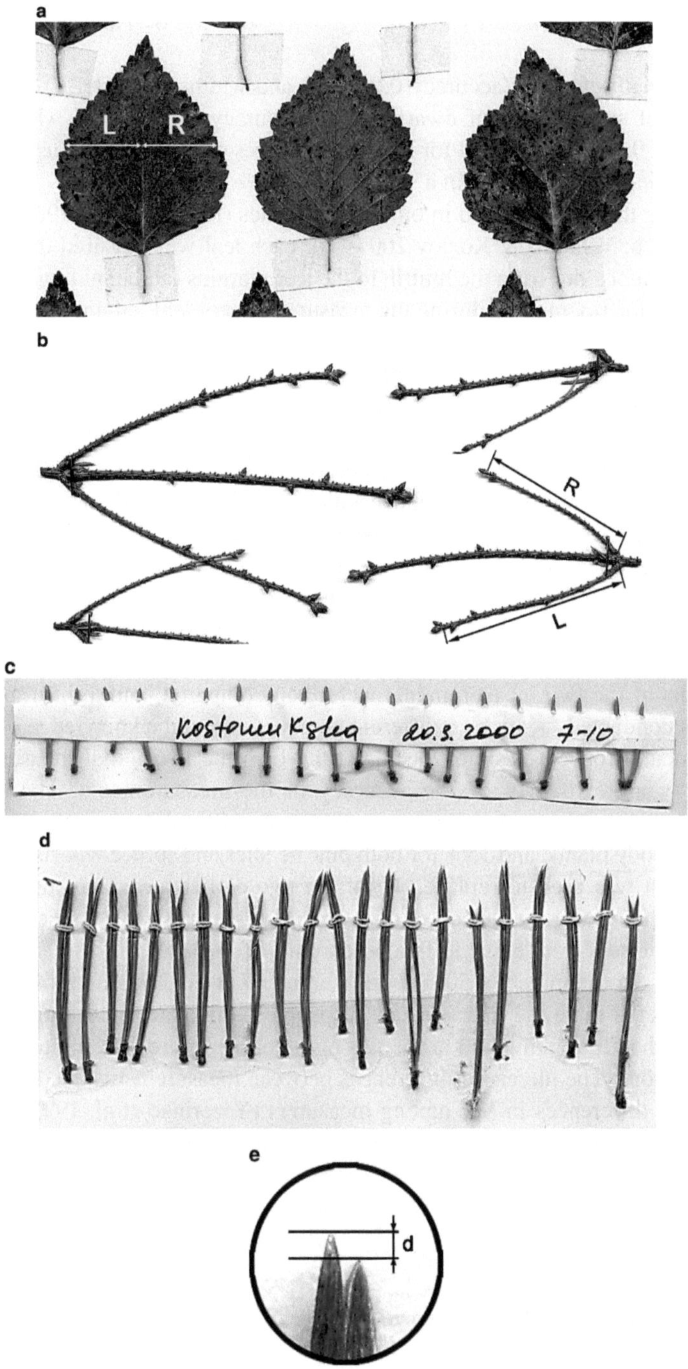

Fig. 5.4 Examples of mounting and measurement schemes used for leaves (**a**), annual whorls of spruce (**b**), and pine needles (**c**-**e**). Arrows indicate the measured characters; d - the absolute difference in length between two needles forming a pair (fascicle); L and R - left and right leaf halves (**a**) or left and right shoots forming annual whorl in spruces (**b**)

in needle length of pines (accuracy 0.05 mm) and widths of the left (L) and right (R) halves of small leaves of dwarf shrubs (accuracy of 0.1 mm), while a ruler (accuracy of 0.5 mm) was used for both larger leaves of other deciduous plants and shoots forming annual whorls in a spruce.

Following the protocol used in our earlier studies (Kozlov et al. 1996c; Zvereva et al. 1997a, b; Valkama & Kozlov 2001), for each leaf we measured the width of the left and right sides from the midrib to the leaf margins (at the midpoint between the base and the tip, marked during the measurements of leaf length) perpendicular to the midrib (Fig. 5.4a). For spruce, following the protocol suggested by Kozlov et al. (2001), we measured the lengths of the left and right second-order lateral shoots from the point of departure from the axis to the base of the terminal bud (Fig. 5.4b). Only the lateral shoots arising at the base of the respective (same-year) apical bud were considered, i.e., epicormic shoots (Fig. 5.2b, c) were disregarded. The measurement protocol for pine needles (developed by Kozlov & Niemelä 1999) differed from the protocols used for leaves and shoots: in pines, we independently measured the difference in the length of the two needles in the fascicle (Fig. 5.4d) and the length of the longer of the two needles.

All measurements were performed twice, by different persons, without a priori knowledge of either plant origin or results of previous measurements. The results of the two measurements were then compared, and a third measurement was conducted (again by a different person, without the knowledge of earlier measurements) for leaves/shoots/needles for which the two initial measurements differed by more than twice the accuracy of the measurements. The average proportions of re-measured objects were 5% for dwarf shrub leaves, 10% for larger leaves of woody plants, and 15% for both pine needles and spruce whorls. The third measurement was used to replace one of the two earlier measurements that was considered erroneous; the detection of the error was usually easy since some two thirds of the errors were equal to five or ten units of the scale used.

In total, approximately 150,000 measurements were obtained from nearly 36,000 objects (Table 5.2). More than 30 people participated in taking measurements, and therefore individual properties of the observer are unlikely to influence our conclusions. The increased differences between measurements due to virtually unavoidable differences in ME among measurers (Yezerinac et al. 1992; Helm & Albrecht 2000) were traded for a decreased probability of systematic error, which is more likely to arise when the repeated measurements are conducted by the same person.

5.2.3 *Analysis and Interpretation*

As a basic measure of FA, we have chosen the absolute difference between the sides divided by the mean value at the level of individual measurement. The need for correction for trait size is justified by significant size-dependence ($r = 0.14$, CI = 0.12 … 0.17, $N = 64$) revealed by meta-analysis of correlations between trait asymmetry $|R - L|$ and trait size $[(R + L)/2]$ across all of our data sets (Table 5.2, except for three samples showing antisymmetry). This index, labelled FA2 by Palmer and Strobeck (2003), is one of the most frequently used, in spite of several drawbacks. For discussion on properties of different indices used to quantify FA, consult Palmer (1994) and Palmer and Strobeck (1986, 2003).

In conducting the analyses, we followed the most recent and rigorous methodology developed by Palmer and Strobeck (2003), accounting also for methodological suggestions by Knierim et al. (2007). The validation analysis of each data set included:

1. Data inspection and testing for outliers. The first step of data inspection, aimed at elimination of errors not related to simple measurement imprecision, is described above (Section 5.2.2). Outliers were checked by examining distributions of signed R – L values. After examining all of the data sets, we accepted the rule of thumb and considered FA2 values exceeding 0.3 for leaf and needle measurements and exceeding 0.6 for shoot measurements as statistical outliers. These values, amounting to ca. 0.2% of all measurements, were excluded from further analyses.
2. Test for size-dependence: Spearman coefficient of rank correlation between trait asymmetry $|R - L|$ and trait size $[(R + L)/2]$, averaged from two repeated measurements.
3. Test for significance of FA relative to ME: the two-way, mixed model ANOVA for each sample (i.e., plant species by polluter) with sides as fixed factors and individuals as random factors. This model also tests for significance of DA (Palmer & Strobeck 1986, 2003; Merilä & Björklund 1995) and allows calculation of repeatability (ME5, according to Palmer & Strobeck 2003).
4. Comparison of DA with FA4a index ($FA4a = 0.798\sqrt{var}\,(R–L)$; consult Palmer and Strobeck [2003] for more details) for each sample, where mixed model ANOVA revealed a significant difference between the right and left sides.

The method used to measure FA in pines does not allow the use of mixed model ANOVA; therefore, we tested repeatability of measurements (the absolute difference in length of two needles comprising a pair) by using one-way ANOVA. We did not test pine samples for the presence of DA, since this would make sampling and measurements extremely time-consuming; the absence of DA was presumed on the basis of tests conducted earlier (Kozlov & Niemelä 1999).

Following data validation, we (a) calculated variation in FA among plant individuals nested within plots, (b) calculated variation in FA between study sites on the basis of plant-specific values, and (c) calculated plot-specific values that served as the basis for meta-analysis.

To account for possible effects of DA, we (by means of meta-analysis) contrasted ESs calculated from 28 data sets where the presence of DA was indicated by a statistically significant difference in the width of two halves of a leaf, and from the remaining 24 data sets showing no difference between the left and right sides of a leaf (Table 5.2). Note that the method used to measure needle asymmetry in pines (see above) made the test for DA impossible; therefore, ESs obtained using this method were separately compared with data from other plants - both those showing the presence and absence of DA.

Table 5.2 Basic statistics on measurements of differences between left and right sides of plant organs

					ANOVA[b]					
					Side		Side*Individual			
Location of the polluter	Plant species	Sample size	Size dependence[a]	DA, mm	*F*	*P*	*F*	Repeatability	Shapiro–Wilk test[c]	Kurtosis[d]
Apatity	*Alnus incana*	500	0.08	0.11	1.88	0.17	16.2	0.885	0.996	0.14
	Betula pubescens	500	0.27*	–0.51	129.9	<0.0001	2.37	0.800	0.991**	0.72**
Bratsk	*Betula pubescens*	500	0.16*	–0.14	3.81	0.05	23.9	0.920	0.988***	0.80**
	Picea abies	232	0.23*	–0.23	0.01	0.94	94.3	0.979	0.978**	1.42**
	Pinus sylvestris	896	–0.02	–	–	–	–	0.973	–	–
	Vaccinium vitis-idaea	500	0.18*	–0.01	0.41	0.52	47.2	0.958	0.992**	0.50*
Harjavalta	*Betula pubescens*	450	0.11*	–0.09	2.21	0.14	3.73	0.577	0.663	0.14
	Picea abies	323	–0.04	–0.17	0.15	0.70	132.3	0.985	0.994*	0.39
	Pinus sylvestris	799	–0.02	–	–	–	–	0.994	–	–
	Salix caprea	498	0.18*	–0.13	3.40	0.07	16.5	0.886	0.996	0.22
	Vaccinium vitis-idaea	300	0.39*	–0.06	3.80	0.05	6.03	0.716	0.984**	0.60*
Jonava	*Betula pendula*	500	0.19*	–0.03	0.51	0.48	12.5	0.852	0.980***	1.26**
	Picea abies	398	0.11*	–0.81	3.78	0.06	75.6	0.973	0.989**	0.97**
	Pinus sylvestris	473	0.10*	–	–	–	–	0.990	–	–
Kandalaksha	*Betula pubescens*	498	0.18*	–0.12	5.12	0.02	20.6	0.907	0.990**	0.69**
	Picea abies	448	0.16*	–0.17	1.96	0.16	23.6	0.919	0.991*	0.25
	Pinus sylvestris	965	–0.03	–	–	–	–	0.995	–	–
	Salix caprea	500	0.15*	–0.06	0.00	0.98	53.0	0.963	0.991**	0.50*
	Vaccinium vitis-idaea	500	0.16*	–0.23	199.9	<0.0001	15.2	0.876	0.988***	0.03
Karabash	*Betula pendula*	500	0.27*	–0.28	23.1	<0.0001	17.6	0.892	0.986***	0.80**
	Pinus sylvestris	773	0.02	–	–	–	–	0.997	–	–
	Vaccinium vitis-idaea	397	0.31*	–0.14	48.2	<0.0001	27.6	0.930	0.990**	0.43
Kostomuksha	*Betula pubescens*	499	0.12*	–0.18	9.97	0.0017	38.2	0.949	0.976***	0.46*
	Picea abies	482	0.09	0.27	0.95	0.33	20.7	0.908	0.993*	0.49*
	Pinus sylvestris	900	0.12	–	–	–	–	0.968	–	–
	Salix caprea	500	0.11*	–0.21	11.3	0.0008	57.0	0.965	0.983***	0.41*
	Vaccinium vitis-idaea	499	0.24*	–0.25	237.2	<0.0001	16.2	0.883	0.989**	0.49*

Krompachy	*Betula pendula*	500	0.25*	0.10	0.77	0.38	29.0	0.933	0.986***	1.04**
	Fagus sylvatica	500	0.20*	–0.09	0.67	0.41	32.4	0.940	0.986***	–0.65⁺⁺
	Picea abies	500	0.16*	0.16	0.06	0.81	26.9	0.928	0.994	0.62**
	Pinus sylvestris	961	0.03	–	–	–	–	0.997	–	–
Monchegorsk	*Betula nana*	500	0.10*	–0.23	114.8	<0.0001	28.6	0.932	0.995	–0.05
	Betula pubescens	500	0.15*	–0.08	2.03	0.16	17.4	0.891	0.989***	0.69**
	Picea abies	500	0.22*	–0.23	7.90	0.0051	12.3	0.850	0.996	0.10
	Pinus sylvestris	1,000	–0.01	–	–	–	–	0.983	–	–
	Populus tremula	500	0.23*	0.01	0.01	0.90	23.7	0.919	0.987***	1.05**
	Salix caprea	500	0.26*	–0.01	0.05	0.83	29.3	0.934	0.994	0.26
	Vaccinium vitis-idaea	500	0.11*	–0.18	2.04	0.15	15.8	0.881	0.988***	0.33
Nadvoitsy	*Betula pubescens*	497	0.08	–0.20	6.02	0.01	4.39	0.629	0.987***	0.67**
	Picea abies	497	0.13*	–0.44	4.60	0.03	46.1	0.958	0.995	0.13
	Pinus sylvestris	998	0.23*	–	–	–	–	0.956	–	–
	Vaccinium vitis-idaea	500	0.04	–0.10	30.5	<0.0001	10.4	0.825	0.996	–0.12
Nikel	*Betulapubescens*	500	0.00	–0.09	2.74	0.10	4.81	0.643	0.991**	0.31
	Salix glauca	450	0.15*	–0.01	0.14	0.71	5.02	0.668	0.981***	0.45*
	Vaccinium vitis-idaea	454	0.15*	–0.01	0.46	0.50	6.86	0.746	0.986***	0.07
Norilsk	*Betula nana*	500	0.23*	–0.24	80.7	<0.0001	22.8	0.916	0.986***	1.14**
	Salix lanata	499	0.13*	0.15	2.53	0.11	34.4	0.944	0.989**	0.72**
	Vaccinium uliginosum	490	0.25*	0.09	33.8	<0.0001	83.7	0.976	0.981***	0.46*
	Vaccinium vitis-idaea	500	0.21*	–0.28	11.9	0.0006	14.0	0.867	0.985***	0.47*
Revda	*Betula pubescens*	500	0.15*	–0.18	9.17	0.026	14.0	0.867	0.985***	1.17**
	Picea abies	500	0.16*	–0.37	3.93	0.05	36.1	0.946	0.991**	0.81**
	Pinus sylvestris	915	0.04	–	–	–	–	0.995	–	–
	Vaccinium vitis-idaea	399	0.16*	–0.08	26.3	<0.0001	16.1	0.883	0.980***	0.25
Straumsvík	*Salix herbacea*	500	0.28*	–0.28	150.4	<0.0001	53.0	0.963	0.979***	1.55**
	Vaccinium uliginosum	498	0.14*	–0.29	0.63	0.43	22.1	0.913	0.989***	0.70**
Sudbury	*Betula papyrifera*	457	0.13*	–0.33	14.3	0.0002	18.0	0.897	0.991**	0.81**
	Pinus resinosa	758	0.12*	–	–	–	–	0.979	–	–

(continued)

Table 5.2 (continued)

Location of the polluter	Plant species	Sample size	Size dependence[a]	DA, mm	ANOVA[b] Side F	ANOVA[b] Side P	ANOVA[b] Side*Individual F	Repeatability	Shapiro–Wilk test[c]	Kurtosis[d]
Volkhov	*Betula pubescens*	495	0.08	0.20	9.15	0.026	11.9	0.845	0.989**	0.59*
	Picea abies	350	0.19*	0.41	4.59	0.03	45.5	0.957	0.991*	0.53
	Pinus sylvestris	587	–0.01	–	–	–	–	0.985	–	–
	Salix caprea	487	0.17*	0.03	0.22	0.64	46.4	0.958	0.996	–0.21
Vorkuta	*Betula nana*	499	0.13*	–0.12	28.1	<0.0001	24.0	0.920	0.993*	0.26
	Salix glauca	499	0.17*	–0.01	0.13	0.72	5.01	0.667	0.981***	0.09
	Vaccinium vitis-idaea	500	0.16*	–0.34	10.6	0.0012	8.81	0.796	0.987***	0.12
Žiar nad Hronom	*Carpinus betulus*	490	0.12*	–0.10	2.17	0.14	7.82	0.773	0.988***	0.69**
	Fagus sylvatica	500	0.20*	0.51	13.0	0.003	43.8	0.955	0.982***	–0.70++
	Quercus petraea	260	0.32*	0.11	1.84	0.18	66.5	0.970	0.973**	–0.94++

[a] Spearman coefficient of rank correlation between the absolute value of asymmetry, FA1=|R–L|, and trait size [(R + L)/2]. An asterisk denotes significance at $P = 0.05$.

[b] Two-factor, mixed model ANOVA of untransformed repeated measurements aimed at testing of the significance of fluctuating asymmetry (Side * Individual interaction; for all tests, significant at $P < 0.0001$) against the measurement error on the basis of repeated measurements of leaves. Significant effect of side hints presence of directional asymmetry, while individual effect (F values not shown; for all tests, $P < 0.0001$) denotes variation in leaf size. Measurement protocol applied to pines does not allow this analysis (for explanations, consult Section 5.2.3).

[c] An asterisk denotes significance at $P = 0.05$, two asterisks – at $P = 0.01$, and three asterisks – at $P = 0.001$.

[d] Calculated by SAS univariate procedure as an 'unbiased estimate', i.e. by using Equation 7 in Palmer and Strobeck (2003) and compared with critical values reported in Table 5 (Palmer & Strobeck 2003). An asterisk denotes deviations towards leptocurtosis (narrow-peaked and long-tailed distribution) significant at $P = 0.05$; two asterisks – at $P = 0.01$; two pluses denote deviation towards platycurtosis (broad-peaked or bimodal distribution) significant at $P = 0.05$.

5.3 Results

5.3.1 Data Validation

The mixed model ANOVA revealed a significant difference between the right and left sides in 28 of 55 data sets (Table 5.2). Analysis of the signed (R – L) difference across all data sets demonstrated that the right side was consistently larger (mean DA = 0.11 mm significantly differs from zero: $t = 3.74$, $N = 55$, $P = 0.0005$). However, in all data sets DA was smaller than the FA4a index (see above, Section 5.2.3): on average, DA comprised approximately 30% of FA, suggesting that DA's contribution to the total variation in |R – L| values is small and can therefore be neglected. DA did not differ significantly between shoots and leaves ($F_{1,53} = 0.65$, $P = 0.42$) or between measurements conducted under a dissecting microscope and those performed using a ruler ($F_{1,53} = 1.53$, $P = 0.22$).

Between-side variation (FA) in all samples was significantly ($P < 0.0001$) greater than measurement error (Table 5.2). Importantly, the repeatability values of FA were rather high (57.7–94.4%, mean value 89.3%); repeatability differed among study objects (leaves: 86.1%; needles: 98.4%; shoots: 94.0%; ANOVA for square-root arcsine-transformed values: $F_{2,61} = 21.9$, $P < 0.0001$), and this variation was at least partially explained by higher repeatability of measurements conducted with higher accuracy (under dissecting microscope: 92.3%; by using ruler: 87.3%; $F_{1,62} = 6.27$, $P = 0.01$).

Distributions of signed (R – L) values deviated significantly from the normal distribution in 45 of 55 data sets (Table 5.2). This is by no means surprising, since half of the samples demonstrated significant DI and the differences between the sides were size-dependent (Table 5.2; Section 5.2.3). Importantly, the majority of departures from the normal distribution, as indicated by kurtosis (Table 5.2), were toward leptokurtosis (narrow-peaked and long-tailed distribution); however, three data sets demonstrated platykurtosis (broad-peaked or bimodal distribution) that may indicate the presence of antisymmetry. Since inspection of scatter plots demonstrated that these three distributions (European beech from Krompachy and Žiar nad Hronom, and durmast oak, *Quercus petraea*, from Žiar nad Hronom) are bimodal, these data sets were excluded from the analysis of FA.

5.3.2 Variation between Plant Individuals and between Study Sites

We detected significant between-plant variation in FA within plots in about two fifths of the data sets. Importantly, this variation was significant in 12 of 13 data sets on pines, while the proportion of significant results among other plants was about one fourth. Between-plot variation (explored by using plant-specific means) was significant in only 14 of 64 data sets (Tables 5.3–5.16).

Only ten of 128 individual correlation coefficients were significant, and only three of these ten significant correlations were consistent with the expected pattern, i.e., increase in FA with increase in pollution.

5.3.3 *Overall Effect of Pollution and Sources of Variation*

Pollution had no overall effect on FA (Fig. 5.5); this result did not depend on the method used to calculate ESs ($Q_B = 0.25$, df = 2, $P = 0.88$). Moreover, this conclusion remained valid when samples showing significant DA were excluded from the analysis, although the mean ES in data sets with significant DA ($d = 0.16$, CI = −0.01 … 0.33, $N = 28$) was higher ($Q_B = 4.85$, df = 1, $P = 0.03$) than in data sets where DA was not detected ($d = -0.10$, CI = −0.27 … 0.07, $N = 27$).

Pollution effect on FA did not differ among individual polluters (Fig. 5.6; $Q_B = 16.2$, df = 16, $P = 0.44$); significant increases in FA were observed only around Vorkuta and Norilsk. However, we found consistency ($r_S = -0.62$, $N = 15$ polluters, $P = 0.01$) in the effects of individual polluters on FA and on the rate of photosynthetic reactions ($T_{1/2}$). Changes in other indices of plant vitality (listed in Chapter 4) did not correlate with changes in FA ($r_S = 0.00$ … −0.43, $N = 9$ …17 polluters, $P = 0.24$ … 0.99).

Changes in FA did not depend on either the type of the polluter (Fig. 5.5; $Q_B = 2.25$, df = 4, $P = 0.69$) or the type of its impact on soil pH (Fig. 5.5; $Q_B = 1.29$, df = 2, $P = 0.52$). The only source of variation identified in this database was the geographical position of polluters (Fig. 5.5; $Q_B = 5.93$, df = 1, $P = 0.02$): FA tended to increase around northern polluters and decrease around southern ones.

Responses of different plant species were generally uniform (Fig. 5.7; $Q_B = 15.1$, df = 8, $P = 0.06$); only grayleaf willow (*Salix glauca*) demonstrated a significant increase in FA with an increase in pollution. The effect did not differ either between Gymnosperms and Angiosperms (Fig. 5.7; $Q_B = 0.49$, df = 1, $P = 0.48$) or between evergreen and deciduous species (Fig. 5.7; $Q_B = 0.24$, df = 1, $P = 0.63$).

We detected significant non-linear responses in eight of 64 data sets (six dome-shaped and two U-shaped), which is more than twice as high as can be expected by chance.

5.4 Discussion

5.4.1 *Plants as Study Objects*

So far, asymmetry studies have mostly been conducted on animals, despite the fact that plants have many properties that make them more handy for studies addressing FA, including the possibility to obtain multiple measurements from a single individual (Freeman et al. 1993). However, high phenotypic plasticity and some

Table 5.3 Fluctuating asymmetry of plants in the impact zones of the power plant at Apatity, Russia and the fertiliser factory at Jonava, Lithuania

	Apatity		Jonava		
	Alnus incana	*Betula pubescens*	*Betula pendula*	*Picea abies*	*Pinus sylvestris*
Site	1997	2006	2005	2005	2005
1–1	0.067	0.084	0.089	–	0.0072
1–2	0.068	0.078	0.084	0.0859	0.0062
1–3	0.081	0.065	0.104	0.1103	0.0079
1–4	0.066	0.058	0.083	0.0969	0.0083
1–5	0.064	0.061	0.073	0.0847	0.0086
2–1	0.072	0.074	0.073	–	0.0087
2–2	0.079	0.080	0.067	0.0957	0.0084
2–3	0.079	0.051	0.074	0.1094	0.0087
2–4	0.070	0.062	0.065	0.1068	0.0069
2–5	0.081	0.078	0.057	0.1076	0.0091
ANOVA: *F/P*	1.03/0.43	1.77/0.10	2.65/0.02	1.12/0.38	0.37/0.94
Dist.: *r/P*	0.10/0.78	–0.62/0.06	–0.26/0.46	0.17/0.68	0.29/0.40
Poll.: *r/P*	0.00/0.99	0.49/0.15	–0.17/0.65	0.02/0.96	–0.32/0.37

ANOVA: test for significance of variation between study sites. Dist.: Pearson linear correlation between site-specific means and log-transformed distance from polluter. Poll.: Pearson linear correlation between site-specific means and concentration of the selected pollutant. Sampling year shown under plant name.

Table 5.4 Fluctuating asymmetry of plants in the impact zone of aluminium smelter at Bratsk, Russia

	Betula pubescens	*Picea abies*	*Pinus sylvestris*	*Vaccinium vitis-idaea*
Site	2002	2004	2002	2002
1–1	0.046	0.113	0.0090	0.068
1–2	0.048	0.175	0.0116	0.063
1–3	0.075	0.150	0.0080	0.071
1–4	0.052	0.118	0.0089	0.069
1–5	0.056	0.118	0.0091	0.049
2–1	0.061	0.124	0.0079	0.066
2–2	0.066	0.130	0.0125	0.065
2–3	0.055	0.168	0.0092	0.073
2–4	0.065	0.171	0.0097	0.058
2–5	0.073	0.124	0.0092	0.076
ANOVA: *F/P*	2.36/0.03	0.31/0.97	0.44/0.90	1.06/0.41
Dist.: *r/P*	0.35/0.32	0.05/0.89	–0.25/0.48	–0.11/0.76
Poll.: *r/P*	–0.14/0.69	–0.26/0.47	–0.30/0.40	0.05/0.89

For explanations, consult Table 5.3.

Table 5.5 Fluctuating asymmetry of plants in the impact zone of the nickel-copper smelter at Harjavalta, Finland

	Betula pubescens	*Picea abies*	*Pinus sylvestris*	*Salix caprea*	*Vaccinium vitis-idaea*
Site	2002	2001	2002	2002	2002
1–1	0.072	–	0.0073	0.088	–
1–2	0.071	–	0.0031	0.069	0.063
1–3	0.079	0.099	–	0.087	0.069
1–4	0.068	0.120	0.0053	0.089	0.053
1–5	0.082	0.151	0.0090	0.087	0.060
2–1	0.073	–	0.0087	0.067	–
2–2	0.071	0.114	0.0084	0.080	–
2–3	0.072	0.190	0.0112	0.086	0.061
2–4	0.075	0.124	0.0042	0.092	0.049
2–5	–	0.147	0.0112	0.076	–
ANOVA: *F*/*P*	0.49/0.86	1.65/0.17	2.63/0.03	0.62/0.77	0.93/0.48
Dist.: *r*/*P*	0.60/0.08	0.18/0.69	0.21/0.58	0.42/0.23	−0.34/0.51
Poll.: *r*/*P*	−0.40/0.28	0.02/0.96	−0.53/0.14	−0.56/0.09	0.28/0.59

For explanations, consult Table 5.3.

Table 5.6 Fluctuating asymmetry of plants in the impact zone of the aluminium smelter at Kandalaksha, Russia

	Betula pubescens	*Picea abies*	*Pinus sylvestris*	*Salix caprea*	*Vaccinium vitis-idaea*
Site	2002	2001	1998	2002	2001
1–1	0.075	0.160	0.0043	0.063	0.092
1–2	0.059	0.142	0.0047	0.089	0.091
1–3	0.062	0.133	0.0100	0.071	0.098
1–4	0.071	0.147	0.0036	0.072	0.103
1–5	0.070	0.110	0.0039	0.101	0.081
2–1	0.082	0.101	0.0057	0.068	0.097
2–2	0.080	0.113	0.0118	0.083	0.091
2–3	0.080	0.180	0.0237	0.065	0.083
2–4	0.075	0.115	0.0060	0.076	0.105
2–5	0.082	0.120	0.0118	0.074	0.045
ANOVA: *F*/*P*	0.66/0.74	1.14/0.36	11.5/<0.0001	1.84/0.09	2.35/0.03
Dist.: *r*/*P*	−0.11/0.77	−0.01/0.98	0.13/0.72	0.37/0.29	−0.48/0.15
Poll.: *r*/*P*	−0.11/0.75	−0.07/0.85	−0.30/0.51	−0.13/0.72	0.40/0.25

For explanations, consult Table 5.3.

Table 5.7 Fluctuating asymmetry of plants in the impact zones of the copper smelter at Karabash, Russia and aluminium smelter at Straumsvík, Iceland

	Karabash			Straumsvik	
	Betula pendula	*Pinus sylvestris*	*Vaccinium vitis-idaea*	*Salix herbacea*	*Vaccinium uliginosum*
Site	2003	2003	2003	2002	2002
1–1	0.056	–	–	0.082	0.059
1–2	0.085	0.0114	0.063	0.066	0.077
1–3	0.077	0.0105	0.068	0.072	0.071
1–4	0.073	0.0118	0.055	0.069	0.063
1–5	0.075	0.0171	0.069	0.072	0.063
2–1	0.059	–	–	0.070	0.065
2–2	0.068	0.0167	0.074	0.061	0.077
2–3	0.078	0.0133	0.069	0.072	0.084
2–4	0.083	0.0102	0.073	0.064	0.081
2–5	0.071	0.0106	0.072	0.066	0.082
ANOVA: *F/P*	0.95/0.49	0.33/0.93	0.79/0.60	0.61/0.78	0.95/0.49
Dist.: *r/P*	0.57/0.09	−0.27/0.51	−0.03/0.95	−0.24/0.51	0.22/0.54
Poll.: *r/P*	−0.80/0.0059	0.51/0.19	0.35/0.40	0.70/0.02	−0.55/0.10

For explanations, consult Table 5.3.

Table 5.8 Fluctuating asymmetry of plants in the impact zone of the iron pellet plant at Kostomuksha, Russia

	Betula pubescens	*Picea abies*	*Pinus sylvestris*	*Salix caprea*	*Vaccinium vitis-idaea*
Site	2002	2001	2002	2002	2001
1–1	0.048	0.160	0.0087	0.074	0.099
1–2	0.054	0.152	0.0069	0.065	0.071
1–3	0.058	0.143	–	0.072	0.107
1–4	0.070	0.136	0.0078	0.075	0.078
1–5	0.071	0.093	0.0055	0.076	0.070
2–1	0.049	0.105	0.0077	0.087	0.090
2–2	0.053	0.115	0.0086	0.090	0.095
2–3	0.062	0.148	0.0064	0.077	0.082
2–4	0.058	0.106	0.0078	0.083	0.058
2–5	0.062	0.130	0.0072	0.066	0.092
ANOVA: *F/P*	1.71/0.12	2.10/0.05	0.92/0.51	0.76/0.65	3.87/0.0014
Dist.: *r/P*	0.83/0.0029	−0.13/0.73	−0.56/0.12	−0.50/0.14	−0.37/0.28
Poll.: *r/P*	−0.52/0.13	−0.29/0.41	0.32/0.39	0.57/0.08	0.24/0.50

For explanations, consult Table 5.3.

Table 5.9 Fluctuating asymmetry of plants in the impact zone of the nickel-copper smelter at Monchegorsk, Russia

	Betula nana	*Betula pubescens*	*Picea abies*	*Pinus sylvestris*	*Populus tremula*	*Salix caprea*	*Vaccinium vitis-idaea*
Site	2001	2002	2002	2002	2002	2002	2001
1–1	0.091	0.058	0.113	0.0087	0.060	0.085	0.060
1–2	0.091	0.073	0.117	0.0073	–	0.053	0.081
1–3	–	–	–	–	0.049	–	–
1–4	0.063	0.059	0.113	0.0102	–	0.078	0.067
1–5	0.078	0.065	0.125	0.0099	–	0.074	0.083
2–1	–	–	0.089	–	–	–	–
2–2	0.066	0.053	0.125	0.0075	–	0.062	0.078
2–3	–	–	–	–	0.053	–	–
2–4	0.069	0.067	0.124	0.0163	–	0.079	0.078
2–5	0.076	0.068	0.116	0.0127	0.052	0.072	0.084
2–6	–	–	–	–	0.058	–	–
2–7	–	–	0.133	–	–	–	–
2–8	0.054	0.058	0.116	0.0083	0.056	0.084	0.051
2–9	0.074	0.057	0.120	0.0117	0.043	0.086	0.072
2–10	0.100	0.051	–	0.0097	0.053	0.070	0.069
2–11	–	–	–	–	0.057	–	–
2–12	–	–	0.120	–	0.048	–	–
ANOVA: *F*/*P*	3.94/0.0012	1.12/0.37	0.69/0.74	1.27/0.28	0.85/0.58	2.00/0.06	2.11/0.05
Dist.: *r*/*P*	−0.15/0.68	−0.25/0.49	0.57/0.05	0.20/0.58	−0.39/0.26	0.22/0.54	−0.10/0.79
Poll.: *r*/*P*	−0.08/0.83	0.17/0.65	0.07/0.82	0.19/0.59	0.23/0.52	−0.22/0.54	0.24/0.50

For explanations, consult Table 5.3.

Table 5.10 Fluctuating asymmetry of plants in the impact zone of the copper smelter at Krompachy, Slovakia

	Betula pendula	*Picea abies*	*Pinus sylvestris*
Site	2004	2002	2002
1–1	0.0644	0.126	0.0167
1–2	0.0699	0.122	0.0139
1–3	0.0898	0.104	0.0099
1–4	0.0807	0.132	0.0116
1–5	0.1027	0.079	0.0095
2–1	0.0635	0.165	0.0096
2–2	0.0869	0.095	0.0117
2–3	0.0953	0.141	0.0110
2–4	0.0797	0.115	0.0124
2–5	0.0921	0.131	0.0111
ANOVA: *F*/*P*	1.92/0.08	2.24/0.005	0.55/0.83
Dist.: *r*/*P*	0.70/0.02	−0.31/0.38	−0.55/0.10
Poll.: *r*/*P*	−0.40/0.14	0.40/0.25	−0.20/0.58

For explanations, consult Table 5.3.

Table 5.11 Fluctuating asymmetry of plants in the impact zones of the aluminium smelter at Nadvoitsy, Russia

	Betula pubescens	*Picea abies*	*Pinus sylvestris*	*Vaccinium vitis-idaea*
Site	2004	2004	2004	2004
1–1	0.075	0.110	0.0101	0.056
1–2	0.077	0.092	0.0180	0.065
1–3	0.065	0.117	0.0159	0.061
1–4	0.067	0.123	0.0176	0.058
1–5	0.073	0.120	0.0165	0.071
2–1	0.089	0.140	0.0163	0.048
2–2	0.079	0.155	0.0135	0.068
2–3	0.060	0.098	0.0123	0.047
2–4	0.076	0.121	0.0130	0.056
2–5	0.074	0.137	0.0170	0.059
ANOVA: *F*/*P*	1.23/0.30	0.87/0.56	1.19/0.33	1.51/0.18
Dist.: *r*/*P*	−0.48/0.16	−0.05/0.89	0.42/0.23	0.35/0.32
Poll.: *r*/*P*	0.71/0.02	0.21/0.56	−0.13/0.73	−0.10/0.78

For explanations, consult Table 5.3.

Table 5.12 Fluctuating asymmetry of plants in the impact zones of the nickel-copper smelter at Nikel and ore-roasting plant at Zapolyarnyy, Russia and the nickel-copper smelter at Sudbury, Canada

	Nikel			Sudbury	
	Betula pubescens	*Salix glauca*	*Vaccinium vitis-idaea*	*Betula papyrifera*	*Pinus resinosa*
Site	2001	2001	2001	2007	2002
1–1	0.058	0.085	–	0.046	–
1–2	0.076	0.055	0.060	0.098	0.0020
1–3	0.075	0.074	0.069	0.071	0.0021
1–4	0.066	0.065	0.071	0.094	0.0019
1–5	0.055	0.067	0.056	0.068	0.0027
2–1	0.062	0.070	–	0.055	–
2–2	0.069	–	0.060	0.064	0.0018
2–3	0.058	0.082	0.052	0.073	0.0021
2–4	0.060	0.083	0.064	0.069	0.0028
2–5	0.079	0.073	0.055	0.074	0.0030
ANOVA: *F*/*P*	1.43/0.21	1.52/0.18	0.71/0.67	2.29/0.04	3.17/0.01
Dist.: *r*/*P*	0.14/0.69	−0.21/0.59	−0.02/0.97	0.39/0.26	0.88/0.0035
Poll.: *r*/*P*	−0.29/0.41	0.32/0.40	−0.19/0.65	−0.55/0.10	−0.65/0.08

For explanations, consult Table 5.3.

Table 5.13 Fluctuating asymmetry of plants in the impact zones of the nickel and copper smelters at Norilsk, Russia

	Betula nana	*Salix lanata*	*Vaccinium vitis-idaea*	*Vaccinium uliginosum*
Site	2002	2002	2002	2002
1–1	0.107	–	0.088	0.083
1–2	0.086	0.064	0.071	0.054
1–3	0.077	0.073	0.077	0.061
1–4	0.081	0.078	0.072	0.069
1–5	0.075	0.077	0.074	0.067
2–1	0.068	0.082	0.097	0.067
2–2	0.067	0.053	0.075	0.080
2–3	0.089	0.090	0.078	0.067
2–4	0.075	0.067	0.060	0.054
2–5	0.075	0.071	0.101	0.085
ANOVA: *F*/*P*	1.35/0.24	1.69/0.14	1.82/0.09	1.21/0.31
Dist.: *r*/*P*	−0.35/0.31	0.10/0.80	−0.22/0.55	−0.14/0.70
Poll.: *r*/*P*	0.49/0.15	0.02/ 0.96	0.39/0.26	0.42/0.23

For explanations, consult Table 5.3.

Table 5.14 Fluctuating asymmetry of plants in the impact zones of the copper smelter at Revda, Russia

	Betula pubescens	*Picea abies*	*Pinus sylvestris*	*Vaccinium vitis-idaea*
Site	2003	2003	2003	2003
1–1	0.061	0.107	0.0127	0.068
1–2	0.054	0.108	0.0158	–
1–3	0.065	0.098	0.0167	–
1–4	0.045	0.090	0.0186	0.049
1–5	0.054	0.080	0.0162	0.062
2–1	0.055	0.108	0.0135	0.051
2–2	0.057	0.087	0.0142	0.054
2–3	0.073	0.137	0.0123	0.055
2–4	0.062	0.098	0.0119	0.060
2–5	0.075	0.098	0.0108	0.057
ANOVA: *F*/*P*	1.31/0.26	1.06/0.41	1.68/0.13	1.20/0.33
Dist.: *r*/*P*	0.10/0.79	−0.41/0.24	0.16/0.67	−0.07/0.87
Poll.: *r*/*P*	0.04/0.90	0.40/0.25	−0.37/0.29	0.17/0.69

For explanations, consult Table 5.3.

Table 5.15 Fluctuating asymmetry of plants in the impact zones of aluminium smelter at Volkhov, Russia

	Betula pubescens	*Picea abies*	*Pinus sylvestris*	*Salix caprea*
Site	2002	2002	1998	2002
1–1	0.065	0.135	–	0.094
1–2	0.076	0.132	–	0.086
1–3	0.069	–	–	0.076
1–4	0.060	0.085	0.0056	0.084
1–5	0.051	0.115	0.0116	0.102
2–1	0.075	0.110	0.0114	0.053
2–2	0.068	0.112	–	0.068
2–3	0.074	0.148	0.0085	0.098
2–4	0.066	0.125	0.0061	0.089
2–5	0.085	0.097	0.0073	0.089
ANOVA: *F/P*	1.12/0.37	0.67/0.72	2.18/0.09	2.22/0.04
Dist.: *r/P*	–0.24/0.51	–0.24/0.53	–0.52/0.29	0.69/0.03
Poll.: *r/P*	0.23/0.53	–0.04/0.92	0.57/0.24	–0.81/0.0048

For explanations, consult Table 5.3.

Table 5.16 Fluctuating asymmetry of plants in the impact zones of the power plant at Vorkuta, Russia and the aluminium smelter at Žiar nad Hronom, Slovakia

	Vorkuta			Žiar nad Hronom
	Betiula nana	*Salix glauca*	*Vaccinium vitis-idaea*	*Carpinus betulus*
Site	2001	2001	2001	2002
1–1	0.073	0.080	0.085	0.088
1–2	0.079	0.083	0.082	0.074
1–3	0.066	0.086	0.069	0.081
1–4	0.090	0.080	0.069	0.078
1–5	0.081	0.081	0.058	0.077
2–1	0.074	0.080	0.068	0.056
2–2	0.091	0.084	0.077	0.065
2–3	0.078	0.083	0.062	0.068
2–4	0.053	0.064	0.057	0.061
2–5	0.087	0.084	0.061	0.053
ANOVA: *F/P*	2.08/0.05	0.43/0.91	1.28/0.28	1.82/0.10
Dist.: *r/P*	0.05/0.90	–0.20/0.59	–0.66/0.04	–0.38/0.28
Poll.: *r/P*	0.06/0.88	0.27/0.45	0.25/0.49	0.60/0.07

For explanations, consult Table 5.3.

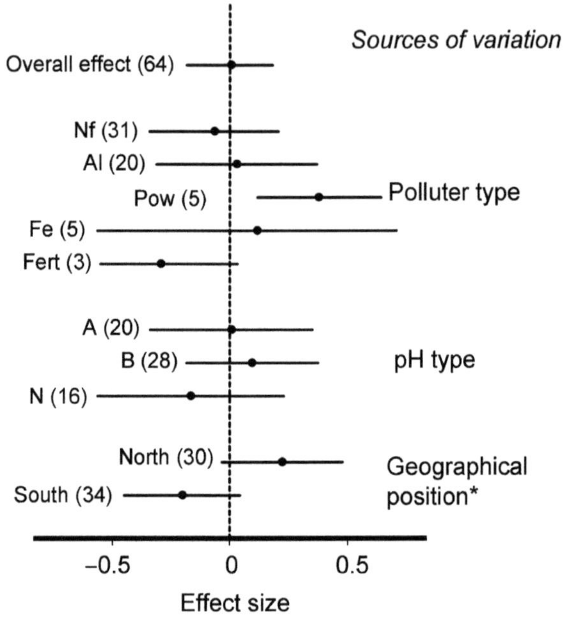

Fig. 5.5 Overall effect and sources of variation in responses of fluctuating asymmetry of woody plants. Horizontal lines denote 95% confidence intervals; sample sizes are shown in brackets; an asterisk denotes significant ($P < 0.05$) between-class heterogeneity. For classifications of polluters and abbreviations consult Table 2.1

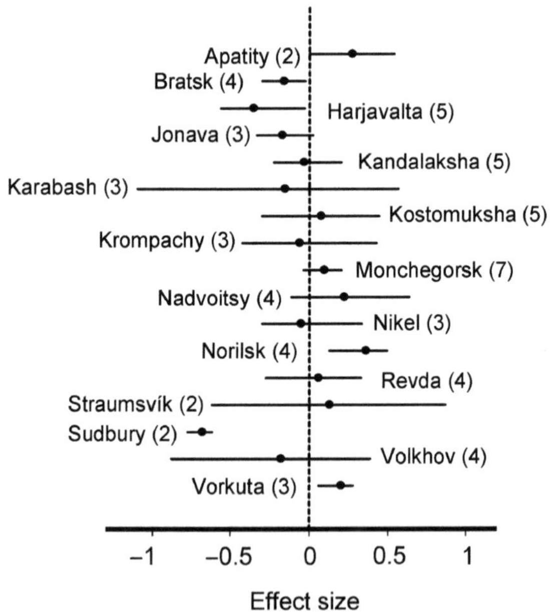

Fig. 5.6 Effects of individual polluters on of fluctuating asymmetry of woody plants. For explanations, consult Fig. 5.5

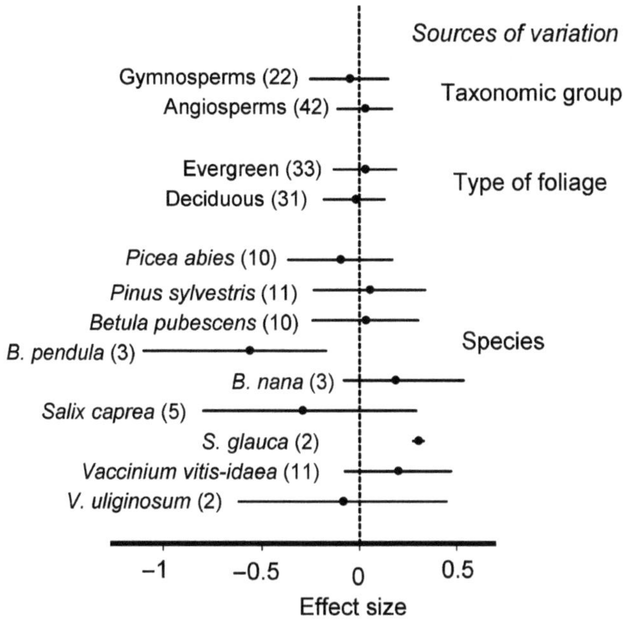

Fig. 5.7 Effects of point polluters on of fluctuating asymmetry of woody plant groups and species. For explanations, consult Fig. 5.5

other plant properties have served as the basis for recent criticism of FA measurements in plants and led Palmer and Strobeck (2003) to the conclusion that departures from symmetry in plants 'seem like unreliable measures of developmental instability'. However, significant between-plant variation detected in about two thirds of our data sets suggests that the sampled plant genets either responded differently to environmental stress (imposed by factors other than pollution) or demonstrated different degrees of genetic variation. Also, the significant leptokurtosis detected in 27 of 51 data sets (Table 5.2) can be seen as evidence of individual differences (see Gangestad & Thornhill 1999). These results, in line with our earlier conclusion (Valkama & Kozlov 2001), suggest that FA in plants may serve an indicator of stress providing that (1) a sufficiently large number of metamers is measured, (2) adequate methods for data validation are used, and (3) the results are interpreted with caution.

5.4.2 Data Quality

Low quality data can easily compromise the results of studies involving fluctuating asymmetry (Kozlov 2001a; Palmer & Strobeck 2003; Van Dongen 2006). Therefore, the utmost care was taken to avoid pitfalls identified by earlier studies.

The rigorous data validation analysis (following the protocol by Palmer & Strobeck 2003) demonstrated the presence of FA in all data sets and suggested that the DA detected in half of our samples was too small to distort our estimates of FA. The latter conclusion was further confirmed by meta-analysis: exclusion of data sets that demonstrated significant DA did not change the results.

DA's independence of both study object and measurement technique suggests that this minor (0.11 mm on average) but significant deviation from perfect symmetry was most likely due to handedness of observers (cf. Helm & Albrecht 2000). In our specific case, we think that handedness may have affected the angle at which the observer looked at the ruler overlapping the leaf.

5.4.3 Pollution Effects on Plant Fluctuating Asymmetry

Meta-analysis of our data demonstrated an absence of pollution effects on the FA of woody plants. This may seem surprising only in relation to published data (Table 5.1), which showed a significant increase in FA with pollution ($z_r = 0.65$, CI = 0.38 ... 0.92, $N = 25$). However, an absence of changes in FA is consistent with other data obtained in the course of our study, including the absence of pollution effects on leaf size (Fig. 4.8), shoot length (Fig. 4.14) and the ratio of variable to maximum chlorophyll fluorescence (Fig. 4.1) in woody plants.

In the following discussion, we will attempt to answer three questions: (1) Does FA reflect the state of plant individuals, or is it just random noise? (2) What are possible reasons for the discrepancy between original and published data? (3) Does pollution affect plant FA?

Since we detected significant between-plot and between-plant variation in a much higher proportion of data sets than expected by chance, we conclude that our estimates of FA reflect the state of the measured plants. This conclusion is supported by the correspondence between changes in chlorophyll fluorescence and in FA: the photochemical reaction was slowed down near the same polluters where we detected an increase in FA. Detection of between-tree differences in one third of the data sets hints that a substantial part of the FA variation is due to a genetic component or to a genotype by environment interaction. This conclusion agrees with an earlier finding on individual responses of FA in birch trees to annual climatic fluctuations (Valkama & Kozlov 2001) and confirms an opinion that it is difficult to separate genetic and environmental components of FA in observational studies (Siikamäki & Lammi 1998; Wilsey & Saloniemi 1999). More generally, we conclude that the effects of genotype in our data set were substantially larger than the effects of pollution, and that individual responses of genotypes to environmental stressors have resulted in zero net effect.

The reasons for the discrepancy between the original and published data cannot be explored directly. However, we suspect that the consistent increase in plant FA with increase in pollution reported in published studies (Table 5.1) is observed for several reasons, including in particular measurement error (discussed by Merilä &

Björklund 1995; Van Dongen et al. 1999), research and publication bias, and confounding factors, such as climatic stress or herbivory.

Since a rigorous data validation protocol was used in only two published studies fitting our selection criteria (Table 5.1), problems associated with measurement error are likely to distort conclusions drawn from these data sets. In particular, an increase in relative measurement error with pollution-induced decrease in leaf size can erroneously be interpreted as an increase in FA with increase in pollution (Kozlov 2001a). Finally, the unavoidable handedness of the observer may result in an increase in DA, which can be confounded with an increase in FA: in our study, the effect of pollution on 'overall asymmetry' in data sets demonstrating significant DA was higher than in data sets showing no DA. To further explore the impact of measurement error on conclusions concerning FA in polluted habitats, we plan to locate and re-measure some samples used in published studies; however, this examination is outside the scope of our book.

Research bias can result from the tendency to explore organisms and/or conditions in which the researcher has a reasonable expectation of detecting statistically significant effects (Gurevitch & Hedges 1999). As in the case of insect herbivores (Section 7.4.1), we suspect that non-random choice of study objects and study sites may have led to overestimation of pollution effects on plant FA. However, shortage of data does not allow for testing this hypothesis in the same manner as with pollution impact on insect herbivory (Zvereva & Kozlov 2009). On the other hand, a funnel plot based on published studies is highly asymmetrical ($\tau_B = 0.38$, $N = 24$, $P = 0.01$), in contrast to the symmetrical pattern based on original data ($\tau_B = -0.004$, $N = 64$, $P = 0.95$). This discrepancy may hint at the existence of considerable publication bias, although other interpretations are also possible (Terrin et al. 2005, and references therein).

Interestingly, the studies published prior to 2000 reported significant increases in FA with pollution ($z_r = 0.99$, CI = 0.57 … 1.40, $N = 13$), while studies published after 2000 reported no overall effect ($z_r = 0.37$, CI = −0.01 … 0.76, $N = 12$). The difference between these two periods was significant ($Q_B = 5.56$, df = 1, $P = 0.02$), hinting that development of the hypothesis linking FA and pollution has shifted in the late 1990s–early 2000s from stage one (supportive evidence, as defined by Leimu & Koricheva 2004) to stage two (accumulation of disconfirming evidence). The magnitude of the effect is likely to be considerably overestimated in the first stage and somewhat underestimated in the second (Leimu & Koricheva 2004). This pattern hints that the hypothesis requires reformulation and restriction of its scope.

Importantly, under certain circumstances, FA indeed increased with pollution (Fig. 5.6). However, we do not believe that this effect can be directly attributed to the toxicity of pollutants. Plant FA can increase in response to climatic stress (Valkama & Kozlov 2001; Kozlov & Niemelä 2003; Hagen et al. 2008) and herbivory (Zvereva et al. 1997a, b; Møller & Shykoff 1999; Kozlov 2005c); both of these factors can be influenced by pollution, leading to the appearance of the expected pattern in some study years (Valkama & Kozlov 2001).

To our knowledge, none of the numerous experiments with industrial pollutants have addressed effects on FA in plants. We strongly support an opinion by Møller and Shykoff (1999) that controlled experimental studies of environmental and genetic effects on FA of plants are badly needed to increase our understanding of the mechanisms causing developmental instability.

5.5 Summary

We did not detect the expected increase of plant FA with increasing levels of pollution: the overall effect did not differ from zero. Absence of effect was consistent across both polluters and plant species. This result contrasts with a meta-analysis of published data, which showed increase in FA with pollution. However, conclusions based on published data may be flawed due to both methodological problems and research and publication biases. Moreover, an increase in FA with pollution may result from the impact of confounding factors, such as climatic stress or herbivory. We conclude that either plants persisting in polluted habitats do not experience environmental stress, or the hypothesis linking FA and environmental stress requires reformulation and restriction of its scope.

Chapter 6
Structure of Plant Communities

6.1 Introduction

Vegetation, i.e., the plant life of a region, not only shapes terrestrial ecosystems but also serves several crucial functions in the biosphere. Due to the importance of vegetation for mankind, deterioration of plant communities under pollution impacts attracted considerable scientific and public attention more than a century ago (Holland 1888; Haselhoff & Lindau 1903; Stoklasa 1923).

Historically, the majority of studies exploring pollution effects on plant communities were conducted in forested areas of Europe and North America. A large body of publications report decreases in forest vitality, often followed by forest decline, at different scales, from local, around point polluters (National Research Council of Canada 1939; Bunce 1979; Symeonides 1979; Sutherland & Martin 1990; Rigina & Kozlov 2000; Aznar et al. 2007), to regional (Pitelka & Raynal 1989; Kandler & Innes 1995; Bussotti & Ferretti 1998; Akselsson et al. 2004; Allen et al. 2007).

The effects of extreme pollution pressure on plant communities are relatively well documented, and forest decline near big smelters is a textbook example of the adverse impact of aerial pollution on terrestrial ecosystems (Treshow 1984; Freedman 1989). However, we still lack an understanding of patterns and processes occurring at lower levels of pollution. Although some responses are common across several community types and across several kinds of polluters, natural variation, the adaptation potential of individual species and intrinsic community differences complicate any interpretation (Armentano & Bennett 1992). Long ago, meta-analysis was suggested as useful tool to fully interpret the literature and to ascertain the likelihood of trends common to ecosystems and pollution regimes (Armentano & Bennett 1992); however, to our knowledge, no attempts were made to achieve this goal. As a result, conclusions on the overall effects of pollution on vegetation are usually made on the basis of a few case studies reporting consequences of the most severe impacts (Gordon & Gorham 1963; Wood & Nash 1976; Freedman & Hutchinson 1980b). Moreover, these conclusions are geographically biased, since almost no data exist on pollution-induced changes in plant communities other than in northern and temperate forests (Zvereva et al. 2008).

M.V. Kozlov et al. *Impacts of Point Polluters on Terrestrial Biota*,
DOI 10.1007/978-90-481-2467-1_6, © Springer Science+Business Media B.V. 2009

Last but not least, forestry-oriented studies often concentrate on losses of economically important products, primarily timber, and neglect changes in other groups of plants (Linzon 1986; Carrier & Krippl 1990; Feng et al. 2002). However, losses of ecosystem services may be much more costly than losses of timber (Costanza et al. 1997). To attempt a better understanding of the ecosystem-level effects of pollution on terrestrial biota (Chapter 9), we have chosen for this study a set of indices reflecting both the structure and function of the selected plant communities.

A primary characteristic of vegetation is its three-dimensional structure, described by the horizontal and vertical distributions of plant biomass. This structure is determined by both environmental and historical factors and species composition. Horizontal distribution, i.e., the pattern of spacing of plant stems on the ground, is not considered in the present study. Vertical distribution of biomass is explored by evaluating the cover of vegetation layers, which is a rough but commonly used approximation of biomass. Additionally, we measured stand basal area and height of the dominant tree species, which are indicative of the aboveground biomass and provide a link to several functional characteristics of the stand.

Our meta-analysis of indices reflecting plant community structure, along with basic goals (search for general patterns and identification of sources of variation), aims at verification of several assumptions concerning pollution impacts on forests, in particular:

(a) Standing biomass, as reflected by height and basal area, decreases with pollution.
(b) Conifers are more sensitive to pollution than deciduous trees, and therefore the proportion of conifers among top-canopy trees decreases with pollution.
(c) Trees are more sensitive to pollution than herbs and grasses, and therefore community destruction under pollution impacts follows a downward pattern (trees decline first, while field layer vegetation is the last to decline).
(d) Natural forest regeneration is suppressed by pollution.

These effects have been observed in only some of the polluted areas, and their generalization (Woodwell 1970; Kozlowski 1980; Smith 1981; Treshow 1984; Freedman 1989) is so far lacking proper statistical support.

Species composition and spatial distribution (reflected by diversity measured at different spatial scales) are inherent parts of plant community structure. Recent meta-analysis, based on 86 individual studies conducted in the impact zones of 60 polluters (Zvereva et al. 2008), demonstrated that plot-specific estimates of species richness of vascular plants (α diversity) generally decreased with pollution. However, the responses were not uniform across the studies. In particular, we have revealed the effects of methodology on the variation of effect sizes (ESs) as well as the publication bias (i.e., studies covered by the ISI database reported adverse effects that were on average twice as large as those reported in other studies). This gives special importance to the analysis of original data that were collected using uniform methods. Furthermore, our data allow testing of the hypothesis that the species composition changes with pollution, a problem that was impossible to resolve by meta-analysis of the published data, which are generally restricted to summary statistics, such as species richness or diversity indices (Zvereva et al. 2008).

On the other hand, limitations imposed by the structure of published data (Zvereva et al. 2008) allowed us to explore only a few potential sources of variation in responses of plant communities to industrial pollution. In particular, the majority of published studies compared species lists, without any attempt to address the effects of overall abundance of plants on species richness. Therefore, in this chapter, in addition to performing meta-analysis of vegetation cover *per se*, we explored relationships between changes in species richness and in vegetation cover that serves as a rough measure of abundance. This was done in order to test the hypothesis (Kozlov et al. 1998) that a decline in species richness (measured per area unit) with an increase in pollution is an artefact caused by an overall decrease in plant abundances.

The latter hypothesis is in line with several publications (Begg et al. 1997; Salminen & Haimi 1999; Hobbs 2003; Chalcraft et al. 2008) that stressed the importance of spatial scales in assessing pollution effects on biodiversity. The structure of our data (replicated lists of species for each study site) allows us to explore the effects of industrial pollution at three spatial scales: small (within plots), intermediate (among plots) and large (within the study site). This approach is necessary to check whether the assessment of α diversity (i.e., mean number of species per study plot) accurately predicts pollution effects on regional biodiversity.

Finally, we were interested in temporal changes in plant communities, and in mechanisms behind these changes. While vegetation damage around industrial enterprises is routinely attributed to pollution (even when pollution is only one of the factors responsible for deterioration of ecosystem health), at a regional level atmospheric pollutants are seen as one of many causal factors related to vegetation changes, primarily forest decline (Whittaker et al. 1974; Cogbill 1977; Field et al. 1992; Duchesne et al. 2002). Elucidating temporal and spatial patterns of the effects imposed by industrial polluters on structure, productivity, and regeneration of plant communities may lead to a better understanding of the role of atmospheric pollutants in regional processes and thus result in a wide array of practical applications.

6.2 Materials and Methods

6.2.1 Study Areas and Sampling Design

The structure of plant communities was assessed around 14 of 18 polluters (Tables 6.1–6.41). We have chosen not to collect information on the abundances and diversity of plants around Apatity, Harjavalta, Krompachy and Volkhov because vegetation in these areas was greatly modified by other kinds of human activities. In particular, primary forests near the power plant in Apatity were

Table 6.1 Vegetation cover (%) in the impact zone of aluminium smelter at Bratsk, Russia (data of 2002)

Site	Top-canopy	Understorey	Field layer	Mosses
1–1	46.7	10.0	60.0	23.3
1–2	76.7	5.0	22.7	23.3
1–3	36.7	18.3	61.7	21.7
1–4	36.7	13.3	60.0	46.7
1–5	56.7	6.0	73.3	33.3
2–1	53.3	16.7	28.7	60.0
2–2	60.0	30.0	23.3	50.0
2–3	66.7	6.0	43.3	30.0
2–4	76.7	11.7	55.0	25.0
2–5	53.3	5.0	63.3	46.7
ANOVA: *F/P*	7.19/0.0001	2.46/0.04	9.40/<0.0001	5.37/0.0009
Dist.: *r/P*	−0.03/0.94	−0.45/0.19	0.70/0.03	−0.09/0.80
Poll.: *r/P*	−0.10/0.79	0.22/0.54	0.46/0.18	0.48/0.16

All data were collected in 2002 from three plots 10 × 10 m size. Cover of bare ground equals to zero in all sites; epigeic lichens found at two sites only (1–5: cover 0.7%; 2–5: cover 0.3%). ANOVA: test for significance of variation between study sites. Dist.: Pearson linear correlation between site-specific means and log-transformed distance from polluter. Poll.: Pearson linear correlation between site-specific means and concentration of the selected pollutant.

Table 6.2 Vegetation cover (%) in the impact zone of the fertiliser factory at Jonava, Lithuania

Site	Top-canopy	Understorey	Field layer	Mosses
1–1	46.7	78.3	24.0	8.8
1–2	63.3	11.7	33.5	59.5
1–3	41.7	16.7	8.6	89.5
1–4	45.0	11.0	12.0	64.0
1–5	50.0	43.3	20.0	49.5
2–1	40.0	21.7	44.5	38.5
2–2	46.7	25.0	7.6	20.0
2–3	50.0	61.7	18.5	58.1
2–4	35.0	71.7	17.0	91.5
2–5	63.3	5.7	16.2	82.5
ANOVA: *F/P*	2.23/0.06	18.0/<0.0001	9.42/<0.0001	12.8/<0.0001
Dist.: *r/P*	0.15/0.69	−0.11/0.77	−0.62/0.06	0.66/0.04
Poll.: *r/P*	−0.52/0.12	0.59/0.07	−0.35/0.32	−0.23/0.52

Cover of top-canopy and understorey estimated in 2002 from three plots 10 × 10 m size, cover of other layers – in 2007 from ten plots 1 × 1 m size. Cover of epigeic lichens and bare ground equals to zero in all sites. For other explanations, consult Table 6.1.

repeatedly felled and burned, and the secondary regrowth is now under severe recreation impact. No primary forests are left near the aluminium smelter in Volkhov, and the entire area has been greatly modified by centuries of agricultural practice. In Harjavalta, Scots pine stands near the smelter were planted in the late 1950s to replace forest killed by pollution; most of the more distant stands were

Table 6.3 Vegetation cover (%) in the impact zone of the aluminium smelter at Kandalaksha, Russia

Site	Top-canopy	Understorey	Field layer	Mosses	Lichens	Bare ground
1–1	66.7	6.3	34.5	60.0	0.6	0
1–2	70.0	2.3	24.3	69.0	0	0
1–3	56.7	0.3	33.5	52.5	0.1	0
1–4	63.3	0.7	55.5	39.5	0	0
1–5	60.0	3.0	60.0	11.5	0	1.0
2–1	60.0	1.0	28.5	11.6	0	10.0
2–2	76.7	1.0	35.7	33.5	0	1.0
2–3	45.0	0	18.7	56.1	0.5	0
2–4	50.0	1.7	42.5	51.5	0	0
2–5	60.0	0	24.1	60.5	4.5	0
ANOVA: *F*/*P*	1.63/ 0.17	1.46/ 0.23	6.41/ <0.0001	7.01/ <0.0001	4.37/ <0.0001	73.0/ <0.0001
Dist.: *r*/*P*	−0.32/0.36	−0.33/0.35	0.34/0.33	0.16/0.65	–	–
Poll.: *r*/*P*	0.51/0.13	0.33/0.36	−0.30/0.40	−0.15/0.68	–	–

Cover of top-canopy and understorey estimated in 2002 from three plots 10 × 10 m size, cover of other layers – in 2007 from ten plots 1 × 1 m size. Absence of correlation coefficients indicates that the data were not used in meta-analyses. For other explanations, consult Table 6.1.

Table 6.4 Vegetation cover (%) in the impact zone of the copper smelter at Karabash, Russia

Site	Top-canopy	Understorey	Field layer	Mosses	Bare ground
1–1	16.7	0	0.2	0	13.3
1–2	23.3	0	3.0	1.7	0
1–3	46.7	2.0	51.5	0	0
1–4	46.7	0.7	53.0	0.9	0
1–5	63.3	7.7	23.4	0.1	0
2–1	10.3	0	0	0	10.0
2–2	30.0	5.0	0.2	2.5	0
2–3	53.3	4.3	19.1	0.0	0
2–4	40.0	3.3	67.5	2.7	0
2–5	60.0	3.0	25.0	15.8	0
ANOVA: *F*/*P*	15.5/<0.0001	2.47/0.04	35.9/<0.0001	8.24/<0.0001	6.87/0.0002
Dist.: *r*/*P*	0.87/0.001	0.48/0.16	0.72/0.02	0.49/0.15	–
Poll.: *r*/*P*	−0.83/0.003	−0.40/0.25	−0.68/0.03	−0.32/0.37	–

Cover of top-canopy and understorey estimated in 2003 from three plots 10 × 10 m size, cover of other layers – in 2007 from ten plots 1 × 1 m size. Cover of epigeic lichens equals to zero in all sites. Absence of correlation coefficients indicates that the data were not used in meta-analyses. For other explanations, consult Table 6.1.

also planted in different years and are intensively managed to maximise timber production. The mountainous landscape around Krompachy, in combination with a long history of human settlements, mining, and intensive agriculture, made selection of comparable study sites nearly impossible. For more details on these and other impact zones, consult Section 2.2.

Table 6.5 Vegetation cover (%) in the impact zone of the iron pellet plant at Kostomuksha, Russia

Site	Top-canopy	Understorey	Field layer	Mosses	Lichens
1–1	46.7	0	76.7	70.0	0
1–2	46.7	0	57.0	95.0	0
1–3	36.7	0	43.3	95.0	0
1–4	33.3	0	74.0	95.0	0
1–5	33.3	1.0	63.3	95.0	0
2–1	30.0	3.0	71.7	68.3	0
2–2	60.0	0	51.3	95.0	0
2–3	50.0	0	45.0	95.0	0
2–4	40.0	0	50.0	94.3	9.0
2–5	40.0	0.3	73.7	95.0	0.3
ANOVA: *F*/*P*	2.19/0.07	4.85/0.0018	2.81/0.03	20.7/<0.0001	2.45/0.05
Dist.: *r*/*P*	−0.19/0.61	−0.49/0.15	−0.19/0.60	0.84/0.002	–
Poll.: *r*/*P*	−0.24/0.50	0.85/0.002	0.41/0.24	−0.80/0.005	–

Cover of top-canopy and understorey estimated in 2002 from three plots 10 × 10 m size, cover of other layers – in 2006 from ten plots 1 × 1 m size. Cover of bare ground equals to zero in all sites. Absence of correlation coefficients indicates that the data were not used in meta-analyses. For other explanations, consult Table 6.1.

Table 6.6 Vegetation cover (%) in the impact zone of the nickel-copper smelter at Monchegorsk, Russia

Site	Top-canopy	Understorey	Field layer	Mosses	Lichens	Bare ground
1–1	0.7	9.3	0.4	10.9	0	90.0
1–2	0	38.3	0	2.0	0	70.0
1–4	13.3	6.7	15.0	6.5	10.2	10.0
1–5	15.0	11.7	40.5	10.1	5.6	10.3
2–3	0	5.0	1.0	2.5	0.9	91.7
2–4	0	11.7	0.7	25.4	5.9	78.3
2–5	43.3	13.3	16.6	4.3	7.0	16.0
2–8	10.0	3.0	48.5	46.0	4.4	4.3
2–9	36.7	11.7	56.0	43.3	2.7	1.7
2–10	36.7	10.0	52.0	63.3	2.2	0
ANOVA: *F*/*P*	10.3/<0.0001	2.77/0.03	24.5/<0.0001	8.28/<0.0001	2.54/0.02	105.4/<0.0001
Dist.: *r*/*P*	0.70/0.02	−0.25/0.48	0.88/0.001	0.72/0.02	0.34/0.34	−0.88/0.001
Poll.: *r*/*P*	−0.53/0.11	−0.11/0.77	−0.70/0.02	−0.46/0.18	−0.33/0.35	0.83/0.003

Cover of top-canopy and understorey estimated in 2001 from three plots 10 × 10 m size, cover of other layers – in 2007 from ten plots 1 × 1 m size. For other explanations, consult Table 6.1.

Table 6.7 Vegetation cover (%) in the impact zone of the aluminium smelter at Nadvoitsy, Russia

Site	Top-canopy	Understorey	Field layer	Mosses
1–1	40.0	0	22.5	8.0
1–2	53.3	3.3	22.9	71.0
1–3	43.3	0.3	37.8	7.0
1–4	35.0	1.3	39.7	67.0
1–5	43.3	0.4	57.5	12.5
2–1	31.7	0.7	21.0	18.2
2–2	56.7	5.0	33.5	26.0
2–3	21.7	10.0	40.0	49.0
2–4	20.0	0	32.5	59.5
2–5	31.7	2.4	45.5	4.0
ANOVA: *F*/*P*	6.23/0.0003	4.33/0.0031	4.21/0.0001	13.3/<0.0001
Dist.: *r*/*P*	−0.25/0.48	−0.05/0.89	0.89/0.0006	0.09/0.80
Poll.: *r*/*P*	0.46/0.18	−0.06/0.87	−0.80/0.006	−0.11/0.76

Cover of top-canopy and understorey estimated in 2004 from three plots 10 × 10 m size, cover of other layers – in 2007 from ten plots 1 × 1 m size. Cover of epigeic lichens and bare ground equals to zero in all sites. For other explanations, consult Table 6.1.

Table 6.8 Vegetation cover (%) in the impact zone of the nickel-copper at Nikel and ore-roasting plant at Zapolyarnyy, Russia

Site	Top-canopy	Understorey	Field layer	Mosses	Lichens	Bare ground
1–1	2.0	15.0	0	0	0	68.3
1–2	2.0	31.7	2.5	3.0	0	53.3
1–3	1.7	36.7	45.2	16.0	0	0.7
1–4	3.3	14.3	33.0	32.0	0	0.3
1–5	12.7	7.3	39.0	40.0	12.7	1.3
2–1	3.7	13.7	1.1	1.7	0	81.7
2–2	4.3	15.0	7.4	7.5	0	46.7
2–3	8.3	30.0	19.7	1.5	1.6	5.0
2–4	8.3	18.3	49.0	20.5	14.0	0.7
2–5	17.7	28.3	25.3	45.5	2.3	2.3
ANOVA: *F*/*P*	1.54/0.20	1.59/0.18	20.5/<0.0001	8.26/<0.0001	3.49/0.0010	14.4/<0.0001
Dist.: *r*/*P*	0.66/0.04	0.07/0.84	0.79/0.006	0.85/0.002	0.51/0.13	−0.91/0.0003
Poll.: *r*/*P*	−0.43/0.22	−0.29/0.42	−0.71/0.02	−0.61/0.06	−0.38/0.27	0.85/0.002

Cover of top-canopy and understorey estimated in 2001 from three plots 10 × 10 m size, cover of other layers – in 2007 from ten plots 1 × 1 m size. For other explanations, consult Table 6.1.

Table 6.9 Vegetation cover (%) in the impact zone of the nickel-copper smelters at Norilsk, Russia

Site	Top-canopy	Understorey	Field layer	Mosses	Lichens	Bare ground
1–1	0.7	6.0	38.3	0	0	20.0
1–2	2.0	61.7	43.3	0	26.7	2.0
1–3	2.7	63.3	48.7	1.0	78.3	0.7
1–4	16.7	50.0	33.3	1.0	56.7	3.0
1–5	6.7	35.0	51.7	6.0	40.0	1.3
2–1	0	11.7	35.0	0	18.3	6.7
2–2	1.3	25.0	31.7	0	70.0	3.7
2–3	31.7	23.3	63.3	0	15.0	1.0
2–4	13.3	38.3	60.0	1.0	93.3	0
2–5	20.0	6.7	73.3	6.7	86.7	0.3
ANOVA: *F/P*	9.17/<0.0001	5.96/0.0004	4.73/0.0018	40.9/<0.0001	7.75/<0.0001	21.8/<0.0001
Dist.: *r/P*	0.56/0.09	0.32/0.36	0.62/0.05	0.73/0.02	0.64/0.05	−0.78/0.008
Poll.: *r/P*	−0.46/0.18	−0.57/0.08	−0.49/0.15	−0.49/0.15	−0.66/0.04	0.93/<0.0001

All data were collected in 2002 from three plots 10 × 10 m size. For other explanations, consult Table 6.1.

Table 6.10 Vegetation cover (%) in the impact zone of the copper smelter at Revda, Russia

Site	Top-canopy	Understorey	Field layer	Mosses
1–1	18.3	0	0.8	41.5
1–2	56.7	5.3	15.0	18.2
1–3	45.0	18.3	39.0	32.5
1–4	56.7	3.0	54.7	1.0
1–5	46.7	43.3	45.5	24.7
2–1	50.0	1.0	7.1	0.3
2–2	50.0	1.0	27.5	3.1
2–3	53.3	11.7	21.3	1.0
2–4	53.3	1.0	54.0	19.5
2–5	38.3	30.0	66.0	67.2
ANOVA: *F/P*	2.02/0.09	5.87/0.0005	20.3/<0.0001	11.2/<0.0001
Dist.: *r/P*	0.26/0.46	0.69/0.03	0.92/0.0001	0.34/0.34
Poll.: *r/P*	−0.47/0.17	−0.56/0.09	−0.83/0.003	−0.23/0.52

Cover of top-canopy and understorey estimated in 2003 from three plots 10 × 10 m size, cover of other layers – in 2007 from ten plots 1 × 1 m size. Cover of epigeic lichens and bare ground equals to zero in all sites. For other explanations, consult Table 6.1.

Table 6.11 Vegetation cover (%) in the impact zone of the aluminium smelter at Straumsvík, Iceland

Site	Understorey	Field layer	Mosses	Lichens	Bare ground
1–1	0	3.6	29.5	13.6	0
1–2	0	42.9	64.0	0.4	0
1–3	0	31.9	64.0	0.6	8.3
1–4	0	13.8	60.0	4.3	2.3
1–5	6.0	52.0	34.0	1.9	2.0
2–1	0	26.4	46.0	2.9	0
2–2	0	14.1	71.5	1.2	1.3
2–3	0	17.5	59.5	5.6	0.5
2–4	0	20.7	76.0	0.6	2.2
2–5	0.7	23.1	76.0	0.2	0.5
ANOVA: *F/P*	4.78/0.0017	4.13/0.0002	3.75/0.0005	5.14/<0.0001	3.50/0.01
Dist.: *r/P*	–	0.42/0.22	0.24/0.50	–0.34/0.33	0.02/0.96
Poll.: *r/P*	–	–0.58/0.08	–0.62/0.06	0.84/0.002	–0.26/0.47

Cover of understorey estimated in 2002 from three plots 10 × 10 m size, cover of other layers – in 2007 from ten plots 1 × 1 m size. Top-canopy cover equals to zero in all sites. Absence of correlation coefficients indicates that the data were not used in meta-analyses. For other explanations, consult Table 6.1.

Table 6.12 Vegetation cover (%) in the impact zone of the nickel-copper at Sudbury, Canada

Site	Top-canopy	Understorey	Field layer	Mosses	Lichens
1–1	3.0	0	13.9	9.9	10.2
1–2	18.3	0	36.2	15.7	5.6
1–3	20.0	4.3	40.5	3.5	2.3
1–4	13.3	16.7	43.8	6.4	1.8
1–5	40.0	0.3	3.5	8.6	0
2–1	3.0	0	10.0	20.7	11.5
2–2	40.0	10.3	5.1	20.1	4.2
2–3	23.3	16.7	43.5	0.8	0
2–4	20.0	10.0	49.5	0.2	0
2–5	36.7	11.0	10.6	0.9	0
ANOVA: *F/P*	8.16/<0.0001	5.02/0.0013	13.7/<0.0001	2.99/0.0037	3.94/0.0003
Dist.: *r/P*	0.66/0.04	0.46/0.18	0.18/0.62	–0.75/0.01	–0.92/0.0002
Poll.: *r/P*	–0.46/0.18	–0.48/0.16	–0.49/0.16	0.83/0.003	0.92/0.0001

Cover of top-canopy and understorey estimated in 2007 from three plots 10 × 10 m size, cover of other layers – in 2007 from ten plots 1 × 1 m size. For other explanations, consult Table 6.1.

Table 6.13 Vegetation cover (%) in the impact zone of the power plant at Vorkuta, Russia

Site	Understorey	Field layer	Mosses	Lichens	Bare ground
1–1	66.0	55.0	73.3	1.7	0
1–2	69.3	33.3	80.0	2.0	0
1–3	52.7	38.3	86.7	4.7	0
1–4	55.3	47.0	78.3	8.7	1.7
1–5	58.7	25.0	85.0	6.7	3.3
2–1	70.3	35.0	73.3	0	0
2–2	62.7	33.3	83.3	5.7	0
2–3	62.3	31.7	86.7	6.7	0
2–4	56.0	25.7	81.7	6.0	2.0
2–5	60.0	27.3	90.0	2.3	0.7
ANOVA: *F/P*	5.29/0.0009	1.42/0.24	4.57/0.0022	1.83/0.12	2.31/0.06
Dist.: *r/P*	−0.75/0.01	−0.50/0.14	0.75/0.01	0.65/0.04	–
Poll.: *r/P*	0.68/0.03	−0.13/0.72	−0.26/0.47	−0.45/0.19	–

All data were collected in 2002 from three plots 10 × 10 m size. Absence of correlation coefficients indicates that the data were not used in meta-analyses. For other explanations, consult Table 6.1.

Table 6.14 Vegetation cover (%) in the impact zone of the aluminium smelter at Žiar nad Hronom, Slovakia

Site	Top-canopy	Understorey	Field layer	Mosses	Bare ground
1–1	91.7	1.7	0	0.3	0
1–2	85.0	0.7	0.2	0.3	0.2
1–3	93.3	0	7.3	0.1	1.3
1–4	80.0	0.3	7.3	0.3	1.7
1–5	88.3	0	15.0	2.0	0.5
2–1	93.3	5.7	3.3	2.7	15.0
2–2	93.3	0	9.3	11.7	6.7
2–3	91.7	1.0	16.7	3.3	1.7
2–4	93.3	0	3.0	2.0	1.0
2–5	85.0	1.7	10.0	4.3	1.7
ANOVA: *F/P*	4.22/0.0035	2.88/0.02	3.01/0.02	9.00/<0.0001	14.4/<0.0001
Dist.: *r/P*	−0.25/0.49	−0.38/0.28	0.71/0.02	0.23/0.53	−0.19/0.60
Poll.: *r/P*	−0.01/0.99	0.09/0.80	09.66/0.04	−0.38/0.28	−0.28/0.44

All data were collected in 2002 from three plots 10 × 10 m size. Cover of epigeic lichens equals to zero in all sites. For other explanations, consult Table 6.1.

Table 6.15 Stand characteristics in the impact zone of aluminium smelter at Bratsk, Russia

Site	Stand characteristics			Seedling density, exx/ha	Proportion (%) of Scots pine	
	Basal area, m^2/ha	Composition[a]	Height, m		Among mature trees	Among seedlings
1–1	20.3	7P2B1A	23.0	245	72.1	57.8
1–2	28.0	8P1B1A	24.3	11	86.0	0
1–3	32.7	8P2B	31.7	37	83.6	8.3
1–4	32.7	8P1B1A	27.0	53	80.8	16.7
1–5	35.0	7P3B	27.7	64	69.8	46.7
2–1	8.7	4L3S3B	16.0	0	4.2	–
2–2	30.3	7P1S1B1A	17.0	43	66.6	8.3
2–3	33.3	7P2L1B	23.0	128	72.9	68.4
2–4	21.7	7P2B1A	27.7	187	70.4	42.4
2–5	26.7	8P2B	26.7	379	75.9	76.8
ANOVA (G_H): $F(\chi2)/P$	13.3/ <0.0001	189.9/ <0.0001	10.2/ <0.0001	5.30/ 0.0009	9.15/ <0.0001	3.96/ 0.0082
Dist.: *r*/*P*	0.54/0.11	–	0.71/ 0.02	0.37/ 0.30	0.41/ 0.23	0.40/ 0.28
Poll.: *r*/*P*	−0.85/ 0.0016	–	−0.67/ 0.03	−0.31/ 0.39	−0.90/ 0.0004	−0.15/ 0.71

Data collected in 2002.

[a] A, European aspen (*Populus tremula*); B, white birch (*Betula pubescens*); L, Siberian larch (*Larix sibirica*); P, Scots pine (*Pinus sylvestris*); S, Siberian spruce (*Picea abies* ssp. *obovata*).

G_H' test for significance of variation in stand composition between study sites. For other explanations, consult Table 6.1

Table 6.16 Stand characteristics in the impact zone of the fertiliser factory at Jonava, Lithuania

Site	Stand characteristics			Seedling density, exx/ha	Proportion (%) of Scots pine among mature trees[b]
	Basal area, m^2/ha	Composition[a]	Height, m		
1–1	44.3	10P	20.3	43	96.3
1–2	41.3	9P1Q	24.7	176	91.1
1–3	41.0	10P	23.0	53	100.0
1–4	38.0	7P2B1S	29.3	133	72.3
1–5	41.0	10P	23.3	219	99.2
2–1	34.3	10P	15.3	53	100.0
2–2	37.3	5P3B1S1Q	27.0	85	54.8
2–3	45.7	10P	25.0	165	100.0
2–4	33.0	10P	17.3	523	98.0
2–5	51.0	7P3S	25.7	528	71.9
ANOVA (G_H): $F(\chi^2)/P$	2.42/0.05	901.4/<0.0001	45.41/<0.0001	6.14/0.0004	39.9/<0.0001
Dist.: *r*/*P*	0.27/0.45	–	0.44/0.20	0.62/0.06	−0.11/0.76
Poll.: *r*/*P*	−0.46/0.18	–	−0.26/0.46	0.12/0.73	−0.14/0.71

Data collected in 2005.

[a] B, common birch (*Betula pendula*); P, Scots pine (*Pinus sylvestris*); Q, English oak (*Quercus robur*); S, Norway spruce (*Picea abies*).

[b] Only one Scots pine seedling had been recorded in the course of the survey (on site 1–4).

For other explanations, consult Tables 6.1 and 6.15

Table 6.17 Stand characteristics in the impact zone of the aluminium smelter at Kandalaksha, Russia

Site	Stand characteristics[a]		Seedling density, exx/ha	Proportion (%) of Scots pine	
	Basal area, m^2/ha	Composition[b]		Among mature trees	Among seedlings
1–1	23.3	10P	544	98.5	15.1
1–2	34.0	10P	885	96.1	9.9
1–3	27.0	10P	85	100.0	21.1
1–4	28.7	9P1B	208	94.2	21.4
1–5	21.7	8P2B	763	84.2	8.8
2–1	23.7	10P	389	100.0	3.9
2–2	29.3	10P	229	100.0	36.7
2–3	13.0	9P1S	117	91.4	91.1
2–4	23.3	9P1S	400	89.9	79.6
2–5	16.0	10P	75	98.4	53.7
ANOVA (G_H): F (χ^2)/P	2.48 0.0434	30.0/0.04	5.51/0.0007	2.21/0.0674	5.44/0.0008
Dist.: r/P	−0.33/0.35	–	−0.20/0.57	−0.54/0.10	0.39/0.27
Poll.: r/P	0.53/0.11	–	0.39/0.25	0.54/0.11	−0.61/0.06

Data collected in 2002.

[a] Stand height had not been measured.

[b] B, white birch (*Betula pubescens*); P, Scots pine (*Pinus sylvestris*); S, Siberian spruce (*Picea abies* ssp. *obovata*).

For other explanations, consult Tables 6.1 and 6.15

Table 6.18 Stand characteristics in the impact zone of the copper smelter at Karabash, Russia

Site	Stand characteristics			Seedling density, exx/ha	Proportion (%) of Scots pine	
	Basal area, m^2/ha	Composition[a]	Height, m		Among mature trees	Among seedlings
1–1	5.7	10B	5.3	16	0	0
1–2	20.0	2P8B	18.0	917	22.8	88.8
1–3	28.0	6P4B	24.7	368	60.8	60.9
1–4	34.7	6P4B	24.0	165	57.7	48.0
1–5	30.0	5P5B	22.0	213	53.3	70.5
2–1	2.7	10B	6.7	0	0	–
2–2	19.7	5P5B	11.3	720	46.7	98.4
2–3	28.3	6P4B	24.3	560	57.6	64.4
2–4	32.0	4P6B	21.3	16	40.9	50.0
2–5	57.7	8P2B	20.3	165	84.7	39.9
ANOVA (G_H): F (χ^2)/P	23.4/ <0.0001	206.9/ <0.0001	75.9/ <0.0001	1.51/ 0.21	3.56/ 0.0086	3.22/ 0.03
Dist.: r/P	0.90/ 0.0005	–	0.79/ 0.0065	−0.19/ 0.60	0.80/ 0.0054	−0.02/ 0.96
Poll.: r/P	−0.83/ 0.0031	–	−0.91/ 0.0003	−0.17/ 0.63	−0.79/ 0.0070	−0.21/ 0.58

Data collected in 2003.

[a] B, common birch (*Betula pendula*); P, Scots pine (*Pinus sylvestris*).

For other explanations, consult Tables 6.1 and 6.15

Table 6.19 Stand characteristics in the impact zone of the iron pellet plant at Kostomuksha, Russia

Site	Stand characteristics[a]		Seedling density, exx/ha	Proportion (%) of Norway spruce	
	Basal area, m^2/ha	Composition[b]		Among mature trees	Among seedlings
1–1	25.7	5S4P1B	91	51.5	0
1–2	34.5	9S1P	96	88.9	0
1–3	36.7	5S4P1B	299	47.5	26.3
1–4	32.3	6S3P1B	69	56.3	11.1
1–5	26.7	8P1S1B	32	14.6	16.7
2–1	29.7	5S4P1B	283	48.0	4.5
2–2	44.0	6S3P1B	91	58.3	14.1
2–3	30.5	7S2P1B	5	66.6	100.0
2–4	20.3	9P1B	240	0	0
2–5	23.3	8P1B1S	53	9.7	0
ANOVA (G_H): F (χ^2)/P	3.35/0.02	399.6/<0.0001	1.23/0.33	14.0/<0.0001	6.11/0.0007
Dist.: r/P	−0.24/0.50	–	−0.33/0.35	−0.45/0.19	0.07/0.85
Poll.: r/P	0.13/0.71	–	0.38/0.29	0.19/0.60	−0.20/0.57

Data collected in 2002.

[a] Stand height had not been measured.

[b] B, white birch (*Betula pubescens*); P, Scots pine (*Pinus sylvestris*); S, Siberian spruce (*Picea abies* ssp. *obovata*).

For other explanations, consult Tables 6.1 and 6.15

Table 6.20 Stand characteristics in the impact zone of the nickel-copper smelter at Monchegorsk, Russia

Site	Stand characteristics			Seedling density, exx/ha	Proportion (%) of Norway spruce	
	Basal area, m^2/ha	Composition[a]	Height, m		Among mature trees	Among seedlings
1–1	0.3	10B	2.3	5	0	0
1–2	0.3	10B	2.1	528	0	0
1–4	2.3	6P4B	2.6	528	0	0
1–5	5.3	5P5B	7.3	736	0	15.3
2–3	0	–	1.9	0	–	–
2–4	0	–	2.0	0	–	–
2–5	2.0	5B3S2P	6.9	907	27.8	5.1
2–8	12.0	9S1B	11.3	1,173	87.2	84.5
2–10	11.0	5B4S1P	11.8	320	36.3	71.6
2–12	18.7	8S2B	11.3	363	83.6	27.1
ANOVA (G_H): F (χ^2)/P	36.6/ <0.0001	94.2/ <0.0001	19.5/ <0.0001	5.85/ 0.0005	23.3/ <0.0001	49.8/ <0.0001
Dist.: r/P	0.87/0.0010	–	0.87/0.0010	0.45/0.20	0.73/0.04	0.63/0.09
Poll.: r/P	−0.62/0.05	–	−0.65/0.04	−0.59/0.07	−0.47/0.24	0.53/0.18

Data collected in 2001.

[a] B, white birch (*Betula pubescens*); P, Scots pine (*Pinus sylvestris*); S, Siberian spruce (*Picea abies* ssp. *obovata*).

For other explanations, consult Tables 6.1 and 6.15

Table 6.21 Stand characteristics in the impact zone of the aluminium smelter at Nadvoitsy, Russia

Site	Stand characteristics			Seedling density, exx/ha	Proportion (%) of Scots pine	
	Basal area, m^2/ha	Composition[a]	Height, m		Among mature trees	Among seedlings
1–1	41.3	9P1B	19.7	2,021	86.9	55.1
1–2	31.0	7P2B1A	17.7	837	72.8	0.4
1–3	36.7	9P1B	24.3	272	86.5	0
1–4	35.7	6P4B	24.7	69	62.4	24.4
1–5	26.7	5B3P1S1A	17.0	341	34.2	7.8
2–1	33.7	7P3B	21.3	1,472	72.0	86.7
2–2	29.0	6P2B1A1S	24.3	192	61.2	0
2–3	27.7	9P1B	24.7	64	90.9	24.4
2–4	27.0	9P1B	17.7	53	92.7	36.7
2–5	40.3	6P3B1A	25.0	309	55.1	4.7
ANOVA (G_H): F (χ^2)/P	1.81/0.12943	260.6/<0.0001	72.3/<0.0001	3.08/0.02	4.87/0.0015	6.45/0.0003
Dist.: r/P	−0.24/0.50	–	0.00/0.99	−0.73/0.02	−0.48/0.16	−0.54/0.11
Poll.: r/P	0.12/0.72	–	−0.14/0.71	0.72/0.02	0.14/0.71	0.47/0.17

Data collected in 2004.

[a]A, European aspen (*Populus tremula*); B, white birch (*Betula pubescens*); P, Scots pine (*Pinus sylvestris*); S, Siberian spruce (*Picea abies* ssp. *obovata*). For other explanations, consult Tables 6.1 and 6.15

Table 6.22 Stand characteristics in the impact zone of the nickel-copper smelter at Nikel and ore-roasting plant at Zapolyarnyy, Russia

Site	Stand characteristics			Seedling density, exx/ha	Proportion (%) of mountain birch	
	Basal area, m^2/ha	Composition[a]	Height, m		Among mature trees	Among seedlings
1–1	0	–	4.2	0	–	–
1–2	0.3	10B	3.3	0	100.0	–
1–3	2.0	10B	6.8	747	100.0	90.5
1–4	0.7	10B	3.7	0	100.0	–
1–5	0	–	3.8	64	–	100.0
2–1	1.7	10B	2.5	0	100.0	–
2–2	1.0	–	3.3	395	0	56.1
2–3	4.3	6P4B	6.7	2,000	38.9	92.6
2–4	2.0	10B	6.0	1,227	100.0	89.4
2–5	11.7	6P4B	7.2	864	38.9	95.1
ANOVA (G_H): F (χ^2)/P	8.49/<0.0001	39.2/<0.0001	11.3/<0.0001	2.59/0.04	6.02/0.0079	2.82/0.09
Dist.: r/P	0.40/0.25	–	0.45/0.19	0.31/0.38	−0.01/0.98	0.85/0.03
Poll.: r/P	−0.26/0.47	–	−0.42/0.23	−0.39/0.27	0.09/0.84	−0.94/0.0044

Data collected in 2001.
[a]B, white birch (*Betula pubescens*); P, Scots pine (*Pinus sylvestris*).
For other explanations, consult Tables 6.1 and 6.15

Table 6.23 Stand characteristics in the impact zone of the nickel-copper smelters at Norilsk, Russia

Site	Stand characteristics			Seedling density, exx/ha	Proportion (%) of Siberian larch	
	Basal area, m^2/ha	Composition[a]	Height, m		Among mature trees	Among seedlings
1–1	0	–	–	0	–	–
1–2	0.3	10L	7.0	0	100.0	
1–3	1.0	10L	6.5	0	100.0	–
1–4	4.0	10L	6.0	5	100.0	100.0
1–5	2.0	8L2A	5.5	16	75.0	100.0
2–1	0	–	–	5	–	100.0
2–2	0.3	10B	2.0	59	0	0
2–3	10.7	6L3B1S	16.0	1,093	58.3	10.2
2–4	5.0	10L	8.3	37	100.0	100.0
2–5	7.0	6L4B	19.3	37	64.3	60.0
ANOVA (G_H): F (χ^2)/P	9.75/<0.0001	41.6/0.007	57.3/<0.0001	20.7/<0.0001	4.52/0.02	5.05/0.03
Dist.: r/P	0.54/0.10	–	0.49/0.21	0.04/0.91	0.39/0.34	0.23/0.63
Poll.: r/P	–0.45/0.20	–	–0.34/0.41	–0.06/0.87	–0.78/0.02	–0.22/0.64

Data collected in 2002.

[a]A, Siberian alder (*Duschekia fruticosa*); B, white birch (*Betula pubescens*); L, Siberian larch (*Larix sibirica*); S, Norway spruce (*Picea abies* ssp. *obovata*). For other explanations, consult Tables 6.1 and 6.15

Table 6.24 Stand characteristics in the impact zone of the copper smelter at Revda, Russia

Site	Stand characteristics			Seedling density, exx/ha	Proportion (%) of Scots pine	
	Basal area, m^2/ha	Composition[a]	Height, m		Among mature trees	Among seedlings
1–1	23.0	4S4F2B	18.7	229	3.2	0
1–2	33.0	4S4F2B	24.0	917	0	0
1–3	41.0	6F2B1S1A	26.3	885	1.7	0
1–4	28.7	5S3B2P	25.0	176	23.1	0
1–5	34.7	8F2S	25.0	336	0	0
2–1	17.0	5P2A2S1F	19.3	1,243	47.7	11.5
2–2	29.3	5S5P	18.3	576	44.1	11.5
2–3	24.7	5B3P1F1S	21.7	288	29.3	5.7
2–4	47.7	8P1S1B	24.7	165	82.6	0
2–5	52.0	6S3P1F	24.0	96	26.7	0
ANOVA (G_H): F (χ^2)/P	9.34/<0.0001	904.6/<0.0001	7.29/0.0001	5.62/0.0007	11.7/<0.0001	3.24/0.01
Dist.: r/P	0.72/0.02	–	0.80/0.0060	−0.58/0.08	−0.03/0.94	−0.59/0.07
Poll.: r/P	−0.73/0.02	–	−0.94/<0.0001	0.28/0.43	0.11/0.77	−0.63/0.05

Data collected in 2003.

[a]A, aspen (*Populus tremula*); B, white birch (*Betula pubescens*); F, Siberian fir (*Abies sibirica*); P, Scots pine (*Pinus sylvestris*); S, Siberian spruce (*Picea abies* ssp. *obovata*).

For other explanations, consult Tables 6.1 and 6.15

Table 6.25 Stand characteristics in the impact zone of the nickel-copper smelter at Sudbury, Canada

Site	Stand basal area, m^2/ha	Stand composition[a]	Proportion (%) of aspen among mature trees[b]
1–1	0	–	–
1–2	4.3	6P2M2B	0
1–3	15.7	4P4O1B1F	0
1–4	11.0	7A2P1B	72.8
1–5	30.7	5A3F1B1O	48.5
2–1	0	–	–
2–2	17.7	8B2M	0
2–3	15.7	8A2P	75.3
2–4	11.3	4A2M2P2B	21.2
2–5	24.0	7A3F	65.2
ANOVA (G_H): F (χ^2)/P	7.48/<0.0001	322.3/<0.0001	9.35/0.0001
Dist.: r/P	0.80/0.0051	–	0.63/0.10
Poll.: r/P	−0.67/0.04	–	−0.64/0.09

Data collected in 2007.

[a] A, quaking aspen (*Populus tremuloides*); B, canoe birch (*Betula papyrifera*); F, balsam fir (*Abies balsamea*); M, red maple (*Acer rubrum*); O, red oak (*Quercus rubra*); P, jack pine (*Pinus banksiana*).

[b] Seedlings were not counted.

For other explanations, consult Tables 6.1 and 6.15

Table 6.26 Stand characteristics in the impact zone of the aluminium smelter at Žiar nad Hronom, Slovakia

Site	Stand characteristics			Seedling density, exx/ha	Proportion (%) of European beech	
	Basal area, m^2/ha	Composition[a]	Height, m		Among mature trees	Among seedlings
1–1	25.0	8B2H	30.0	416	81.8	65.7
1–2	24.3	5H4P1B	23.3	629	7.3	8.3
1–3	30.7	6H4B	26.3	4,149	36.9	89.6
1–4	40.0	9B1F	33.7	2,800	94.3	77.8
1–5	33.0	9B1O	33.3	5,728	91.5	65.2
2–1	27.0	6B3H1P	30.0	6,149	62.8	63.3
2–2	27.3	10B	29.0	6,736	100.0	99.3
2–3	25.3	6B2H1M1L	32.0	5,771	55.4	29.5
2–4	44.0	9B1F	35.3	1,691	87.0	98.2
2–5	29.0	7B1A1H1F	36.3	9,989	79.3	3.3
ANOVA (G_H): F (χ^2)/P	6.90/0.0002	604.5/<0.0001	6.46/0.0003	4.72/0.0019	67.2/<0.0001	12.9/<0.0001
Dist.: r/P	0.61/0.06	–	0.78/0.0074	0.56/0.09	0.54/0.11	0.08/0.82
Poll.: r/P	–0.44/0.20	–	–0.47/0.17	–0.67/0.03	–0.29/0.41	–0.18/0.62

Data collected in 2002.

[a] A, European ash (*Fraxinus excelsior*); B, European beech (*Fagus sylvatica*); F, silver fir (*Abies alba*); H, European hornbeam (*Carpinus betulus*); M, Norway maple (*Acer platanoides*); O, durmast oak (*Quercus petraea*); L, small-leaved lime (*Tilia cordata*).

For other explanations, consult Tables 6.1 and 6.15.

Table 6.27 Occurrences of vascular plants (numbers of sampling plots, out of three, on which the species was recorded) in the impact zone of the aluminium smelter at Bratsk, Russia

Species	Life form	Occurrences of species on sampling plots									
		1–1	1–2	1–3	1–4	1–5	2–1	2–2	2–3	2–4	2–5
Abies sibirica	w	0	0	0	0	0	0	0	0	1	0
Achillea millefolium	h	0	1	0	3	0	0	2	0	1	0
Aconitum volubile	h	0	0	0	0	0	0	0	0	0	0
Adenophora coronopifolia	h	2	2	0	0	0	0	0	0	0	0
Adoxa moschatellina	h	1	0	2	0	2	2	3	0	1	0
Agrimonia pilosa	h	0	0	2	2	0	1	0	3	0	0
Angelica sylvestris	h	0	1	2	0	2	1	0	0	2	2
Antennaria dioica	h	1	0	0	0	0	0	0	1	0	0
Artemisia latifolia	h	0	0	0	0	3	0	0	0	0	0
Betula pendula	w	0	0	3	3	0	0	0	3	0	0
Betula pubescens	w	3	3	1	2	3	3	3	0	3	3
Calamagrostis arundinacea	g	2	1	1	1	3	1	1	3	2	1
Calamagrostis purpurea	g	1	0	1	0	0	0	2	0	0	0
Carex globularis	g	1	2	1	2	2	3	1	3	1	2
Cirsium helenioides	h	1	1	0	1	0	0	0	0	0	2
Clematis alpina ssp. *sibirica*	ds	0	1	3	2	2	3	0	3	3	1
Conioselinum tataricum	h	3	2	0	2	2	2	3	0	0	1
Cotoneaster niger	w	0	0	0	0	0	1	0	1	0	0
Crepis sibirica	h	0	1	1	0	1	0	0	0	1	0
Cypripedium guttatum	h	1	1	0	1	0	0	0	0	0	1
Dendranthema zawadskii	h	0	0	0	0	0	0	0	1	0	0
Duschekia fruticosa	w	0	0	3	3	0	2	1	3	2	0
Epilobium angustifolium	h	3	1	3	3	1	3	3	0	2	1
Equisetum pratense	h	0	0	0	0	0	0	0	0	3	3
Euphorbia jenisseiensis	h	1	3	0	0	0	0	0	0	0	0
Festuca gigantea	h	0	0	3	0	0	0	0	0	0	0
Festuca ovina	g	0	0	0	0	0	0	1	0	0	0
Fragaria vesca	h	0	0	0	0	0	3	0	0	0	0
Galium boreale	h	3	2	0	1	3	3	0	1	3	2
Geranium albiflorum	h	2	3	0	2	0	3	3	0	2	0
Goodyera repens	h	0	0	0	0	1	0	0	0	0	0
Gymnocarpium dryopteris	h	0	0	1	1	2	0	0	0	3	0
Hieracium sp.	h	1	0	0	0	0	0	0	0	0	0
Hieracium umbellatum	h	0	1	1	0	0	0	0	0	1	0
Lactuca sibirica	h	1	1	3	0	1	1	0	0	1	1
Larix sibirica	w	3	0	2	2	1	2	3	3	2	3
Lathyrus sp.	h	3	3	1	3	3	1	1	3	1	2
Lathyrus vernus	h	0	0	1	3	3	0	0	0	3	3
Ledum palustre	ds	3	0	1	0	3	1	3	0	0	3
Lilium pilosiusculum	h	0	0	0	0	0	0	0	0	1	0
Linnaea borealis	h	0	1	3	3	3	0	2	2	3	2
Lonicera altaica	w	1	0	0	0	0	0	0	1	1	0
Lonicera caerulea ssp. *pallasii*	w	0	0	0	1	2	0	0	1	0	2
Luzula pilosa	g	0	0	0	0	0	0	0	0	1	0
Lycopodium annotinum	h	0	0	0	0	0	0	0	0	0	2
Maianthemum bifolium	h	2	0	2	3	2	0	3	2	2	3
Melica nutans	g	0	0	0	0	0	0	0	0	1	0
Orthilia secunda	h	1	0	0	0	1	0	0	3	0	0

(continued)

Table 6.27 (continued)

Species	Life form	Occurrences of species on sampling plots									
		1–1	1–2	1–3	1–4	1–5	2–1	2–2	2–3	2–4	2–5
Paris quadrifolia	h	0	0	0	2	0	0	0	0	0	0
Pedicularis labradorica	h	0	0	0	0	0	0	1	0	0	0
Pedicularis resupinata	h	0	0	0	0	1	0	1	0	0	0
Peucedanum palustre	h	1	0	0	0	0	0	0	0	0	0
Picea abies ssp. *obovata*	w	1	1	0	0	0	3	1	2	2	3
Pinus sibirica	w	0	0	0	0	0	0	0	3	3	0
Pinus sylvestris	w	3	3	3	3	3	2	3	3	3	3
Pleurospermum uralense	h	0	1	0	1	1	0	1	0	0	1
Polemonium racemosum	h	0	1	0	0	0	0	0	0	0	0
Populus tremula	w	2	2	2	2	1	1	3	0	3	3
Pulmonaria mollis	h	0	1	2	1	1	0	0	0	1	0
Pulsatilla flavescens	h	2	2	1	1	2	1	0	0	0	1
Pyrola rotundifolia	h	3	0	3	1	2	0	0	2	0	3
Ranunculus acris ssp. *borealis*	h	0	1	0	0	0	0	0	0	0	0
Rosa acicularis	w	2	2	3	3	2	3	3	2	3	3
Rubus matsumuranus	w	0	0	1	2	0	1	0	0	0	0
Rubus saxatilis	h	1	3	3	3	3	2	3	3	3	3
Rumex aquaticus	h	0	0	0	0	0	1	0	0	0	0
Salix caprea	w	0	0	1	0	0	1	0	1	1	0
Salix hastata	w	0	1	0	0	1	1	0	0	0	1
Salix phylicifolia	w	0	0	0	0	1	2	0	0	0	0
Salix taraikensis	w	0	0	0	0	1	0	0	0	0	0
Sanguisorba officinalis	h	0	3	3	2	0	1	1	0	0	0
Saposhnikovia divaricata	h	0	1	0	0	0	0	0	0	0	0
Saussurea sp.	h	0	2	0	0	0	0	0	0	0	0
Saxifraga nelsoniana	h	0	0	0	1	0	0	0	0	0	0
Senecio sp. (cf. *integrifolius*)	h	0	0	0	0	0	0	0	2	1	0
Solidago daurica	h	0	0	0	1	0	0	0	0	0	0
Sorbus aucuparia	w	0	0	3	3	1	2	0	0	1	0
Spiraea media	w	3	3	0	2	2	3	2	1	3	0
Tanacetum bipinnatum	h	0	2	0	1	0	2	0	0	0	0
Tanacetum vulgare	h	0	0	0	0	0	0	1	0	0	0
Thalictrum minus	h	0	3	2	3	2	1	1	0	3	3
Trientalis europaea	h	0	0	2	2	0	0	1	0	2	3
Trifolium lupinaster	h	0	1	0	0	1	0	0	1	2	3
Trollius asiaticus	h	3	3	0	3	3	3	3	3	3	3
Vaccinium myrtillus	ds	0	0	0	0	0	0	0	0	0	0
Vaccinium uliginosum	ds	2	0	0	0	2	0	3	0	0	0
Vaccinium vitis-idaea	ds	3	3	3	3	3	1	3	3	3	3
Veratrum sp.	h	0	0	0	0	0	0	2	0	0	0
Vicia cracca	h	3	0	0	0	1	1	1	0	2	3
Viola canina	h	1	0	0	0	2	3	1	0	0	1
Viola epipsiloides	h	0	0	0	1	0	0	0	0	1	0
Viola uniflora	h	3	2	1	3	1	1	1	0	1	3

Data collected 2–4.8.2002. Life forms: ds - dwarf shrubs, g - grasses, h - herbs, w - trees and shrubs.

Table 6.28 Occurrences of vascular plants (numbers of sampling plots, out of three, on which the species was recorded) in the impact zone of the fertilising factory at Jonava, Lithuania

Species	Life form	Occurrences of species on sampling plots									
		1–1	1–2	1–3	1–4	1–5	2–1	2–2	2–3	2–4	2–5
Acer negudo	w	1	0	0	0	0	0	0	0	0	0
Acer platanoides	w	1	2	0	0	0	1	3	3	3	0
Betula pendula	w	0	1	0	2	0	0	3	0	0	1
Bilderdykia convolvulus	h	0	0	0	0	0	3	0	0	0	0
Calamagrostis arundinacea	g	0	0	0	1	0	0	0	1	3	3
Calamagrostis canescens	g	0	0	3	3	3	0	0	0	0	0
Carduus crispus	h	0	1	0	0	0	0	0	0	0	0
Carpinus betulus	w	0	0	0	0	0	0	1	0	0	0
Chelidonium majus	h	2	3	0	0	0	3	3	0	0	0
Convollaria majalis	h	0	1	0	0	1	0	0	0	0	0
Corylus avellana	w	2	3	3	3	2	0	3	1	0	3
Cystopteris fragilis	h	0	0	2	2	1	0	0	3	0	0
Dryopteris carthusiana	h	0	0	0	0	0	0	0	0	0	3
Dryopteris filix-mas	h	1	1	3	0	3	2	3	3	2	1
Epilobium angustifolium	h	0	0	0	0	0	2	0	2	1	0
Equisetum fluviatile	h	0	0	0	0	1	0	0	0	0	0
Equisetum pratense	h	1	0	0	0	0	0	0	0	0	0
Erodium cicutarium	h	0	0	1	0	0	0	0	0	0	0
Euonymus verrucosus	w	0	0	0	0	0	0	1	0	0	0
Fragaria vesca	h	1	2	2	3	3	0	0	3	3	2
Frangula alnus	w	1	3	3	3	3	0	2	3	3	3
Galeopsis tetrahit	h	2	2	0	2	1	3	0	2	2	0
Galium album	h	0	0	0	0	0	0	0	0	2	0
Galium boreale	h	0	1	0	0	0	0	0	0	0	0
Galium palustre	h	0	1	0	0	0	0	0	0	0	0
Geum urbanum	h	0	3	0	0	0	0	0	0	2	0
Hieracium laevigatum	h	0	0	0	2	0	0	0	0	0	0
Hieracium mixopolium	h	0	0	0	0	0	0	0	0	0	0
Hieracium silvularum	h	0	0	1	0	1	0	0	1	0	0
Humulus lupulus	h	3	0	0	0	0	0	0	0	0	0
Hypericum perforatum	h	0	0	0	1	1	0	0	0	2	0
Impatiens parviflora	h	0	3	2	1	3	0	3	0	0	0
Knautia arvensis	h	0	0	0	2	0	0	0	1	3	0
Lonicera xylosteum	w	1	0	0	0	0	0	0	0	0	0
Luzula pilosa	g	0	0	0	3	2	0	0	0	0	1
Lysimachia vulgaris	h	0	0	0	0	3	0	0	0	0	0
Maianthemum bifolium	h	0	1	0	0	0	0	0	0	0	1
Malus domestica	w	0	0	0	0	0	0	0	2	0	0
Malus sylvestris	w	0	1	0	0	0	0	0	0	0	0
Melampyrum pratense	h	0	0	0	0	0	0	0	0	0	0
Melica nutans	g	0	3	0	0	0	0	0	0	0	0
Moehringia trinervia	h	1	3	3	2	3	3	3	0	0	1
Mycelis muralis	h	1	2	3	0	3	3	0	3	2	3
Oxalis acetosella	h	3	3	3	1	3	0	3	0	3	3
Paris quadrifolia	h	0	0	0	0	0	0	0	0	0	1
Phragmitis australis	g	0	0	0	0	2	0	0	0	0	0
Picea abies	w	0	1	1	3	3	0	3	2	2	3
Pinus sylvestris	w	3	3	3	3	3	3	3	3	3	3

(continued)

Table 6.28 (continued)

Species	Life form	Occurrences of species on sampling plots									
		1–1	1–2	1–3	1–4	1–5	2–1	2–2	2–3	2–4	2–5
Populus tremula	w	0	0	0	0	0	0	0	1	2	2
Prunus domestica	w	1	0	1	0	0	1	0	0	0	0
Prunus serotina	w	0	0	0	0	0	0	0	2	0	0
Pyrola rotundifolia	h	0	0	1	0	0	0	0	0	0	0
Pyrus communis	w	0	1	0	0	0	0	0	0	0	0
Quercus robur	w	3	3	3	3	3	3	2	3	3	3
Ribes uva-crispa	w	0	0	0	0	0	1	0	0	0	0
Rubus caesius	w	3	0	0	0	0	0	0	0	0	0
Rubus idaeus	w	3	3	3	3	3	3	3	3	3	3
Rumex acetosa	h	0	0	0	1	0	0	0	0	1	0
Rumex acetosella	h	0	0	0	3	0	0	0	0	0	0
Salix cinerea	w	0	1	0	0	0	0	0	0	0	0
Sambucus racemosa	w	0	1	1	0	0	0	0	0	0	0
Silene vulgaris	h	0	0	0	0	0	0	0	0	0	0
Solidago virgaurea	h	0	2	0	1	1	0	0	0	0	1
Sorbus aucuparia	w	2	2	3	3	3	1	3	3	2	3
Stellaria holostea	h	0	0	0	0	2	0	0	0	0	0
Tilia cordata	w	1	0	0	0	1	0	0	0	1	0
Urtica dioica	h	3	3	0	0	0	3	2	1	2	0
Vaccinium myrtillus	ds	0	0	0	3	0	0	0	0	0	2
Veronica chamaedrys	h	0	0	0	0	0	0	0	2	0	0
Veronica officinalis	h	0	0	0	0	0	0	0	0	2	0
Viburnum opulus	w	0	1	0	0	1	0	0	0	0	0
Vicia sepium	h	0	1	0	0	0	0	0	0	0	0
Viola riviniana	h	0	0	0	3	0	0	0	0	0	0

Data collected 3–5.9.2005. Life forms: ds - dwarf shrubs, g - grasses, h - herbs, w - trees and shrubs.

Table 6.29 Occurrences of vascular plants (numbers of sampling plots, out of three, on which the species was recorded) in the impact zone of the aluminium smelter at Kandalaksha, Russia

Species	Life form	Occurrences of species on sampling plots									
		1–1	1–2	1–3	1–4	1–5	2–1	2–2	2–3	2–4	2–5
Andromeda polifolia	ds	0	0	0	0	0	0	0	1	0	0
Arctostaphylos uva-ursi	ds	0	0	0	0	0	0	0	0	0	2
Betula nana	w	0	0	0	0	0	0	0	2	0	0
Betula pendula	w	1	0	0	0	0	0	1	0	0	0
Betula pubescens	w	3	3	3	3	3	2	3	3	1	2
Calluna vulgaris	ds	0	0	0	0	1	0	0	3	2	2
Carex globularis	g	0	0	1	0	0	0	0	2	0	0
Deschampsia cespitosa	g	0	0	0	0	0	1	0	0	0	0
Deschampsia flexuosa	g	3	1	3	3	3	2	3	1	2	0
Diphasium complanatum	h	0	0	0	0	2	0	0	0	0	0
Empetrum nigrum	ds	2	3	3	3	3	3	3	3	3	3
Epilobium angustifolium	h	1	2	1	2	3	3	3	0	1	1
Gymnocarpium dryopteris	h	0	0	0	1	0	0	0	0	0	0
Hieracium sp.	h	0	0	0	0	1	1	0	0	0	0
Juniperus communis ssp. *nana*	w	1	1	3	3	2	1	1	0	1	1
Ledum palustre	ds	3	3	2	2	3	1	3	3	3	0
Linnaea borealis	h	3	3	2	3	3	3	3	0	3	0
Luzula pilosa	g	2	0	0	3	2	2	2	0	2	0
Lycopodium annotinum	h	0	1	0	1	0	0	0	0	0	0
Lycopodium clavatum	h	0	1	0	0	0	0	0	0	0	0
Melampyrum sylvaticum	h	0	1	3	3	3	0	2	0	3	0
Moneses uniflora	h	1	0	0	0	0	3	1	0	0	0
Picea abies ssp. *obovata*	w	2	3	2	3	3	1	1	3	3	3
Pinus sylvestris	w	3	3	3	3	3	3	3	3	3	3
Poa pratensis	g	0	0	0	0	0	1	1	0	0	0
Populus tremula	w	3	3	0	0	3	2	0	0	0	0
Salix caprea	w	3	1	3	1	2	3	3	0	0	1
Salix phylicifolia	w	2	0	0	0	1	0	0	0	0	0
Solidago virgaurea	h	2	3	2	3	3	3	2	0	0	0
Sorbus aucuparia	w	3	3	1	3	1	3	3	0	1	0
Trientalis europaea	h	3	0	0	2	0	2	3	0	1	0
Vaccinium myrtillus	ds	3	3	3	3	3	3	3	3	3	3
Vaccinium uliginosum	ds	3	1	2	2	3	0	3	3	2	2
Vaccinium vitis-idaea	ds	3	3	3	3	3	3	3	3	3	3

Data collected 26.6.2002. Life forms: ds - dwarf shrubs, g - grasses, h - herbs, w - trees and shrubs.

Table 6.30 Occurrences of vascular plants (numbers of sampling plots, out of three, on which the species was recorded) in the impact zone of the copper smelter at Karabash, Russia

Species	Life form	Occurrences of species on sampling plots									
		1–1	1–2	1–3	1–4	1–5	2–1	2–2	2–3	2–4	2–5
Abies sibirica	w	0	1	0	0	0	0	0	1	0	1
Achillea millefolium	h	0	0	0	0	3	0	1	0	2	1
Aconitum septentrionale	h	0	0	1	0	2	0	0	0	0	0
Adenophora lilifolia	h	0	0	1	1	0	0	0	3	0	2
Aegopodium podagraria	h	0	0	0	3	3	0	0	0	3	1
Agrimonia pilosa	h	0	0	0	1	2	0	0	0	1	1
Agrostis capillaris	g	0	2	0	0	2	0	0	0	3	0
Agrostis gigantea	g	0	0	0	0	0	0	1	0	0	0
Ajuga reptans	h	0	0	3	2	3	0	0	2	3	0
Alchemilla sp.	h	0	0	0	0	0	0	0	0	3	0
Alnus incana	w	0	0	0	0	0	0	2	0	0	0
Angelica sylvestris	h	0	0	0	2	0	0	0	2	0	1
Antennaria dioica	h	0	0	0	0	0	0	0	0	0	2
Asarum europaeum	h	0	0	1	0	0	0	0	0	2	0
Athyrium filix-femina	h	0	0	0	1	1	0	0	0	1	0
Betula pendula	w	3	2	1	3	3	3	3	1	3	0
Betula pubescens	w	0	3	0	0	0	0	0	3	1	3
Brachypodium pinnatum	g	0	1	0	0	1	0	0	1	1	0
Bupleurum longifolium ssp. *aureum*	h	0	0	0	0	0	0	0	0	0	1
Calamagrostis arundinacea	g	0	3	3	0	0	0	2	3	1	0
Calamagrostis obtusata	g	0	1	3	3	0	0	0	3	0	0
Calamagrostis purpurea ssp. *langsdorfii*	g	0	0	0	0	0	0	0	0	0	3
Campanula glomerata	h	0	0	1	1	2	0	0	0	1	0
Campanula sp.	h	0	0	0	0	0	0	0	1	0	1
Campanula stevenii ssp. *wolgensis*	h	0	0	2	0	0	0	0		0	0
Carduus nutans	h	0	0	0	0	2	0	0	0	0	0
Carex montana	g	0	0	0	0	0	0	0	0	1	0
Carex sp.	g	0	0	0	0	0	0	0	0	0	1
Cerastium fontanum ssp. *triviale*	h	0	0	0	0	0	0	0	0	1	0
Cerastium pauciflorum	h	0	0	0	2	0	0	0	0	0	0
Chamaecytisus ruthenicus	w	0	2	3	3	1	0	0	3	2	3
Conioselinum tataricum	h	0	0	0	0	2	0	1	1	3	0
Cotoneaster niger	w	0	0	0	0	0	0	0	0	1	0
Crataegus sanguinea	w	0	0	0	1	1	0	0	0	0	0
Dactylis glomerata	g	0	0	0	0	2	0	0	0	1	0
Deschampsia cespitosa	g	0	2	0	3	3	0	0	0	3	0
Digitalis grandiflora	h	0	0	2	1	0	0	0	0	0	1
Elymus repens	g	0	1	0	0	0	0	0	0	0	0
Epilobium angustifolium	h	0	0	0	0	0	0	0	0	0	2
Epipactis atrorubens	h	0	0	1	0	1	0	0	0	0	0
Epipactis helleiborine	h	0	0	1	0	0	0	0	0	0	0
Equisetum pratense	h	0	2	0	0	0	0	0	0	0	0
Euphorbia sp.	h	0	0	0	0	0	0	0	1	0	0
Festuca rubra	g	0	2	0	0	0	0	1	2	1	0
Festuca valesiaca	g	0	0	0	0	0	0	0	1	0	0

(continued)

Table 6.30 (continued)

Species	Life form	Occurrences of species on sampling plots									
		1–1	1–2	1–3	1–4	1–5	2–1	2–2	2–3	2–4	2–5
Filipendula ulmaria	h	0	0	0	0	1	0	0	0	1	0
Filipendula vulgaris	h	0	0	2	2	2	0	0	0	1	1
Fragaria vesca	h	0	0	3	3	3	0	0	3	3	3
Galeopsis bifida	h	0	0	0	0	0	0	0	0	0	1
Galium boreale	h	0	0	3	2	3	0	1	3	2	3
Galium odoratum	h	0	0	0	0	0	0	0	0	1	0
Galium ruthenicum	h	0	0	0	0	0	0	2	0	0	0
Geranium sylvaticum ssp. *pseudosibiricum*	h	0	0	0	0	2	0	0	0	0	0
Geranium sylvaticum ssp. *sylvaticum*	h	0	0	3	3	3	0	0	3	3	1
Geum rivale	h	0	0	0	0	3	0	0	0	0	0
Geum urbanum	h	0	0	0	2	0	0	0	0	3	0
Glehoma hederacea	h	0	0	0	0	0	0	0	0	2	0
Hieracium caespitosum ssp. *brevipilum*	h	0	0	0	0	2	0	0	0	1	0
Hieracium pilosella	h	0	0	0	0	0	0	0	0	1	0
Hieracium umbellatum	h	0	0	1	0	0	0	0	2	0	0
Hypericum maculatum	h	0	0	0	0	2	0	0	0	0	0
Hypochoeris maculata	h	0	0	0	0	0	0	1	2	0	0
Inula salicina	h	0	0	2	0	0	0	0	0	0	0
Larix sibirica	w	0	0	0	0	0	0	0	3	0	0
Lathyrus gmelinii	h	0	0	0	0	0	0	0	0	1	2
Lathyrus pisiformis	h	0	0	2	1	1	0	0	0	0	0
Lathyrus pratensis	h	0	0	0	1	1	0	0	0	0	0
Lathyrus vernus	h	0	0	3	3	2	0	0	2	2	3
Leucanthemum vulgare	h	0	0	0	0	1	0	0	0	3	0
Lilium martagon	h	0	0	2	2	0	0	0	2	2	1
Lonicera sp.	w	0	0	0	0	3	0	1	0	0	0
Luzula pilosa	g	0	0	0	1	0	0	0	0	0	0
Maianthemum bifolium	h	0	1	1	3	0	0	0	3	3	2
Melampyrum pratense	h	0	0	2	1	1	0	0	3	0	1
Melica nutans	g	0	0	0	0	1	0	0	0	3	0
Moneses uniflora	h	0	0	0	1	0	0	0	0	0	2
Neottia nudus-avis	h	0	0	0	0	0	0	0	0	0	1
Neottianthe cucullata	h	0	0	0	1	0	0	0	0	0	0
Origanum vulgare	h	0	0	1	1	0	0	0	0	0	0
Orthilia secunda	h	0	0	2	3	1	0	1	3	1	3
Phleum phleoides	g	0	0	0	0	1	0	0	0	0	0
Picea abies ssp. *obovata*	w	0	0	0	0	0	0	0	1	0	1
Pimpinella saxifraga	h	0	0	3	2	0	0	2	2	0	2
Pinus sylvestris	w	0	3	3	3	3	0	3	3	3	3
Plantago major	h	0	0	0	0	0	0	0	0	3	0
Plantago media	h	0	0	0	1	0	0	0	0	0	0
Pleurospermum uralense	h	0	0	0	0	0	0	0	2	0	1
Polygonatum officinale	h	0	0	2	1	0	0	0	3	1	3
Polygonum bistorta	h	0	0	0	0	0	0	0	0	2	0
Populus tremula	w	0	3	3	3	1	0	0	1	0	0
Potentilla erecta	h	0	0	1	0	0	0	0	2	0	0

(continued)

Table 6.30 (continued)

Species	Life form	Occurrences of species on sampling plots									
		1–1	1–2	1–3	1–4	1–5	2–1	2–2	2–3	2–4	2–5
Primula veris ssp. *macrocalix*	h	0	0	1	0	3	0	0	0	1	0
Prunella vulgaris	h	0	0	0	2	1	0	0	0	3	0
Prunus padus	w	0	0	0	2	1	0	0	0	2	2
Pteridium aquilinum	h	0	1	2	0	3	0	0	2	0	0
Pulmonaria mollis	h	0	0	0	0	1	0	0	0	0	0
Pulmonaria obscura	h	0	0	2	3	2	0	0	0	2	2
Pyrola media	h	0	0	0	2	0	0	0	1	0	1
Pyrola rotundifolia	h	0	0	0	3	0	0	0	1	3	1
Ranunculus acris	h	0	0	0	0	0	0	0	0	1	0
Ranunculus auricomus	h	0	0	0	0	0	0	0	0	1	0
Ranunculus cassubicus	h	0	0	0	2	2	0	0	0	3	0
Ranunculus monophyllus	h	0	0	0	0	0	0	0	0	2	0
Ranunculus polyanthemos	h	0	0	0	0	1	0	0	0	2	0
Ranunculus repens	h	0	0	0	0	0	0	0	0	0	1
Ranunculus sp.	h	0	0	0	0	2	0	0	0	1	0
Rosa canina	w	0	0	1	3	3	0	0	2	3	0
Rubus idaeus	w	0	0	0	0	1	0	0	0	2	0
Rubus saxatilis	h	0	0	3	3	3	0	0	3	3	3
Salix caprea	w	0	0	0	0	1	0	0	0	0	0
Sanguisorba officinalis	h	0	0	2	2	2	0	3	3	1	3
Saussurea controversa	h	0	0	0	0	1	0	0	0	0	3
Silene nutans	h	0	0	0	1	0	0	0	1	0	2
Silene repens	h	0	0	1	0	0	0	0	0	0	0
Silene wolgensis	h	0	0	0	0	0	0	2	0	0	0
Solidago virgaurea	h	0	0	1	1	2	0	0	2	0	1
Sorbus aucuparia	w	0	0	1	3	3	0	1	2	0	3
Stellaria graminea	h	0	0	0	2	0	0	0	0	1	0
Succisa pratensis	h	0	0	1	0	0	0	0	1	0	1
Taraxacum sp.	h	0	0	0	0	1	0	0	0	3	1
Thalictrum minus	h	0	0	0	0	1	0	0	0	1	1
Tilia cordata	w	0	0	0	0	0	0	0	0	0	1
Trientalis europaea	h	0	0	0	1	0	0	0	2	1	2
Trifolium lupinaster	h	0	0	2	3	3	0	3	2	2	3
Trifolium medium	h	0	0	3	3	3	0	0	1	1	2
Trifolium pratense	h	0	0	0	0	0	0	0	0	3	0
Trifolium repens	h	0	0	0	3	2	0	0	0	3	0
Trollius europaeus	h	0	0	0	0	0	0	0	0	2	3
Tussilago farfara	h	0	0	0	1	3	0	0	0	0	1
Urtica dioica	h	0	0	0	0	2	0	0	0	2	0
Vaccinium myrtillus	ds	0	1	1	1	0	0	0	3	0	0
Vaccinium vitis-idaea	ds	0	2	3	3	1	0	0	3	2	3
Veronica chamaedrys	h	0	0	3	3	3	0	0	1	3	3
Vicia cracca	h	0	0	3	0	2	0	0	0	0	0
Vicia sepium	h	0	0	0	3	1	0	0	0	2	0
Vicia sylvatica	h	0	0	0	0	0	0	0	0	1	3
Viola canina	h	0	1	1	1	0	0	0	2	1	1
Viola hirta	h	0	0	1	1	0	0	0	1	3	2
Viola mirabilis	h	0	0	0	2	0	0	0	0	0	0

Data collected 23–25.7.2003. Life forms: ds - dwarf shrubs, g - grasses, h - herbs, w - trees and shrubs.

Table 6.31 Occurrences of vascular plants (numbers of sampling plots, out of three, on which the species was recorded) in the impact zone of the iron pellet plant at Kostomuksha, Russia

Species	Life form	Occurrences of species on sampling plots									
		1–1	1–2	1–3	1–4	1–5	2–1	2–2	2–3	2–4	2–5
Andromeda polifolia	ds	0	0	0	1	0	0	0	0	0	0
Betula pendula	w	1	0	0	0	0	0	0	0	0	1
Betula pubescens	w	3	2	3	3	2	3	2	3	3	3
Calamagrostis arundinacea	g	2	1	0	3	0	3	1	0	0	1
Calamagrostis purpurea ssp. *phragmitoides*	g	1	0	0	0	0	0	0	0	0	0
Calluna vulgaris	ds	0	0	0	0	0	0	0	0	0	1
Carex globularis	g	0	1	1	1	0	0	0	0	0	0
Deschampsia flexuosa	g	3	3	3	3	1	3	3	3	0	2
Empetrum nigrum	ds	0	0	0	2	0	0	0	0	3	3
Epilobium angustifolium	h	3	0	1	0	0	3	1	1	1	0
Equisetum sylvaticum	h	0	1	0	1	0	0	0	0	0	0
Geranium sylvaticum ssp. *sylvaticum*	h	0	0	0	0	0	3	0	0	0	0
Goodyera repens	h	0	0	1	1	0	1	0	1	0	0
Gymnadenia conopsea	h	0	0	0	0	0	1	0	0	0	0
Gymnocarpium dryopteris	h	0	1	1	0	0	2	0	0	0	0
Hieracium caespitosum	h	0	0	0	0	0	1	0	0	0	0
Hieracium murorum	h	0	0	0	0	0	1	0	0	0	0
Juniperus communis ssp. *nana*	w	0	0	3	0	1	2	1	0	0	3
Ledum palustre	ds	0	0	0	2	3	0	0	0	3	3
Linnaea borealis	h	3	3	3	0	0	3	3	0	0	0
Listera cordata	h	0	2	0	0	0	0	0	0	0	0
Luzula pilosa	g	3	0	3	0	0	1	2	3	0	1
Lycopodium annotinum	h	0	1	0	0	0	2	2	0	0	0
Maianthemum bifolium	h	0	1	2	0	0	2	0	0	0	0
Melampyrum pratense	h	0	0	0	0	0	1	0	0	0	0
Melampyrum sylvaticum	h	2	2	3	2	3	3	2	3	1	3
Orthilia secunda	h	2	1	3	0	0	0	1	0	0	0
Picea abies ssp. *obovata*	w	3	3	3	3	2	3	3	3	1	3
Pinus sylvestris	w	3	0	3	2	3	0	3	2	3	3
Populus tremula	w	1	0	0	1	2	3	2	0	2	0
Rubus chamaemorus	h	0	0	0	2	0	0	0	0	0	0
Salix caprea	w	2	0	0	0	0	0	1	1	0	1
Salix phylicifolia	w	0	0	0	0	0	0	0	1	0	0
Solidago virgaurea	h	3	1	3	0	0	3	2	2	0	0
Sorbus aucuparia	w	3	3	3	2	1	3	3	1	0	2
Trientalis europaea	h	3	1	1	0	0	3	0	1	0	0
Vaccinium myrtillus	ds	3	3	3	3	3	3	3	3	3	3
Vaccinium uliginosum	ds	0	0	0	1	3	0	0	1	3	3
Vaccinium vitis-idaea	ds	3	3	3	3	3	3	3	3	3	3

Data collected 18.7.2002. Life forms: ds - dwarf shrubs, g - grasses, h - herbs, w - trees and shrubs.

Table 6.32 Occurrences of vascular plants (numbers of sampling plots, out of three, on which the species was recorded) in the impact zone of the nickel-copper smelter at Monchegorsk, Russia

Species	Life form	Occurrences of species on sampling plots									
		1–1	1–2	1–4	1–5	2–3	2–5	2–7	2–8	2–9	2–12
Andromeda polifolia	ds	0	0	1	0	0	2	0	0	0	0
Arctostaphylos alpinus	h	0	0	0	0	0	1	0	0	0	0
Betula nana	w	1	1	2	3	0	3	1	2	1	1
Betula pubescens	w	0	2	2	3	3	2	2	3	2	3
Calamagrostis lapponica	g	0	2	0	1	0	0	0	0	0	0
Calluna vulgaris	ds	0	0	2	2	0	0	0	0	1	0
Carex brunnescens	g	0	0	0	1	0	0	0	0	0	0
Carex cespitosa	g	0	0	0	0	0	0	0	1	0	0
Carex dioica	g	0	0	1	0	0	1	0	1	0	0
Carex flava	g	0	0	0	0	0	0	0	1	0	0
Carex globularis	g	0	0	0	1	0	0	0	0	0	0
Carex juncella	g	0	0	0	0	0	0	0	1	0	0
Carex nigra	g	0	0	0	0	0	1	0	0	0	0
Carex rostrata	g	0	0	0	0	0	1	0	0	0	0
Carex vaginata	g	0	0	1	0	0	0	0	1	0	0
Cirsium helenioides	h	0	0	0	0	0	0	0	1	0	0
Comarum palustre	h	0	0	0	0	0	0	0	0	0	1
Cornus suecica	h	0	0	0	0	0	0	0	1	0	0
Crepis paludosa	h	0	0	1	0	0	0	0	0	0	1
Dactylorhiza maculata	h	0	0	0	0	0	0	0	0	1	0
Deschampsia cespitosa	g	0	0	0	1	0	0	0	0	0	0
Deschampsia flexuosa	g	0	1	1	1	0	0	3	2	2	2
Drosera rotundifolia	h	0	0	0	0	0	0	0	1	0	0
Eleocharis quinqueflora	g	0	0	0	0	0	0	0	1	0	0
Empetrum nigrum	ds	1	1	3	2	1	1	3	3	3	2
Epilobium angustifolium	h	0	0	1	2	1	0	2	1	1	1
Equisetum fluviatile	h	0	0	0	0	0	0	0	0	0	1
Equisetum palustre	h	0	0	1	1	0	2	0	1	0	0
Equisetum sylvaticum	h	0	1	1	1	0	0	0	1	0	0
Eriophorum angustifolium	g	0	0	1	0	0	2	0	1	0	0
Eriophorum scheuchzeri	g	0	0	1	0	0	0	0	0	0	0
Eriophorum vaginatum	g	0	1	0	1	0	1	0	1	0	0
Festuca ovina	g	0	0	0	0	0	0	0	1	0	0
Geranium sylvaticum ssp. *sylvaticum*	h	0	0	0	0	0	0	0	0	1	0
Juniperus communis ssp. *nana*	w	0	0	0	1	0	0	1	2	0	2
Ledum palustre	ds	0	1	3	3	0	3	2	1	3	0
Linnaea borealis	h	0	0	0	0	0	0	2	1	2	0

(continued)

Table 6.32 (continued)

Species	Life form	Occurrences of species on sampling plots									
		1–1	1–2	1–4	1–5	2–3	2–5	2–7	2–8	2–9	2–12
Luzula pilosa	g	0	0	0	0	0	0	0	0	1	1
Lycopodium annotinum	h	0	0	0	0	0	0	1	0	0	0
Melampyrum sylvaticum	h	0	0	0	0	0	0	0	1	1	2
Menyanthes trifoliata	h	0	0	0	0	0	1	0	0	0	0
Molinia caerulea	g	0	0	0	0	0	0	0	1	0	0
Orthilia secunda	ds	0	0	1	0	0	0	0	0	1	0
Parnassia palustris	h	0	0	0	0	0	0	0	1	0	0
Picea abies ssp. *obovata*	w	0	1	2	1	0	2	2	3	3	3
Pinguicula vulgaris	h	0	0	0	0	0	0	0	1	0	0
Pinus sylvestris	w	0	1	3	3	0	2	1	0	1	0
Poa pratensis	g	0	1	0	0	0	0	0	0	0	0
Populus tremula	w	0	0	0	0	0	0	0	0	1	0
Potentilla erecta	h	0	0	0	0	0	0	0	2	0	0
Rubus chamaemorus	h	0	1	0	2	0	2	0	2	0	0
Salix borealis	w	1	1	3	2	0	0	0	2	1	2
Salix caprea	w	0	1	3	1	2	0	1	1	0	1
Salix glauca	w	0	0	1	0	0	0	0	0	1	0
Salix myrsinites	w	0	0	0	1	0	0	0	0	0	0
Salix phylicifolia	w	0	1	1	1	1	0	0	0	1	2
Saussurea alpina	h	0	0	0	0	0	0	0	1	0	0
Scirpus hudsonianus	g	0	0	0	0	0	0	0	1	0	0
Solidago virgaurea	h	0	0	0	0	0	0	0	1	0	2
Sorbus aucuparia	w	0	0	0	0	0	0	0	0	1	0
Tofieldia pusilla	h	0	0	0	0	0	0	0	1	0	0
Trientalis europaea	h	0	0	0	0	0	0	0	2	0	2
Vaccinium microcarpum	ds	0	0	1	0	0	1	0	1	0	0
Vaccinium myrtillus	ds	0	0	2	0	1	0	3	3	3	2
Vaccinium uliginosum	ds	0	1	3	3	0	2	0	2	2	3
Vaccinium vitis-idaea	ds	0	0	2	3	2	1	3	2	3	3
Viola epipsila	h	0	0	0	0	0	0	0	1	0	0

Data collected 17.6–6.7.1997. Life forms: ds - dwarf shrubs, g - grasses, h - herbs, w - trees and shrubs.

Table 6.33 Occurrences of vascular plants (numbers of sampling plots, out of three, on which the species was recorded) in the impact zone of the aluminium smelter at Nadvoitsy, Russia

Species	Life form	Occurrences of species on sampling plots									
		1–1	1–2	1–3	1–4	1–5	2–1	2–2	2–3	2–4	2–5
Achillea millefolium	h	0	0	0	0	0	1	0	0	0	0
Alnus incana	w	0	2	2	0	0	0	2	0	2	3
Antennaria dioica	h	1	0	0	0	0	0	0	0	0	0
Betula nana	w	0	1	0	0	1	0	0	0	0	0
Betula pendula	w	3	0	2	2	0	3	1	1	2	2
Betula pubescens	w	3	3	1	3	3	3	3	3	0	3
Calamagrostis arundinacea	g	0	0	1	0	0	0	0	0	0	0
Calamagrostis canescens	g	0	1	0	0	0	0	0	0	0	0
Calluna vulgaris	ds	3	0	2	1	0	3	0	3	2	1
Carex brunnescens	g	0	0	0	0	0	0	1	0	0	0
Carex curta	g	0	1	0	0	0	0	0	0	0	0
Carex globularis	g	0	0	0	0	1	0	0	0	0	1
Carex magellanica	g	0	1	0	0	0	0	0	0	0	0
Deschampsia cespitosa	g	0	1	0	1	0	0	2	0	0	0
Deschampsia flexuosa	g	0	1	3	3	0	3	1	3	0	3
Dryopteris carthusiana	h	0	1	0	0	0	0	1	0	0	0
Empetrum nigrum	ds	3	1	1	1	2	3	0	3	3	0
Epilobium angustifolium	h	0	1	3	2	0	3	2	0	0	0
Equisetum arvense	h	0	3	0	0	0	0	0	0	0	0
Equisetum sylvaticum	h	0	0	0	2	0	0	1	0	0	0
Festuca ovina	g	0	0	0	0	0	1	0	0	0	0
Fragaria vesca	h	0	0	0	0	0	0	1	0	0	0
Geranium sylvaticum ssp. *sylvaticum*	h	0	0	0	0	0	0	1	0	0	0
Geum rivale	h	0	1	0	0	0	0	0	0	0	0
Hieracium laevigatum	h	0	0	2	0	0	3	0	0	0	0
Hieracium murorum	h	0	0	3	1	0	2	2	0	0	2
Juncus filiformis	g	0	0	0	0	1	0	0	0	0	0
Juniperus communis ssp. *nana*	w	0	0	0	3	0	1	0	3	0	3
Ledum palustre	ds	1	2	1	3	3	0	0	2	3	2
Linnaea borealis	h	0	0	1	1	0	0	2	0	0	0
Luzula pilosa	g	0	0	3	1	0	0	2	0	0	3
Lycopodium annotinum	h	0	0	1	1	0	0	0	0	0	0
Maianthemum bifolium	h	1	1	0	0	0	0	3	1	0	1
Melampyrum pratense	h	1	0	3	2	2	0	1	2	2	3
Menyanthes trifoliata	h	0	3	0	0	0	0	0	0	0	0
Moneses uniflora	h	1	1	0	0	0	0	0	0	0	0
Orthilia secunda	h	0	2	0	0	0	0	0	0	0	0
Oxalis acetosella	h	0	0	0	0	0	0	3	0	0	0
Picea abies ssp. *obovata*	w	3	3	0	2	2	3	3	0	0	2
Pinus sylvestris	w	3	3	3	3	3	3	3	3	3	3
Platanthera bifolia	h	0	1	0	0	0	0	0	0	0	0

(continued)

Table 6.33 (continued)

Species	Life form	Occurrences of species on sampling plots									
		1–1	1–2	1–3	1–4	1–5	2–1	2–2	2–3	2–4	2–5
Populus tremula	w	2	0	1	0	3	3	0	2	2	3
Pyrola minor	h	0	2	0	0	0	1	2	0	0	0
Ranunculus repens	h	0	1	0	0	0	0	2	0	0	0
Rubus chamaemorus	h	0	2	0	0	0	0	0	0	0	1
Rubus idaeus	w	0	0	0	0	0	0	3	0	0	0
Rubus saxatilis	h	0	0	0	0	0	0	1	0	0	0
Salix aurita	w	0	0	0	0	1	0	0	0	0	0
Salix caprea	w	1	0	3	0	1	3	3	0	2	1
Salix myrsinifolia	w	0	1	0	1	0	0	0	0	0	0
Salix phylicifolia	w	0	0	0	0	0	0	0	0	0	1
Solidago virgaurea	h	0	0	0	0	0	1	0	0	0	0
Sorbus aucuparia	w	1	2	3	2	0	3	3	0	3	3
Taraxacum sp.	h	0	0	0	0	0	1	0	0	0	0
Trientalis europaea	h	0	1	0	0	0	0	3	0	0	0
Tussilago farfara	h	0	0	0	0	0	0	1	0	0	0
Urtica dioica	h	0	0	0	0	0	0	1	0	0	0
Vaccinium microcarpum	ds	0	2	0	0	0	0	0	0	0	0
Vaccinium myrtillus	ds	2	3	3	3	3	3	3	3	3	3
Vaccinium uliginosum	ds	0	1	0	1	2	0	0	3	3	2
Vaccinium vitis-idaea	ds	3	3	3	3	3	3	3	3	3	3
Veronica chamaedrys	h	0	0	0	0	0	0	2	0	0	0
Vicia sepium	h	0	0	0	0	0	0	1	0	0	0
Viola riviniana	h	0	1	0	0	0	0	0	0	0	0

Data collected 25–27.7.2002. Life forms: ds - dwarf shrubs, g - grasses, h - herbs, w - trees and shrubs.

Table 6.34 Occurrences of vascular plants (numbers of sampling plots, out of three, on which the species was recorded) in the impact zone of the of the nickel-copper smelter at Nikel and ore-roasting plant at Zapolyarnyy, Russia

Species	Life form	Occurrences of species on sampling plots									
		1–1	1–2	1–3	1–4	1–5	2–1	2–2	2–3	2–4	2–5
Alnus incana	w	0	0	0	0	0	1	0	0	0	0
Andromeda polifolia	ds	0	0	3	3	1	1	2	1	1	1
Arctostaphylos alpinus	h	0	0	3	3	0	0	0	0	2	0
Arctostaphylos uva-ursi	ds	0	0	0	0	0	0	1	0	0	0
Bartsia alpina	h	0	0	0	1	0	0	0	0	1	0
Betula nana	w	0	3	3	3	3	0	0	3	3	3
Betula pubescens	w	2	1	3	2	2	2	3	3	3	3
Calamagrostis lapponica	g	0	0	0	0	0	0	0	2	0	0
Calamagrostis purpurea	g	1	0	0	0	0	0	0	0	0	0
Calamagrostis stricta	g	0	0	0	0	0	0	0	0	1	0
Calluna vulgaris	ds	0	0	0	0	0	0	2	1	0	0
Caltha palustris	h	0	0	0	0	0	0	0	0	1	0
Campanula rotundifolia	h	0	0	0	0	0	0	1	0	0	0
Carex aquatilis	g	1	0	0	0	0	0	0	0	0	0
Carex bigelowii	g	0	0	0	0	1	0	0	0	0	0
Carex brunnescens	g	0	0	0	0	0	0	0	0	0	1
Carex curta	g	0	0	1	0	0	0	0	0	0	0
Carex dioica	g	0	0	1	0	0	0	0	0	0	0
Carex nigra	g	0	0	1	3	2	0	0	0	0	0
Carex rostrata	g	1	0	1	0	0	0	0	0	0	0
Carex rotundata	g	0	0	0	0	0	0	0	0	0	1
Carex vaginata	g	0	0	0	3	1	0	2	0	2	0
Cirsium helenioides	h	0	0	0	0	0	0	1	0	1	0
Cornus suecica	h	0	0	3	3	1	0	0	0	3	0
Dactylorhiza traunsteineri	h	0	0	0	1	0	0	2	0	1	0
Deschampsia cespitosa	g	2	0	0	0	0	1	0	1	0	0
Deschampsia flexuosa	g	0	3	3	3	3	2	3	3	3	3
Empetrum nigrum ssp. *hermaphroditum*	ds	2	3	3	3	3	1	3	3	3	3
Epilobium angustifolium	h	0	0	0	0	0	0	2	0	0	0
Equisetum arvense	h	0	1	0	0	0	2	2	0	2	0
Equisetum fluviatile	h	1	0	1	0	0	0	0	0	0	0
Equisetum palustre	h	0	0	1	0	0	0	0	0	0	0
Equisetum sylvaticum	h	0	0	3	0	1	0	1	3	0	2
Eriophorum angustifolium	g	1	0	0	0	1	0	0	0	0	0
Eriophorum russeolum	g	1	0	0	0	0	0	0	0	0	0
Eriophorum scheuchzeri	g	1	0	0	0	0	0	0	0	0	0
Eriophorum vaginatum	g	0	0	2	0	1	0	1	0	1	2

(continued)

Table 6.34 (continued)

Species	Life form	Occurrences of species on sampling plots									
		1–1	1–2	1–3	1–4	1–5	2–1	2–2	2–3	2–4	2–5
Geranium sylvaticum ssp. *sylvaticum*	h	0	0	0	1	0	0	1	0	2	0
Geum rivale	h	0	0	0	0	0	0	0	0	1	0
Gymnadenia conopsea	h	0	0	0	0	0	0	1	0	0	0
Hieracium alpinum	h	0	0	0	0	1	0	0	0	0	0
Hieracium vulgatum	h	0	0	0	0	0	0	0	0	1	0
Juncus trifidus	g	0	0	0	2	2	0	0	0	0	0
Juniperus communis ssp. *nana*	w	0	1	1	1	0	0	1	0	1	1
Ledum palustre	ds	0	2	3	1	0	2	3	3	3	2
Linnaea borealis	h	0	0	0	0	0	0	0	0	2	1
Lychnis alpina	h	0	0	0	0	0	1	0	0	0	0
Lycopodium dubium	h	0	0	0	1	0	0	0	0	0	0
Melampyrum pratense	h	0	0	1	2	1	0	0	0	0	0
Melica nutans	g	0	0	0	0	0	0	1	0	0	0
Molinica caerulea	g	0	0	0	1	1	0	0	0	0	0
Pedicularis lapponica	h	0	0	0	2	0	0	0	0	0	0
Phyllodoce caerulea	ds	0	0	0	3	1	0	0	0	0	0
Pinguicula vulgaris	h	0	0	0	1	1	0	0	0	0	0
Pinus sylvestris	w	0	1	0	0	0	1	2	3	1	2
Poa angustifolia	g	1	0	0	0	0	0	0	0	0	0
Poa pratensis	g	1	0	0	0	0	1	0	0	0	0
Potentilla erecta	h	0	0	0	1	0	0	1	0	1	0
Pyrola minor	h	0	0	0	0	0	0	1	0	1	0
Rubus chamaemorus	h	0	1	3	3	1	0	2	3	2	2
Rubus saxatilis	h	0	0	0	0	0	0	1	0	0	0
Rumex acetosa	h	0	0	0	0	0	0	1	0	1	0
Salix borealis	w	1	0	1	1	0	2	2	1	1	0
Salix caprea	w	0	0	1	0	0	0	0	0	0	0
Salix glauca	w	2	1	2	3	1	2	0	1	0	1
Salix lapponum	w	0	0	1	1	1	0	0	0	0	0
Salix myrsinites	w	0	0	0	0	0	2	2	0	2	0
Salix phylicifolia	w	2	3	3	0	0	0	0	2	0	1
Saussurea alpina	h	0	0	0	1	0	0	0	0	1	0
Solidago virgaurea	h	0	0	2	3	1	0	1	0	3	0
Sorbus aucuparia	w	0	0	0	0	0	0	0	0	3	0
Taraxacum sp.	h	0	0	1	1	0	0	0	0	0	0
Tofieldia pusilla	h	0	0	0	1	0	0	1	0	0	0
Trientalis europaea	h	0	0	0	1	0	0	0	0	1	1
Vaccinium myrtillus	ds	0	3	3	3	3	1	3	3	3	3
Vaccinium uliginosum	ds	1	2	3	3	0	1	3	3	2	3
Vaccinium vitis-idaea	ds	0	3	3	3	3	1	3	3	3	3

Data collected 17–18.7.2001. Life forms: ds - dwarf shrubs, g - grasses, h - herbs, w - trees and shrubs.

Table 6.35 Occurrences of vascular plants (numbers of sampling plots, out of three, on which the species was recorded) in the impact zone of the of the nickel-copper smelters at Norilsk, Russia

Species	Life form	Occurrences of species on sampling plots									
		1–1	1–2	1–3	1–4	1–5	2–1	2–2	2–3	2–4	2–5
Achillea impatiens	h	0	0	1	0	0	0	0	2	0	0
Aconitum septentrionale	h	0	0	1	0	0	0	0	0	0	0
Aconitum volubile	h	0	0	1	0	0	0	0	0	0	0
Allium schoenoprasum	h	1	0	0	0	0	0	0	0	0	0
Andromeda polifolia	ds	3	1	0	1	0	3	0	0	3	0
Anthoxanthum alpinum	g	0	0	0	0	1	0	0	3	0	0
Arctagrostis latifolia	g	0	1	0	0	2	1	0	0	0	0
Arctostaphylos alpinus	ds	0	0	1	3	3	0	0	0	0	0
Betula nana	w	1	3	3	3	3	3	3	3	3	1
Betula pubescens	w	0	0	0	0	0	0	1	3	0	3
Calamagrostis purpurea ssp. *purpurea*	g	0	0	0	0	0	0	2	2	0	0
Calamagrostis stricta	g	0	0	0	0	0	1	2	0	0	0
Cardamine macrophylla	h	0	0	2	0	0	0	0	0	3	0
Cardaminopsis petraea	h	0	0	1	0	0	0	0	0	0	0
Carex globularis	g	0	0	0	0	0	0	0	0	0	3
Carex parallela ssp. *redowskiana*	g	0	1	0	0	0	0	0	0	0	0
Carex pediformis	g	0	1	0	0	0	0	0	0	2	0
Carex rotundata	g	0	2	0	1	0	1	0	0	1	0
Carex vaginata	g	0	2	3	1	0	0	0	0	2	0
Cassiope tetragona	ds	1	0	0	0	0	0	0	0	0	0
Cirsium helenioides	h	0	0	0	0	0	0	0	1	0	0
Clematis alpina ssp. *sibirica*	ds	0	0	0	0	0	0	0	2	0	0
Conioselinum tataricum	h	0	0	0	0	0	0	0	1	0	0
Dryas octopetala	h	3	3	1	2	2	0	0	0	1	0
Duschekia fruticosa	w	3	3	1	3	3	3	0	3	3	2
Empetrum nigrum	ds	3	3	1	3	3	3	3	3	3	2
Epilobium angustifolium	h	0	2	0	0	0	0	0	0	0	0
Epilobium davuricum	h	0	1	0	0	0	0	0	0	0	0
Equisetum arvense	h	0	1	1	0	1	1	1	0	0	0
Equisetum sp.	h	3	3	3	3	3	2	3	3	3	3
Eriophorum scheuchzeri	g	1	0	0	0	0	1	1	0	0	0
Festuca ovina	g	1	0	3	3	2	0	0	0	0	0
Festuca rubra	g	0	0	2	1	0	0	0	0	3	0
Festuca vivipara	g	0	2	0	0	0	0	0	0	0	0
Galium boreale	h	0	0	1	0	0	0	0	2	0	0
Geranium sylvaticum ssp. *pseudosibiricum*	h	0	0	1	0	0	0	0	2	0	0
Hedysarum hedysaroides ssp. *arcticum*	h	2	1	0	0	0	0	0	0	2	0
Hierochloe alpina	g	0	0	1	0	1	0	0	0	0	0
Juniperus communis ssp. *nana*	w	2	0	0	0	0	0	0	3	1	0
Lagotis minor	h	0	0	0	0	0	0	0	0	2	0
Larix sibirica	w	2	2	2	3	3	2	0	3	3	3
Ledum decumbens	ds	0	2	2	3	3	3	3	0	3	3
Linnaea borealis	h	0	0	0	0	0	0	0	3	0	1
Luzula arctica	g	0	0	0	0	1	0	0	0	0	0
Luzula confusa	g	0	0	0	1	0	0	0	0	0	0

(continued)

Table 6.35 (continued)

Species	Life form	Occurrences of species on sampling plots									
		1–1	1–2	1–3	1–4	1–5	2–1	2–2	2–3	2–4	2–5
Luzula tundricola	g	0	0	0	0	1	0	0	0	0	0
Lychnis sibirica ssp. *samojedorum*	h	0	1	0	0	0	0	0	0	0	0
Lycopodium annotinum	h	0	0	0	0	0	0	0	1	0	0
Minuartia stricta	h	0	1	0	0	0	0	0	0	0	0
Oxytropis campestris ssp. *sordida*	h	2	0	0	0	0	0	0	0	0	0
Parnassia palustris	h	0	0	0	0	0	0	0	0	2	0
Pedicularis labradorica	h	0	0	1	3	2	1	0	2	1	0
Pedicularis oederi	h	0	1	0	0	0	0	0	0	0	0
Pedicularis sudetica	h	0	0	0	1	0	1	0	0	0	0
Petasites frigidus	h	0	1	1	3	2	0	0	0	0	0
Petasites sibiricus	h	0	0	0	0	0	0	0	0	1	0
Picea abies ssp. *obovata*	w	0	0	0	0	0	0	3	3	1	3
Poa arctica	g	2	3	2	3	1	2	0	2	0	0
Poa pratensis	g	3	2	0	2	3	1	3	1	0	0
Polygonum bistorta	h	3	3	2	0	0	0	0	1	2	0
Polygonum viviparum	h	1	3	2	3	2	0	1	2	3	0
Potentilla sp.	h	0	0	0	0	1	0	0	0	0	0
Pyrola minor	h	0	1	1	1	0	0	0	2	3	0
Pyrola rotundifolia	h	1	1	2	1	0	0	0	3	0	0
Rubus arcticus	ds	0	0	0	0	0	0	0	2	0	0
Rubus chamaemorus	h	0	0	0	0	0	2	3	0	0	3
Rumex thyrsiflorus	h	0	0	0	0	0	0	1	0	0	0
Salix arctica	w	3	3	2	0	0	1	0	3	3	0
Salix bebbiana	w	1	2	2	3	3	3	3	2	3	0
Salix cinerea	w	0	0	0	0	0	1	0	0	0	0
Salix hastata	w	0	3	1	2	2	0	0	0	0	0
Salix lanata	w	1	3	2	1	3	2	3	0	3	0
Salix phylicifolia	w	0	0	0	1	0	0	0	0	0	0
Salix recurvigemmis	w	2	1	2	0	0	3	0	0	0	0
Saussurea alpina	h	1	3	2	0	0	0	1	3	3	0
Saxifraga nelsoniana	h	0	0	0	1	0	0	0	0	2	0
Saxifraga serpyllifolia	h	2	0	0	0	0	0	0	0	0	0
Schoenus ferrugineus	g	0	1	0	0	0	0	0	0	0	0
Senecio integrifolius	h	0	2	0	0	1	0	0	0	1	0
Seseli condensatum	h	0	0	1	0	0	0	0	1	1	0
Silene wahlbergella	h	0	1	0	0	0	0	0	0	0	0
Solidago virgaurea	h	0	0	1	0	0	0	0	3	0	0
Sorbus aucuparia	w	0	0	0	0	0	0	0	1	0	0
Stellaria palustris	h	0	1	3	2	3	0	2	0	2	0
Thalictrum alpinum	h	0	1	2	0	0	0	0	0	1	0
Thalictrum minus	h	0	0	1	0	0	0	0	1	0	0
Tofieldia pusilla	h	0	0	0	0	0	0	0	0	1	0
Trollius asiaticus	h	3	0	1	0	0	0	0	2	1	0
Vaccinium microcarpum	ds	0	0	0	0	0	0	0	0	1	0
Vaccinium myrtillus	ds	0	0	0	0	0	0	0	3	0	0
Vaccinium uliginosum	ds	3	3	3	3	3	3	3	3	3	2
Vaccinium vitis-idaea	ds	0	3	3	3	3	2	1	3	3	3
Valeriana capitata	h	0	0	1	0	0	0	0	0	2	0
Veratrum sp.	h	0	0	0	0	0	0	0	3	0	0
Vicia sp.	h	0	1	0	0	0	0	0	0	0	0

Data collected 23–28.7.2002. Life forms: ds - dwarf shrubs, g - grasses, h - herbs, w - trees and shrubs.

Table 6.36 Occurrences of vascular plants (numbers of sampling plots, out of three, on which the species was recorded) in the impact zone of the of the copper smelter at Revda, Russia

Species	Life form	Occurrences of species on sampling plots									
		1–1	1–2	1–3	1–4	1–5	2–1	2–2	2–3	2–4	2–5
Abies sibirica	w	3	3	3	1	3	3	3	3	3	3
Achillea millefolium	h	0	0	0	0	0	0	0	0	1	0
Aconitum septentrionale	h	0	0	1	3	3	0	0	2	0	0
Actaea spicata	h	0	0	1	0	2	0	0	0	0	0
Aegopodium podagraria	h	0	1	1	3	3	3	1	3	3	3
Agrostis capillaris	g	0	0	0	0	0	0	3	0	1	0
Ajuga reptans	h	0	0	0	3	3	0	0	3	1	1
Alchemilla acutiloba	h	0	0	0	0	0	0	0	1	0	1
Alchemilla murbeckiana	h	0	0	0	0	0	0	0	0	1	0
Alchemilla sarmatica	h	0	0	0	2	0	0	0	0	0	0
Alhemilla wichurae	h	0	0	0	0	0	0	0	0	1	0
Alnus incana	w	0	0	0	0	0	3	0	2	0	0
Angelica sylvestris	h	0	0	0	0	0	0	0	0	1	0
Asarum europaeum	h	0	0	3	0	2	0	0	1	1	3
Athyrium filix-femina	h	0	1	3	0	3	0	0	0	0	2
Betula pendula	w	3	0	0	1	0	3	2	1	3	0
Betula pubescens	w	3	3	0	3	2	3	1	2	3	1
Brachypodium pinnatum	g	0	0	0	2	1	0	0	0	0	0
Cacalia hastata	h	0	0	2	0	2	0	0	0	1	0
Calamagrostis obtusata	g	3	3	2	2	2	3	1	3	3	1
Calamagrostis purpurea ssp. *langsdorfii*	g	1	1	2	0	0	0	0	0	0	0
Campanula glomerata	h	0	0	0	0	0	0	0	1	0	0
Carex digitata	g	0	0	0	0	0	0	0	0	1	0
Carex montana	g	0	0	2	1	0	0	0	0	0	0
Carex sp.	g	1	0	0	0	0	0	0	0	0	0
Centaurea sp.	h	0	0	0	0	0	0	0	1	0	0
Cerastium fontanum ssp. *triviale*	h	0	0	0	0	2	0	0	0	0	0
Cerastium pauciflorum	h	0	2	3	2	0	0	1	0	1	3
Chamaecytisus ruthenicus	h	0	0	0	0	0	1	2	0	3	0
Chrysosplenium alternifolium	h	0	0	0	1	1	0	0	0	0	0
Circaea alpina	h	0	0	1	0	3	0	0	0	0	3
Cirsium helenioides	h	0	1	0	0	0	0	0	1	0	0
Cirsium oleraceum	h	0	0	0	0	0	0	0	1	0	0
Cirsium sp.	h	0	0	0	1	0	0	0	0	1	0
Clematis alpina ssp. *sibirica*	h	1	0	0	0	0	0	0	0	1	0
Corydalis solida	h	0	0	0	0	1	0	0	0	0	0
Crepis paludosa	h	0	1	1	0	1	0	0	0	0	0
Crepis sibirica	h	0	0	1	0	0	0	0	0	0	0
Daphne mezereum	w	0	0	0	1	1	0	0	0	1	0
Deschampsia cespitosa	g	0	2	0	2	0	2	0	2	1	3
Digitalis grandiflora	h	0	0	0	0	0	0	0	1	0	0
Dryopteris assimilis	h	0	0	0	0	1	0	0	0	0	0
Dryopteris carthusiana	h	0	1	0	1	2	0	0	0	0	1
Dryopteris filix-mas	h	0	1	3	0	0	0	0	2	0	0
Elymus repens	g	0	0	0	0	0	0	1	0	0	0
Epilobium angustifolium	h	1	0	1	0	0	1	3	1	0	0
Epilobium palustre	h	0	0	0	0	1	0	0	0	0	0

(continued)

Table 6.36 (continued)

Species	Life form	Occurrences of species on sampling plots									
		1–1	1–2	1–3	1–4	1–5	2–1	2–2	2–3	2–4	2–5
Equisetum pratense	h	0	0	0	0	0	1	0	0	0	0
Equisetum sylvaticum	h	0	3	1	3	3	2	0	0	0	3
Euphorbia sp.	h	0	0	0	0	0	0	0	0	0	3
Filipendula ulmaria ssp. *denudata*	h	0	0	0	2	0	0	0	1	0	0
Fragaria vesca	h	0	0	0	3	3	0	0	1	3	3
Galium boreale	h	0	0	0	3	0	3	0	2	3	0
Galium odoratum	h	0	0	0	0	3	0	0	0	0	3
Geranium sylvaticum ssp. *sylvaticum*	h	0	1	2	3	0	0	0	3	3	3
Geum rivale	h	0	1	0	0	3	0	0	2	0	2
Glechoma hederacea	h	0	0	0	0	1	0	0	1	0	0
Gymnocarpium dryopteris	h	0	2	1	0	1	0	0	0	0	0
Heracleum sibiricum	h	0	0	2	0	1	0	0	1	0	0
Hieracium sp.	h	0	0	1	0	0	0	0	0	0	0
Hieracium umbellatum	h	0	0	0	0	0	0	0	0	1	0
Hypericum maculatum	h	0	0	0	1	0	0	0	0	0	0
Impatiens noli-tangere	h	0	0	0	0	3	0	0	0	0	3
Juniperus communis	w	0	0	0	0	0	1	0	0	2	0
Lathyrus gmelinii	h	0	3	1	1	2	0	0	3	0	0
Lathyrus vernus	h	0	1	1	3	3	0	0	0	3	1
Lilium martagon	h	0	0	0	0	0	0	0	0	2	0
Lonicera tatarica	w	0	0	1	0	2	0	0	0	3	0
Luzula pilosa	g	0	0	1	3	1	0	0	0	3	0
Lysimachia vulgaris	h	0	0	0	0	0	0	0	2	0	0
Maianthemum bifolium	h	0	3	3	3	2	0	0	3	3	0
Melampyrum pratense	h	0	0	1	3	0	0	0	1	2	0
Melica nutans	g	1	1	1	0	1	0	0	1	1	0
Milium effusum	g	0	0	0	0	2	0	0	0	0	1
Moneses uniflora	h	0	0	2	0	0	0	0	0	0	0
Myosotis sylvatica	h	0	0	1	1	3	0	0	0	0	2
Orthilia secunda	h	0	0	1	3	0	0	0	0	2	0
Oxalis acetosella	h	0	2	3	2	3	0	0	1	1	3
Padus racemosa	w	0	1	0	1	0	0	0	3	1	0
Paris quadrifolia	h	0	0	2	0	3	0	0	0	0	0
Picea abies ssp. *obovata*	w	3	3	3	3	3	3	3	3	3	3
Pinus sylvestris	w	2	0	0	3	0	3	3	3	3	1
Plantago major	h	0	0	0	0	0	0	0	0	1	0
Platanthera bifolia	h	0	0	0	0	0	0	0	0	1	0
Poa palustris	g	0	0	0	0	0	0	0	0	0	1
Populus tremula	w	0	0	1	2	0	2	3	2	0	0
Potentilla erecta	h	0	0	0	0	0	0	0	1	0	0
Prunella vulgaris	h	0	0	0	3	1	0	0	0	1	1
Pteridium aquilinum	h	0	0	0	0	0	0	0	1	0	0
Pulmonaria mollis	h	0	0	1	1	3	0	0	0	1	0
Pulmonaria obscura	h	0	0	0	1	0	0	0	0	0	0
Pyrola media	h	0	2	2	1	0	0	0	0	1	1
Pyrola minor	h	0	0	0	0	0	0	0	1	0	0

(continued)

Table 6.36 (continued)

Species	Life form	Occurrences of species on sampling plots									
		1–1	1–2	1–3	1–4	1–5	2–1	2–2	2–3	2–4	2–5
Pyrola rotundifolia	h	0	0	0	1	0	0	0	2	2	0
Ranunculus acris ssp. *borelais*	h	0	0	0	0	0	0	0	0	1	0
Ranunculus cassubicus	h	0	0	0	0	0	0	0	1	1	0
Ranunculus polyanthemos	h	0	0	0	0	0	0	0	0	1	0
Ranunculus repens	h	0	0	0	3	0	0	0	0	1	3
Ribes nigrum	w	0	1	2	0	0	1	0	0	2	0
Rosa acicularis	w	0	0	0	0	0	1	0	0	0	0
Rosa canina	w	0	3	0	3	3	0	1	3	3	0
Rubus idaeus	w	0	2	3	1	3	1	2	3	1	3
Rubus saxatilis	h	0	0	2	3	2	0	0	3	3	2
Rumex acetosa	h	0	0	0	0	0	1	0	0	0	0
Salix caprea	w	2	1	0	0	0	2	2	2	1	1
Sambucus racemosa	w	0	0	1	0	2	0	0	0	0	0
Sanquisorba officinalis	h	1	0	0	0	0	3	1	3	2	0
Senecio nemorensis	h	0	0	1	0	1	0	0	0	0	0
Solidago virgaurea	h	0	0	1	2	3	0	0	0	2	0
Sorbus aucuparia	w	2	3	3	3	3	0	1	2	2	3
Stachys officinalis	h	0	0	0	1	0	0	0	0	0	0
Stellaria bungeana	h	0	0	0	0	1	0	0	0	0	0
Stellaria graminea	h	0	0	0	0	0	0	1	0	0	0
Stellaria holostea	h	0	0	0	2	3	0	0	0	0	0
Stellaria longifolia	h	0	0	0	0	0	0	0	0	1	0
Stellaria media	h	0	0	0	0	0	0	1	0	0	0
Succisa pratensis	h	0	0	0	0	0	0	0	1	0	0
Taraxacum officinale	h	0	0	0	1	0	0	1	0	0	0
Thalictrum flavum	h	0	0	0	0	3	0	0	0	0	0
Thalictrum minus	h	0	0	1	1	0	0	0	3	1	1
Tilia cordata	w	0	0	3	0	0	0	0	0	0	0
Trientalis europaea	h	0	0	2	3	1	0	0	0	3	0
Trifolium lupinaster	h	0	0	0	0	0	0	0	0	2	0
Trifolium medium	h	0	0	0	0	0	1	1	0	0	0
Trifolium pratense	h	0	0	0	0	0	0	0	0	1	0
Trifolium repens	h	0	0	0	0	0	0	0	0	1	0
Trollius europaeus	h	0	0	0	2	0	0	0	2	1	3
Tussilago farfara	h	0	0	0	1	0	0	3	2	0	0
Urtica dioica	h	0	0	0	0	1	1	2	0	0	3
Vaccinium myrtillus	df	3	0	0	2	0	0	1	0	3	0
Vaccinium vitis-idaea	df	3	0	0	0	0	1	1	1	3	0
Valeriana wolgensis	h	0	0	0	0	1	0	0	0	0	0
Veratrum lobelianum	h	0	2	0	1	0	0	0	3	0	0
Veronica chamaedrys	h	0	0	0	3	0	0	1	0	2	1
Viburnum opulus	w	0	1	0	0	1	0	0	3	0	0
Vicia sepium	h	0	0	0	2	0	0	0	0	3	0
Vicia silvatica	h	0	0	0	0	1	0	1	0	0	1
Viola canina	h	0	0	0	2	0	0	1	0	2	0
Viola mirabilis	h	0	0	0	0	2	0	0	0	0	0
Viola selkirkii	h	0	0	1	0	1	0	0	0	0	0

Data collected 17–21.7.2003. Life forms: ds - dwarf shrubs, g - grasses, h - herbs, w - trees and shrubs.

Table 6.37 Occurrences of vascular plants (numbers of sampling plots, out of three, on which the species was recorded) in the impact zone of the of the aluminium smelter at Straumsvík, Iceland

Species	Life form	Occurrences of species on sampling plots									
		1–1	1–2	1–3	1–4	1–5	2–1	2–2	2–3	2–4	2–5
Agrostis capillaris	g	0	0	0	0	0	0	0	1	0	0
Agrostis vinealis	g	0	3	1	2	1	2	2	2	3	1
Alchemilla acutiloba	h	0	0	0	0	1	0	0	0	0	0
Alchemilla alpina	h	0	3	2	2	3	3	3	3	3	0
Anthoxanthum alpinum	g	0	0	0	0	0	0	0	0	0	1
Anthoxanthum odoratum	g	0	0	1	0	0	0	0	0	0	0
Arctostaphylos uva-ursi	ds	0	0	0	0	3	0	0	2	0	2
Armeria maritima	h	0	0	3	1	1	0	1	0	1	0
Athyrium filix-femina	h	0	0	0	0	0	2	0	0	0	0
Betula pubescens	w	0	0	0	0	3	0	0	0	0	1
Calluna vulgaris	ds	0	0	2	1	3	3	3	3	1	3
Cardaminopsis petraea	h	0	2	2	3	0	2	3	3	1	1
Carex bigelowii	g	0	1	1	1	1	0	1	0	0	2
Carex sp.	g	0	0	0	1	1	0	0	0	1	0
Cerastium alpinum	h	0	0	0	1	1	0	0	1	1	0
Cerastium arcticum	h	0	0	1	1	0	1	0	0	0	0
Cystopteris fragilis	h	3	1	0	0	0	2	1	1	1	0
Dactylorhiza maculata	h	0	0	0	1	1	1	0	1	0	0
Deschampsia flexuosa	g	0	0	0	1	1	0	1	1	0	1
Draba incana	h	0	1	0	1	0	0	0	0	0	0
Dryas octopetala	h	0	3	2	1	1	2	3	3	2	0
Empetrum nigrum	ds	3	3	3	3	3	3	3	3	3	3
Erigeron borealis	h	0	1	1	0	0	2	2	1	2	2
Eriophorum scheuchzeri	g	0	0	0	0	0	0	0	1	0	0
Erysimum hieracifolium	h	1	0	0	0	0	0	0	0	0	0
Festuca richardsonii	g	0	2	2	3	1	1	1	2	2	0
Festuca vivipara	g	3	3	3	1	3	3	3	3	3	1
Galium boreale	h	3	3	3	3	3	3	3	3	3	2
Galium normanii	h	0	0	0	0	0	0	0	1	0	0
Galium uliginosum	h	2	1	0	0	1	0	0	0	0	0
Galium verum	h	0	3	3	3	3	3	3	3	3	2
Geranium sylvaticum	h	0	0	0	0	1	0	0	1	0	2
Geum rivale	h	0	0	0	0	0	0	0	1	0	0
Gypsophila fastigiata	h	0	2	1	0	0	0	0	0	0	0
Hieracium atratum s.l.	h	0	0	0	0	0	0	0	0	1	0
Hieracium pilosella s.l.	h	0	0	0	0	0	0	0	1	0	0
Hieracium praealtum	h	0	0	0	0	0	0	0	1	0	0
Hieracium sp.	h	2	0	1	0	3	0	0	0	0	0
Hieracium × floribundum	h	0	0	0	0	0	1	0	0	0	0
Juncus trifidus	g	3	2	3	3	3	3	3	3	3	3
Kobresia myosuroides	g	0	0	0	0	0	1	0	1	0	0
Luzula spicata	g	3	1	2	3	2	2	2	2	3	3
Luzula sudetica	g	0	3	1	1	2	2	3	2	3	1
Lychnis viscaria	h	0	0	0	0	0	0	1	0	0	0
Oxyria digyna	h	3	0	0	0	0	0	0	0	0	1
Pinguicula vulgaris	h	0	0	0	1	0	1	0	0	0	0
Plantago maritima	h	0	1	0	0	1	2	1	0	0	0
Poa glauca	g	1	0	0	0	0	1	1	0	0	1

(continued)

Table 6.37 (continued)

Species	Life form	Occurrences of species on sampling plots									
		1–1	1–2	1–3	1–4	1–5	2–1	2–2	2–3	2–4	2–5
Poa pratensis	g	1	1	0	0	0	0	0	0	1	0
Polygonum viviparum	h	0	2	2	0	3	3	2	3	1	2
Potentilla crantzii	h	0	2	2	2	1	1	1	3	2	1
Ranunculus acris	h	0	2	0	0	1	2	1	1	0	1
Rhinantus minor	h	0	0	0	0	0	0	1	0	0	0
Rubus saxatilis	h	0	0	0	0	2	1	2	1	1	2
Rumex acetosa	h	2	1	0	0	2	3	0	0	0	1
Salix herbacea	w	2	3	3	3	1	3	3	3	3	0
Salix lanata	w	0	0	0	0	0	1	0	0	0	0
Salix phylicifolia	w	0	0	0	0	0	0	0	0	0	3
Saxifraga aizoides	h	0	0	0	0	1	0	0	1	0	0
Saxifraga cespitosa	h	1	1	0	0	0	0	0	0	0	1
Saxifraga hirculus	h	0	0	0	1	0	0	1	0	0	0
Sedum villosum	h	0	0	0	0	0	0	1	0	0	0
Sesleria albicans	g	0	0	0	0	0	0	0	1	0	0
Silene acaulis	h	0	2	3	3	2	0	2	3	1	1
Taraxacum officinale s.l.	h	0	0	0	1	0	0	0	0	0	0
Taraxacum sp.	h	0	1	0	0	1	2	2	2	2	2
Thalictrum alpinum	h	0	3	2	2	2	2	3	3	3	1
Thelypteris phegopteris	h	0	0	0	0	0	0	0	1	0	0
Thymus praecox ssp. *arcticus*	h	2	3	3	3	3	3	3	3	3	2
Trisetum spicatum	g	0	0	0	0	0	0	1	0	0	0
Vaccinium uliginosum	ds	2	1	3	1	3	3	3	3	3	3
Veronica officinalis	h	0	0	0	0	0	0	1	0	0	0

Data collected 11–13.7.2002. Life forms: ds - dwarf shrubs, g - grasses, h - herbs, w - trees and shrubs.

Table 6.38 Occurrences of vascular plants (numbers of sampling plots, out of three, on which the species was recorded) in the impact zone of the of the nickel-copper smelter at Sudbury, Canada

Species	Life form	Occurrences of species on sampling plots									
		1–1	1–2	1–3	1–4	1–5	2–1	2–2	2–3	2–4	2–5
Abies balsamea	w	0	0	2	0	3	0	0	1	0	3
Acer rubrum	w	0	2	2	3	2	0	3	1	3	3
Acer spicatum	w	0	0	0	0	0	0	1	0	0	3
Achillea millefolium	w	0	0	0	3	0	0	0	0	0	0
Actaea sp.	h	0	0	0	0	0	0	0	0	0	1
Agrostis gigantea	h	2	0	0	0	0	2	1	0	0	0
Agrostis hyemalis var. *scabra*	g	2	3	3	2	0	3	2	0	1	0
Alnus incana	w	0	0	0	1	0	0	0	0	0	0
Alnus viridis	w	0	0	0	0	0	0	0	0	0	0
Amelanchier sp.	w	0	0	1	0	0	0	0	0	0	1
Anaphalis margaritacea	h	1	0	0	1	0	0	1	0	0	0
Anemone quinquefolia	h	0	0	0	0	0	0	0	0	0	1
Antennaria neglecta	h	0	0	0	1	0	0	0	0	0	0
Aster macrophyllus	h	0	0	0	3	0	0	0	3	3	1
Betula alleghaniensis	w	0	0	0	0	0	0	0	0	0	0
Betula papyrifera	w	3	3	3	2	3	3	3	1	3	1
Calamagrostis sp.	g	0	2	0	3	0	0	0	1	0	2
Carex scoparia	g	3	0	0	0	0	2	2	0	0	0
Carex sp.	g	0	0	0	3	1	0	0	3	1	3
Cirsium sp.	h	1	0	0	1	0	0	0	0	0	0
Clintonia borealis	h	0	0	2	1	0	0	0	2	0	3
Comptonia peregrina	w	0	3	0	2	0	0	0	0	1	0
Coptis trifolia	h	0	0	0	1	0	0	0	3	3	0
Cornus canadensis	h	0	0	3	2	1	0	0	1	1	0
Cornus sericea	w	0	0	0	0	0	0	0	0	0	1
Corylus cornuta	w	0	0	0	0	3	0	0	2	0	0
Crataegus sp.	w	0	0	0	0	0	0	0	0	0	1
Danthonia spicata	g	0	0	0	2	0	0	0	0	0	0
Deschampsia cespitosa	g	3	3	0	0	0	3	0	0	0	0
Diervilla lonicera	w	0	0	0	3	1	0	0	3	3	2
Dryopteris carthusiana	h	0	0	0	0	1	0	0	0	1	3
Epigaea repens	w	0	0	2	0	0	0	0	0	0	0
Epilobium angustifolium	h	0	0	0	1	0	0	0	1	0	0
Erigeron strigosus	h	2	0	0	0	0	0	0	0	0	0
Fragaria virginiana	h	0	0	0	2	0	0	0	0	0	0
Gallium triflorum	h	0	0	0	1	0	0	0	2	0	2
Gaultheria procumbens	ds	0	0	1	2	0	0	0	0	3	0
Heiracium aurantiacum	h	2	0	0	1	0	0	0	0	0	0
Heiracium caespitosum	h	2	0	0	2	0	0	2	0	1	0
Kalmia angustifolia	w	0	0	0	0	0	0	0	0	0	0
Ledum groenlandicum	df	0	0	0	0	0	0	0	0	0	0
Lotus corniculatus	h	2	0	0	0	0	0	0	0	0	0
Maianthemum canadense	h	0	0	2	0	0	0	0	0	2	1
Matteuccia struthiopteris	h	0	0	0	0	0	0	0	0	0	1
Monotropa hypopithys	h	0	0	1	0	0	0	0	0	0	0
Nemopanthus mucronatus	w	0	0	0	0	1	0	0	0	0	0
Osmunda claytoniana	h	0	0	0	0	0	0	2	0	0	0
Phleum pratense	g	0	0	0	0	0	0	0	0	0	0

(continued)

Table 6.38 (continued)

Species	Life form	Occurrences of species on sampling plots									
		1–1	1–2	1–3	1–4	1–5	2–1	2–2	2–3	2–4	2–5
Picea glauca	w	0	0	2	0	0	0	0	2	1	0
Picea mariana	w	0	0	0	0	0	0	0	0	0	0
Pinus banksiana	w	0	2	2	1	0	0	0	1	1	0
Pinus resinosa	w	1	0	0	2	0	0	0	0	0	0
Pinus strobus	w	0	3	0	0	0	0	0	0	1	0
Polygala paucifolia	h	0	0	0	0	0	0	0	1	1	3
Populus balsamifera	w	2	0	0	1	0	0	0	0	0	0
Populus grandidentata	w	2	0	0	0	0	0	0	0	0	0
Populus tremuloides	w	3	1	0	3	3	1	1	3	1	3
Prunus pensylvanica	w	0	0	0	0	0	0	1	0	0	0
Pteridium aquilinum	h	0	3	3	3	0	0	1	3	3	1
Quercus rubra	w	0	0	3	0	0	0	3	0	3	3
Ribes glandulosum	w	0	0	0	0	1	0	0	0	0	0
Ribes sp.	w	0	0	0	0	0	0	0	1	0	1
Rosa acicularis	w	0	0	0	3	0	0	1	2	0	0
Rubus idaeus	w	0	0	0	1	0	0	0	2	0	1
Rubus pubescens	ds	0	0	0	0	0	0	0	0	0	1
Rumex acetosella	h	3	0	0	0	0	2	2	0	0	0
Salix sp.	w	3	2	1	2	0	0	2	2	1	0
Smilacina racemosa	h	0	0	0	0	0	0	0	0	1	0
Solidago canadensis	h	0	0	0	0	0	0	0	0	0	0
Solidago rugosa	h	2	0	0	1	0	0	0	0	0	0
Solidago uliginosa	h	0	0	0	1	0	0	0	0	0	0
Taraxacum officinale	h	0	0	0	0	0	0	0	0	0	0
Thuja occidentalis	w	0	0	0	0	0	0	0	0	0	0
Trientalis borealis	h	0	0	0	0	1	0	0	0	0	2
Trifolium sp.	h	1	0	0	0	0	0	0	0	0	0
Ulmus americana	w	0	0	0	0	0	0	0	0	0	2
Vaccinium angustifolium	df	0	3	3	3	1	0	2	0	2	0
Vaccinium myrtilloides	df	0	2	3	2	0	0	1	2	2	0
Verbascum thapsus	h	1	0	0	0	0	0	0	0	0	0
Vicia cracca	h	0	0	0	2	0	0	0	0	0	0
Viola sp.	h	0	0	0	1	0	0	0	0	0	1

Data collected 19–21.10.2007. Life forms: ds - dwarf shrubs, g - grasses, h - herbs, w - trees and shrubs.

Table 6.39 Occurrences of vascular plants (numbers of sampling plots, out of three, on which the species was recorded) in the impact zone of the of the power plant at Vorkuta, Russia

Species	Life form	Occurrences of species on sampling plots									
		1–1	1–2	1–3	1–4	1–5	2–1	2–2	2–3	2–4	2–5
Achillea millefolium	h	1	3	3	2	0	3	0	2	0	3
Adoxa moschatellina	h	0	0	0	0	0	0	0	0	0	2
Angelica decurrens	h	0	1	0	0	0	0	0	0	0	1
Arctostaphylos alpinus	h	2	0	0	0	1	0	0	1	0	1
Artemisia vulgaris	h	0	0	0	2	0	1	0	1	0	0
Barbarea vulgaris	h	0	2	0	3	0	0	0	1	0	0
Betula nana	w	3	3	3	3	3	3	3	3	3	3
Calamagrostis langsdorfii ssp. *langsdorfii*	g	0	0	0	1	0	0	0	0	0	1
Cardamine pratensis	h	0	0	0	0	0	3	0	0	0	1
Carex nigra	g	0	0	1	0	1	3	0	3	0	3
Carum carvi	h	0	0	0	0	0	0	0	0	0	3
Cerastium cerastoides	h	0	0	0	0	0	1	0	0	0	0
Cerastium jenisejense	h	2	1	3	2	1	1	3	3	0	3
Chrysosplenium alternifolium	h	1	0	0	0	0	0	0	0	0	0
Empetrum nigrum	ds	0	0	0	0	3	0	1	0	1	1
Epilobium angustifolium	h	1	0	0	0	0	2	1	1	0	0
Equisetum arvense	h	2	0	0	1	0	2	0	0	0	1
Equisetum palustre	h	1	0	0	0	0	0	0	0	0	0
Eriophorum scheuchzeri	g	0	0	1	0	0	0	0	0	0	0
Erysimum cheiranthoides	h	0	0	0	0	0	0	2	1	0	0
Festuca ovina	g	3	3	3	3	3	3	3	3	3	3
Geranium albiflorum	h	3	0	0	1	0	0	0	0	0	0
Geum rivale	h	1	0	0	0	0	0	0	0	0	0
Hedysarum hedysaroides ssp. *arcticum*	h	0	0	0	0	0	2	0	0	0	0
Hieracium sp.	h	0	0	0	0	0	1	0	1	0	0
Ledum decumbens	ds	0	0	0	0	2	0	0	0	0	0
Luzula campestris ssp. *frigida*	g	0	0	0	1	0	0	0	1	0	0
Myosotis laxa ssp. *caespitosa*	h	1	0	0	0	0	0	0	0	0	0
Parnassia palustris	h	0	1	0	0	0	1	0	0	0	0
Pedicularis lapponica	h	2	2	2	2	0	2	3	1	2	1
Petasites frigidus	h	0	0	3	3	2	2	2	1	0	0
Pleurospermum uralense	h	0	1	0	0	0	0	0	0	0	0
Poa alpigena	g	0	0	0	0	0	0	0	1	0	1
Poa pratensis	g	3	3	3	3	3	3	3	3	3	3
Polemonium acutiflorum	h	1	1	1	2	0	0	0	0	0	2
Polygonum bistorta	h	3	3	0	1	1	1	3	1	0	0
Polygonum viviparum	h	1	0	1	0	1	3	0	1	0	0
Pyrola minor	h	3	0	2	3	0	3	1	1	1	1
Ranunculus acris ssp. *borealis*	h	0	0	0	0	0	1	0	0	0	0

(continued)

Table 6.39 (continued)

Species	Life form	Occurrences of species on sampling plots									
		1–1	1–2	1–3	1–4	1–5	2–1	2–2	2–3	2–4	2–5
Ranunculus acris ssp. *glabriusculus*	h	1	0	0	1	0	1	0	0	0	2
Rubus arcticus	h	2	3	3	3	2	3	1	1	3	3
Rubus chamaemorus	h	0	0	2	0	1	0	0	1	2	0
Salix glauca	w	3	3	3	3	3	3	3	3	3	3
Salix hastata	w	0	0	0	0	0	2	0	0	0	0
Salix lanata	w	3	3	3	3	3	3	3	1	2	3
Salix lapponum	w	0	1	0	0	1	3	0	2	0	2
Salix phylicifolia	w	3	2	1	3	1	3	3	2	0	3
Salix reticulata	w	0	0	0	0	0	1	1	0	0	0
Salix × *dasyclados*	w	3	3	3	3	0	3	3	2	2	3
Sanguisorba officinalis	h	0	0	0	0	0	1	0	0	0	0
Saussurea alpina	h	1	0	2	0	0	0	0	0	0	0
Senecio integrifolius	h	0	0	0	0	1	0	0	0	0	1
Solidago virgaurea	h	0	2	1	2	2	0	0	1	0	0
Stellaria fennica	h	0	0	0	0	1	0	0	0	0	1
Stellaria palustris	h	0	0	0	0	0	0	2	0	0	0
Tanacetum bipinnatum	h	0	1	0	0	0	0	0	0	0	0
Taraxacum croceum	h	0	0	0	1	0	1	0	1	0	0
Taraxacum perfiljevii	h	0	0	0	0	0	1	0	0	0	0
Trientalis europaea	h	0	0	0	0	0	0	0	0	0	0
Vaccinium myrtillus	ds	0	0	0	0	3	0	0	0	0	0
Vaccinium uliginosum	ds	3	3	3	2	3	3	3	3	3	3
Vaccinium vitis-idaea	ds	3	3	3	3	3	3	3	3	3	3
Valeriana capitata	h	1	0	0	0	0	0	0	0	0	1
Veratrum lobelianum	h	0	3	0	1	0	0	0	0	0	0
Veronica longifolia	h	1	1	2	0	0	2	0	2	0	0

Data collected 8–10.7.2001. Life forms: ds - dwarf shrubs, g - grasses, h - herbs, w - trees and shrubs.

Table 6.40 Occurrences of vascular plants (numbers of sampling plots, out of three, on which the species was recorded) in the impact zone of the of the aluminium smelter at Žiar nad Hronom, Slovakia

Species	Life form	Occurrences of species on sampling plots									
		1–1	1–2	1–3	1–4	1–5	2–1	2–2	2–3	2–4	2–5
Abies alba	w	0	0	0	2	2	0	0	1	2	1
Acer campestre	w	0	0	3	0	3	0	1	3	0	0
Acer platanoides	w	0	0	0	0	3	0	0	3	0	2
Actaea spicata	w	0	0	1	0	0	0	0	1	0	0
Ajuga reptans	h	0	1	1	0	2	0	0	0	0	3
Asarum europaeum	h	0	0	0	0	0	0	0	3	1	3
Athyrium filix-femina	h	1	0	0	1	0	0	0	0	3	2
Ballota nigra	h	0	0	0	0	1	0	0	0	0	0
Bilderdykia dumetorum	h	0	0	0	0	0	0	0	0	0	1
Cardamine impatiens	h	0	0	0	0	0	0	0	0	0	1
Carex muricata	g	0	0	0	0	0	1	0	0	0	0
Carex pilosa	g	0	0	2	0	1	0	3	1	3	0
Carex rhizina	g	0	1	0	0	0	0	0	0	0	0
Carex sylvatica	g	0	0	0	0	1	0	0	0	0	0
Carpinus betulus	w	2	3	3	0	1	3	1	3	0	1
Corylus avellana	w	0	0	0	0	0	0	0	0	2	0
Crataegus monogyna	w	2	3	0	0	0	0	0	0	0	0
Cystopteris fragilis	h	0	0	0	0	0	0	1	3	0	1
Deschampsia flexuosa	g	0	0	0	0	0	2	3	0	0	0
Dryopteris filix-mas	h	0	0	0	0	1	1	1	3	3	0
Elymus caninus	g	0	1	0	0	0	0	0	1	0	0
Epilobium collinum	h	0	0	0	0	0	0	3	0	0	0
Epilobium montanum	h	0	0	0	0	0	2	0	2	1	0
Euphorbia amygdaloides	h	0	0	1	0	0	0	0	0	0	0
Fagus sylvatica	w	2	3	3	3	3	3	3	3	3	3
Fragaria vesca	h	0	0	0	0	0	1	1	0	0	0
Fraxinus excelsior	w	0	0	0	0	0	0	0	1	0	3
Galeobdolon luteum	h	0	0	0	0	1	0	0	1	0	1
Galeopsis ladanum	h	0	0	0	0	0	0	0	0	0	1
Galium aparine	h	0	0	0	0	0	3	2	0	0	0
Galium boreale	h	0	0	0	0	0	0	0	0	0	0
Galium odoratum	h	0	0	3	0	3	0	3	3	1	3
Geranium robertianum	h	0	0	0	0	1	0	0	0	0	1
Glechoma hederacea	h	1	0	0	0	2	0	0	2	0	2
Gymnocarpium dryopteris	h	0	0	0	0	0	0	0	0	1	0
Hieracium murorum	h	1	0	0	0	0	0	0	0	0	0
Impatiens glandulifera	h	0	0	0	0	0	0	0	2	0	0
Lamium album	h	0	0	1	0	0	0	0	0	0	0
Lathyrus vernus	h	0	0	0	0	0	0	1	0	0	0
Ligustrum vulgare	w	0	0	0	0	0	3	0	0	0	0
Luzula luzuloides	g	0	0	0	3	0	3	3	0	3	2
Maianthemum bifolium	h	1	0	0	0	0	0	0	0	0	0

(continued)

Table 6.40 (continued)

Species	Life form	Occurrences of species on sampling plots									
		1–1	1–2	1–3	1–4	1–5	2–1	2–2	2–3	2–4	2–5
Mycelis muralis	h	0	0	0	0	0	0	1	0	1	0
Oxalis acetosella	h	0	0	0	0	2	0	3	2	1	2
Picea abies	w	0	0	0	0	0	1	**0**	0	0	0
Pinus sylvestris	w	0	3	0	0	0	0	0	0	0	0
Polygonatum latifolium	h	0	0	1	0	0	0	0	0	0	0
Populus tremula	w	0	0	0	0	0	0	0	0	0	0
Primula elatior	h	0	0	1	0	0	0	0	0	0	0
Pulmonaria obscura	h	0	0	0	0	0	0	0	3	0	0
Quercus petraea	w	1	3	1	0	1	3	2	0	0	0
Quercus rubra	w	0	0	0	0	0	2	0	0	0	0
Ribes uva-crispa	w	0	0	0	0	0	0	0	1	0	0
Rosa sp.	w	0	1	0	0	0	0	0	0	0	0
Rubus caesius	w	1	0	0	2	0	0	0	0	0	1
Rubus idaeus	w	0	0	0	0	0	0	0	1	0	0
Sorbus aucuparia	w	2	1	0	1	0	0	0	0	0	0
Stellaria media	h	0	0	0	0	0	0	0	0	1	2
Tilia cordata	w	0	0	0	0	0	0	0	3	0	0
Urtica dioica	h	0	0	0	0	0	0	0	0	0	2
Veronica chamaedrys	h	0	0	0	0	0	2	0	0	0	0
Viola reichenbachiana	h	0	2	3	0	3	2	0	3	1	0

Data collected 29.8–1.9.2002. Life forms: ds - dwarf shrubs, g - grasses, h - herbs, w - trees and shrubs

Table 6.41 Results of statistical analyses of the data on diversity of vascular plants: between-site variation and correlations with distances and pollution loads

Polluter	Statistics[a]	All vascular plants	Trees and shrubs	Grasses	Herbs	Field layer
Bratsk	ANOVA: *F/P*	2.72/0.03	3.47/0.01	0.61/0.78	3.03/0.02	3.74/0.007
	Dist.: *r/P*	0.63/0.05	0.23/0.52	0.02/0.96	0.48/0.16	0.53/0.12
	Poll.: *r/P*	−0.34/0.34	0.22/0.53	−0.10/0.77	−0.37/0.29	−0.47/0.17
Jonava	ANOVA: *F/P*	5.19/0.0011	5.22/0.001	17.7/<0.0001	2.92/0.02	4.60/0.002
	Dist.: *r/P*	0.56/0.09	0.36/0.30	0.78/0.007	0.14/0.70	0.47/0.17
	Poll.: *r/P*	−0.12/0.75	0.12/0.75	−0.28/0.43	−0.05/0.89	−0.21/0.55
Kandalaksha	ANOVA: *F/P*	7.32/0.0001	3.63/0.008	3.17/0.02	13.0/<0.0001	8.81/<0.0001
	Dist.: *r/P*	−0.36/0.31	−0.42/0.22	−0.43/0.21	−0.38/0.28	−0.28/0.44
	Poll.: *r/P*	0.42/0.22	0.56/0.09	0.29/0.41	0.50/0.14	0.28/0.43
Karabash	ANOVA: *F/P*	55.5/<0.0001	21.3/<0.0001	9.63/<0.0001	55.4/<0.0001	56.3/<0.0001
	Dist.: *r/P*	0.92/0.0002	0.80/0.005	0.55/0.10	0.92/0.0001	0.75/0.01
	Poll.: *r/P*	−0.87/0.0010	−0.88/0.001	−0.74/0.01	−0.82/0.0035	−0.69/0.03
Kostomuksha	ANOVA: *F/P*	5.55/0.0007	1.61/0.18	5.55/0.0007	20.1/<0.0001	7.11/0.0001
	Dist.: *r/P*	−0.64/0.05	−0.40/0.25	−0.63/0.05	−0.73/0.02	−0.63/0.05
	Poll.: *r/P*	0.48/0.16	0.52/0.12	0.52/0.12	0.39/0.27	0.39/0.27
Monchegorsk	ANOVA: *F/P*	8.01/<0.0001	3.33/0.01	2.42/0.05	5.73/0.0006	8.19/<0.0001
	Dist.: *r/P*	0.48/0.16	0.36/0.31	0.16/0.67	0.58/0.08	0.50/0.14
	Poll.: *r/P*	−0.68/0.03	−0.67/0.04	−0.51/0.14	−0.49/0.15	−0.64/0.04
Nadvoitsy	ANOVA: *F/P*	5.29/0.0009	6.71/0.0002	4.18/0.004	8.84/<0.0001	4.27/0.003
	Dist.: *r/P*	−0.44/0.21	−0.21/0.55	0.14/0.71	−0.43/0.21	−0.44/0.21
	Poll.: *r/P*	0.58/0.08	0.35/0.32	−0.01/0.97	0.59/0.07	0.55/0.10
Nikel	ANOVA: *F/P*	2.97/0.0204	1.47/0.23	1.35/0.28	2.54/0.04	3.28/0.01
	Dist.: *r/P*	0.54/0.10	0.37/0.30	0.36/0.31	0.42/0.22	0.54/0.11
	Poll.: *r/P*	−0.66/0.04	−0.54/0.11	−0.15/0.68	−0.51/0.14	−0.63/0.05
Norilsk	ANOVA: *F/P*	14.4/<0.0001	3.09/0.02	5.55/0.0007	13.5/<0.0001	33.7/<0.0001
	Dist.: *r/P*	0.56/0.12	0.33/0.39	0.56/0.11	0.38/0.31	0.58/0.10
	Poll.: *r/P*	−0.61/0.08	−0.37/0.33	−0.63/0.07	−0.42/0.26	−0.61/0.08

Revda	ANOVA: *F/P*	15.2/<0.0001	6.90/0.0002	0.90/0.54	17.6/<0.0001	15.0/<0.0001
	Dist.: *r/P*	0.75/0.01	−0.07/0.84	0.54/0.11	0.91/0.0003	0.89/0.0005
	Poll.: *r/P*	−0.77/0.01	0.07/0.85	−0.65/0.04	−0.83/0.003	−0.82/0.004
Straumsvík	ANOVA: *F/P*	3.07/0.02	1.00/0.47	1/32/0.29	3.13/0.02	3.12/0.02
	Dist.: *r/P*	0.04/0.91	0.53/0.12	0.14/0.70	−0.20/0.58	0.00/0.99
	Poll.: *r/P*	−0.64/0.05	−0.62/0.06	−0.66/0.04	−0.51/0.14	−0.60/0.07
Sudbury	ANOVA: *F/P*	10.4/<0.0001	8.32/<0.0001	5.82/0.0005	7.71/<0.0001	8.21/<0.0001
	Dist.: *r/P*	0.45/0.19	0.80/0.005	−0.64/0.05	0.40/0.25	0.19/0.59
	Poll.: *r/P*	−0.50/0.14	−0.82/0.04	−0.40/0.25	0.63/0.05	−0.25/0.48
Vorkuta	ANOVA: *F/P*	8.10/<0.0001	5.27/0.001	4.94/0.0014	6.30/0.0003	6.11/0.0004
	Dist.: *r/P*	−0.50/0.14	−0.63/0.05	0.19/0.59	−0.58/0.08	−0.43/0.22
	Poll.: *r/P*	0.41/0.24	0.55/0.10	0.15/0.58	0.39/0.27	0.33/0.35
Žiar nad Hronom	ANOVA: *F/P*	3.46/0.01	5.06/0.001	4.31/0.003	4.43/0.0027	3.96/0.005
	Dist.: *r/P*	0.34/0.34	−0.26/0.48	0.06/0.88	0.56/0.09	0.55/0.10
	Poll.: *r/P*	−0.50/0.14	0.10/0.79	−0.43/0.21	−0.57/0.08	−0.66/0.004

Distance was log-transformed prior the correlation analysis. Pollution load was measured as foliar concentrations of fluorine (Bratsk, Kandalaksha, Nadvoitsy, Straumsvík, Žiar nad Hronom), strontium (Jonava), iron (Kostomuksha, Vorkuta), nickel (Monchegorsk, Nikel/Zapolyarnyy, Sudbury, Norilsk), and copper (Karabash, Revda) (for actual concentrations, consult Tables 2.27–2.43). In Norilsk, plot 2–5 was excluded from the analysis

The majority of information on the structure of plant communities was collected from three plots, 10 × 10 m in size, selected for each study site. These plots were not marked, and they were surveyed only once. However, after processing of the first data sets, we have recognised that this spatial scale is only marginally suitable to measure the cover of field layer vegetation, mosses and epigeic lichens. Therefore, for ten of 14 surveyed polluters, we have additionally assessed the cover of these plant groups, as well as of bare ground and surface stones (the latter index is reported in Chapter 3) in ten plots, 1 × 1 m in size, selected at 10 m intervals along a line crossing at least two of three larger plots.

6.2.2 Vegetation Cover

Vegetation cover was estimated visually by the same observer (M.V.K.). Repeated measurements indicated sufficient accuracy of these estimates: the differences between two measurements were below 5% for absolute values of vegetation cover not exceeding 50%, and below 10% for larger cover values.

We separately estimated and analysed cover of the following layers: (a) the top-canopy layer formed by mature woody plants (this layer is absent in tundra sites around Straunsvík and Vorkuta); (b) the understorey, i.e., the intermediate layer between the top-canopy and the field layer in forested habitats (the understorey is formed by woody plants, including both mature low-stature species and juvenile individuals of the top-canopy species); (c) the field layer, consisting of herbs, grasses and sedges, and dwarf shrubs; (d) mosses; and (e) epigeic lichens. Simultaneously, we estimated the proportion of bare ground and surface stone cover; the latter data are reported and discussed in Chapter 3 (Tables 3.2–3.15).

To test the hypothesis that trees are more sensitive to pollution than herbs and grasses, we introduced an additional response variable, the ratio between site-specific cover estimates of top-canopy and field layers. A decrease of this variable with pollution will support the hypothesis mentioned above.

6.2.3 Stand Characteristics

Stand characteristics are averaged from three point samplings, conducted from centres of the same plots (10 × 10 m size) that were used for assessment of tree cover (see above, Section 6.2.2). Stand basal area was measured by a relascope, as described by West (2003). Stand composition, expressed as relative abundances of forest-forming species (rounded to the nearest 10% in Tables 6.15–6.26), was calculated from the same records as stand basal area. The average height of the stand is based on three plot-specific values, each obtained by measuring heights of five trees forming the top-canopy layer by the angle of elevation method.

Response variables used in the analyses were (a) stand basal area, (b) stand height, (c) the proportion of the main forest-forming tree species (determined by the type of forest in which the study plots were selected), and (d) the proportion of conifers.

6.2.4 Regeneration of Dominant Woody Plants

Seedlings of woody plants, saplings and young trees (defined as trees that have not yet reached 5% of the average height of mature trees at this study site; generally less than 120 cm tall) were counted in each 10 × 10 m plot. To account for small seedlings (less than 5 cm tall) that are not easy to recognise in dense field layer vegetation, we (whenever necessary) carefully checked five to ten 1 × 1 m subplots within each plot and multiplied the average number of small seedlings by 100 to obtain a plot-specific estimate.

Response variables used in the analyses were (a) total number of seedlings, saplings and young trees, (b) the proportion of the dominant tree species, and (c) the proportion of conifers among seedlings, saplings and young trees.

6.2.5 Diversity of Vascular Plants

Every effort was made to record all species of vascular plants within each 10 × 10 m plot. Easily recognisable species were recorded *in situ* by using pre-printed forms. Vouchers of other species were determined with the assistance of professional botanists (listed in the Acknowledgements section); these vouchers are now deposited in the herbarium of the University fo Turku. In total, we were able to provide species names for about 99% of the collected specimens; the abbreviation 'sp.' in Tables 6.27–6.40 refers to non-flowering individuals whose identity cannot be revealed with certainty.

The nomenclature of plants generally follows Tutin (1964, 1968, 1972, 1976, 1980). Siberian species absent from Europe are given according to Baikov (2005), and North American species are given according to Gleason and Cronquist (1993).

To allow direct comparison of our conclusions with the meta-analysis of the published data (Zvereva et al. 2008), we considered the following partially subordinated groups: grasses and sedges (grasses hereafter), herbs, field layer vegetation (grasses and sedges, herbs and dwarf shrubs), shrubs and trees (woody plants hereafter), and all vascular plants pooled. In our analysis, we included dwarf shrubs, which are woody plants, in the field layer vegetation. This combination seemed more relevant in terms of vegetation structure because for many life-history traits, dwarf shrubs are more similar to perennial herbs than to top-canopy species.

Our data allowed calculating α, β, and γ diversity of vascular plants. For each study site, α diversity was measured as the mean number of plant species within each of three replicate 10 × 10 m plots. These values were calculated for (a) the overall species richness of vascular plants, (b) trees and shrubs, (c) herbs, (d) grasses, and (e) field layer vegetation. Analyses reported below are based on effect sizes, calculated from correlations of site-specific means with log-transformed distances from polluters (Table 6.41).

Since three methods of ES calculations (described in Section 2.5.2.2) for α diversity yielded the same conclusions, and overall effects on different groups of plants were uniform (see below, Section 6.3.4), we explored pollution effects on β and γ diversity of all vascular plants by only one method, i.e., by contrasting species richness in the two most and two least polluted study sites around each of the polluters.

Following the protocol described by Chalcraft et al. (2008), we measured site-specific β diversity as the average pairwise Jaccard distance in plant species composition among three replicate plots within the site. The two aspects of spatial variation combined in Jaccard distance (Koleff et al. 2003) can be separated by calculating β_{gl}, which measures variation in species composition attributable to spatial variation in diversity (i.e., some localities contain more species than other localities) and β_{sim}, which measures spatial variation in species composition after adjusting for differences in α diversity (i.e., some localities contain species that are absent in other localities) (Lennon et al. 2001; Koleff et al. 2003). For a given pair of plots, these indices were estimated by:

$$\beta_{gl} = 2 \times \mathrm{abs}(b - c) / [2a + b + c] \tag{6.1}$$

$$\beta_{sim} = \min(b,c) / [\min(b,c) + a] \tag{6.2}$$

where:

a = the number of species that both plots have in common

b = the number of species that are found in the first plot only and

c = the number of species that are found in the second plot but not in the first one

Since we surveyed three 10 × 10 m plots, our site-specific estimates of both β_{gl} and β_{sim} were each based on three pairwise values. We estimated γ diversity as the total number of plant species found in all three replicate plots within the site.

6.2.6 *Species Composition of Vascular Plants*

As has been pointed out earlier, problems in comparative analysis of data sets from different floristic regions complicate exploration of pollution-induced changes in species composition by meta-analysis (Zvereva et al. 2008). However, using original data (Tables 6.27–6.40), we can test the hypotheses that pollution affects species composition of vascular plants, either by selective species removal or colonisation by tolerant species, or both, against the null hypothesis that between-site variation in species composition is random (Table 6.42). This can be done by combining data on species richness with estimates of between-site similarities, calculated as the Jaccard index (i.e., the number of common species divided by the total number of species recorded at both sites). We assumed that any non-random (e.g., pollution-induced) change in plant species composition will result in a smaller similarity between polluted and unpolluted sites within the same pollution gradient, compared to the similarity between two unpolluted sites (Table 6.42).

Table 6.42 Relationships between characteristics of plant communities in polluted and unpolluted sites expected under different assumptions concerning pollution effects on species composition of vascular plants

Assumptions	Species richness[a]	Between-site similarity[b]	Uncommon species[c]
Variation in species composition is random, i.e., probabilities of both extinction and colonization are equal for all species	NP = NC	SPC = SC	UP = UC
Pollution causes only selective removal of sensitive species	NP < NC	SPC < SC	UP < UC
Pollution causes only colonization by tolerant species	NP > NC	SPC < SC	UP > UC
Pollution causes species replacement, i.e. selective species removal followed by colonization by tolerant species	NP = NC	SPC < SC	UP = UC

[a] NP, mean number of species in polluted sites; NC, mean number of species in clean sites.
[b] SPC, average similarity (Jaccard index) between polluted and clean sites around the same polluter; SC, similarity between two clean sites around the same polluter.
[c] UP, proportion of species present in polluted site but absent in clean site of the same gradient (relative to species number in polluted site); UC, proportion of species present in clean site but absent in polluted site of the same gradient (relative to species number in clean site).

Since estimates of species richness may be affected by plant abundances (see below, Section 6.4.3) for both heavily polluted and control plots, we estimated the proportion of uncommon species, i.e., species that were absent in a site on the opposite end of the same pollution gradient. These proportions were averaged for two gradients around the same polluter and then compared across all polluters using the Kruskal-Wallis test. If pollution only removes sensitive species, then flora of the most polluted site should represent a subset of flora of an unpolluted site, with all species shared with the unpolluted site. And *vice versa* if pollution only facilitates colonization by tolerant species, then flora of the unpolluted site should represent a subset of flora of the most polluted site (Table 6.42).

6.3 Results

6.3.1 Vegetation Cover

Variation in the cover of different plant groups between study sites was significant in 53 of 63 data sets. This variation was always significant in mosses (14 data sets), while canopy cover showed the lowest variation (significant in seven of 12 data sets). However, only 44 of 118 individual correlation coefficients (with both distance and pollution) were significant (Tables 6.1–6.14).

Vegetation layers responded to pollution impacts in an uncoordinated manner; an average pairwise correlation between the cover of different layers (site-specific values standardised by polluter) did not differ from zero ($z_r = 0.01$, CI = −0.06 … 0.08, $N = 10$ pairwise correlations). This result indicates that separate analyses of changes in the cover of different layers were not redundant. Since the proportion of bare ground negatively correlated with vegetation cover ($z_r = -0.30$, CI = −0.49 … −0.11, $N = 5$ pairwise correlations), it was excluded from the meta-analysis.

Pollution effects on canopy cover (Fig. 6.1), understorey vegetation, including shrubby vegetation in treeless areas (Fig. 6.2), and ground lichens ($z_r = -0.08$, CI = −0.66 … 0.63, $N = 6$) did not differ from zero; adverse effects were detected for field layer vegetation (Fig. 6.3) and mosses (Fig. 6.4). These conclusions did not depend on the method used to calculate ES (canopy cover: $Q_B = 0.37$, df = 2, $P = 0.83$; field layer cover: $Q_B = 0.26$, df = 2, $P = 0.88$).

The absence of overall effects of pollution on canopy and understorey covers is due to contrasting responses to impacts of different polluters: a decline around non-ferrous smelters and an increase around aluminium smelters (Figs. 6.1 and 6.2). Consistently, we detected significant negative effects of acidifying polluters and significant positive effects of alkalysing polluters on both the top-canopy and the understorey plants. Pollution effects were similar around both southern and northern polluters (Figs. 6.1 and 6.2).

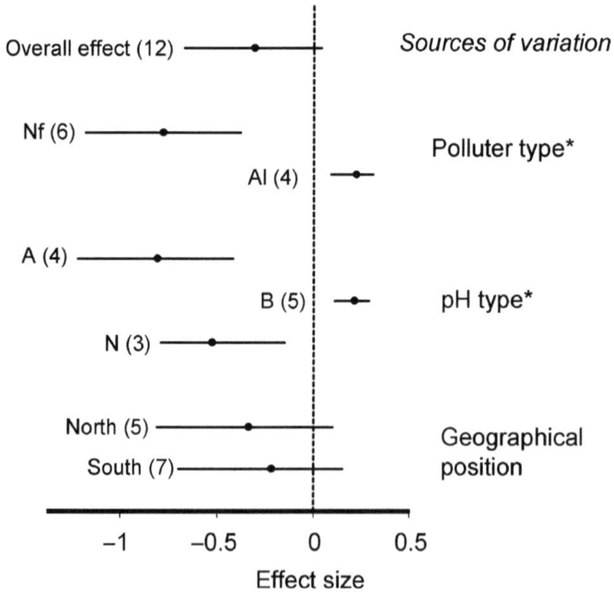

Fig. 6.1 Overall effect and sources of variation in responses of canopy cover. Horizontal lines denote 95% confidence intervals; sample sizes are shown in brackets; an asterisk denotes significant ($P < 0.05$) between-class heterogeneity. For classifications of polluters consult Table 2.1

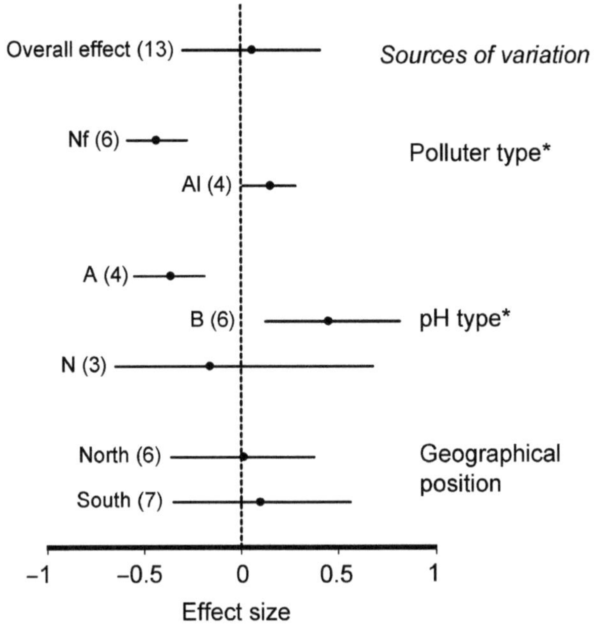

Fig. 6.2 Overall effect and sources of variation in responses of understorey vegetation cover (including shrubby vegetation in treeless areas). For explanations, consult Fig. 6.1

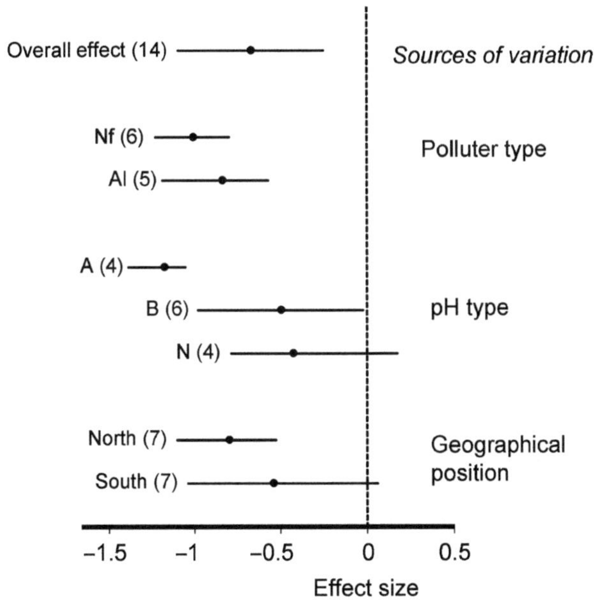

Fig. 6.3 Overall effect and sources of variation in responses of field layer vegetation cover. For explanations, consult Fig. 6.1

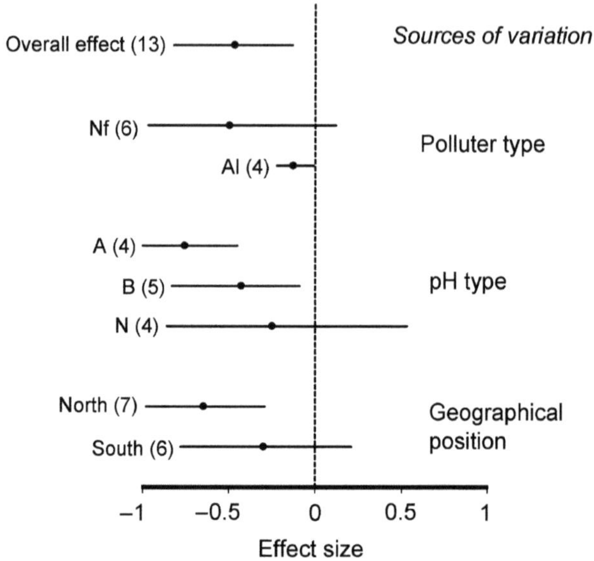

Fig. 6.4 Overall effect and sources of variation in responses of moss cover. For explanations, consult Fig. 6.1

The adverse effects of pollution on the cover of either field layer vegetation or mosses did not vary with polluter type; both acidification and alkalinisation resulted in a significant reduction of both these groups (Figs. 6.3 and 6.4). Adverse effects were significant only around the northern polluters (Figs. 6.3 and 6.4). Changes in the cover of epigeic lichens showed no variation in respect to explored categorical variables (data not shown).

Top-canopy and understorey plants generally demonstrated a weaker decline in cover than field layer vegetation; the ratio between site-specific cover estimates of these plant layers increased with pollution ($z_r = 0.42$, CI = 0.05 … 0.78, $N = 13$). This effect was independent of either the type or geographical position of the polluter ($Q_B = 0.31$, df = 1, $P = 0.56$ and $Q_B = 0.12$, df = 1, $P = 0.72$, respectively), but it changed with the pollution impact on soil pH ($Q_B = 20.9$, df = 2, $P = 0.01$). A decline in the field layer relative to top-canopy plants around acidifying polluters ($z_r = 0.96$, CI = 0.78 … 1.15, $N = 4$) was stronger than around alkalysing polluters ($z_r = 0.44$, CI = 0.07 … 0.78, $N = 6$). Around polluters that did not change soil pH, the effect was the opposite: trees and the understorey declined faster than field layer vegetation ($z_r = -0.35$, CI = −0.40 … −0.29, $N = 2$).

We have detected significant non-linear responses in 11 of 59 data sets (three dome-shaped and eight U-shaped).

6.3.2 Stand Characteristics

Stand characteristics generally showed pronounced between-site variation. This variation was significant in 11 of 12 data sets on basal area, nine of nine data sets on tree height, 12 of 12 data sets on species composition, and 11 of 12 data sets on the proportion of dominant tree species in the stand. However, only a few individual correlation coefficients (with both distance and pollution; nine of 24 for basal area, nine of 18 for tree height, and five of 24 for proportion of dominant tree species in the stand) were significant (Tables 6.15–6.26).

Stand height and basal area (values standardised by impact zone) significantly correlated to each other ($r = 0.63$, $N = 88$ sites, $P < 0.0001$) and showed a uniform response to pollution (correlation between ESs: $r = 0.91$, $N = 9$ impact zones, $P = 0.0008$); both these indices significantly decreased near point polluters (Figs. 6.5 and 6.6). These conclusions did not depend on the method used to calculate ES (basal area: $Q_B = 0.56$, df = 2, $P = 0.76$; tree height: $Q_B = 1.61$, df = 2, $P = 0.45$).

The general negative effect of pollution on stand basal area was mostly due to the significant effects of non-ferrous smelters, whereas the effects of aluminium plants did not differ from zero. A decrease in basal area was pronounced only in the

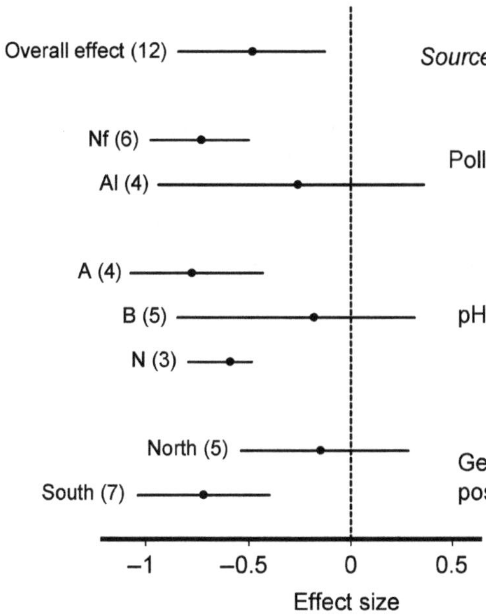

Fig. 6.5 Overall effect and sources of variation in responses of stand basal area. For explanations, consult Fig. 6.1

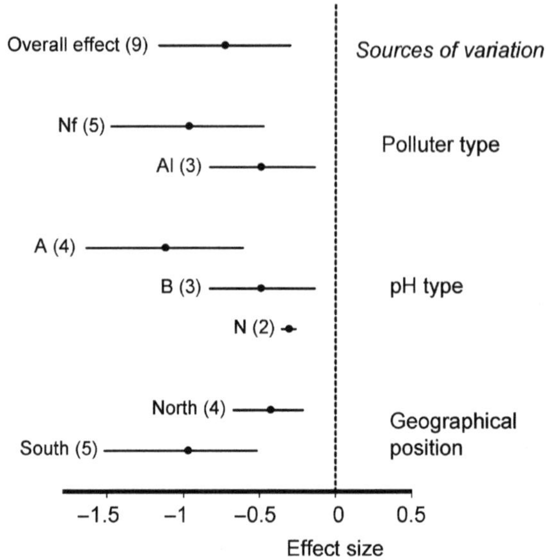

Fig. 6.6 Overall effect and sources of variation in responses of stand height. For explanations, consult Fig. 6.1

impact zones of southern polluters (Fig. 6.5). At the same time, we identified no sources of variation in stand height responses to pollution; within each group of categorical variables, all effects were significantly negative (Fig. 6.6).

Although pollution effects on stand basal area and canopy cover closely correlated to each other ($r = 0.79$, $N = 12$ impact zones, $P = 0.002$), an overall adverse effect was detected for basal area only (Fig. 6.5). This discrepancy is primarily due to contrasting impacts of aluminium smelters on these two variables: canopy cover tended to increase (Fig. 6.1), while basal area tended to decrease near these polluters (Fig. 6.5).

Neither proportion of dominant tree species ($z_r = -0.34$, CI = −0.75…0.07, $N = 12$) nor proportion of conifers (Fig. 6.7) changed under pollution impacts. However, the absence of overall effects on the latter index was due to counterbalancing impacts of non-ferrous smelters and aluminium plants (Fig. 6.7).

We have detected significant non-linear responses in two of 21 data sets (one dome-shaped and one U-shaped).

6.3.3 *Regeneration of Dominant Woody Plants*

Between-site variation in the total number of seedlings, saplings, and young trees was significant in nine of 11 data sets, and in the proportion of dominant tree species

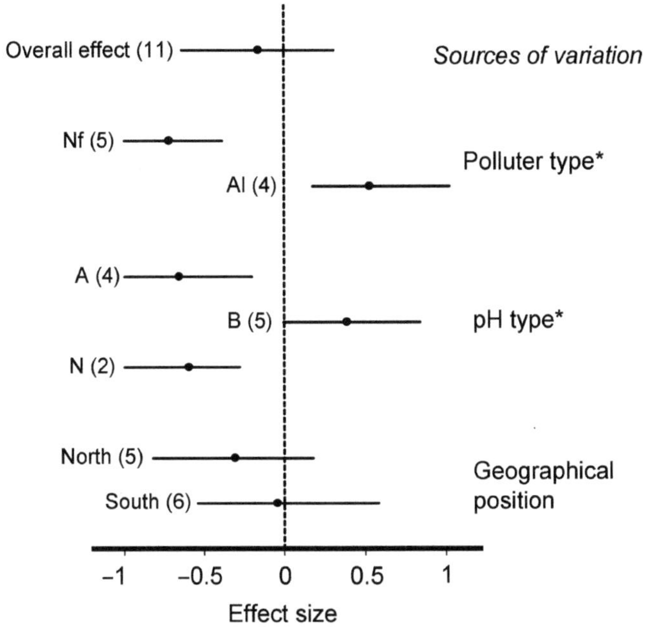

Fig. 6.7 Overall effect and sources of variation in responses of stand composition, measured by proportion of conifers in stand basal area. For explanations, consult Fig. 6.1

in another nine of 11 data sets. However, only three of 22 individual correlation coefficients (with both distance and pollution) were significant for each of these two variables (Tables 6.15–6.26); there was no concordance between data sets in which individual effects were significant.

Pollution did not affect forest regeneration, as can be concluded from the absence of effects on the number of tree seedlings (Fig. 6.8), the proportion of dominant tree species among seedlings ($z_r = -0.22$, CI = −0.66…0.22, $N = 10$), or the proportion of conifers ($z_r = -0.21$, CI = −0.59 … 0.17, $N = 11$). The conclusion on the absence of an overall effect on the number of seedlings did not depend on the method used to calculate ES ($Q_B = 0.22$, df = 2, $P = 0.90$).

The effects of pollution on the composition of mature (top-canopy) and young trees (i.e., seedlings and saplings) did not differ from each other (proportion of dominant tree species: $Q_B = 0.17$, df = 1, $P = 0.68$; proportion of conifers: $Q_B = 0.02$, df = 1, $P = 0.88$). None of the characteristics describing abundance and diversity of seedlings, saplings and young trees varied with either type or geographical position of polluters, or their impact on soil pH. Pollution effects on the abundance of regrowth were generally independent from effects on stand basal area (correlation between ESs: $r = 0.44$, $N = 11$, $P = 0.18$).

We have detected significant non-linear responses in two of 11 data sets (one dome-shaped and one U-shaped).

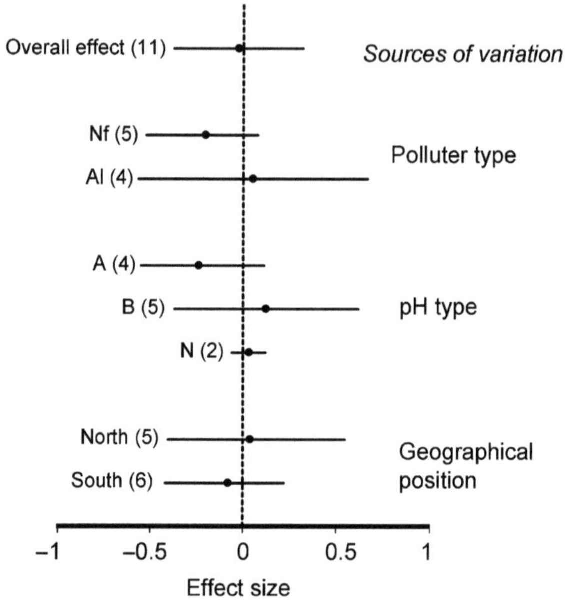

Fig. 6.8 Sources of variation in responses of stand natural regeneration, measured by number of seedlings, saplings, and young trees. For explanations, consult Fig. 6.1

6.3.4 Diversity of Vascular Plants

6.3.4.1 Local (Plot-Specific) Species Richness (α Diversity)

Between-site variation in species richness of different plant groups was significant in 63 of 70 data sets. However, only 35 of 140 individual correlation coefficients (with both distance and pollution) were significant (Tables 6.27–6.40).

The overall effect of pollution on the species richness of vascular plants did not differ from zero (Fig. 6.9). This conclusion did not depend on the method used to calculate ES ($Q_B = 1.63$, df = 2, $P = 0.44$).

Changes in α diversity were generally consistent among all groups of plants (Fig. 6.10). Responses of trees and shrubs did not differ from responses of field layer vegetation ($Q_B = 0.36$, df = 1, $P = 0.55$); grasses and herbs also showed uniform responses to pollution ($Q_B = 0.24$, df = 1, $P = 0.62$).

The absence of an overall effect is due to significant differences between effects caused by different types of polluters; species richness of all explored plant groups, as well as total species richness of vascular plants, decreased around non-ferrous smelters but did not change around aluminium smelters (Figs. 6.9 and 6.10). Accordingly, a decrease in species richness was recorded only near acidifying

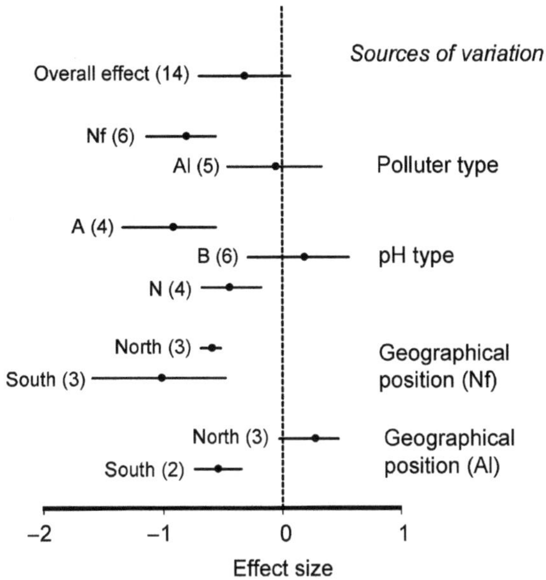

Fig. 6.9 Overall effect and sources of variation in responses of species richness of vascular plants. For explanations, consult Fig. 6.1

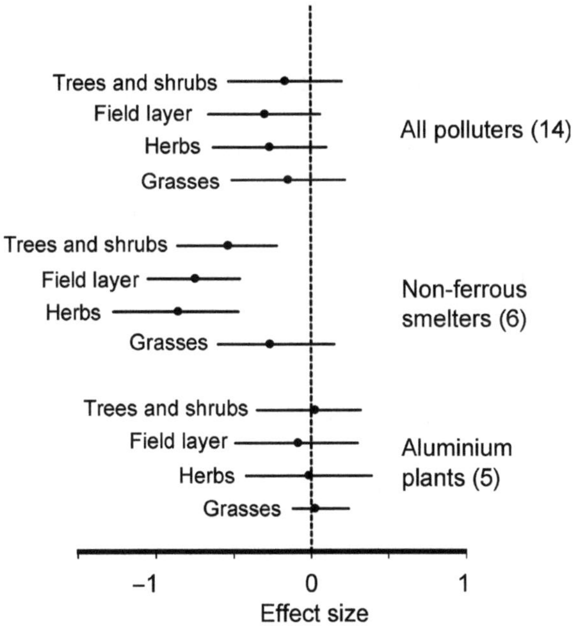

Fig. 6.10 Overall effect and sources of variation in responses of species richness of different, partially subordinated, groups of vascular plants. For explanations, consult Fig. 6.1

and neutral polluters (Fig. 6.9). Non-ferrous smelters adversely affected all plant groups except for grasses, while the effects of aluminium smelters on grasses did not differ from effects on other plant groups (Fig. 6.10). Changes in α diversity were only weakly related to changes in cover (field layer vegetation: $r = 0.50$, $N = 14$ impact zones, $P = 0.07$).

The pollution effects on the species richness of vascular plants were stronger in southern regions (Fig. 6.9). The geographical difference was significant between aluminium smelters ($Q_B = 4.76$, df = 1, $P = 0.03$) but did not reach the significance level between non-ferrous smelters ($Q_B = 2.61$, df = 1, $P = 0.11$).

We have detected significant non-linear responses in nine of 70 data sets (eight dome-shaped and one U-shaped).

6.3.4.2 Spatial Variation in Diversity (β_{gl} Diversity)

Pollution did not cause changes in spatial variation in species numbers, measured by β_{gl} ($d = 0.07$, CI = −0.75 … 0.88, $N = 14$). This effect was consistent among all groups of categorical variables and uniform across all impact zones ($Q_T = 16.2$, df = 13, $P = 0.24$). The effects of pollution on β_{gl} were independent of the effects on other measures of diversity (correlations between ESs: $r_S = -0.02$ … −0.32, $N = 14$ impact zones, $P = 0.27$ … 0.95).

6.3.4.3 Spatial Variation in Species Composition (β_{sim} Diversity)

Pollution did not cause changes in spatial variation in species composition, measured by β_{sim} ($d = -0.25$, CI = −0.96 … 0.47, $N = 14$). This effect was consistent among all groups of categorical variables and uniform across all impact zones ($Q_T = 14.9$, df = 13, $P = 0.31$). The effects of pollution on β_{sim} were independent of the effects on other measures of diversity (correlations between ESs: $r_S = -0.02$ … 0.25, $N = 14$ impact zones, $P = 0.39$ … 0.95).

6.3.4.4 Regional (Site-Specific) Species Richness (γ Diversity)

Pollution tended to reduce the species richness at a site-specific level, although the effect remained non-significant ($d = -0.55$, CI = −1.27 … 0.19, $N = 14$). This effect was consistent among all groups of categorical variables and uniform across all impact zones ($Q_T = 14.6$, df = 13, $P = 0.33$). The effects of pollution on γ diversity were consistent with effects on α diversity (correlations between ESs: $r_S = 0.75$, $N = 14$ impact zones, $P = 0.002$) but independent of the effects on either β_{gl} ($r_S = -0.32$, $N = 14$, $P = 0.27$) or β_{sim} ($r_S = 0.25$, $N = 14$, $P = 0.39$).

6.3.5 Species Composition of Vascular Plants

An average similarity (Jaccard index) between polluted and control sites (mean ± S.E.: 0.297 ± 0.045) did not differ ($\chi^2 = 2.74$, df = 1, $P = 0.10$) from the similarity between two control sites (0.419 ± 0.033) for the entire sample of 14 polluters. However, when polluter types were analysed separately, the difference did not appear significant for aluminium smelters ($\chi^2 = 0.01$, df = 1, $P = 0.92$), while for non-ferrous smelters, the average similarity between polluted and control sites (0.169 ± 0.048) was smaller ($\chi^2 = 5.79$, df = 1, $P = 0.02$) than the similarity between two control sites (0.309 ± 0.028). The overall pollution effects on the similarity between polluted and control sites were not significant for northern polluters, but they were marginally significant for southern polluters ($\chi^2 = 0.20$, df = 1, $P = 0.65$ and $\chi^2 = 2.98$, df = 1, $P = 0.08$, respectively).

Heavily polluted sites, on average, contained about the same ($\chi^2 = 1.19$, df = 1, $P = 0.28$) proportion of species that were absent on the opposite end of the pollution gradient (mean ± S.E.: 0.501 ± 0.055) as control sites (0.589 ± 0.064). This conclusion remained valid for both non-ferrous and aluminium smelters analysed separately ($\chi^2 = 2.56$, df = 1, $P = 0.11$ and $\chi^2 = 0.27$, df = 1, $P = 0.60$, respectively). The results did not differ between northern and southern polluters ($\chi^2 = 0.69$, df = 1, $P = 0.41$ and $\chi^2 = 1.47$, df = 1, $P = 0.22$, respectively).

6.4 Discussion

6.4.1 Vegetation Structure and Productivity

Pollution impacts on vegetation are far from being simply detrimental. One of the most interesting findings of our study is the diversity of the responses of plant communities to the impacts of point polluters. Both direction and magnitude of responses depend on many factors, including characteristics of both the polluter and plant communities, and presumably also pollution history. Moreover, different indices of plant community structure responded to pollution impacts in an uncoordinated manner.

First and most importantly, vegetation responses depend on the polluter type. In particular, decreases in the cover of top-canopy plants (Fig. 6.1) and the understorey (Fig. 6.2), in the proportion of conifers among top-canopy plants (Fig. 6.7), and in stand basal area (Fig. 6.5) were detected only near non-ferrous smelters. In contrast, near aluminium smelters we observed increases in the cover of top-canopy plants (Fig. 6.1) and the understorey (Fig. 6.2), and in the proportion of conifers among top-canopy plants (Fig. 6.7), while stand basal area did not change (Fig. 6.5). Reports on forest damage around alkalysing polluters, such as aluminium smelters and magnesite plants, are rare; acute damages were generally observed in the past, prior to implementation of strict control for emissions of gaseous fluorine and HF (Bohne 1971; Gilbert 1975). Recently, forest deterioration around aluminium smelters has mostly been associated with giant industrial enterprises in Siberia (Lyubashevsky et al. 1996; Mikhailova & Berezhnaya 2000; Mikhailova 2003). Our data support conclusion by Freedman (1989) that the adverse effects of aluminium industries on stands are less acute than the effects of non-ferrous smelters.

An absence of overall effects of pollution on the cover of top-canopy and understorey plants seems to contradict the general belief that forests decline with pollution, both around point polluters (Gordon & Gorham 1963; Freedman & Hutchinson 1980b; Innes & Oleksyn 2000) and on a regional scale (Pitelka & Raynal 1989; Bussotti & Ferretti 1998; Akselsson et al. 2004). However, our results indicate that it does not make any sense to discuss an overall ES calculated by meta-analysis because the sign and magnitude of this overall effect will mostly depend on the relative numbers of different polluters included in the sample. In particular, a zero effect on the covers of top-canopy (Fig. 6.1) and understorey plants (Fig. 6.2), as well as on the proportion of conifers (Fig. 6.7), resulted from averaging significant negative effects observed around non-ferrous smelters with significant positive effects observed around aluminium industries.

The only stand characteristics that decreased around all types of polluters were basal area (Fig. 6.5) and height (Fig. 6.6). This conclusion is consistent with the large body of forestry literature (Kozlowski 1980; Smith 1981), as well as with the results of meta-analyses of plant growth in polluted areas (Chapter 4, Roitto & Kozlov 2007, Roitto et al. 2009). More generally, our data confirm the historically accepted opinion (Woodwell 1970; Kozlowski 1980; Smith 1981; Treshow 1984; Freedman 1989) that standing biomass, as reflected by height and basal area, decreases with pollution. Since plant biomass is often used as a proxy for productivity (Clements & Newman 2002), and trees form the larger part of vegetation biomass in forest ecosystems, we (in line with Odum 1985) conclude that aboveground productivity generally decreases with pollution.

At the same time, both field layer vegetation and mosses showed similar negative responses to pollution by different industries (Figs. 6.3 and 6.4). Although a decline in field layer vegetation near point polluters is widespread (Freedman & Hutchinson 1980b; Salemaa et al. 2001; Taylor & Fox 2001), mechanisms behind this effect remain unclear. An absence of regeneration on heavily toxic soils explains the steady decline of field layer vegetation in industrial barrens (Zverev et al. 2008), but we still lack an understanding of processes occurring at lower levels of pollution load. In particular, repeated attempts to experimentally reproduce the reduction of field layer vegetation generally failed; its cover did not change or even increased following applications of acid rain and heavy metals (Nygaard & Abrahamsen 1991; Shevtsova & Neuvonen 1997; Zobel et al. 1999). Of course, these experiments could be too short to mimic effects that became evident following decades of pollution impacts. Moreover, in our opinion, these small-scale experiments were initially condemned to failure due to the impossibility of reproducing landscape-level effects, such as modifications of microclimate (discussed in Section 9.1.1).

Another important finding of our analysis is geographical variation in vegetation responses to pollution. In particular, the cover of both field layer and mosses declines only around the northern polluters (Fig. 6.3), while stand basal area declines only around the southern polluters (Fig. 6.5). We suggest several explanations for this phenomenon. First, dense canopies of more productive southern forests intercept the larger part of pollutant deposition (Nieminen et al. 1999; Neal 2002), thus providing better shelter to field layer vegetation than sparse subarctic stands. Second, even slight effects on stand density in southern forests enhance light availability to forest floor vegetation, favouring growth of grasses and herbs (McClenahen 1978; Vacek et al. 1999) and thus alleviating direct negative effects of pollution. Responses to pollution may also depend on initial diversity and productivity of affected communities, as well as on climatic effects on mobility and toxicity of pollutants (Odum 1985; Chalcraft et al. 2008; Zvereva et al. 2008).

In our opinion, the inability to create a phenomenological model explaining the detected variability of plant community responses results, in particular, from a shortage of information about the pollution impacts on plant–plant interactions. Long ago it became apparent that the responses of trees to pollution may be quite different under competitive conditions in a forest stand from what would be expected from experiments conducted with single individuals or single species (West et al. 1980; Auerbach 1981). However, almost no research on this problem has been done since then; only 14% of presentations delivered at the fifteenth to twenty-first biennial International IUFRO Meetings for Specialists in Air Pollution Effects on Forest Ecosystems (Tesche & Feiler 1992; Cox et al. 1996; Bussotti et al. 1996; Anonymous 1998, 2000, 2004; Maňkovská 2002) reported the effects of pollution on biotic interactions, and of those, most described insect–plant and plant–mycorrhyza relationships. Thus, single-species studies still dominate in pollution ecology research.

The general theory predicts that in harsh abiotic environments, plant–plant interactions will be mostly facilitative, in contrast to the dominance of competition in optimal abiotic environments (Brooker & Callaghan 1998), and this pattern was recently observed in some studies conducted around the nickel-copper smelter in Monchegorsk (Zvereva & Kozlov 2004, 2007; Eränen & Kozlov 2007). The existence of positive interactions can partially explain why the effects of pollution on vegetation are expressed to a lesser extent than expected; adverse effects may be ameliorated by positive interactions between plants as well as between plants and mycorrhizal fungi (Zvereva & Kozlov 2004, 2007; Eränen & Kozlov 2007; Ruotsalainen et al. 2007, 2008).

To conclude, responses of plant communities to pollution are diverse, and the outcome of pollution impacts on vegetation strongly depends on both the polluter and the structure of the affected community. The effects of pollution on vegetation have frequently been overestimated due to generalization of patterns that were observed around non-ferrous smelters, which impose the most acute impacts on forests.

6.4.2 *Stand Regeneration*

Regeneration (or its absence) depends on (a) regeneration sources, i.e., seed production by extant plants and their accumulation in soil seedbanks, (b) seed germinability and seedling survival, and (c) growing space for regrowth. Although pollution has adverse overall effects on plant reproduction (Roitto & Kozlov 2007; Roitto et al. 2009), several studies revealed the presence of germinable seeds even in the most contaminated study sites (Komulainen et al. 1994; Huopalainen et al. 2000; Winterhalder 2000; Salemaa & Uotila 2001). And, at least in some situations, the number of germinable seeds did not change with the distance from the polluter (Salemaa & Uotila 2001). While revegetation from seed banks is often hampered by soil toxicity (Kozlov 2005b; Salemaa & Uotila 2001), an overall decrease in stand basal area and field layer cover with pollution (Figs. 6.3 and 6.5) may favour recruitment of some tree species in less toxic soils (Eränen & Kozlov 2009; Zverev 2009). Another important issue for forest regeneration is the effect of air pollution on seedling competitive ability (Merino et al. 2008), but we are not aware of studies explicitly addressing this problem (except for Eränen & Kozlov 2009).

Data on the density and diversity of seedlings and young trees were only rarely monitored along pollution gradients (Lehvavirta & Rita 2002; Zverev 2009). In our data set, the overall effect of pollution on forest regeneration did not differ from zero and was consistent among all groups of categorical variables (Fig. 6.8). This result contradicts observations conducted in industrial barrens, where natural regeneration is suppressed (Kozlov & Zvereva 2007a); however, industrial barrens are extremes that were only observed in some of our pollution gradients. On the other hand, the relative contribution of stochastic factors to regeneration processes may be rather high (up to 83% in a study by Kubota & Hara 1996), and therefore extreme spatial variation in both density and diversity of regrowth may have prevented us from detecting pollution effects, should they exist.

To conclude, our results disagree with conclusions that natural forest regeneration is always suppressed by pollution (Kozlowski 1980; Smith 1981; Treshow 1984; Freedman 1989). An absence of regeneration occurs only in industrial barrens that have developed around non-ferrous smelters.

6.4.3 *Diversity of Vascular Plants*

As mentioned in Chapter 1, ecologists still have no unequivocal answer to an eternal question: Does industrial pollution always result in lower biodiversity? It has long been accepted that undisturbed communities have the highest species richness (Margalef 1968; Odum 1985). An alternative hypothesis suggests that species richness is maximised at intermediate levels of disturbance (Grime 1973; Connell 1978) because superior competitors and disturbance-tolerant species may coexist only at these conditions. It was also suggested that disturbance may increase diversity in communities, the initial diversity of which is low (Odum 1985).

There is no doubt that pollution is one of the factors contributing to destruction of natural habitats (Barbault 2001). On the other hand, habitats deteriorated by pollution may serve as refugia of rare and endangered species (reviewed by Kozlov & Zvereva 2007a). The absence of correlation between pollution load and diversity has been reported for different groups of insects from impact zones of several polluters (Kozlov 1997; Kozlov & Zvereva 1997; Butovsky & Gongalsky 1999; Brandle et al. 2001; Kozlov & Whitworth 2002; Ermakov 2004; Kozlov et al. 2005b), and a meta-analysis of published data yielded zero overall effect of pollution on diversity of terrestrial arthropods (Zvereva & Kozlov 2009). Thus, the validity of the widespread opinion that polluted habitats generally display a reduction in diversity (Magurran 1988) can be questioned.

The overall effect on species richness (α diversity) of vascular plants in our sample of 14 point polluters did not differ from zero (Fig. 6.9). This result contradicts the robust adverse effect of pollution on floristic diversity detected by a meta-analysis of published data (Zvereva et al. 2008) that yielded an effect two times stronger ($z_r = -0.78$; Zvereva et al. 2008) than meta-analysis of the original data ($z_r = -0.31$; Fig. 6.9).

The difference in outcomes of these two meta-analyses may indicate that the choice of polluters by authors of the published papers was biased; the polluters with evident changes in plant communities were preferentially selected to study effects on plant diversity (object selection bias). Combined with the previously discovered publication bias (journals tended to publish studies that agree with the general paradigm, i.e., adverse effects of pollution on biodiversity; Zvereva et al. 2008), our results indicate that a negative effect of pollution on plant diversity is overestimated. This gives special importance to an exploration of factors contributing to high variation in response patterns around different polluters; observed effects (Table 6.41) varied from strongly negative (e.g., around Karabash and Revda) to neutral or even positive (e.g., around Kostomuksha and Vorkuta).

A positive correlation between changes in cover and in species richness confirms the hypothesis (Kozlov et al. 1998) that the magnitude of decline in species richness with an increase in pollution is overestimated due to a confounding decrease in plant abundances. Similar overestimation of adverse effects on species richness was recently discovered for terrestrial arthropods (Zvereva & Kozlov 2009). Thus, methodologies need to be developed and additional data collected to clearly separate effects on plant diversity from effects on plant abundance.

Furthermore, a pronounced discrepancy in mean effects of aluminium smelters between published studies ($z_r = -1.45$; $N = 4$ polluters; Zvereva et al. 2008) and original data ($z_r = -0.06$; $N = 5$ polluters; Fig. 6.9) can be seen as an indication of research bias acting via selection of 'representative' study sites in such a way that presumed adverse effects are most evident. On the other hand, the discrepancy in conclusions on the effects of aluminium smelters may have resulted from changes in environmental regulations. Data used in an earlier meta-analysis were collected between 1962 and 1989, when emissions of pollutants were generally higher (see Sections 1.2 and 2.2.2) than in the 2000s, when the original data were collected.

In agreement with earlier conclusions, the strongest negative effects were detected around non-ferrous smelters. All six smelters included in our analysis caused dramatic changes in plant communities, including development of industrial barrens that represent an extreme state of pollution-induced ecosystem deterioration (Kozlov & Zvereva 2007a). A strong effect of these smelters on vegetation may be explained by a combination of soil acidification, accumulation of heavy metals, and landscape-level changes leading to loss of topsoil (Chapter 3) and unfavourable changes in microclimate (see Section 9.1.1 for discussion).

Although the geographic distribution of surveyed polluters (Fig. 2.1) is not as extensive as in the meta-analysis of published data (Fig. 1 in Zvereva et al. 2008), conclusions on geographic variation in the magnitude of plant community responses to pollution are consistent between the published and original data sets. Our data confirmed a stronger negative impacts of polluters located in warmer climates for both non-ferrous smelters and aluminium plants (Fig. 6.9).

The existence of geographical variation in responses of plant communities to pollution is one of the most interesting findings of our meta-analyses. This result is especially intriguing because it is in contrast to the general opinion on the higher sensitivity (fragility) of northern ecosystems to different kinds of human-induced disturbances. A lower sensitivity of high-latitude plant communities to pollution impacts may result from several factors, including both community structure and behaviour of pollutants.

In a meta-analysis of published data, we linked stronger responses of southern plant communities with their higher diversity, because the magnitude of species loss under pollution impacts increased with the species richness of undisturbed communities (Zvereva et al. 2008). This result is consistent with theoretical predictions by Odum (1985), who expected lower or even positive effects of disturbances on communities with lower initial diversity. Although causal relationships cannot be inferred from our data, we suggest two possible explanations for the observed pattern. First, species living in more predictable southern environments (where species richness is higher) are less able to tolerate stress than species living in less predictable northern environments (Clements & Newman 2002), which may have evolved preadaptations (Rapport et al. 1985). Second, longer vegetation periods at lower latitudes increase the exposure of plants to pollutants, while higher temperatures and increased precipitation enhance mobility and increase the toxicity of pollutants (Cairns et al. 1975; Klein 1989; Tipping et al. 1999).

Data on the pollution effects on β and γ diversity are scarce; we are only aware of publications reporting the effects of experimental applications of acid rain, heavy metals and nitrogen deposition on plant diversity on different spatial scales (Zobel et al. 1999; Chalcraft et al. 2008). These publications demonstrated increases in between-plot variation in the floristic composition of experimental plots, i.e., increases in β diversity. However, this experimental result contradicts observations of decreases in the spatial variability of structural and functional characteristics of forest litter with pollution (Bringmark & Bringmark 1995; Vorobeichik 1997). Although we did not detect pollution effects on β and γ diversity, an absence of correlations between patterns observed on different spatial scales suggests that extrapolation of the results obtained at the lowest hierarchical level may substantially bias our conclusions on the impact of pollution on biodiversity on larger spatial scales.

Finally, comparison of our results with *a priori* predictions (Table 6.42) demonstrated, that only non-ferrous smelters changed the composition of affected communities, acting via selective removal of some (presumably the most sensitive) species; both species richness and the similarity between polluted and clean sites decreased with pollution. Polluted sites around large non-ferrous smelters, such as Karabash (Table 6.30) and Monchegorsk (Table 6.32), did not contain any species that were absent in controls, suggesting an absence of colonisation. In contrast to non-ferrous industries, effects of aluminium smelters were minor relative to random variation.

To conclude, our data suggest that adverse effects of pollution on plant diversity are generally overestimated. We observed decreases in α diversity and changes in species composition only around non-ferrous smelters; moreover, the magnitude of these effects may appear smaller when a decline in plant abundances is accounted for. Adverse effects were better expressed in southern regions with higher initial diversity.

6.4.4 Temporal Changes in Plant Community Structure

What is the fate of polluted ecosystems? This question is vital for forestry worldwide, as the proportion of forested areas affected by relatively high levels of pollutant deposition is predicted to substantially increase by 2050 (Fowler et al. 1999). Maintenance of forests in polluted areas requires more intensive management than in unpolluted areas, involving 'soft' techniques and highly skilled manual labour. Regular curative measures, forming the basis for silviculture in polluted areas, should be preventive, improving the ecological stability of stands in such a way that they will better resist unavoidable pollution impacts (Kozlov 2004). Therefore, understanding pollution effects on the development of plant communities is badly needed to develop management practices for sustainable development of polluted regions.

Our understanding of changes in vegetation structure and productivity under pollution impacts is hampered by a shortage of long-term observations, documenting both the decline and recovery of vegetation following increases or decreases in pollution loads. This gives special importance to studies conducted in the Sudbury area (Anand et al. 2005) and in other areas recovering after closure of polluters (Wagner 2004) or substantial emission declines (Zverev 2009). However, temporal changes in plant communities can be inferred from static succession analysis, i.e., by comparing simultaneously collected data from study sites that presumably are at different succession stages.

Gordon and Gorham (1963) described pollution-induced changes of vegetation as peeling off the layers of forest structure: first the trees, followed by tall shrubs, and finally, under the severest conditions, the short shrubs and herbs. The similarity of this process with vegetation changes caused by chronic irradiation allowed Woodwell (1970) to suggest the generality of this 'downward' pattern of ecosystem destruction, later called 'the syndrome of spatial decline of the vertical strata of the terrestrial vegetation' (Freedman 1989). However, our meta-analysis demonstrated that this sequence, although recorded in some case studies, cannot be seen as a rule. Moreover, we have found the general pattern to be exactly opposite: field layer vegetation declines with pollution more strongly than top-canopy and understorey plants (Figs. 6.1–6.3). Also, the appearance of heavily polluted sites indicates that on many occasions, the field layer suffers first, disappearing while trees continue to grow on bare or nearly bare ground (please see color plates 22, 36 and 54 in Appendix II).

Both the severity and duration of pollution impacts may have contributed to variable outcomes of studies addressing temporal changes in polluted communities. Extreme levels of environmental contamination, existing near large point polluters, acted as selection factors eliminating not only sensitive species but also sensitive genotypes of more tolerant species; progenies of survivors showed increased tolerance to pollution (Bradshaw & McNeilly 1981; Macnair 1997; Kozlov 2005b; Eränen 2008; Eränen et al. 2009). On the other hand, inertia existing in plant communities (Milchunas & Lauenroth 1995; Zverev 2009) decreases the rate of pollution-induced changes in the abundance of long-lived plants, explaining their prevalence in severely stressed communities of industrial barrens (Kozlov & Zvereva 2007a). This pattern contradicts the hypothesis by Odum (1985) that the proportion of opportunistic species should increase with stress.

Woodwell (1967) suggested that the changes that occur in forest communities with severe disturbances (more specifically, under chronic radiation) tend to be just the reverse of those occurring during a normal (e.g., post-fire) succession. Later on (Woodwell 1970), he found many similarities between plant community deterioration under chronic radiation (Woodwell 1967) and chronic pollution (Gorham & Gordon 1963). Odum (1985) listed a reversal of succession among the general ecosystem responses to abiotic stress. This conclusion, accepted by a number of ecologists (Sigal & Suter 1987; Treshow & Anderson 1989), seems to be based on the following effects observed in stressed communities: (a) lower diversity, (b) poor stratification due to the elimination of woody plants, and (c) selection for rapid growth forms (partially due to the elimination of woody plants). However, our analyses cast doubts on the generality of all these phenomena, and plant communities affected by chronic pollution (please see color plates 9–11, 20–23, 29, 32–39, 45–47, 49, 53, 54 and 59 in Appendix II) differ substantially from communities representing early stages of post-fire succession. A decrease in productivity with pollution (Section 6.4.1) also contradicts an assumption on succession reversal, since productivity generally decreases in the course of succession (Odum 1969b). Moreover, Liu et al. (2007) concluded that sulphur dioxide, by damaging more sensitive pioneer species, can accelerate succession rather than reverse it.

Our data also demonstrated that pollution did not change the species composition of seedlings of top-canopy species. This result is interesting in view of the paradigm of decreasing similarity between seed banks and vegetation as succession proceeds (Thompson 2000). Although this paradigm has repeatedly been questioned, and temporal patterns of similarity between seed banks and vegetation differed widely among studies (Wagner et al. 2006), our data can still be seen as an indication of the relatively stable state of polluted forest ecosystems.

To conclude, we have not found any indication of succession reversal under pollution impacts. Moreover, we did not detect any dynamic process that may lead to further shifts in vegetation structure, e.g., by selective species removal or by species replacement due to altered composition of regrowth. All explored polluters have been functioning for decades, and therefore transition of plant communities from an unpolluted to a polluted state seems to be already completed. An absence of a transitional zone is most likely the result of relatively stable or decreasing amounts of emissions.

6.5 Summary

Although point polluters may drastically change both the structure and diversity of surrounding vegetation, pollution effects are not uniform and therefore cannot be generalized. Responses of plant communities demonstrated significant variation in respect to both type and geographical position of the affected community. Effects of non-ferrous smelters/acidifying polluters on most community characteristics (vegetation cover, stand basal area, proportion of conifers among top-canopy plants, diversity of vascular plants) were significantly negative, while the effects of aluminium smelters/alkalysing polluters were generally neutral or even positive. Geographical variation was inconsistent among explored community characteristics; stand basal area and plant diversity decreased around southern polluters, while the cover of field layer vegetation decreased around the northern polluters. We conclude that the adverse effects of point polluters on vegetation have been overestimated due to research and publication biases, as well as due to the tendency of both narrative reviews and textbooks to use the most striking examples of community deterioration around non-ferrous smelters to illustrate pollution impacts on terrestrial biota.

Chapter 7
Insect Herbivory

7.1 Introduction

Since highly productive terrestrial ecosystems sustain a larger level of herbivory per unit of primary production than less productive ecosystems (McNaughton et al. 1989), pollution-induced changes in the structure and productivity of plant communities (Chapter 6) are likely to change the pattern of herbivory. On the other hand, herbivores consume a large proportion of primary production, contributing to regulation of plant biomass in many environments (Cyr & Pace 1993; Gruner et al. 2008, and references therein). Thus, knowledge of herbivore responses to pollution is crucial for understanding ecosystem-level effects and predicting changes in ecosystem structure and functions under pollution impacts (Lavelle et al. 2006; Maleque et al. 2006; Wolf et al. 2008).

Both actual losses in plant biomass and subsequent decreases in productivity caused by events of mass occurrences (outbreaks) of herbivorous insects are relatively well documented. In contrast, biomass consumption during years when herbivore density is low (termed endemic herbivory) remains almost unexplored, although this gap in knowledge was identified long ago (Schowalter et al. 1986). Several studies suggest that in subarctic, boreal and temperate forests, insect herbivores consume 1–10% of the foliage of dominant tree species (Bray 1964; Haukioja et al. 1973; Glasov 1986; Bogacheva 1990; Kozlov 2008); an average of 7.1% is reported for temperate broad-leaved forests on the basis of 13 studies (Coley & Barone 1996). Even fewer data are available on endemic herbivory in polluted areas; the majority of conclusions are based on estimations of herbivore densities, while direct measurements of losses in plant foliage along pollution gradients are surprisingly scarce. In particular, we were able to include in our meta-analysis of published data only 17 ESs for plant damage caused by insect herbivores, compared with 142 ESs for abundances of insect herbivores (Zvereva & Kozlov 2009).

The first documented observations on changes in the abundance of leaf rollers damaging Norway spruce around a German iron foundry were conducted in 1832–1833 (Riemer & Whittaker 1989). Scientific reports on pollution impacts on forest pests appeared in the late 1940s–early 1950s; their number increased steadily until the mid-1990s and then started to decline (Fig. 1.2). Several narrative reviews

M.V. Kozlov et al. *Impacts of Point Polluters on Terrestrial Biota*,
DOI 10.1007/978-90-481-2467-1_7, © Springer Science+Business Media B.V. 2009

(Alstad et al. 1982; Selikhovkin 1988; Riemer & Whittaker 1989; Kozlov 1990, 1997; Heliövaara & Väisänen 1993; Rusek & Marshall 2000) provided a summary of field observations and laboratory experiments. Although extraordinary diversity of both insects and polluters and variability of the responses made generalizations difficult, several frequently observed phenomena were considered general regularities. In particular, these reviews concluded that herbivory tends to increase with pollution.

A quantitative research synthesis of the published data (Zvereva & Kozlov 2009) demonstrated a positive association between pollution and herbivory. However, while analysis of earlier publications (1970s–1980s) yielded a high positive effect, intensive studies conducted in the 1990s found a non-significant overall effect (Fig. 1 in Zvereva & Kozlov 2009). This temporal trend is consistent with the development of several other research paradigms; earlier studies tend to overestimate the effect by preferentially documenting the patterns that are consistent with the prevailing theory (Leimu & Koricheva 2004). On the other hand, this temporal pattern may have resulted from an overall decrease in emissions or from adaptations of the affected communities. Exploration of the original data may help to distinguish between these hypotheses, since the applied sampling design allowed us to avoid many of the pitfalls that have affected the published data.

7.2 Materials and Methods

7.2.1 Densities of Birch-Feeding Herbivores

7.2.1.1 Selection of the Model Object

For records of population densities, we have chosen insects feeding on white/mountain birch foliage. This tree species was abundant around many polluters, allowing comparison of data collected from different impact zones. Birch forms the basis for important food chains comprising herbivorous mammals, birds, mites and insects. The number of leaf-chewing and leaf-mining insect species damaging birch is estimated at 240 in Russia (Sinadsky 1973) and 270 in the British Isles (Atkinson 1992).

Insects feeding on birches have been studied by our team for decades (Kozlov 1984, 1985, 1997; Ruohomäki et al. 1996; Zvereva & Kozlov 2006b; Kozlov 2008), and therefore both the methodology of censuses and expertise necessary to identify birch-feeding herbivores and their feeding marks were readily available for this study.

7.2.1.2 Identification of Recognisable Taxonomic Units

The species level is in some sense the natural level to construct an operational measure of diversity, and this is the lowest taxonomic level at which insect fauna can be surveyed (Trueman & Cranston 1994). However, some practical limitations, such as a large number of sampled specimens, limited time for data processing, and absence of the required taxonomic expertise, often make it impossible to provide all specimens with the correct scientific names. Attempts to reduce the level of identification by sorting samples into groups that look homogeneous to the researcher (parataxonomist) are sometimes formalised by delimiting Recognisable Taxonomic Units (RTUs) or morphospecies (Cranston 1990). Indeed, some of these groups (RTUs) may include several species, while different stages (sexes, castes) of the same species may be classified as different RTUs. Then the number of RTUs and the abundance of each RTU are assessed with no attempt to name the taxa. Although this concept has many drawbacks, indices of relative abundance and diversity based on RTUs may be used at least in inter-ecosystem comparisons (Bolger et al. 2000; Fabricius et al. 2003; Das et al. 2005; Fontaine et al. 2007). In pollution-related studies, the RTU concept has been applied to assess the diversity of flies (Diptera: Brachycera) and birch-feeding herbivores in the surroundings of the nickel-copper smelter in Monchegorsk (Kozlov 1997).

The following RTUs were sufficiently abundant to be included in our meta-analysis: (1) rolls of leaf rolling weevils, *Deporaus betulae* (L.) (Coleoptera: Rhynchitidae); (2) mines of *Eriocrania* spp. (Lepidoptera: Eriocraniidae); (3) serpentine mines of *Stigmella* spp. (Lepidoptera: Nepticulidae); (4) a gallery, later expanded to a blotch mine, of *Ectoedemia minimella* (Zett.) (Lepidoptera: Nepticulidae); (5) mines and downward folds of the leaf margin made by larvae of *Parornix* spp. (Lepidoptera: Gracillariidae) - care was taken to count the number of mines, not the number of folds, since one larva often makes more than one fold; (6) upperside blotch mine of *Phyllonorycter coryfoliellus* (Hbn.); (7) *Coleophora* spp. (Lepidoptera: Coleophoridae) - since our censuses were conducted after larvae completed their development, here we counted the number of mines, not the number of larvae (one larva produces multiple mines); (8) nests of leaf rollers (Lepidoptera: Tortricidae) - as in the case with *Coleophora* spp., we counted nests, not larvae (one larva can produce several nests during its lifetime); (9) mines of sawflies (Hymenoptera: Tenthredinidae) that include several species from the genera *Fenusa, Profenusa, Messa* and *Scolioneura*; and (10) larvae of leaf-chewing sawflies (Hymenoptera: Tenthredinidae). These ten RTUs represent three feeding guilds, among which miners (moths and sawflies) and rollers (Tortricidae and *Deporaus betulae*) were represented by sufficient numbers of RTUs to allow their separate analyses.

7.2.1.3 Censusing and Estimation of Population Intensities

We conducted herbivore censusing on 10 or 25 (depending on average herbivore densities) randomly chosen birch trees (aged 10–50 years, height 50–300 cm); care was taken to select birches of all size classes at all study sites, thus avoiding a possible size-related bias in density estimation. Censusing was conducted in impact zones of nine polluters, generally over several years.

The stem of each censused birch was measured (to the nearest mm) at the base of the tree crown to serve as an estimate of the efficient diameter, which is proportional to the amount of foliage in the tree crown (Kozlov & Sokolova 1984). If the tree crown had dead branches (a relatively common situation in severely contaminated study sites), according to the pipe model theory (Shinozaki et al. 1964a, b) the efficient diameter was calculated by subtracting cross-sectional areas of dead branches from cross-sectional areas of the tree trunk. In practice, experienced observers (M.V.K. and V.E.Z.) were able to visually estimate the efficient diameters of birches when the trunk diameter was less than 50 mm.

7.2.1.4 Data Processing and Statistical Analysis

Leaf area (S, m^2) of each censused birch was estimated from its efficient diameter (d, cm) using the following equation: $S = 0.1234 \times d^2$ (Kozlov & Sokolova 1984), and herbivore densities were expressed as the number of individuals or RTUs per 1 m^2 of leaf area. This approach allowed us to account for variation in tree size and vitality, which may change with pollution. Note that leaf size in white birch was not affected by pollution (Fig. 4.4); thus, our measure of leaf area is roughly proportional to leaf number and therefore consistent with the number of individuals per 100 leaves (used by Koricheva & Haukioja 1992) or per 100 shoots (used by Bylund & Tenow 1994). More generally, the index used in this study corresponds to the population intensity (density per available resources) *sensu* Southwood (1978).

We did not select certain RTUs for censusing, thus avoiding one of the most common research biases identified by Zvereva and Kozlov (2009), namely, the selective investigation of species that were most abundant in heavily polluted sites. However, since some of the RTUs were extremely infrequent, we decided to disregard the data sets (RTU by polluter by study year) in which the selected RTU was present in only one or two of ten study sites. This decision was based on an assumption that our density estimates for these species are not sufficiently accurate to reveal pollution-related patterns. If two or more data sets were available for the same RTU by polluter combination, then the study year with the highest overall abundance was chosen for meta-analysis. Altogether, we obtained data on ten RTUs, with three to nine RTUs per polluter (Tables 7.1–7.9), yielding 47 ESs in total.

Since distributions of tree-specific intensities were highly skewed, with a large number of zeros, the Kruskal-Wallis test was used to explore variation among study sites. Site-specific intensities (total number of individuals divided by summed area of censused birches) were used to calculate Spearman rank correlations with both distances from polluter and concentrations of pollutants.

7.2.2 *Foliar Damage to Woody Plants*

7.2.2.1 Indices of Foliar Damage

There are several ways to estimate foliar damage by insect herbivores. In particular, Alliende (1989) used two indices: proportion of damaged leaves in a sample of

Table 7.1 Densities (exx/m[2] of leaf area) of birch-feeding herbivores in the impact zone of the power plant at Apatity, Russia

Site	*Eriocrania* spp.	*Parornix* spp.	Tortricidae	Mining Tenthredinidae
1–1	0.514	0.514	1.285	0
1–2	3.915	0	0.652	0
1–3	1.083	0.271	0	0
1–4	3.683	0.230	0.230	0.230
1–5	0.602	0.402	0.201	0.603
2–1	1.228	0	0.491	0.491
2–2	0.461	0.461	0.461	0
2–3	1.772	0.295	1.772	0
2–4	0.448	0	0.672	0.224
2–5	0.281	1.407	1.688	0.281
χ^2/P	13.4/0.14	12.8/0.17	7.26/0.61	9.28/0.41
Dist.: r_S/P	−0.26/0.46	0.18/0.62	−0.25/0.49	0.55/0.10
Poll.: r_S/P	0.42/0.22	−0.20/0.58	−0.04/0.91	−0.71/0.02

Censuses are based on ten trees per site. For description of RTUs consult Section 7.2.1.2. Date of censusing: 30.7.2006. χ^2: test for significance of variation between study sites. Dist.: Spearman rank correlation between site-specific means and distance from polluter. Poll.: Spearman rank correlation between site-specific means and concentration of the selected pollutant.

Table 7.2 Densities (exx/m[2] of leaf area) of birch-feeding herbivores in the impact zone of aluminium smelter at Bratsk, Russia

Site	*Eriocrania* spp.	*Stigmella* spp.	Tortricidae	Mining Tenthredinidae
1–1	0.514	0.514	1.285	0
1–2	3.915	0	0.652	0
1–3	1.084	0.271	0	0
1–4	3.684	0.230	0.230	0.230
1–5	0.603	0.402	0.201	0.603
2–1	1.228	0	0.491	0.491
2–2	0.461	0.461	0.461	0
2–3	1.772	0.295	1.772	0
2–4	0.448	0	0.672	0.224
2–5	0.281	1.407	1.688	0.281
χ^2/P	27.0/0.0014	20.6/0.01	10.2/0.33	14.2/0.12
Dist.: r_S/P	−0.28/0.42	0.13/0.72	0.02/0.96	0.51/0.13
Poll.: r_S/P	0.37/0.29	−0.28/0.44	−0.41/0.24	−0.41/0.24

Censuses are based on ten trees per site. For description of RTUs consult Section 7.2.1.2. Dates of censusing: 2–4.8.2002. For other explanations, consult Table 7.1.

100 leaves and proportion of leaf area removed by herbivores. Although the first index may seem oversimplified, it is often used in ecological research (Pook et al. 1998; Stapel et al. 1998; Greenberg et al. 2000; Ornelas et al. 2004; Crutsinger & Sanders 2005).

Table 7.3 Densities (exx/m² of leaf area) of birch-feeding herbivores in the impact zone of aluminium smelter at Kandalaksha, Russia

Site	*Deporaus betulae*	*Eriocrania* spp.	*Stigmella* spp.	Tortricidae
1–1	0.487	1.298	0.162	0.649
1–2	6.393	0.417	0	0.139
1–3	3.299	0.367	0	0
1–4	1.578	0	0	0
1–5	1.738	0.372	0	0
2–1	1.824	0	0.140	0.281
2–2	1.907	0.293	0.147	1.173
2–3	0	0.194	0.194	0
2–4	0.815	0.326	0.163	0.652
2–5	2.170	0.114	0.457	0
χ^2/P	12.7/0.12	7.38/0.50	15.9/0.04	17.5/0.03
Dist.: r_S/P	0.02/0.96	−0.17/0.65	0.08/0.83	−0.62/0.05
Poll.: r_S/P	0.24/0.51	0.16/0.66	−0.44/0.21	0.40/0.26

Censuses are based on 25 trees per site. For description of RTUs consult Section 7.2.1.2. Dates of censusing: 16–17.7.2002. For other explanations, consult Table 7.1.

Table 7.4 Densities (exx/m² of leaf area) of birch-feeding herbivores in the iron pellet plant at Kostomuksha, Russia

Site	*Deporaus betulae*	*Eriocrania* spp.	Tortricidae
1–1	1.364	0	1.169
1–2	2.267	0	2.591
1–3	1.145	0.164	0.982
1–4	2.809	0	0.766
1–5	0.925	1.542	0.463
2–1	0	0.190	0.569
2–2	0	0.167	2.165
2–3	0.602	0.201	2.607
2–4	0	0	0.891
2–5	0.134	0.134	1.873
χ^2/P	16.0/0.07	6.12/0.73	16.0/0.07
Dist.: r_S/P	0.15/0.67	0.00/0.99	−0.21/0.56
Poll.: r_S/P	0.07/0.85	0.13/0.73	−0.27/0.45

Censuses are based on 25 trees per site. For description of RTUs consult Section 7.2.1.2. Date of censusing: 18.7.2002. For other explanations, consult Table 7.1.

7.2.2.2 Consumption of Foliage

The average proportion of consumed foliar biomass is an important parameter reflecting an outcome of insect-plant relationships. This index is a product of (a) proportion of damaged leaves, which is easy to estimate and (b) average consumption per damaged leaf, an accurate measurement of which (according to protocol suggested by Williams & Abbott 1991) requires much more effort. Since average consumption per damaged leaf showed little variation, even across large geographical regions (Kozlov 2008), we hypothesised that pollution had no effect on this index. This hypothesis was verified using seven plant species in impact zones of six polluters (ten data sets in total).

Table 7.5 Densities (exx/m^2 of leaf area) of birch-feeding herbivores in the impact zone of the nickel-copper smelter at Monchegorsk, Russia

Site	*Deporaus betulae*	*Eriocrania* spp.	*Parornix* spp.	*Phyllonorycter coryfoliellus*	Tortricidae	Mining Tenthredinidae	Chewing Tentredinidae
1–1	0.207	5.373	1.701	0.409	2.044	1.839	0.204
1–2	2.893	0.286	0.142	0.179	3.585	0.538	0.717
1–4	4.641	4.761	0.947	1.202	3.907	1.352	0.150
1–5	3.912	2.418	0.885	6.766	0.531	0.398	0
2–3	1.672	4.612	0.848	0.112	16.12	0.336	18.02
2–4	0.591	9.009	0.664	1.794	3.948	1.615	0
2–5	0.346	1.811	1.02	0.436	0.727	0.145	0.145
2–8	0	0.955	1.44	0.284	0.568	0	0
2–10	4.334	3.029	1.938	0	0.171	0.171	0
2–12	0	1.737	1.548	0	0.151	0	0
χ^2/P	16.6/0.06	23.3/0.0056	20.3/0.02	98.3/<0.0001	96.5/<0.0001	29.1/0.0006	151.5/<0.0001
Dist.: r_S/P	–0.12/0.75	–0.38/0.28	0.49/0.15	–0.28/0.43	–0.80/0.0061	–0.78/0.0072	–0.80/0.006
Poll.: r_S/P	–0.16/0.68	0.43/0.21	–0.48/0.16	0.26/0.48	0.84/0.0022	0.47/0.17	0.59/0.07

Censuses are based on 25 trees per site. For description of RTUs consult Section 7.2.1.2. Dates of censusing: 22–27.6.2000 (*Eriocrania* spp.), 6–12.7.2003 (*Deporaus betulae*), 21–27.8.2004 (*Phyllonorycter coryfoliellus*, Tortricidae, mining and chewing Tenthredinidae) and 7–13.8.2006 (*Parornix* spp.). For other explanations, consult Table 7.1.

Table 7.6 Densities (exx/m^2 of leaf area) of birch-feeding herbivores in the impact zone of the aluminium smelter at Nadvoitsy, Russia

Site	*Deporaus betulae*	*Eriocrania* spp.	*Stigmella* spp.	*Ectoedemia minimella*	*Parornix* spp.	*Coleophora* spp.	Tortricidae	Mining Tenthredinidae
1–1	0	2.707	4.060	0	0.677	0	9.812	2.369
1–2	8.772	3.759	1.253	0	1.671	0	1.671	1.671
1–3	6.663	6.664	1.481	0	0.740	0	1.851	0.370
1–4	0	0.339	8.470	16.16	3.049	20.57	8.808	4.743
1–5	9.808	6.13	0.817	0	3.678	0	8.991	4.904
2–1	2.217	1.109	3.695	1.550	3.695	0	15.89	2.217
2–2	4.608	2.127	1.063	0	0	0.437	1.418	0
2–3	11.76	0.619	0.619	0	0.619	0	1.857	0
2–4	19.43	3.643	0	0.476	6.881	11.91	3.238	6.881
2–5	2.197	9.155	0.366	0	0	0	4.028	2.197
χ^2/P	22.5/0.0074	13.9/0.13	23.3/0.0055	33.9/<0.0001	22.6/0.0072	18.4/0.03	31.0/0.0003	14.8/0.10
Dist.: r_s/P	0.27/0.46	0.32/0.36	−0.49/0.15	0.01/0.97	0.10/0.79	0.19/0.61	0.01/0.97	0.37/0.29
Poll.: r_s/P	−0.25/0.49	−0.43/0.21	0.56/0.09	0.05/0.89	−0.05/0.88	−0.11/0.76	−0.05/0.88	−0.42/0.23

Censuses are based on ten trees per site. For description of RTUs consult Section 7.2.1.2. Dates of censusing: 25.7.2004 (*Ectoedemia minimella*, *Coleophora* spp.) and 23.8.2005 (all other RTUs). For other explanations, consult Table 7.1.

Table 7.7 Densities (exx/m^2 of leaf area) of birch-feeding herbivores in the impact zone of the nickel-copper smelter at Nikel and ore-roasting plant at Zapolyarnyy, Russia

Site	*Deporaus betulae*	*Eriocrania* spp.	Tortricidae
1–1	0	0.264	80.06
1–2	4.088	2.518	8.490
1–3	15.03	3.333	1.878
1–4	0.680	0.317	0.340
1–5	17.16	1.761	0.980
2–1	2.436	0.739	35.07
2–2	0.334	5.899	8.518
2–3	0.842	9.230	6.948
2–4	10.32	2.204	5.822
2–5	6.105	1.333	7.069
χ^2/P	28.3/0.0008	23.3/0.0056	50.8/<0.0001
Dist.: r_s/P	0.67/0.03	0.02/0.96	−0.84/0.0022
Poll.: r_s/P	−0.64/0.05	−0.02/0.96	0.93/0.0001

Censuses are based on 25 trees per site in 2001 and on ten trees per site in 2003. For description of RTUs consult Section 7.2.1.2. Dates of censusing: 17–18.7.2001 (*Eriocrania* spp.) and 17–18.8.2003 (*Deporaus betulae*, Tortricidae). For other explanations, consult Table 7.1.

Table 7.8 Densities (exx/m^2 of leaf area) of birch-feeding herbivores in the impact zone of the copper smelter at Revda, Russia

Site	*Deporaus betulae*	*Eriocrania* spp.	*Stigmella* spp.	*Parornix* spp.	Tortricidae	Mining Tenthredinidae
1–1	36.77	11.64	0	0.931	8.378	0
1–2	0	1.630	2.445	0.407	9.778	0
1–3	46.20	2.862	1.635	5.724	11.86	1.227
1–4	14.63	0.488	1.463	4.876	6.826	4.876
1–5	2.321	9.283	5.570	2.321	14.85	0
2–1	12.94	0.563	0	1.126	0	1.688
2–2	19.14	3.828	1.094	1.094	3.281	0.547
2–3	24.46	0	0.612	0.612	18.35	0
2–4	18.16	5.010	0	0.626	8.141	0
2–5	45.07	7.000	0.875	3.501	20.57	3.063
χ^2/P	17.1/0.05	37.4/<0.0001	22.2/0.0083	7.80/0.55	35.4/<0.0001	19.4/0.02
Dist.: r_s/P	0.07/0.84	0.15/0.69	0.44/0.21	0.46/0.19	0.58/0.08	0.25/0.49
Poll.: r_s/P	0.15/0.69	−0.07/0.85	−0.62/0.06	−0.50/0.14	−0.43/0.22	−0.24/0.50

Censuses are based on ten trees per site. For description of RTUs consult Section 7.2.1.2. Dates of censusing: 19–21.7.2003. For other explanations, consult Table 7.1.

A meta-analysis of the pollution effect on average consumption per damaged leaf was based on contrasting the two most polluted and two least polluted study sites (Table 7.10). In each site, we selected the first five mature trees (at least 3 m high) with accessible twigs and harvested the first ten leaves bearing feeding marks by herbivores. These leaves were then classified visually, according to the scale used by Alliende (1989) and Bach (1990), into the following seven groups: intact, 0.01–1%, 1.1–5%, 6–25%, 26–50%, 51–75% and 76–100% of leaf lamina removed by herbivores.

Table 7.9 Densities (exx/m² of leaf area) of birch-feeding herbivores in the impact zone of the aluminium smelter at Volkhov, Russia

Site	*Deporaus betulae*	*Eriocrania* spp.	*Stigmella* spp.	*Parornix* spp.	*Coleophora* spp.	Tortricidae	Mining Tenthredinidae
1–1	14.91	85.37	1.242	12.43	21.99	11.77	2.353
1–2	2.450	23.51	8.575	9.800	0.500	26.11	0.842
1–3	0	30.81	0	4.514	9.903	4.477	2.686
1–4	19.24	51.47	0	16.28	0	17.59	13.19
1–5	36.24	149.6	0.747	17.93	90.10	3.884	31.85
2–1	0	24.77	0	10.13	22.38	6.919	0.769
2–2	0	15.06	1.339	2.679	25.61	0.939	0.939
2–3	0	91.64	0	0	57.60	9.670	8.703
2–4	53.74	53.09	0	12.68	17.70	9.358	1.871
2–5	69.26	77.51	0	44.71	14.68	21.94	1.371
χ^2/P	32.1/0.0002	13.7/0.13	36.4/<0.0001	32.2/0.0002	43.1/<0.0001	27.4/0.0012	28.3/0.0008
Dist.: r_S/P	0.72/0.02	0.53/0.12	−0.38/0.28	0.59/0.07	0.01/0.99	0.05/0.88	0.54/0.11
Poll.: r_S/P	−0.69/0.03	−0.50/0.14	0.38/0.28	−0.65/0.04	0.28/0.42	−0.21/0.56	−0.72/0.02

Censuses are based on ten trees per site. For description of RTUs consult Section 7.2.1.2. Dates of censusing: 18.9.1994 (*Deporaus betulae*, *Parornix* spp., *Stigmella* spp.), 5.9.1999 (Tortricidae, mining Tenthredinidae) and 9.6.2002 (all other RTUs). For other explanations, consult Table 7.1.

Table 7.10 Average proportion of leaf surface consumed by insect herbivores by the end of the growth season

	Harjavalta	Karabash		Monchegorsk		Nadvoitsy	Nikel	Revda	Sudbury	Volkhov
Study sites and statistics	*Salix caprea*	*Betula pendula*	*Alnus incana*	*Salix borealis*	*Salix caprea*	*Betula pubescens*	*Betula pubescens*	*Populus tremula*	*Betula papyrifera*	*Betula pubescens*
1–1	0.038	0.062	0.413	0.189	0.028	0.199	0.078	0.084	0.082	0.071
2–1	0.063	0.152	0.201	0.084	0.044	0.247	0.090	0.122	0.074	0.073
1–5	0.059	0.143	0.213	0.052	0.051	0.146	0.099	0.072	0.133	0.146
2–5	0.035	0.175	0.218	0.039	0.010	0.109	0.096	0.193	0.049	0.174
ANOVA: F/P	0.03/0.87	1.17/0.39	0.74/0.48	2.96/0.23	0.06/0.83	9.85/0.09	0.19/0.70	0.22/0.68	0.10/0.78	38.2/0.03

Data collected in 2007, except for Monchegorsk (1999) and Harjavalta (1997). ANOVA: test for significance of variation between study sites.

All these censuses were conducted by the same observer (M.V.K.). The numbers of leaves in each class were recorded, and an average proportion of the leaf area removed by herbivores was, according to Lincoln and Mooney (1984), calculated for each tree using frequency weighted average of mid-points.

7.2.2.3 Proportion of Damaged Leaves

Since pollution had no overall effect on the average consumption per damaged leaf (see below, Section 7.3.2.1), the proportion of leaves bearing any signs of damage imposed by leaf chewers or leaf miners was adopted as an operational index of foliar damage.

The proportion of damaged leaves was assessed in the same plant individuals that were used for measuring vitality indices (Chapter 4) and fluctuating asymmetry (Chapter 5), with five individuals per study site. In trees, we selected one twig with approximately 120–150 leaves at a height of 1–3 m and counted damaged leaves among the first 100 leaves starting from the tip of the branch. In dwarf shrubs, the counts were conducted in the laboratory using samples collected for shoot length measurements. Nearly all counts were conducted by the same observer (V.E.Z.).

The proportion of damaged leaves was assessed in 18 species of woody plants belonging to four families (Betulaceae, Salicaceae, Fagaceae, and Ericaceae), with one to six species per polluter, yielding 51 ES in total (Tables 7.11–7.19).

7.3 Results

7.3.1 Densities of Birch-Feeding Herbivores

Variation in herbivore densities between study sites was significant in 31 of 46 data sets. However, only 15 of 92 individual correlations (with both distance and pollution) were significant (Tables 7.1–7.9).

Pollution had no overall effect on densities of birch-feeding herbivores (Fig. 7.1); this result did not depend on the method used to calculate ES ($Q_B = 0.29$, df = 2, $P = 0.87$). Pollution effects on herbivore densities did not differ among individual polluters (Fig. 7.2; $Q_B = 10.3$, df = 8, $P = 0.24$); a significant density decrease was observed only around Revda (Fig. 7.2).

The effect of pollution on herbivore densities did not depend on either the type of the polluter or its impact on soil pH (Fig. 7.1). The only source of variation identified in this database was the geographical position of polluters (Fig. 7.1; $Q_B = 4.92$, df = 1, $P = 0.03$); southern polluters negatively affected herbivore densities, whereas northern polluters caused slight, although non-significant, increases in herbivore densities.

Responses of different RTUs were generally similar (Fig. 7.3; $Q_B = 9.54$, df = 6, $P = 0.15$); only *Parornix* spp. demonstrated a significant decrease in density with increases in pollution. Miners and rollers did not differ in their responses to pollution (Fig. 7.3; $Q_B = 1.21$, df = 1, $P = 0.27$).

Table 7.11 Proportion (%) of leaves bearing feeding marks of insect herbivores in the impact zones of the power plant at Apatity, Russia, aluminium smelter at Bratsk, Russia and nickel-copper smelter at Harjavalta, Finland

	Apatity			Bratsk		Harjavalta	
	Alnus incana	*Betula pubescens*	*Vaccinium vitis-idaea*	*Betula pubescens*	*Vaccinium vitis-idaea*	*Betula pubescens*	*Salix caprea*
Site	2005	2005	2005	2002	2002	2002	2002
1–1	14.8	27.0	–	8.4	3.0	54.4	77.2
1–2	18.4	26.6	3.4	8.2	5.7	24.4	32.2
1–3	20.6	29.2	7.4	10.2	6.7	64.0	85.2
1–4	16.4	25.6	1.4	16.8	10.3	95.2	73.8
1–5	9.2	16.6	1.4	29.4	19.2	70.4	79.0
2–1	22.0	36.6	0.4	4.4	0.2	75.8	40.1
2–2	24.4	29.2	3.2	15.4	8.0	74.3	70.6
2–3	41.6	32.4	4.0	12.6	12.4	58.8	39.4
2–4	16.2	32.6	4.2	21.6	15.6	55.4	56.9
2–5	37.2	28.6	2.0	49.6	7.2	53.2	61.2
ANOVA: *F*/*P*	0.88/0.55	1.78/0.10	1.53/0.18	6.98/<0.0001	7.18/<0.0001	5.90/<0.0001	3.27/0.0046
Dist.: *r*/*P*	0.10/0.77	–0.40/0.25	0.02/0.96	0.77/0.0091	0.75/0.01	0.14/0.70	0.40/0.25
Poll.: *r*/*P*	–0.07/0.85	0.21/0.56	0.30/0.44	–0.51/0.13	–0.71/0.02	–0.68/0.03	–0.55/0.10

Sampling year shown under the plant name. ANOVA: test for significance of variation between study sites. Dist.: Pearson linear correlation between site-specific means and log-transformed distance from polluter. Poll.: Pearson linear correlation between site-specific means and concentration of the selected pollutant.

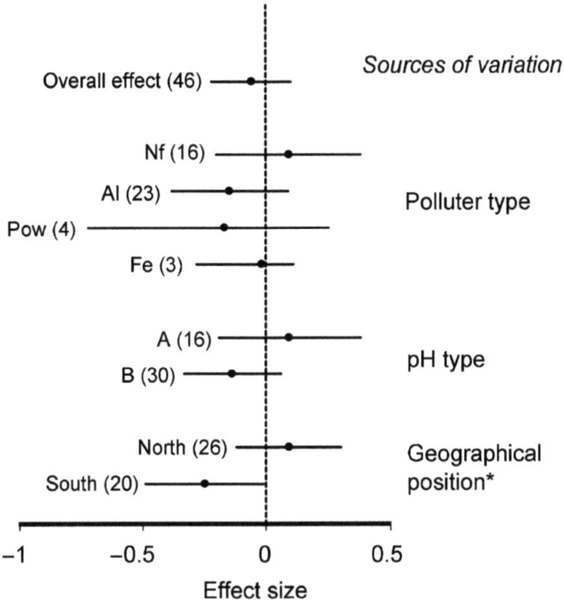

Fig. 7.1 Overall effect and sources of variation in responses of birch-feeding herbivore densities. Horizontal lines denote 95% confidence intervals; sample sizes are shown in brackets; an asterisk denotes significant ($P < 0.05$) between-class heterogeneity. For classifications of polluters consult Table 2.1

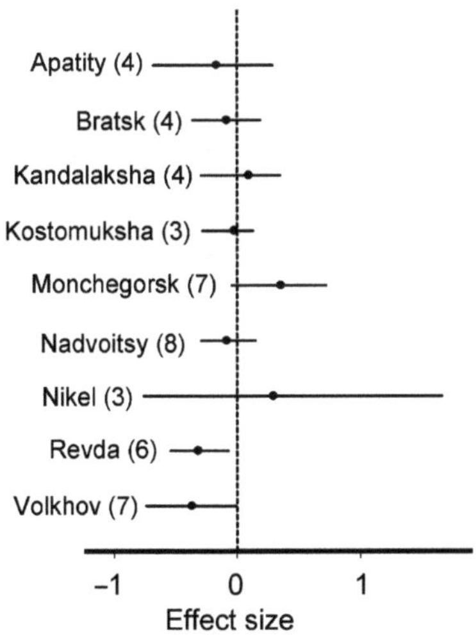

Fig. 7.2 Effects of individual polluters on densities of birch-feeding herbivores. For explanations, consult Fig. 7.1

Table 7.12 Proportion (%) of leaves bearing feeding marks of insect herbivores in the impact zones of the fertilising factory at Jonava, Lithuania, and aluminium smelter at Nadvoitsy, Russia

	Jonava			Nadvoitsy		
	Betula pendula	*Betula pubescens*	*Betula pubescens*	*Vaccinium myrtillus*	*Vaccinium vitis-idaea*	*Vaccinium uliginosum*
Site	2005	2004	2005	2005	2004	2005
1–1	84.8	55.6	51.0	3.0	3.6	8.7
1–2	96.8	23.0	54.8	3.4	5.2	3.4
1–3	95.4	23.8	77.4	3.4	4.4	3.6
1–4	94.0	52.8	70.0	10.4	8.6	5.6
1–5	87.6	25.8	89.0	2.5	6.2	7.2
2–1	95.0	33.8	60.2	4.8	3.2	6.4
2–2	97.6	30.0	41.6	4.8	19.8	4.8
2–3	86.6	15.0	57.8	8.6	4.4	5.2
2–4	94.0	28.6	68.4	9.7	25.0	14.6
2–5	98.4	25.0	86.8	4.4	4.8	17.4
ANOVA: *F/P*	4.06/0.0009	5.45/<0.0001	2.15/0.05	1.73/0.12	6.37/<0.0001	3.17/0.0061
Dist.: *r/P*	0.04/0.91	−0.29/0.41	0.81/0.0039	0.25/0.49	0.13/0.72	0.38/0.28
Poll.: *r/P*	−0.11/0.77	0.24/0.50	−0.72/0.02	−0.40/0.25	−0.12/0.74	−0.42/0.23

For explanations, consult Table 7.11.

Table 7.13 Proportion (%) of leaves bearing feeding marks of insect herbivores in the impact zones of the aluminium smelter at Kandalaksha, Russia and iron pellet plant at Kostomuksha, Russia

	Kandalaksha				Kostomuksha		
	Betula pubescens	*Salix caprea*	*Vaccinium myrtillus*	*Vaccinium uliginosum*	*Betula pubescens*	*Betula pubescens*	*Salix caprea*
Site	2002	2002	2005	2005	2002	2006	2002
1–1	21.4	16.4	34.8	13.8	17.6	11.0	3.2
1–2	51.6	10.7	36.6	8.6	35.6	14.2	23.3
1–3	17.4	2.9	14.8	13.6	23.4	15.2	6.0
1–4	26.4	4.2	40.4	13.8	13.6	21.6	15.0
1–5	16.2	2.8	2.8	0.6	8.8	32.2	8.9
2–1	8.2	5.8	9.2	15.6	8.6	9.2	2.5
2–2	12.2	8.2	2.8	4.6	36.6	16.4	18.5
2–3	16.0	4.6	3.4	2.6	57.4	41.2	21.4
2–4	14.6	8.0	6.8	3.2	41.3	31.8	18.7
2–5	8.8	6.7	5.2	2.0	16.8	37.2	11.0
ANOVA: *F/P*	4.19/0.0007	3.05/0.0073	7.55/<0.0001	4.35/0.0005	8.35/<0.0001	3.38/0.0036	7.22/<0.0001
Dist.: *r/P*	−0.08/0.83	−0.47/0.17	−0.23/0.52	−0.65/0.04	0.05/0.89	0.70/0.02	0.35/0.33
Poll.: *r/P*	0.33/0.36	0.45/0.20	0.37/0.29	0.66/0.04	−0.42/0.23	−0.60/0.07	−0.51/0.13

For explanations, consult Table 7.11.

Table 7.14 Proportion (%) of leaves bearing feeding marks of insect herbivores in the impact zones of the copper smelters at Karabash, Russia and Krompachy, Slovakia

	Karabash				Krompachy		
	Alnus incana	*Betula pendula*	*Betula pendula*	*Vaccinium vitis-idaea*	*Betula pendula*	*Fagus sylvatica*	*Fagus sylvatica*
Site	2007	2003	2007	2003	2004	2002	2004
1–1	–	63.8	17.0	–	35.0	46.2	50.4
1–2	100.0	76.2	79.2	16.6	26.4	33.6	25.2
1–3	99.4	90.6	96.0	7.0	10.2	21.0	13.6
1–4	96.6	86.0	99.6	6.0	44.4	19.0	20.4
1–5	95.6	92.4	88.2	26.6	13.8	25.4	6.8
2–1	91.0	34.0	9.2	–	62.0	34.8	23.0
2–2	97.6	48.4	79.8	5.0	33.2	38.2	19.4
2–3	98.4	68.4	95.2	10.	61.6	65.0	25.8
2–4	94.2	82.4	95.6	6.4	42.8	59.6	39.2
2–5	98.2	82.6	89.4	7.0	68.8	25.6	52.8
ANOVA: *F*/*P*	1.92/0.09	9.41/<0.0001	31.0/<0.0001	2.41/0.04	10.5/<0.0001	5.02/0.0002	7.57/<0.0001
Dist.: *r*/*P*	0.20/0.60	0.84/0.0025	0.77/0.009	0.16/0.70	−0.01/0.98	−0.45/0.19	−0.16/0.66
Poll.: *r*/*P*	−0.57/0.11	−0.92/0.0002	−0.91/0.0002	−0.27/0.52	0.09/0.81	−0.10/0.78	−0.21/0.55

For explanations, consult Table 7.11.

Table 7.15 Proportion (%) of leaves bearing feeding marks of insect herbivores in the impact zone of the nickel-copper smelter at Monchegorsk, Russia

	Betula nana	*Betula pubescens*	*Populus tremula*	*Salix borealis*	*Salix borealis*	*Salix caprea*	*Vaccinium uliginosum*
Site	2002	2002	2002	2002	2005	2002	2005
1–1	2.6	9.6	12.6	91.4	32.6	4.2	0.0
1–2	5.0	46.2	–	78.8	71.2	6.4	0.6
1–3	–	–	3.4	–	–	–	–
1–4	6.0	21.6	–	63.0	30.4	8.4	0.8
1–5	11.8	17.6	–	24.6	48.4	4.6	1.2
2–2	3.6	12.2	–	58.6	12.4	5.6	0.2
2–3	–	–	90.2	–	–	–	–
2–4	0.6	14.0	–	55.4	12.2	0.0	0.2
2–5	1.8	9.8	6.4	85.2	10.8	9.0	0.0
2–6	–	–	42.0	–	–	–	–
2–8	4.0	12.0	8.4	65.2	51.6	44.4	0.8
2–9	2.4	18.6	2.8	27.6	14.0	3.8	0.8
2–10	6.4	16.8	1.6	38.8	8.0	2.2	0.6
2–11	–	–	33.7	–	–	–	–
2–12	–	–	1.0	–	–	–	–
ANOVA: *F/P*	1.94/0.007	4.49/0.0004	11.9/<0.0001	9.58/<0.0001	5.92/<0.0001	3.32/0.0041	0.68/0.72
Dist.: *r/P*	0.25/0.43	–0.07/0.84	–0.30/0.40	–0.65/0.04	–0.20/0.59	0.30/0.41	0.58/0.08
Poll.: *r/P*	–0.57/0.08	–0.20/0.58	0.82/0.0034	0.48/0.16	–0.24/0.50	–0.28/0.44	–0.79/0.0066

For explanations, consult Table 7.11.

Table 7.16 Proportion (%) of leaves bearing feeding marks of insect herbivores in the impact zone of the nickel-copper smelter at Nikel and ore-roasting plant at Zapolyarnyy, Russia

	Betula nana	*Betula pubescens*	*Betula pubescens*	*Salix glauca*	*Salix glauca*	*Vaccinium myrtillus*	*Vaccinium uliginosum*
Site	2001	2001	2005	2001	2005	2005	2005
1–1	0.4	7.2	5.4	25.6	31.6	–	–
1–2	4.2	5.8	21.4	14.4	23.2	0	2.0
1–3	1.0	17.6	2.6	29.4	58.2	0.4	2.2
1–4	1.6	7.2	6.6	12.4	77.2	4.4	1.8
1–5	1.0	6.4	6.6	12.6	25.6	0.6	1.4
2–1	21.8	4.0	1.2	–	15.2	0.2	0.6
2–2	2.0	8.0	3.2	14.4	13.0	0.2	0.2
2–3	0.8	7.0	1.6	3.8	–	0.6	1.8
2–4	0.4	14.4	3.4	10.0	30.0	0.2	2.6
2–5	1.4	15.6	13.8	6.4	23.6	1.0	1.8
ANOVA: *F/P*	6.82/<0.0001	1.62/0.14	2.17/0.05	1.76/0.12	3.23/0.0083	4.77/0.0005	0.92/0.51
Dist.: *r/P*	−0.42/0.23	0.42/0.22	0.13/0.71	−0.53/0.14	0.35/0.36	0.46/0.22	0.53/0.15
Poll.: *r/P*	0.47/0.17	−0.41/0.24	−0.21/0.56	0.50/0.17	−0.32/0.40	−0.32/0.40	−0.63/0.07

For explanations, consult Table 7.11.

Table 7.17 Proportion (%) of leaves bearing feeding marks of insect herbivores in the impact zones of the nickel-copper smelters at Norilsk, Russia and copper smelter at Revda, Russia

	Norilsk				Revda		
	Betula nana	*Salix lanata*	*Vaccinium vitis-idaea*	*Vaccinium uliginosum*	*Betula pubescens*	*Populus tremula*	*Vaccinium vitis-idaea*
Site	2002	2002	2002	2002	2003	2007	2003
1–1	0.4	11.0	9.0	6.8	78.6	41.4	10.2
1–2	0	10.7	0.6	4.5	70.2	25.0	–
1–3	0.4	11.7	1.2	0.2	85.6	24.8	–
1–4	0	5.7	1.6	1.0	88.2	76.0	9.6
1–5	0	9.8	0	0.8	62.2	18.4	4.2
2–1	0	13.1	12.6	0.2	51.8	23.2	9.8
2–2	1.6	5.3	2.4	6.7	69.6	81.6	3.6
2–3	0	10.2	1.0	0.8	68.6	69.6	5.4
2–4	0.6	3.8	0	0	81.6	14.0	9.2
2–5	–	–	–	–	93.8	55.4	16.8
ANOVA: *F*/*P*	2.46/0.02	0.86/0.55	2.23/0.05	1.30/0.27	3.35/0.0039	8.73/<0.0001	2.31/0.05
Dist.: *r*/*P*	–0.23/0.55	–0.42/0.27	–0.84/0.0046	–0.59/0.10	0.51/0.13	0.00/0.99	0.22/0.60
Poll.: *r*/*P*	0.17/0.66	0.38/0.32	0.84/0.0046	0.59/0.10	–0.48/0.16	0.17/0.63	–0.16/0.70

For explanations, consult Table 7.11.

Table 7.18 Proportion (%) of leaves bearing feeding marks of insect herbivores in the impact zones of the aluminium smelters at Straumsvík, Iceland and Volkhov, Russia, nickel-copper smelter at Sudbury, Canada and power plant at Vorkuta, Russia

	Straumsvík		Sudbury	Volkhov		Vorkuta	
	Salix herbacea	*Vaccinium uliginosum*	*Betula papyrifera*	*Betula pubescens*	*Salix caprea*	*Betula nana*	*Salix glauca*
Site	2002	2002	2007	2002	2002	2002	2002
1–1	2.0	2.7	94.0	79.0	69.7	19.8	55.8
1–2	3.7	1.6	65.0	84.4	64.6	46.8	44.2
1–3	13.1	2.0	86.8	80.4	89.1	45.2	61.0
1–4	1.0	5.2	81.8	86.4	78.1	–	–
1–5	2.0	2.6	92.4	73.6	38.0	26.4	30.6
2–1	5.4	6.7	96.6	75.2	40.0	11.4	47.4
2–2	4.5	1.0	98.0	60.6	36.1	60.8	71.6
2–3	7.6	5.1	92.8	90.7	98.4	56.0	45.8
2–4	6.7	1.4	94.6	94.0	96.4	51.2	28.6
2–5	6.4	26.2	98.4	88.7	95.3	48.2	40.0
ANOVA: *F/P*	1.58/0.15	1.14/0.36	7.25/<0.0001	1.18/0.33	6.23/<0.0001	2.89/0.01	0.43/0.91
Dist.: *r/P*	−0.01/0.99	0.51/0.13	0.08/0.83	0.45/0.19	0.47/0.17	0.53/0.14	−0.53/0.014
Poll.: *r/P*	−0.33/0.35	−0.17/0.64	0.15/0.69	−0.46/0.18	−0.57/0.09	−0.08/0.84	0.32/0.41

For explanations, consult Table 7.11.

Table 7.19 Proportion (%) of leaves bearing feeding marks of insect herbivores in the impact zone of the aluminium smelter at Žiar nad Hronom, Slovakia

	Carpinus betulus	*Carpinus betulus*	*Fagus sylvatica*	*Fagus sylvatica*	*Quercus petraea*
Site	2002	2004	2002	2004	2002
1–1	98.6	93.0	97.6	96.0	76.2
1–2	85.5	96.2	37.4	68.2	71.8
1–3	95.9	93.4	87.6	72.8	–
1–4	99.4	–	42.8	69.8	84.2
1–5	97.1	97.0	54.2	78.8	–
2–1	89.8	86.8	68.0	48.6	54.4
2–2	98.3	90.2	66.8	70.2	77.4
2–3	70.0	86.8	46.0	55.8	–
2–4	95.2	81.6	66.0	90.2	–
2–5	93.6	84.0	82.6	77.8	73.1
ANOVA: *F*/*P*	1.75/0.11	2.24/0.05	5.64/<0.0001	4.44/0.0005	1.19/0.35
Dist.: *r*/*P*	0.06/0.87	−0.44/0.24	−0.20/0.58	0.04/0.91	0.36/0.48
Poll.: *r*/*P*	0.11/0.77	0.46/0.21	0.29/0.42	0.43/0.21	0.09/0.86

For explanations, consult Table 7.11.

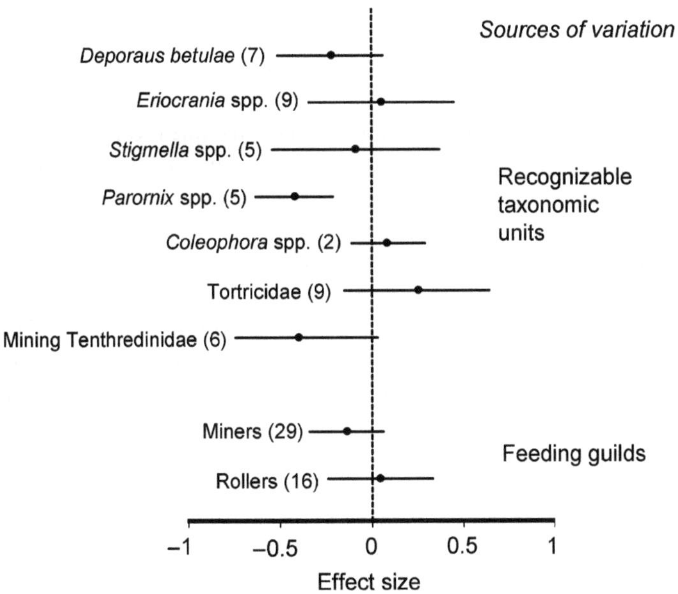

Fig. 7.3 Effects of point polluters on densities of recognizable taxonomic units and feeding guilds of birch-feeding herbivores. For explanations, consult Fig. 7.1

We did not detect significant non-linear responses in any of the 47 density by distance data sets.

7.3.2 *Foliar Damage to Woody Plants*

7.3.2.1 Consumption of Foliage

Pollution did not influence average consumption per damaged leaf in nine of ten data sets (Table 7.10). Consistently, an overall effect did not differ from zero ($d = -0.07$, CI = −0.88 … 0.74, $N = 10$). This result in particular ensures the use of the proportion of damaged leaves as an adequate measure of plant damage imposed by herbivorous insects.

7.3.2.2 Proportion of Damaged Leaves

Variation in herbivore intensities between study sites was significant in 43 of 61 data sets. However, only 21 of 122 individual correlations (with both distance and pollution) were significant (Tables 7.11–7.19).

Correlation between plot-specific values obtained for the same plant species in different years was generally positive ($z_r = 0.31$, CI = 0.02 … 0.60, $N = 9$ polluters), confirming that the pattern of pollution effects on foliar damage was similar among study years.

Foliar damage decreased with pollution (Fig. 7.4); this result did not depend on the method used to calculate ES ($Q_B = 0.27$, df = 2, $P = 0.88$). Individual polluters differentially affected the proportion of damaged leaves (Fig. 7.5: $Q_B = 57.4$, df = 15, $P < 0.00001$); foliar damage decreased near seven polluters, increased near five polluters, and did not change around four polluters.

The effect of pollution on foliar damage by herbivores did not depend on either the type of polluter or its impact on soil pH (Fig. 7.4). Consistent with the effects on herbivore densities (Fig. 7.1) foliar damage decreased around southern polluters but did not change around northern polluters (Fig. 7.4; $Q_B = 7.11$, df = 1, $P = 0.008$).

Pollution had a significant effect on damage imposed by insect herbivores on two of 11 plant species: damage to white/mountain birch decreased, and damage to European aspen increased, while damage to other species did not change with pollution (Fig. 7.6). However, this variation was far from significant ($Q_B = 7.40$, df = 10, $P = 0.69$). Similarly, we found no differences between pollution effects on herbivore damage to trees, shrubs, and dwarf shrubs ($Q_B = 0.37$, df = 2, $P = 0.85$).

We have detected significant non-linear responses in eight of 60 data sets (seven dome-shaped and one U-shaped).

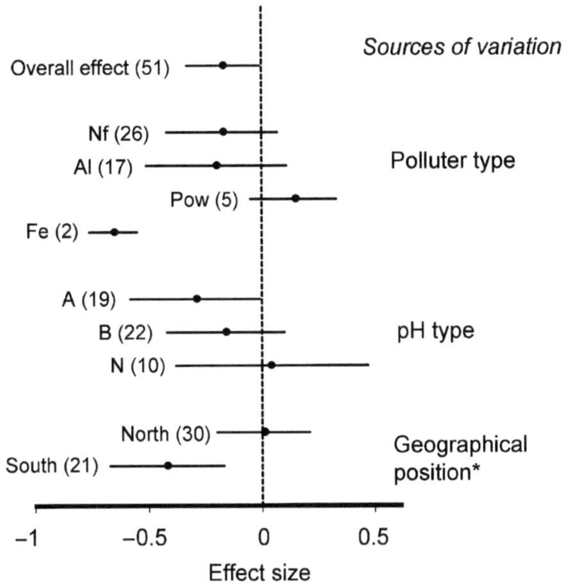

Fig. 7.4 Overall effect and sources of variation in foliar damage to woody plants caused by insect herbivores. For explanations, consult Fig. 7.1

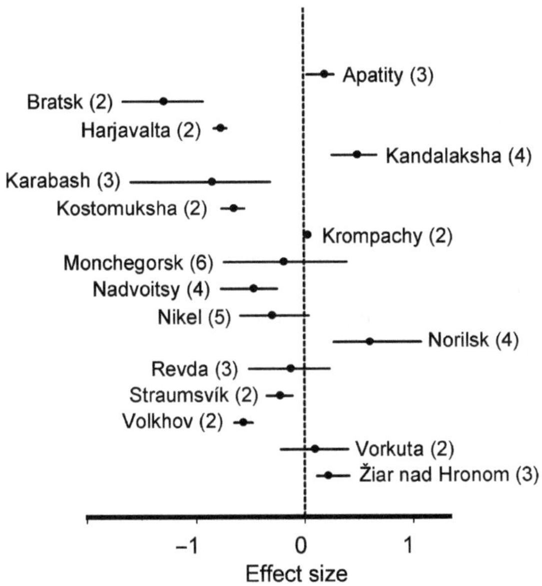

Fig. 7.5 Effects of individual polluters on foliar damage to woody plants caused by insect herbivores. For explanations, consult Fig. 7.1

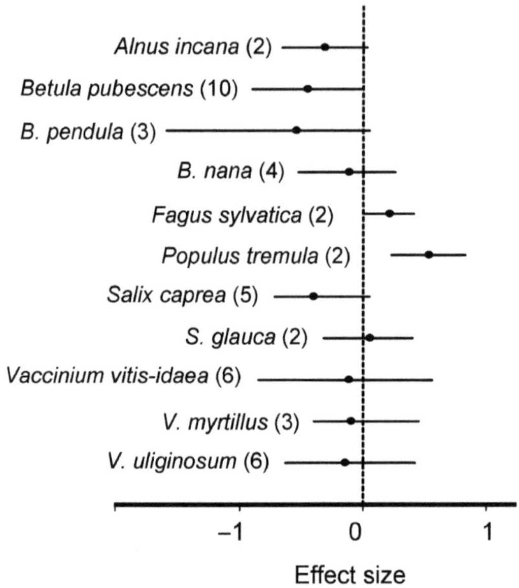

Fig. 7.6 Effects of point polluters on foliar damage to woody plant species caused by insect herbivores. For explanations, consult Fig. 7.1

7.4 Discussion

7.4.1 Herbivory in Polluted Areas

Our meta-analyses of either population density of birch-feeding herbivores or foliar damage imposed by herbivores on several host plants did not confirm the conclusion of several case studies and narrative reviews (Berger & Katzensteiner 1994; Baltensweiler 1985; Führer 1985; Jones 2001; Jones & Paine 2006) on a general increase of herbivory under pollution impact. This result, however, strongly supports our conclusion that significant increase in herbivore abundances near point polluters, revealed by meta-analysis of published data (Zvereva & Kozlov 2009), is an artefact caused by research biases that have led to overestimation of this effect.

The research bias may result from the tendency to explore organisms and/or conditions in which the researcher has a reasonable expectation of detecting significant effects (Gurevitch & Hedges 1999). In particular, studies of pollution impacts on herbivore populations are likely to start in the year and on the site when herbivore damage in a polluted area is apparent, and the species that causes this damage is usually selected as the study object. Importantly, in meta-analysis of published studies, mean ES based on years randomly chosen from multiyear data sets did not differ from zero, while single-year studies yielded significant effects (Zvereva & Kozlov 2009). Further bias may be introduced by selection of a pair of 'typical' study sites (one polluted and one control): since between-site variation was significant in approximately three-quarters of the data sets, the risk of detection of the expected pattern increases with a decrease in the number of study sites.

The data collected in the course of our study do not suffer from these biases; we censused all herbivore species feeding on birch and used ten study sites selected without *a priori* knowledge of the level of herbivory at these sites. Thus, our results indicate that increase in herbivory near point polluters is an infrequent phenomenon rather than general regularity.

While changes in the densities of birch-feeding herbivores were surprisingly uniform (Fig. 7.2), the effects of individual polluters on foliar damage were variable (Fig. 7.5). However, we have not discovered any differences either between classes of polluters (Figs. 7.1 and 7.4) or among woody plant species (Fig. 7.6). The effects of pollution on herbivore densities assessed both directly and indirectly (i.e., by measuring foliar damage) were significantly more detrimental around southern polluters (Figs. 7.1 and 7.4).

7.4.2 Pollution Effects on Insect–Plant Relationships

Injury imposed by herbivorous insects makes the leaf less palatable, and subsequent feeding is more likely to occur in intact leaves, resulting in a dispersion of damage (Valladares & Hartley 1994; Fisher et al. 1999; Staley & Hartley 2002). Since the area consumed per damaged leaf did not differ between polluted and control sites (Section 7.3.2.1), we concluded that pollution did not affect plant ability to develop rapid inducible responses. However, we previously (Zvereva & Kozlov 2000b) detected adverse effect of pollution on delayed inducible responses, i.e., decrease of plant quality next season after damage, which is one of the important feedbacks regulating the population dynamics of herbivores. This contradiction emphasizes the need in further studies of pollution impact on plant responses to herbivory.

7.4.3 Dome-Shaped Responses

Narrative reviews (Führer 1985; Selikhovkin 1988; Riemer & Whittaker 1989; Kozlov 1990, 1997; Heliövaara & Väisänen 1993) refer to several dozen studies claiming discovery of a dome-shaped pattern of density changes along pollution gradients, which was therefore considered a typical response of herbivorous insects to pollution. However, statistical support for this conclusion was found only in 5.48% of publications covered by our meta-analysis (Zvereva & Kozlov 2009). Since analysis of our original data revealed nearly the same proportion of data sets demonstrating a dome-shaped response (6.5%), we conclude that dome-shaped patterns of density changes are neither typical nor even frequent among insect herbivores.

7.4.4 *Insect Feeding Guilds*

Although earlier reviews (Larsson 1989; Koricheva et al. 1998) concluded that insect feeding guild is the most important determinant of herbivore responses to host plant stress, including stress imposed by pollution, we have not detected any pollution effects on either miners or rollers when analysing both published (Zvereva & Kozlov 2009) and original (Fig. 7.3) data. However, the effects of pollution on free-living defoliators, as well as on foliar damage imposed mostly by this guild, were positive in a meta-analysis of published data (Zvereva & Kozlov 2009) but negative in the original data (Fig. 7.4). This hints that an overall pollution effect on free-living defoliators is close to zero and is therefore similar to effects on miners and rollers.

The discrepancy between the detected effects on free-living defoliators may be partially explained by a research bias that affected the published data (see above). On the other hand, in the current study we have not surveyed coniferous plants which, according to published data, suffer more pronounced increases in herbivore damage under pollution impacts than deciduous plants (Zvereva & Kozlov 2009). Last but not least, our data concern only endemic herbivory, while the meta-analysis of published data included also measurements performed during outbreaks of certain species. This may also have contributed to contrasting outcomes of the two meta-analyses, since pollution effects on some herbivores were expressed only in high-density years (Kozlov 2003; Zvereva & Kozlov 2006b).

An absence of significant interguild variation in both meta-analyses may have resulted from substantial differences in duration of stress impacts between observational and experimental studies. Differential responses of feeding guilds have been detected on experimentally stressed plants (Larsson 1989; Koricheva et al. 1998), while plants naturally growing in the polluted environment frequently do not exhibit (and presumably do not experience) any effects of stress (Chapters 4 and 5, and references therein) due to both acclimation and evolved resistance (Bradshaw & McNeilly 1981; Macnair 1997; Kozlov 2005b; Eränen 2008). Our results in particular suggest that some of the regularities discovered by means of manipulative studies are of low predictive value when applied to natural ecosystems.

7.5 Summary

We have not discovered the expected increase of herbivory in polluted areas: the overall effects were either negative (foliar damage) or absent (herbivore densities) for all studied feeding guilds. The dome-shaped patterns in herbivory along pollution gradients were rare and occasional. Types of polluters do not differ in their effects on plant damage by herbivorous insects; however, pollution tended to suppress herbivory in areas with a warmer and more humid climate. We conclude that increase in herbivory at polluted sites is an infrequent phenomenon rather than a general pattern.

Chapter 8
Methodology of Pollution Ecology: Problems and Perspectives

8.1 Importance of Observational Studies

Charles J. Krebs (1989) started his famous book *Ecological Methodology* with the following statement: 'Ecologists collect data and, as in other fields in biology, the data they collect are to be used for testing hypotheses'. This is true, but where do the new hypotheses originate?

The roots of many theories, including the Newton's law of universal gravitation and Darwin's evolutionary theory, emerge from systematic or occasional observations. Accumulation of data on rainfall acidity, measurements of which for a long time were driven by scientific curiosity, in the 1950s and 1960s allowed scientists to determine the origin of acidity and recognise the damaging effects of acid rain. The importance of observational studies (sometimes termed mensurative experiments) still remains high, especially for environmental sciences, which often face novel problems associated with the rapid development of our civilisation.

Environmental scientists have often been blamed for preferring a narrative approach to hypothesis testing. For example, only 20% of papers published in the *Journal of Applied Ecology* in 1999 explicitly stated clearly testable hypotheses (Ormerod et al. 1999). However, in our opinion, this situation reflects a shortage of relevant hypotheses rather than the incompetence of environmental scientists or their reluctance to use the hypothesis-testing approach. Our survey of several dozens of publications summarising the knowledge on pollution effects on biota (Smith 1974; Miller & McBridge 1975; Kozlowski 1980; Auerbach 1981; Alstad et al. 1982; Newman & Schreiber 1984; Odum 1985; Rapport et al. 1985; Schindler 1987; Sigal & Suter 1987; Bååth 1989; Freedman 1989; Riemer & Whittaker 1989; Treshow & Anderson 1989; Barker & Tingey 1992; Heliövaara & Väisänen 1993; Clements & Newman 2002) turned up quite a few testable predictions.

The advancement of pollution ecology requires further accumulation of observational data on changes in landscapes, ecosystems, communities, populations and individual organisms occurring around industrial polluters. This information is necessary for exploratory analyses leading to the generation of specific hypotheses, which can be tested using field and laboratory experiments.

M.V. Kozlov et al. *Impacts of Point Polluters on Terrestrial Biota*,
DOI 10.1007/978-90-481-2467-1_8, © Springer Science+Business Media B.V. 2009

8.2 Interpretation of Experimental Results

A substantial part of our knowledge of responses of organisms, populations, and communities to pollution originates from experimental studies, mostly conducted in artificial (laboratory) environments. For example, 42 of 50 relevant recent (2008) publications (25 from *Environmental Pollution* and another 25 from *Water, Air, and Soil Pollution*) referred to laboratory experiments, compared to 11 papers that reported field manipulations; only four papers combined these approaches. Six papers reported use of micro- or mesocosms, while the remaining majority of experimental systems were obviously oversimplified in terms of both abiotic and biotic environments.

More generally, ecosystem-level and community-level field experiments with industrial pollutants remain as rare as they were 2 decades ago (Schindler 1987), and most of these experiments are conducted in aquatic systems. The limited number of large-scale manipulations with terrestrial biota is considered a significant shortcoming of ecotoxicology (Clements & Newman 2002). Harvesting the results of 'unintentional pollution experiments' (Lee 1998) established by industries long ago (Section 1.4) can partially overcome this problem.

Laboratory studies, by eliminating a substantial part of natural variability, are likely to produce biased results, in particular due to (a) investigation of only a few 'model' species, with preferences for short-living and easy-to-handle organisms, (b) unrealistic environments, including the use of closed chambers, unnatural growth media for plants, the absence of mutualists, e.g., mycorrhizal fungi, competitors and benefactors, (c) unrealistic forms in which pollutants are applied to organisms, (d) unrealistic demographic structures of experimental populations, e.g., preferential use of seedlings to explore responses of woody plants, and (e) the short duration of the experiments relative to decades or even centuries of ecosystem exposure to industrial pollutants (Patterson & Olson 1983; Stenström 1991; Sandermann et al. 1997; Saxe et al. 1998; Weltje 1998; Ahonen-Jonnarth et al. 2000; Koster et al. 2006). Importantly, the outcomes of field and laboratory studies addressing pollution effects on biota have not, to our knowledge, been compared systematically (except for some specific ecotoxicological tests: Hose et al. 2006), and therefore the biases introduced by experimental methodology remain insufficiently known.

The importance of indirect effects, which are usually neglected in manipulative studies with industrial pollutants, can be demonstrated by a seemingly paradoxical increase in plant performance, repeatedly observed in the heavily polluted sites near the nickel-copper smelter in Monchegorsk. For example, in industrial barrens, some dwarf shrubs grow and reproduce better than in unpolluted forests, especially when they are sheltered by mountain birch trees (Zvereva & Kozlov 2004, 2005). The leaf size and shoot growth of boreal willow, *Salix borealis* increased with an increase in pollution (Zvereva et al. 1997a). Similarly, birch seedlings planted in metal-contaminated bare ground after 3 years of exposure were taller, had longer leaves and had a higher survival rate than those in the unpolluted forest (Eränen & Kozlov 2009). In our opinion, these 'positive' effects observed in heavily polluted environments resulted from the absence of competing vegetation that declined decades ago. In general, the effects of pollution on competitors or benefactors of the organism under study may be so strong that they counterbalance direct toxic effects of pollutants.

Additional uncertainties in interpretation of experimental results are introduced by using organisms from populations that have not been exposed to pollutants prior the experiments. One of the examples of an obvious discrepancy between observational and manipulative studies, presumably resulting from this approach, concerns the effects of industrial pollutants on the growth of herbaceous plants. Experimental studies, conducted both in fully controlled environments and under field conditions, usually report the adverse effects of different pollutants on herbaceous plants (Brun et al. 2003; Tuma 2003; Hassan 2004; Rämö et al. 2006; Ryser & Sauder 2006). Consistently, many ecotoxicological tests are based on measurements of plant size or biomass (Rajput & Agrawal 2005; An 2006; Everhart et al. 2006; Rombke et al. 2006; Rooney et al. 2006). At the same time, herbaceous plants naturally occurring near big polluters only rarely differed from plants collected from unpolluted environments in terms of growth characteristics (Kozlov & Zvereva 2007b; Figs. 4.11, 4.12, 4.17 and 4.18). We hypothesise that micro-evolution, often leading to the development of pollution tolerance (Bradshaw & McNeilly 1981; Shaw 1990; Macnair 1997; Barnes et al. 1999; Medina et al. 2007), is a plausible explanation of the discrepancy between the results of controlled experiments and field-collected data (Kozlov & Zvereva 2007b).

We conclude that short-term experiments with non-adapted organisms in over-simplified laboratory environments are likely to overestimate the adverse effects of industrial pollutants. Results of these experiments can be used for estimation of the relative toxicity of different substances, but they are of limited value for both explaining and predicting effects observed in polluted environments. Even more importantly, community responses cannot be inferred from the results of single-species experiments. Thus, both dedicated observational studies and field experiments remain of critical importance for the development of pollution ecology, in particular as a tool to validate the results of laboratory tests.

8.3 The Amount of Reliable Information

It seems that many ecologists consider exploration of the effects of industrial pollution on biota outdated and believe that this research field is nearly exhausted. Both impressions are far from correct.

First and most importantly, inputs of industrial pollutants, primarily sulphur dioxide, into the atmosphere remain extremely high and continue to contribute to local air pollution, smog, acid rain, dry deposition, and global climate change. Although the global SO_2 emissions at the beginning of the 2000s were reduced by 22% relative to the peak value observed at the end of the 1980s, still they remain at the level of the mid-1960s (Stern 2006). Sulphur dioxide was recently identified as the air pollutant of the highest national concern for India (Agrawal 2005). Thus, there is no doubt that sulphur dioxide and many other pollutants, especially metals and fluorine accumulated in soils, continue to affect terrestrial biota. Even under the most optimistic emission scenarios, these effects will remain an issue of importance for at least several decades (Barcan 2002a; Fowler et al. 1999; Karnosky et al. 2003a).

Second, even the acute effects of pollution are not documented properly. In spite of the high number of publications, the amount of reliable data published in a form suitable for meta-analyses is surprisingly low (Fig. 1.2). Moreover, researchers tend to focus on the impacts imposed by 'pollution superstars', such as large non-ferrous smelters, while the effects caused by smaller industries (emitting 1,000 to 10,000 t SO_2 annually) remain almost unexplored. This shortage of information decreases reliability of estimations of dose-response relationships at lower levels of pollution, i.e., in concentration ranges which are most important for predicting pollution effects at regional scales.

Third, researchers tended to explore pollution effects on species of the highest economic importance, primarily forest-forming conifers and agricultural crops. However, incorporation of pollution effects into biogeochemical models, such as LPJ-GUESS (General Ecosystem Simulator: Smith et al. 2001; Sitch et al. 2003), requires parameterisation of growth responses to pollution (including changes in competitive abilities as well) for all cohorts of plant functional types included. Absence of this information hampers the modelling of regional and global effects of industrial pollution on vegetation structure. Importantly, we are likely to underestimate the importance of both 'low' (i.e., not exceeding the critical loads) depositions of pollutants and 'minor' differences in pollution impacts on plant species or functional groups. For example, the introduction of a small biotic disturbance (insect herbivory) into the LPJ-GUESS model demonstrated that relatively minor damage to birch (annual removal of 1–10% of foliage) changes predictions of future forest composition (Wolf et al. 2008).

Thus, there is acute need for reliable quantitative information concerning the responses of different groups of biota to industrial pollution. This information can be immediately utilised in both building phenomenological models and adjusting existing ecosystem simulators in order to improve our predictions of regional to global pollution effects on terrestrial ecosystems.

8.4 Quality of Information

8.4.1 Design of Impact Studies

For a long time, researchers exploring pollution effects on biota were advised to pay special attention to both experimental design and statistical analysis, because they 'are as important as the choice of monitoring parameters and techniques' (Sigal & Suter 1987). However, both recent meta-analyses (Ruotsalainen & Kozlov 2006; Zvereva et al. 2008; Zvereva & Kozlov 2009; Roitto et al. 2009) and narrative review of studies addressing forest health (Percy & Ferretti 2004) clearly demonstrated that the majority of primary studies suffer from a number of methodological shortcomings.

Different approaches have been suggested to monitor chronic, local environmental impacts (Stewart-Oaten & Bence 2001). The most well known include intervention analysis, which compares time series before and after an onset of impact at the affected site, so-called BACI design (Before-After, Control-Impact comparisons), and impact versus reference sites. The latter approach estimates error variation among sites, while intervention analysis and BACI estimate error variation over time (Stewart-Oaten & Bence 2001). All approaches require the selection of adequate temporal and spatial scales (Hewitt et al. 2001), and the results of comparisons should always be interpreted with caution, especially in terms of the causality of the observed differences (Section 9.2.3).

Except for dendrochronological analysis, comparison between observations conducted before and after the beginning of impact is only rarely used in studies of biotic effects caused by industrial pollutants. This is mostly due to an absence of adequately collected information about the state of the impacted ecosystems before the onset of pollution. As a rule, the error term in testing for the significance of the effect is obtained from comparisons among study sites. Therefore the quality of information, obtained by comparison between polluted (treatment) and unpolluted (control) habitats depends critically on the number of study sites. For the following analysis, we used a database of approximately 2,000 primary studies that describe responses of terrestrial biota to industrial pollution and the fit criteria listed in Section 5.1.2.

Generally, the sampling design of published studies was poorly replicated. The median number of study sites in a random sample of 1,000 publications was five (Fig. 8.1). Even more importantly, 35% of publications were based on two or

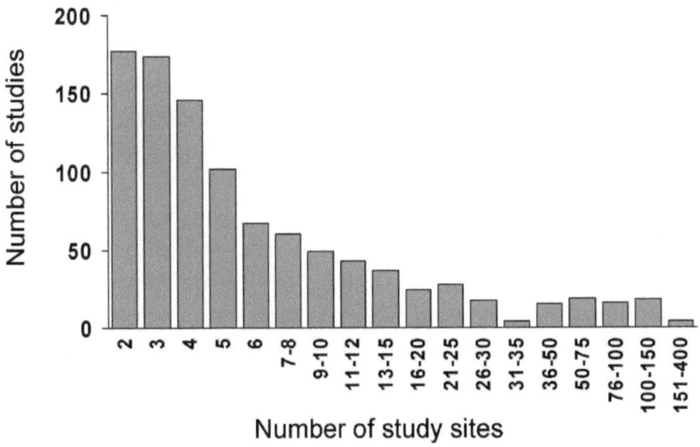

Fig. 8.1 Frequencies of studies based on different numbers of study sites in a random sample of 1,000 publications reporting biotic responses to industrial pollution. For criteria used to select these publications consult Section 5.1.2

three study sites, which means that either 'treatment' or 'control' or both were not replicated. The use of statistical analysis in this situation yields the pseudoreplication problem (Hurlbert 1984; Kozlov 2003, 2007; Kozlov & Hurlbert 2006). Because of repeated attempts to defend the use of non-replicated experimental design (Tatarnikov 2005; Veličković 2007b), we briefly explain the problems arising from comparisons between one polluted and one unpolluted study site.

A statement on the similarity or dissimilarity of two groups of objects is valid only when between-group differences are compared with within-group variation. Very importantly, the level at which the variation is measured within a group is critical. It is also obvious that assessment of within-group variation is only possible when the group consists of more than one object. When there is only a single experimental unit under each treatment, and within-group variation is calculated from measurements made on multiple samples or evaluation units within a single experimental unit (for definitions, consult Kozlov & Hurlbert 2006), then simple pseudoreplication is committed (Hurlbert 1984; Kozlov & Hurlbert 2006). Variation within study sites represents another level, one step down in the hierarchical analysis. This variation can only be used to test for differences between sites, but such tests cannot provide a statistical ground for attributing the differences to pollution (Kozlov 2007).

Earlier meta-analyses (Zvereva & Kozlov 2006a, 2009) demonstrated that effect sizes calculated on the basis of pseudoreplicated studies may be both higher and the same as those based on properly replicated studies. Pseudoreplicated studies allow only calculation of Hedge's d effect size, which increases with both decrease in within-group variation and increase in sample size (Rosenberg et al. 2000). The higher effect sizes obtained by pseudoreplicated studies on insect and plant responses to pollution result from both (a) the generally higher number of samples collected within a site, relative to the number of study sites in properly replicated studies, and (b) generally lower within-site variation relative to between-site variation. Research bias associated with the selection of two 'typical' study sites, one polluted and one unpolluted (control), is likely to further overestimate the effect (Zvereva & Kozlov 2009). Thus, when exclusion of pseudoreplicated studies from meta-analysis results in too low a number of suitable studies (Ruotsalainen & Kozlov 2006; Zvereva & Kozlov 2006a, 2009), the outcomes of these studies should be contrasted with the results based on properly replicated studies to account for the effects of methodology.

To conclude this part of the data quality assessment, we strongly recommend that the impact versus reference sites design includes at least two polluted and two control plots. This design allows correct testing of H0 (no differences between polluted sites and unpolluted sites) by using ANOVA based on site-specific mean values of the character under study. Furthermore, this design allows use of the data in meta-analyses employing both Hedge's d and correlation coefficients as measures of the effect size. However, although four study sites allow the correct use of statistical methods, both the accuracy of the effect estimate and the power of the analysis based on four study sites are rather low (Section 8.4.3).

Another insufficiently explored potential source of problems is the spatial arrangement of study sites. First, this information is reported much more seldom than the number of study sites: it was missing in 493 of 1,000 publications randomly selected from our database. The majority of studies that reported sampling design in sufficient details either selected each study site in a different direction from the polluter (38%) or positioned study sites along a single gradient (transect) starting from the polluter (37%). The latter design is potentially dangerous due to the possibility to erroneously attribute effects caused by another environmental variable to the effects of pollution. For example, the selection of all study sites to the South of the nickel-copper smelter in Monchegorsk, with control sites located up to 200 km from the polluter (Norin & Yarmishko 1990; Kabirov 1993; Zhirov et al. 1993; Evdokimova 2000; Lukina et al. 2005), does not allow discrimination of the effects of pollution from climatic (latitudinal) variation. Choosing study sites in several directions from the polluter (Figs. 2.2–2.19) is likely to minimise impact of confounding variables on the outcome of data analyses.

8.4.2 Signal to Noise Ratio

Most of studies addressing pollution impacts on biota were initiated when effects were already evident, and papers describing severely degraded ecosystems around major smelters (Gordon & Gorham 1963; Wood & Nash 1976; Freedman & Hutchinson 1980b) are most frequently used to illustrate consequences of pollution impact on biota. However, it would be risky to rely on these studies alone to predict ecosystem responses to contamination (Schindler 1987).

Wolterbeek et al. (1996) criticised the literature on biomonitoring for focussing on the detected changes in monitored parameters (signal) and neglecting the information about the range of natural variation (noise). The latter information is indeed essential to test whether the monitored character began to vary outside its normal range, thus indicating that the ecosystem is perturbed or stressed (Hurlbert 1984; Schindler 1987; Hewitt et al. 2001). The problem is that we usually do not know the normal range for any variable, at least for time periods greater than a few years (Schindler 1987). The only exception is dendrochronology, routinely accounting for natural variation in tree growth over decades or even centuries. Even exploration of insect population dynamics is commonly (Turchin 1990; Berryman 1994) based on time series shorter than 30–40 generations (which are considered ideal for detecting factors that influence population dynamics: Royama 1992), because longer-term data are rare (Hunter & Price 1998). Cyclic fluctuations in many population characteristics always pose a risk of erroneous interpretation of the results of short-term studies: different characteristics of populations from polluted and unpolluted habitats may result from asynchronous density changes in spatially isolated populations rather than from pollution impact on population dynamics (Zvereva et al. 2002; Zvereva & Kozlov 2006b, 2009).

In our study we detected significant between-site variation in 73.8% of 782 statistical tests (Chapters 3–7); however, only 20.4% of 1,446 individual correlation coefficients demonstrated significant relationships between the explored characteristics and pollution load. The proportion of significant correlations increased with the proportion of significant tests of between-site variation ($r_S = 0.66$, $N = 19$ characters, $P = 0.0019$), hinting that the contribution of detectable pollution-related effects to the overall spatial variation in all explored characteristics was more or less the same.

Of course, the proportion of significant tests is a very rough measure of the frequency of detectable effects associated with pollution, because this proportion depends on both the accepted level of the significance and on the statistical power of the tests (Section 8.4.3). Still, we were able to detect pollution 'signal' in only about one fourth of situations when the between-plot variation was significant. Furthermore, the detected signal was rather small: on average, only about 15% of variability among our study sites, which always included both most and least polluted habitats (Figs. 2.2–2.19), was explained by pollution load. This conclusion further stresses the need to properly account for variation in the characteristics under study in order to partition pollution effects on biota from variation caused by other factors.

There is no doubt that regional increases in pollutant concentrations are generally below the levels of environmental contamination observed near industrial polluters. Since the localised and relatively strong environmental contamination generally causes biotic effects of low magnitude, accurate identification of the consequences of regional pollution may require more effort than commonly thought. Moreover, the low signal to noise ratio increases the probability of erroneous attribution of natural spatial or temporal variation to pollution. Properly replicated sampling design, selection of adequate temporal and spatial scales, careful use of statistics and exploration of causal relationships behind the observed effects are essential to overcome this problem.

8.4.3 Power of Correlation Analyses

Data collected around the industrial polluters are most frequently analysed by calculating correlations between the measured parameters and either distance from the polluter or concentration of some pollutant. In a random sample of 1,000 publications from our collection, correlation analysis was used in 32% of studies properly reporting the use of statistical methods. Together with publications that employed regression analysis (15%), they comprise nearly a half of all studies.

Analysis of 1,446 individual correlation coefficients calculated from our data (Chapters 3–7) demonstrated that the average absolute value of correlation was relatively small (mean ± S.E.: 0.395 ± 0.006), corresponding to the effect size (z-transformed correlation coefficient) of 0.42. According to classification by Cohen (1988), our sample included 24.9% of small effects (≤0.20), 24.3% of medium effects (0.20–0.40) and 50.8% of large effects (>0.40).

Assuming that the effect size of 0.4 is sufficiently representative for correlations between the measures of pollution load and biotic variables, we conclude that the statistical power ($\alpha = 0.05$, one-tailed test) of each individual correlation analysis based on ten study sites was on average only 34%. In other words, two thirds of the individual analyses committed the Type II statistical error, i.e., acceptance of the null hypothesis (variation of measured variable is independent from variation in pollution load), while in fact H1 (variation of measured variable is associated with variation in pollution load) was true.

The average statistical power of 40–47% (to detect a medium effect size) of tests published in behavioural journals was considered 'distressingly low' by Jennions and Møller (2003). The situation in pollution ecology is even worse. The median number of study sites in a random sample of papers that have used correlation analysis to detect the biotic effects of pollution was seven. Thus, on average only 25% of the published tests had a chance to detect an effect size of 0.4. Power of these published tests was only sufficient (80%) to detect very large effects ($|r| \geq 0.83$, corresponding to $z_r = 1.19$), which are extremely infrequent (5% of our sample).

The most straightforward solution to this problem is to increase sample sizes. To increase the statistical power of correlation analysis to the recommended level of 80% (to detect an effect size of 0.4), the number of study sites should be at least 36. It may be argued (Jennions & Møller 2003) that logistic, ethical, conservation, and financial constraints make this impossible: only 7.2% of 1,000 publications from our database reported data sampling from 38 or more study sites (Fig. 8.1). Moreover, since many researchers tend to design their studies by analogy with previously published work, it seems unlikely that the number of study sites in field studies addressing pollution effects on biota will increase rapidly.

The low statistical power of tests based on small samples demonstrates the need to use meta-analysis for generalization of the accumulated data. Although this need was recognised long ago (Armentano & Bennett 1992), little was done to fully interpret the literature and ascertain the likelihood of trends common to ecosystems and pollutant regimes (Ruotsalainen & Kozlov 2006; Zvereva et al. 2008; Zvereva & Kozlov 2009; Roitto et al. 2009). The experience of medical sciences demonstrated that meta-analysis of numerous small-scale studies (only few of which are likely to detect significant trends) may provide a more cost-effective way of assessing the value of a treatment than investing in a few large-scale studies (Song et al. 2000).

8.4.4 *Correlation with Pollution or Correlation with Distance?*

It has been frequently argued that comparisons of the results from the literature are complicated by varying methods and objectives (Armentano & Bennett 1992; Glasziou & Sanders 2002). One of specific sources of variation in methodology of pollution-oriented studies is the choice of explanatory variables for correlation analyses. Authors of primary studies correlate their results with a variety of parameters, including distance from the polluter and concentrations of different pollutants in ambient air, snow, soils, or different organisms. Some authors constructs so-called 'toxicity indices', usually computed as linear combinations of absolute or relative concentrations of several pollutants (Rühling & Tyler 1973; Vorobeichik et al.

1994; Mumtaz et al. 1997; Mowat & Bundy 2002; Simonsen et al. 2008). Consequently, the validity of a meta-analysis that combines studies based on correlations with different explanatory variables can potentially be questioned.

Averaging the absolute values of Pearson's linear coefficients across the entire study (Chapters 3–7) demonstrated that correlations with both log-transformed distance from the polluter and foliar concentration of one of the pollutants yielded apparently the same overall conclusions (z-transformed values: $F_{1,\ 1444} = 0.00$, $P = 0.95$). This is not surprising, since concentrations of all pollutants in all media (ambient air, soils, plant and animal tissues) decrease proportionally with increasing distance from the smelter (Freedman & Hutchinson 1980a; Barkan 1993; Ruohomäki et al. 1996; Kozlov 2005a). Thus, the selection of the measure of the pollution load depends primarily on the study goals. If the study aims to demonstrate the effect, then distance from the polluter may be the best proxy of pollution load. Even in medical studies, population exposure to pollution is often estimated on the basis of distance to pollution source (Gottlieb et al. 1982; Biggeri et al. 1996; Monge-Corella et al. 2008). Of course there are numerous exceptions related to local orography and meteorology (Vorobeichik et al. 1994; Kozlov et al. 1995), which may be critically important for case studies, but are of relatively little value for meta-analyses. Moreover, exploration of dose–response relationships requires the use of pollutant concentration (usually log-transformed) instead of distance.

Emissions of any polluter consist of dozens of substances, many of which are toxic. Since it is impossible to attribute the effects observed in the course of field studies to any of the individual pollutants, the best solution is to correlate the results with one of the 'main' pollutants, which may serve as an indicator of pollution load. Concentrations of metals, fluorine, or sulphur dioxide have been the most frequently used indicators in published primary studies. Importantly, estimates of pollutant concentrations in both ambient air and plant foliage show substantial temporal variation (Vorobeichik et al. 1994; Kozlov et al. 1995; Kozlov 2005a). Therefore, we recommend using the concentrations of individual pollutants (but not the toxicity indices) in media that accumulate pollutants during months (snow) or decades (soils and litter).

8.4.5 Gradient Approach or the Planned Contrast?

Keeping in mind that the median number of study sites in a random sample of 1,000 published primary studies was five (Section 8.4.1), we checked whether a meta-analysis based on the contrast between the two most polluted and the two least polluted study sites (Hedge's d effect size) yielded the same conclusion as a meta-analysis based on z-transformed correlation coefficients calculated from ten study sites.

Three methods of data analysis (correlation with distance, correlation with pollution, and contrast between two most and two least polluted sites) after appropriate transformations produced similar effect sizes in all individual analyses (reported in Chapters 3–7) and in a pooled data set ($F_{2,\ 1824} = 1.36$, $P = 0.26$). However, this result should not be interpreted as an excuse for using low number of study sites: although the average effect size was not affected, variance of each individual effect based on the contrast between the two most and two least polluted sites was on average seven times higher than for effect sizes based on ten study sites, thus affecting the significance of the overall effect and hampering identification of the parameters explaining variation in effect sizes.

8.4.6 Importance of Supplementary Information

Information on the polluters, as well as on their environmental impacts, was poorly reported in most of the primary studies used in meta-analyses of the effects caused by industrial pollution (Ruotsalainen & Kozlov 2006; Zvereva et al. 2008; Zvereva & Kozlov 2009; Roitto et al. 2009). As a result, the values of many explanatory variables necessary for our analyses had to be found in additional publications, on the Internet and by personal contacts with authors, regional authorities or company representatives. This was the most difficult and time-consuming part of data collection; still, in some cases we were unable to obtain the data of critical importance, which affected the number of studies involved in individual analyses.

Integrating our experiences over the course of this work, we suggest the *minimum minimorum* list of characteristics that need to be reported in each study addressing impact of industrial polluters on biota in order to allow its efficient use in subsequent comparative studies an\d meta-analyses.

Emission source
- General information
  - Name of the polluter
  - Type of the polluter (e.g., copper smelter, coal-burning power plant)
- Geographical information
  - Country, administrative region within the country
  - Position relative to large settlement(s)
  - Latitude, longitude, and altitude above sea level
- Historical information
  - The year of establishment (beginning of pollution impact)
- Chemical information
  - List of most important pollutants
  - Amount of emissions during the study year(s)

Impact zone
- Geographical information
  - Presence of other point polluters with similar or larger environmental impacts within the impact zone of the polluter under study
- Landscape information
  - Biome
  - Landscape characteristics (plain, hilly, river valley, rocky mountain slopes)

Ecological information
- Pre-industrial vegetation (type, physiognomy)
- Soil type and basic characteristics (including pre-industrial pH of topsoil)

Historical information
- Brief history of environmental impact
- References to most important studies reporting pollution loads and environmental effects

Chemical information
- Spatial pattern of pollutant distribution (maximum concentrations, size and shape of the contaminated area)

8.5 Research and Publication Biases

Bias is a term used to describe a tendency or preference towards a particular perspective, ideology or result, especially when the tendency interferes with the ability to be impartial, unprejudiced, or objective. Investigation of biases and prevention of their impacts on general conclusions is of specific importance for meta-analyses (Begg 1994; Palmer 1999; Murtaugh 2002; Leimu & Koricheva 2004; Delgado-Rodríguez 2006; Formann 2008).

The accuracy of conclusions made on the basis of meta-analyses may suffer from the research bias - the tendency to perform experiments on organisms or under conditions in which one has a reasonable expectation of detecting statistically significant effects (Gurevitch & Hedges 1999). For example, studies of herbivore populations are likely to start in the year and on the site when herbivore damage is apparent (Zvereva & Kozlov 2009). Similarly, studies of pollution impact on vegetation are generally initiated following the appearance of visible damage (Anderson 1966; Linzon 1966; Murtha 1972; Tikkanen & Niemelä 1995) and are often confined to sites where the damage is most prominent. Some of these impact studies were even lacking controls (Wong 1978; Banásová et al. 1987). On the other hand, the decline of forests near point polluters resulted in the exclusion of barren areas from forestry-oriented projects, reports of which (Wotton & Hogan 1981; Scale 1982) therefore provided no information on the consequences of the most severe impacts on vegetation. The non-random selection of a pair of 'typical' study sites (one polluted vs. one control site) tends to overestimate the effect through a bias towards intuitive results (Zvereva & Kozlov 2009). Last but not least, the majority of pollution studies originated from Europe and, to a lesser extent, from North America, while other regions remain almost unexplored (Ruotsalainen & Kozlov 2006; Zvereva et al. 2008; Zvereva & Kozlov 2009; Roitto et al. 2009). Especially critical for the generalization of the results is the absence of information on subtropical and tropical regions, housing the largest part of terrestrial biodiversity.

If the probability of publication depends on factors other than the quality of the research, then we may face the problem of publication bias. Studies that disagree with the prevailing trend will not be published, especially when they are based on relatively small samples (Light & Pillemer 1984; Begg 1994), or their publication will be delayed (Møller & Jennions 2001), or they will appear in less visible journals (Leimu & Koricheva 2004). Preferential publication of studies with significant outcomes that confirm the general paradigm may result in a correlation between effect size and sample size. In our meta-analyses of the published results, effect sizes were generally independent of sample size (Ruotsalainen & Kozlov 2006; Zvereva et al. 2008; Zvereva & Kozlov 2009), except for studies on plant fluctuating asymmetry (Section 5.4.3). We also found bias in reporting the data when quantitative information was provided only for species fitting the research hypothesis (density increase), while species that did not show density changes were only briefly mentioned (Selikhovkin 1986; Shelukho 2002).

Due to the sensitivity of the research topic, some studies may remain unpublished due to efforts by industry to protect itself from unwanted examination of its impact on the environment and surrounding population (discussed by Moffatt et al. 2000). The best known example is the governmental policy of the former USSR, where publication of 'negative' information was prohibited for decades (Komarov 1978). We are aware of several recent examples, based on financial rather than legal restrictions. The experience of epidemiological research showed that opposite situations may also well exist, when negative or inconclusive evidence is likely to be discounted (Balshem 1991; Moffatt et al. 2000). This source of dissemination bias deserves special investigation.

The studies included in ISI databases reported stronger negative effects of pollution on diversity of vascular plants (Zvereva et al. 2008) and abundance of herbivorous insects (Zvereva & Kozlov 2009) but weaker effects on plant growth and reproduction (Roitto et al. 2009) than other publications. The discovery of this bias emphasises the need to account for studies published in less visible data sources (book chapters, conference proceedings, journals published in national languages, and 'grey' literature) in order to obtain less biased estimates of effect size in meta-analytical research syntheses (Murtaugh 2002).

The temporal trend in the magnitude of the reported effects is a general phenomenon in ecology (Jennions & Møller 2002; Leimu & Koricheva 2004). We detected a decrease in the estimates of pollution effects on herbivore population density from 1965–1989 to 1990–2008 (Zvereva & Kozlov 2009), which we tentatively attribute to the improvement in research methodology. The hypothesis on the increase in herbivory under pollution impact came through developmental stages that are typical for any ecological hypothesis (Leimu & Koricheva 2004): from supportive evidence of the newly formulated hypothesis to the accumulation of disconfirming evidence. This accumulation of non-supporting evidence during the period of most intensive studies in the 1990s resulted in a shift from highly significant positive effects to non-significant effects and caused doubts about the generality of this phenomenon. Meta-analysis of published data contributes to the third stage in the development of the hypothesis - the restriction of its scope by discovering sources of variation in herbivore responses to pollution (Zvereva & Kozlov 2009).

Comparison of original data (Chapters 3–7) with the outcomes of earlier studies (Table 8.1) demonstrated that published data generally overestimate (by a factor of 5 on average) the magnitude of the effects of industrial pollution on terrestrial biota.

Table 8.1 Comparison between mean effects sizes based on published studies and on the original data

	Published data			Original data		
Character	*r*	*N*	Reference	*r*	*N*	Reference
Leaf/needle size	−0.48	204	Roitto et al. 2009	−0.11	88	Fig. 4.7
Shoot length	−0.47	164	Roitto et al. 2009	−0.14	111	Fig. 4.13
Radial increment	−0.59	40	Roitto et al. 2009	−0.35	10	Fig. 4.19
Plant fluctuating asymmetry	0.31	25	Section 5.4.3, Table 5.1	0.00	61	Fig. 5.5
Plant diversity	−0.64	45	Zvereva et al. 2008	−0.33	12	Fig. 6.9
Density of insect herbivores	0.44	142	Zvereva & Kozlov 2009	−0.06	45	Fig. 7.1
Plant damage by insect herbivores	0.64	17	Zvereva & Kozlov 2009	−0.23	59	Fig. 7.4

All effect sizes were converted to correlation coefficients.

Furthermore, for both density of insect herbivores and plant damage by these insects, the original data indicated either no effect or a slight decrease with pollution, while published data reported strong increase with pollution (Table 8.1). Of course, it can be argued that we collected our data in the 2000s, when emissions from industrial enterprises were generally lower than in previous decades. However, we think that an overall decline of emissions explains only a minor part of the detected differences between the published and original data. A substantial part of the effects observed near the point polluters is due to a large pool of pollutants (metals and fluorine-containing substances) accumulated in soils (Haidouti et al. 1993; Lyubashevsky et al. 1996; Giller et al. 1998; Nahmani & Lavelle 2002; Kozlov & Zvereva 2007a). Natural leaching of these pollutants will continue for decades or centuries before they approach pre-industrial levels (Tyler 1978; Barcan 2002a), and only the very first signs of vegetation recovery were observed around some of the polluters explored by our team (Eeva & Lehikoinen 2000; Chernenkova et al. 2001). Therefore, we tend to attribute the differences detected between the effect sizes calculated from published and original data (Table 8.1) not to emission decline but to the biases discussed above.

Since we revealed a marked diversity of pollution effects on terrestrial biota, we strongly encourage researchers and editors to publish results that are unexpected or seem strange. The occasional observation of truly surprising phenomena is the norm in ecology, not the exception (Doak et al. 2008). Elimination of these results at the pre-publication stage (decision not to submit the manuscript) or by reviewers (frequently due to disagreement with the prevailing paradigm) is likely to bias our estimates of overall effects, sometimes leading to the dominance of incorrect or exaggerated opinions (partially discussed in Section 9.2). Even more importantly, these ‘outliers’ are critically important for the exploration of the sources of variation in responses of organisms and ecosystems to pollution.

8.6 Summary

Further accumulation of reliable observational data remains of critical importance for the development of pollution ecology. Results of short-term experiments with non-adapted organisms in over-simplified laboratory environments should be interpreted cautiously, since they are likely to overestimate adverse effects. Proper replication of sampling at all hierarchical levels, selection of adequate temporal and spatial scales, and careful use of statistics are the key factors assuring the quality of information on the effects of pollution on terrestrial biota. This information, summarised by meta-analyses or other appropriate procedures, is necessary for exploratory research followed by the generation of specific hypotheses, which can be tested using field and laboratory experiments. In research syntheses, the utmost care should be taken to recognise biases affecting the outcomes of individual studies and mitigate their impacts on our knowledge.

Chapter 9
Effects of Industrial Polluters: General Patterns and Sources of Variation

9.1 The State of Knowledge

We identified about 2,500 data sources that report impacts of point polluters on terrestrial biota and fit criteria listed in Section 5.1.2. We think that our collection is sufficiently representative and covers at least a half of existing publications. Although we have started a systematic review of these publications, we estimate that the completed meta-analyses (Ruotsalainen & Kozlov 2006; Zvereva et al. 2008; Zvereva & Kozlov 2009; Roitto et al. 2009) still cover less than 25% of the available information. The aim of Section 9.1 is to briefly evaluate knowledge of changes of landscapes, ecosystems, communities, populations and individual organisms in impact zones of industrial enterprises in order to reveal critical shortcomings and data gaps.

9.1.1 Landscape-Level Effects

The emissions of point polluters can act in such a way that the consequences of pollution-induced changes become the dominant feature of the landscape (Bohne 1971; Broto et al. 2007). However, integrated studies of polluted landscapes are infrequent (Gilbert 1975; Doncheva 1979; Kozlov & Zvereva 2007a), and they only rarely include the societal dimension (Broto et al. 2007). Investigation of pollution impact on biota requires a reduction of scale that may not be appropriate for the study of landscapes; therefore, landscape-level effects were not explored in the course of our project. However, these effects are of critical importance for understanding the indirect impacts of industrial pollution on biota, and we would like to stress the need for exploration of specific landscapes that have evolved around point polluters.

Landscape-level effects are easily recognisable around many of polluters considered in this book. Big polluters dominate both land and sky, especially when puffs of smoke make them visible from large distances (please see color plates 18, 50, 61, 64 and 65 in Appendix II), and are often surrounded by extensive areas of disturbed

M.V. Kozlov et al. *Impacts of Point Polluters on Terrestrial Biota*,
DOI 10.1007/978-90-481-2467-1_9, © Springer Science+Business Media B.V. 2009

land (please see color plates 20–23, 32–38, 45–47, 53 and 54 in Appendix II). The visual perception of this disturbance is, in the first line, linked with vegetation damage. Forest death (please see color plates 32 and 33 in Appendix II) is often seen as the most important change, possibly also due to the symbolic connotations of the forest. This phenomenon attracted emotional public attention: 'Bushes shrank and vanished. Grasses died away. Blighted land replaced the forest. All around us dead hills, red, raw, ribbed by erosion, stood stark in the sunshine. Hardly two miles from dense woodland, we were in the midst of a moonscape on earth.... We were in the southeast corner of Tennessee, in the Ducktown Desert of the Copper Basin' (Teale 1951). On the other hand, the visual appearance of vegetation in heavily polluted treeless habitats (e.g., tundra) does not differ from unpolluted sites (please compare color plates 66 and 67 in Appendix II) and therefore causes less public awareness. Last but not least, low levels of pollution may cause no visible changes (please compare color plates 56 and 57 in Appendix II), or slight effects of pollution may be masked by consequences of other, stronger impacts associated with human settlements, such as cutting of timber, littering, and recreational activities (please compare color plates 2 and 3, 16 and 17, 40 and 41 in Appendix II).

Extreme depositions of airborne pollutants, especially of sulphur dioxide accompanied by heavy metals, resulted in the appearance of industrial barrens - bleak open landscapes, with only small patches of vegetation surrounded by bare land (please see color plates 9, 21, 29, 34–36, 54 and 59 in Appendix II). These landscapes appeared as a by-product of human activities about a century ago and were first reported in the North America. Recently, we collected information on about 40 industrial barrens worldwide (Kozlov & Zvereva 2007a, and unpublished). Among polluters, the effects of which are documented in the present book, industrial barrens occur around non-ferrous smelters at Harjavalta, Karabash, Krompachy, Monchegorsk, Norilsk, Nikel, Revda, and Sudbury. The extent of industrial barrens varies from few hectares (Harjavalta, Krompachy) to several hundreds of square kilometres (Norilsk, Monchegorsk). In spite of a general reduction in biodiversity, industrial barrens still support a variety of life, including regionally rare and endangered species, as well as populations that evolved specific adaptations to the harsh and toxic environments (Kozlov & Zvereva 2007a).

Pollution often changes the visual perception of landscapes, making them unattractive - especially when dead trees, which remain standing for decades (please see color plates 32, 33, 47 and 49 in Appendix II), cut the skyline (Gilbert 1975). However, standing trees, even when they are dead, maintain climatic and biotic stability of contaminated habitats, in particular by ameliorating microclimate (Wołk 1977) and preventing soil erosion. The importance of standing trees was proven experimentally: the introduction of dead birches to tundra created favourable microclimate for field layer vegetation in spring and early summer (Molau 2003). Cutting of dead trees has been performed in many severely polluted areas (Gilbert 1975; Kozlov & Zvereva 2007a); this action improved visual attractiveness of landscapes but accelerated secondary damage. The old clearcuts under severe pollution impact near both Monchegorsk and Nikel have been rapidly transformed to industrial barrens, while in the adjacent uncut areas some vegetation is still alive (Kozlov 2004).

9.1.2 Ecosystem-Level Effects

Ecosystem-level processes form a focal point in the development of the theoretical basis of both 'basic' and 'applied' ecology. Appreciation of the value of ecosystem services for human well-being (Costanza et al. 1997) further stresses the need to explore basic principles governing ecosystem development in a changing world (Kremen & Ostfeld 2005; Mokany et al. 2006; Grimm et al. 2008). However, exploration of ecosystem-level effects is a difficult task, and therefore some predictions of ecosystem changes due pollution impacts (Newman & Schreiber 1984; Odum 1985; Rapport et al. 1985) remain too general to derive testable hypotheses.

A decrease in net primary productivity is considered one of most important ecosystem-level responses to different kinds of disturbances (Odum 1985; Rapport et al. 1985; Sigal & Suter 1987; Treshow & Anderson 1989; Barker & Tingey 1992; Armentano & Bennett 1992; Newman & Schreiber 1984; Bezel 2006). Since biomass or standing crop is a primary variable measured for determining productivity (Sigal & Suter 1987), pollution effects on productivity can be deducted from overall decreases in stand basal area and stand height, as well as from decreases in cover of field layer vegetation and mosses found in our study (Chapter 6). Combined with reduction of shoot length, radial increment, and root growth (Roitto et al. 2009; Chapter 4), these data suggest that a decrease in net primary productivity with pollution is likely to be a general ecosystem-level phenomenon.

Decreased productivity is a pre-requisite of changes in nutrient cycling. Decreases in both basal area and stand height as well as in cover of field layer vegetation (Chapter 6) are indicative of lower litter input and, consequently, lower forest floor thickness (Keane 2008; Merino et al. 2008). A decrease in topsoil depth around non-ferrous smelters (Chapter 3) is consistent with this conclusion. On the other hand, pollution adversely affects soil microbiota (Bååth 1989; Fritze 1992; Ruotsalainen & Kozlov 2006; Chapter 3) and detritophagous arthropods (Zvereva & Kozlov 2009), substantially decreasing the biological decomposition rate and breakdown of the litter (Rühling & Tyler 1973; Prescott & Parkinson 1985; Zwolinski 1994; McEnroe & Helmisaari 2001).

The decreases in both production and decomposition of organic matter are so far the only ecosystem-level responses to pollution, the generality of which is supported by the available information. These effects are in line with the prediction (Odum 1985) that polluted ecosystems become more open: reduction of internal nutrient cycling increases the importance of both input and output environments.

9.1.3 Community-Level Effects

An excellent book by Clements and Newman (2002) provides a broad perspective for bridging ecological and toxicological approaches in the study of community-level effects of pollution on biota. One of the most important conclusions made by these authors concerns the need to pay more attention to the indirect effects of contaminants, mediated by changes in species interactions, such as competition, facilitation, predation and mutualism.

Species diversity is the most frequently measured community parameter (Sigal & Suter 1987). However, the results of case studies are often over-generalized, and therefore they are discussed in Section 9.2.2. Insufficient knowledge of pollution impacts on many groups of terrestrial biota further restricts discussion to changes in vegetation structure and in arthropod communities.

Armentano and Bennett (1992) listed the following responses of plant communities to chronic air pollution: reduced photosynthesis, reduced labile carbohydrate pool, reduced growth of root tips and new leaves, decreased leaf area, differences in species growth performance, reduced community canopy cover, reduced reproductive capacity, shifts in interspecific competitive advantage, alteration of community composition, change in species diversity, and change in community structure (physiognomy). However, our data provided no support for most of these predictions. In particular, overall effect of pollution on leaf/needle size and shoot length of woody plants did not differ from zero (Figs. 4.8 and 4.14). Individual species of woody plants similarly responded to pollution in terms of both leaf/needle size and shoot growth (Figs. 4.10 and 4.16). Canopy cover also did not change with pollution (Fig. 6.1), and no changes in either species richness of vascular plants (Fig. 6.9) or physiognomy of vegetation were observed around many of the studied polluters (please compare color plates 2 and 3, 16 and 17, 40 and 41, 56 and 57, 66 and 67 in Appendix II).

The differential responses of some taxonomic and functional groups of both plants and arthropods to pollution (Zvereva et al. 2008; Zvereva & Kozlov 2009; Chapters 4–5) support the opinion that industrial pollutants affect the trophic structure of ecosystems (Sigal & Suter 1987). In particular, pollution generally imposes larger detrimental effects on predators than on herbivorous arthropods, thus creating so-called enemy-free spaces for herbivores. This phenomenon was documented by several case studies (Zvereva & Kozlov 2000a, b) and confirmed by meta-analysis of published data (Zvereva & Kozlov 2009). At the same time, the available information remains insufficient to support the hypothesis on simplification of ecosystem structure (Odum 1985; Rapport et al. 1985; Freedman 1989; Armentano & Bennett 1992) or on the 'loss of ecosystem integrity' (Barker & Tingey 1992) under pollution impact.

Another frequently repeated statement (Newman & Schreiber 1984; Bezel 2006) is an overall decrease in the size of organisms with pollution. This hypothesis was suggested by Odum (1985), who referred to mesocosm experiments with plankton (that were presumably unpublished at that time), and by Rapport et al. (1985), who based his generalizations on observational studies of vegetation decline due to pollution (Gordon & Gorham 1963) and irradiation (Woodwell 1967). Surprisingly, we found almost no attempts to verify this hypothesis (but see Braun et al. 2004), and therefore its validity and generality remain unknown.

Thus, although some responses to industrial pollutants are common for the communities explored so far, natural variation, rapid development of adaptations and intrinsic community differences complicate any generalization. Importantly, our inability to reveal consistent and significant effect of pollution on most of the explored community characteristics results not from the absence of effects, but from the diversity in responses often yielding a zero overall effect.

9.1.4 Population-Level Effects

Changes in both population structure and some attributes of population dynamics are reported for a number of species. However, only a few species were explored in sufficient detail, and only a limited number of parameters were studied in more than a few species. This naturally limits our understanding of changes in population behaviour under pollution impacts.

Pollution influences the genetic structure of populations, often leading to the development of pollution tolerance. These effects, referred as micro-evolution due to pollution (Medina et al. 2007), are best documented in microbiota; changes in the genetic structure of plant populations were reported more frequently than in animal (mostly arthropod) populations. The heavy metal tolerance of plants, especially grasses, has been studied extensively and provides a well-documented example of rapid evolutionary adaptation (Bradshaw & McNeilly 1981; Shaw 1990; Macnair 1997). Even long-lived trees may develop pollution resistance to both sulphur dioxide and heavy metals (Rachwal & Wit-Rzepka 1989; Turner 1994; Utriainen et al. 1997; Eränen 2008), presumably through survival selection (Kozlov 2005b).

Micro-evolution due to pollution is likely to be more common than has been documented (Barker & Tingey 1992; Kozlov & Zvereva 2007b; Eränen 2008). Therefore, a currently identified (Medina et al. 2007) gap between studies addressing changes in the genetic structure of populations and those assessing effects at higher levels of biological organisation should be overcome to improve understanding of ecosystem responses to pollution. Bridging molecular and community-level studies may revitalise exploration of phenotypic differentiation of populations living in contaminated environments (Kozlov 1990; Newman et al. 1992; Zvereva et al. 2002).

Surprisingly, we have not found any study documenting changes in the spatial structure of plant populations. For animals, fragmentary data exist on insect herbivores (Kozlov 1990, 2003) and small mammals (Mukhacheva 2007). The only conclusion that can be made at the moment is that pollution may change small-scale distribution patterns of organisms, although the mechanisms behind these changes remain nearly unknown.

Studies addressing the pollution impact on population demography conclude that contaminated populations usually differ from control populations in age structure (Zverev et al. 2008). The impacted populations may be both younger (Kucken et al. 1994; Zhuikova et al. 2001) and older (Deyeva & Maznaya 1993; Dmowski et al. 1998; Zverev et al. 2008; Zverev 2009) than the control ones. Since morphological and physiological parameters of plants and animals generally change with age, some of the differences between populations from contaminated and control sites result from changes in age structure rather than from direct effects of pollutants on individual fitness (Zverev et al. 2008).

Even less is known on sex ratio in the affected populations. Although the medical literature suggests that industrial pollution may change sex ratios (Williams et al. 1992; Lloyd et al. 1984, 1985), some studies found no convincing evidence of this effect (Williams et al. 1995; Kozlov 1999). Isolated observations on animal populations in contaminated areas also produced diverse results. The proportion of females of two insect species increased in surroundings of chemical factories (Birg 1989; Chumakov 1989). Male local survival of the pied flycatcher (*Ficedula hypoleuca*) was higher near the non-ferrous smelters at Harjavalta relative to unpolluted habitats (Eeva et al. 2006b), potentially resulting in sex ratio changes along pollution gradients. Finally, no links between pollution load and sex ratio was found in populations of two vole species (*Clethrionomys rufocanus* and *C. glareolus*) near non-ferrous smelters located at Monchegorsk and Revda, respectively (Kataev 1984; Vorobeichik et al. 1994). Variation in sex ratio in soil nematode communities around the Almalyk industrial complex, reported by Pen-Mouratov et al. (2008), was independent of copper concentration in soils of the study sites.

Although abundance is one of the most frequently measured population characteristics, pollution effects on long-term density fluctuations have only rarely been studied (Zvereva et al. 1997b, 2003; Kozlov 2003; Zvereva & Kozlov 2004). First and most importantly, the majority of research projects last 1 to 3 years, thus yielding estimates of population density rather than monitoring population dynamics. So far, only a few species or species groups have been monitored along a pollution gradient uninterruptedly for 10 or more years. Furthermore, pollution effects on a suite of population parameters driving population dynamics, such as death rate, life expectancy, survival and fecundity, are properly documented for a very few species of terrestrial biota.

Especially surprising is absence of information on wildlife mortality in pollution gradients, possibly because direct measurements of death rates seem nearly impossible. Although observations on mortality of cattle and wildlife attributable to air pollution have repeatedly been summarised (Newman 1979; Newman et al. 1992), they almost exclusively concern extreme levels of exposure or catastrophic local events. So far, only indirect data, derived from measurements of abundance of necrophagous beetles, suggest that mortality of vertebrates increases near point polluters (Kozlov et al. 2005a). It seems that modelling of pollution effects on population dynamics (Kidd 1991; Dubey 2004, and references therein) is developed much better than validation of these models with observational data.

Review of available knowledge allowed us to identify a few species for which the amount of available information is sufficient or nearly sufficient to approach understanding of pollution effects on population dynamics. The list includes white birch, *Betula pubescens*; Scots pine, *Pinus sylvestris*; bilberry, *Vaccinium myrtillus*; willow feeding leaf beetle, *Chrysomela lapponica*; Great tit, *Parus major*; pied flycatcher, *Ficedula hypoleuca*; grey-sided vole, *Clethrionomys rufocanus*; and bank vole, *C. glareolus*.

Importantly, comparison of population abundances between polluted and unpolluted study sites can provide little information about pollution effects on population dynamics. The same species can demonstrate different patterns of density changes not only around different polluters, but also in different years around the same polluter. For example, the population density of the leaf beetle, *Chrysomela lapponica*, around a nickel-copper smelter at Monchegorsk demonstrated a significant dome-shaped pattern in 1994 (Zvereva et al. 1995), but not in 1998, when it gradually decreased from the most to the least most polluted sites (Zvereva et al. 2002) nor in 2006–2007 when the density of this species did not change with distance from the polluter (E.L.Z. & M.V.K., personal observation, 2007). Similar variation in density patterns along the Monchegorsk pollution gradient was observed in *Eriocrania* leafminers during 1994–2005 (Zvereva & Kozlov 2006b). Several other studies reporting multiyear density data (Selikhovkin 1995, 1996; Kozlov 2003) also suggest that all kinds of pollution-effect relationships can be observed for the same species during different study years, as a result of asynchronous density fluctuations in spatially isolated populations (Selikhovkin 1995; Zvereva et al. 2002; Kozlov 2003; Zvereva & Kozlov 2006b). Therefore, classification of species on the basis of density changes along pollution gradients, widely accepted in narrative reviews on insect responses to pollution (Alstad et al. 1982; Führer 1985; Kozlov 1990; Heliövaara & Väisänen 1993), is hardly reasonable. However, this does not mean that further accumulation of short-term data on population abundances along pollution gradients is useless. Although short-term observations are of limited value for understanding pollution impacts on the population dynamics of individual species, meta-analysis of these observations can yield important conclusions on overall effects of pollution on population densities of different taxonomic or functional groups of biota.

9.1.5 Organism-Level Effects

A substantial number of observational studies conducted around point polluters focused on responses of individual organisms. Diversity of both organisms and approaches hampers generalization of these data. However, the critical problem, identified two decades ago (Sigal & Suter 1987), is a common absence of linkages between the recorded individual responses and processes occurring in the affected populations and communities.

A number of publications reported increases in the visible injury of plants with pollution (Jacobson & Hill 1970; Sigal & Suter 1987; Treshow & Anderson 1989; Flagler 1998). However, this field of knowledge, the oldest one in the scientific documentation of the effects caused by pollution, has mostly developed in isolation from studies addressing changes in plant growth and reproduction. A variety of symptoms, subjectivity in estimation of damage, common absence of statistical analysis, especially in older publications, and absence of links between visible injury and changes in fitness made interpretation of the accumulated data rather vague (Percy & Ferretti 2004). This also concerns various indices of tree health used mostly by Russian scientists (Alexeyev 1995) and visual estimates of crown defoliation and discoloration, serving as the basis of large-scale forest monitoring program (UN-ECE 2006). Although all these indices are potentially useful for monitoring purposes, we need to stress the low value of publications based on various estimations of visible injury for the development of pollution ecology.

Body size or weight is the only organism-level characteristic that was reported in a sufficient number of studies. Meta-analysis of data on terrestrial arthropods demonstrated that body size significantly decreased with pollution (Zvereva & Kozlov 2009). Observations on birds and mammals (Bezel 2006; Veličković 2007a), although not yet summarised, seem to show a similar trend. Thus, a decrease in body size is likely to be a general regularity for animals.

9.2 Myths of Pollution Ecology

Profound concern for environmental quality, amplified by media attention, resulted in both diminution and exaggeration of particular problems by both representatives of industries and environmental activists. These processes not only biased the ideas of the general public on effects caused by pollution, but also unavoidably influenced scientists. Prejudgement and misinterpretation presumably imposed even stronger adverse impacts on the development of pollution ecology than the methodological shortcomings discussed in Section 8.4. In short, there is no doubt that industrial pollution was the primary cause of extreme environmental deterioration in many areas (Gordon & Gorham 1963; Wood & Nash 1976; Freedman & Hutchinson 1980b; Kozlov & Zvereva 2007a), but these severe effects are neither general nor widespread. Only some authors (Barker & Tingey 1992; Clements & Newman 2002) appreciate unequivocally that the consequences of air pollution to biota and the resulting impact on ecosystem structure and functions (services) are not clearly known. We strongly support this opinion.

9.2.1 Generality of Responses

It has been repeatedly claimed that there are commonly observed patterns of ecosystem responses to stress (Woodwell 1970; Lugo 1978; Auerbach 1981; Odum 1985; Rapport et al. 1985; Freedman 1989), while variation in responses is rarely appreciated (Rapport & Whitford 1999; Clements & Newman 2002). This approach results in an exaggerated impression of both the generality and the uniformity of the effects caused by pollution.

A random sample of 100 publications from our data base (including 50 papers in Russian and 50 papers by native English speakers) demonstrated that the investigated point polluter was explicitly named (or associated with a certain town) in the titles of only ten publications. Titles of another 16 publications, although not pointing out at the polluter, identified the geographical region where the impact zone was located. Thus, titles of three quarters of publications reported just 'pollution impact on…' or something similar, unconsciously assuming that the effects are general and uniform across both polluters and impacted (stressed) communities.

'Stress' is used loosely by ecologists to describe the relative (deviating from the presumed optimum) circumstances affecting species and communities (Lortie & Callaway 2006). As a result, 'low-stress' and 'high-stress' environments are often labelled arbitrarily, on the basis of the researcher's assumptions about which conditions are favourable or unfavourable for the organisms or communities under study. However, whether these organisms or communities really experience stress in 'high-stress' environments is only rarely questioned. Instead, any differences between 'stressful' and 'benign' environments are routinely labelled as 'stress responses'. Generalization of these 'stress responses' is likely to produce misleading results, especially when the differences between the 'low-stress' and 'high-stress' environments are not accounted for (Lortie & Callaway 2006).

Grime (1979) defined stress in terms of productivity. Accordingly, stressful environments are defined as those in which producers are limited in their ability to convert energy to biomass. This approach is commonly used by ecologists to examine stress across communities (Underwood 1989; Goldberg et al. 1999; Parker et al. 1999; Kammer & Mohl 2002; Lortie & Callaway 2006). Thus, an overall decrease in productivity (Section 9.1.2) can be seen as a proof of ecosystem-level stress caused by industrial pollution. However, this does not necessarily mean that all organisms and populations experience stress under pollution impact. What is stressful for one species may be optimal for another species, or even for another population of the same species, because of differences in adaptation to particular environments (Lortie et al. 2004).

Pollution can sometimes even improve environmental conditions for certain species relative to the benign (unpolluted) habitats. For example, the absence of competitors resulted in faster growth of birch seedlings on metal-contaminated soils relative to unpolluted forests with a dense cover of field layer vegetation (Eränen & Kozlov 2009). Similarly, decreased pressure from natural enemies in polluted habitats and sometimes also improvements in host plant quality favoured density increase of some herbivorous insects, in spite of adverse effects of pollutants on the individual performance of these insects (Kozlov et al. 1996b; Zvereva & Kozlov 2000a, b, 2009).

We conclude that a widespread opinion on the existence of a general pattern of ecosystem responses to pollution results mostly from earlier generalizations based on the few most impressive case studies that were available that time (Haywood 1907; National Research Council of Canada 1939; Gorham & Gordon 1960a, b; Gordon & Gorham 1963; Hutchinson & Whitby 1974; Wood & Nash 1976; Freedman & Hutchinson 1980a, b; Amiro & Courtin 1981). We suggest that efforts should be directed primarily at exploring sources of variation in responses to pollution, rather than at searching for the presumed general effects.

9.2.2 Uniformity of Responses

The majority of the existing reviews claim that pollution effects on organisms, populations, communities and ecosystems are generally detrimental (Alstad et al. 1982; Newman & Schreiber 1984; Sigal & Suter 1987; Bååth 1989; Riemer & Whittaker 1989; Treshow & Anderson 1989; Barker & Tingey 1992; Fritze 1992; Heliövaara & Väisänen 1993; Rusek & Marshall 2000; Bezel 2006). Even if some organisms benefit from pollution, this effect is usually interpreted as the sign of an overall decrease in ecosystem vitality or integrity (e.g., pests benefit from plant weakening by pollution and accelerate destruction of plant communities by imposing additional damage: Wentzel & Ohnesorge 1961; Carlson et al. 1977; Baltensweiler 1985; Führer 1985; Berger & Katzensteiner 1994). Although in many situations environmental pollution obviously harms organisms, communities, and ecosystems, this blackening perception is not always true. The adaptation potential of individual organisms and entire ecosystems should not be underestimated.

Decreases in net primary productivity (discussed in Section 9.1.2) and in diversity are most frequently mentioned among the general effects of pollution on ecosystem properties (Odum 1985; Rapport et al. 1985; Sigal & Suter 1987; Treshow & Anderson 1989; Barker & Tingey 1992; Armentano & Bennett 1992; Newman & Schreiber 1984; Bezel 2006). However, we are not aware of any study documenting ecosystem-level diversity along pollution gradients.

The pollution effects on diversity of different groups of biota are not uniform. While the diversity of soil microfungi (Ruotsalainen & Kozlov 2006), vascular plants (Zvereva et al. 2008; Chapter 6), and soil arthropods (Zvereva & Kozlov 2009) generally decreased with pollution, the species richness of insect herbivores increased (Zvereva & Kozlov 2009). Moreover, the reported adverse effects on diversity may appear overestimated, since the researchers generally did not account for pollution effects on species abundance (Zvereva & Kozlov 2009). As we demonstrated earlier, the confounding effect of density may change even the sign of the pollution effect on diversity (Kozlov 1997). Thus, both methodological problems and individualistic responses of different groups of biota make generalization of pollution effects on ecosystem-level diversity premature.

We discovered the diversity in responses of the studied components of terrestrial ecosystems by meta-analyses of both published (Ruotsalainen & Kozlov 2006; Zvereva et al. 2008; Zvereva & Kozlov 2009; Roitto et al. 2009) and original data (Chapters 3–7). We consider this discovery as one of the most interesting and important findings of our research. Identification of factors explaining this variation (partially discussed in Section 9.4) is crucial for understanding and predicting pollution-induced changes in affected ecosystems.

9.2.3 Causality of Responses

It seems that 'presumption of guilt' is commonly accepted by pollution ecologists, and the majority of changes observed around industrial polluters is unthinkingly attributed to the toxicity of pollutants. This is especially common for adverse effects, like the death of cattle, crop damage, or forest decline. Causality is discussed only rarely (Courtin 1994), mostly in situations when effects are somewhat unexpected, such as better plant growth near the polluter (Zvereva et al. 1997a), or absence of the pollutant at the locality when damage occurs (Tikkanen & Niemelä 1995).

Reviews and methodological papers addressing pollution impact on biota usually do not even mention the need to carefully explore the causes of the phenomena observed in polluted environments (Schindler 1987; Sigal & Suter 1987; Freedman 1989; Treshow & Anderson 1989; Barker & Tingey 1992; Bezel 2006) or briefly advertise experimental studies as a tool to demonstrate cause-and-effect relationships (Clements & Newman 2002). However, although the randomised experiment is often invoked as one method for unambiguously determining causality (Clements 2004), even that is questionable when outcomes are stochastic and we rely on statistical interpretation (Fabricius & De'ath 2004).

Pollution ecologists are not the first scientists to face the attribution of causality problem. The concept of causality has a long and complex history. Its everyday usage is often straightforward, but as a philosophical or scientific concept, its definition and use are often contentious (for more details and references consult Fabricius & De'ath 2004). One of the best known examples of rules for attributing the effect (disease) to its presumed cause (microbe) are the Koch's postulates (Koch 1880). A much closer example concerns epidemiology, which deals with issues of complexity comparable to the ecological and environmental sciences and has similar requirements of scientific rigour and a limited ability to use the experimental approach.

Epidemiologists have long ago developed criteria to pass from statistically confirmed association 'to a verdict of causation' (Hill 1965). These criteria include: strength and consistency of observed association, its specificity and temporality (a logical time sequence of events), existence of the relationship between the dose (the putative cause) and the response, coherence with known biological facts, plausibility, experimental support and analogy with similar, better known, situations (Hill 1965; Susser 1986; Fabricius & De'ath 2004). These criteria, with few modifications, were also suggested for studies addressing effects of pollutants on biota (Fox 1991). Importantly, none of the criteria are taken as indicative by themselves, but equally, none are seen as absolutely necessary to evaluate the causal significance of associations (Roth et al. 1982). The more criteria that are satisfied and the stronger the association, the more confidence we should have in our judgement that the association is causal (Fabricius & De'ath 2004).

While the use of these criteria in an individual case study may be difficult or even impossible (but see Schroeter et al. 1993), their applicability to generalized results of multiple studies casts no doubt. Formalising the conclusions of our review on industrial barrens (Kozlov & Zvereva 2007a), we attribute the development of these specific landscapes to the effects of aerial emissions of big non-ferrous smelters because:

1. Most of known industrial barrens (33 of 35) are associated with non-ferrous smelters (consistence and specificity criteria).
2. In all documented situations, industrial barrens developed after the beginning of smelting (temporality criterion).
3. The size of the barren area is roughly proportional to the amount of aerial emissions, and its extent is usually consistent with the extent of territory with excessive soil contamination by toxic metals (dose dependence criterion).
4. Adverse effects of sulphur dioxide and toxic metals on biota were demonstrated in a number of experimental studies (coherence criterion).
5. The phenomenological model describing the development of industrial barrens does not contradict any of basic regularities commonly accepted by ecologists (plausibility criterion).

However, while most barrens are associated with non-ferrous smelters, not every smelter is surrounded by industrial barrens. Thus, we need to restrict the scope of our conclusions, but a shortage of information prevents verification of the hypotheses (Kozlov & Zvereva 2007a) that the detected causal link is valid only for mountainous or hilly landscapes and/or regions with relatively harsh climate.

Meta-analysis was recently suggested as a tool to evaluate whether association between an increase in mortality and air pollution is strong enough to be interpreted as evidence for causality (Bellini et al. 2007). However, to our knowledge, there are no guidelines to judge the strength of evidence by the size of the observed effect, and thus the interpretation of the results remains subjective. In contrast, the significance of the overall effect can be seen as the proof of consistency of the observed association.

An overall effect was significant in nine of 25 meta-analyses of original data collected during the course of our project (Figs. 3.1–3.5, 4.1, 4.4, 4.8, 4.11, 4.14, 4.17, 4.19, 4.20, 5.5, 6.1–6.9, 7.1, 7.4). This can indeed be seen as a weak proof of causal association between pollution load and the changes of measured characteristics of terrestrial ecosystems. On the other hand, zero overall effect can signal that our meta-analyses are combining diverse, sometimes even opposite, effects. The search for sources of variation demonstrated significant effects of pollution in at least one subset of our data classified according to three categorical variables (classes of polluters, their geographical position and effects on soil pH) in 12 of 16 meta-analyses with zero overall effect. The heterogeneity was mostly due to differences between classes of polluters and, consequently, between groups of polluters causing soil acidification and alkalinisation. Thus, although association between the detected changes in terrestrial biota and ‘pollution in general’ was weak, the strength of association increased when we restrict the search of causal relationships to either non-ferrous smelters or aluminium plants.

Exploration of causality is often complicated by indirect (Section 8.2) or secondary effects of pollutants. Pollution frequently triggers secondary effects that may enhance the primary disturbance in a positive feedback fashion (Courtin 1994; Kozlov & Zvereva 2007a). For example, forest thinning results in increased injury from gaseous pollutants (Norokorpi & Frank 1993) and in higher wind speed (Kozlov 2002), which increases climatic stress both directly and by changes in snowpack structure and also imposes mechanical damage to plants (Kozlov 2001b). A thin and compact snow layer explains lower soil temperatures recorded in industrial barrens during the wintertime (Kozlov & Haukioja 1997). Changes in microclimate in combination with a pollution-induced decrease in the cold hardiness of conifers (Sutinen et al. 1996) increase the probability of death of extant trees from freezing injury. Accumulation of woody debris, in combination with proximity to large human settlements, increases the risk of occasional fires. The disappearance of trees increases albedo, and thus lowers April-June temperatures, creating less favourable conditions for plant growth (Molau 2003) and causing delays in phenology (Kozlov et al. 2007). In subarctic regions, changes in microclimate following deforestation are often so strong that they may hamper recovery of vegetation even in absence of additional stressors (Arseneault & Payette 1997; Vajda & Venäläinen 2005). Forest decline reduces recovery of topsoil (Chapter 3) and facilitates soil erosion (please see color plates 20, 29, 34, 46 and 54 in Appendix II), with adverse consequences for all components of biota.

The effects mentioned above are apparent only at the landscape scale, and therefore their importance is often underestimated: all changes observed around point polluters are routinely attributed to toxicity of pollutants, while many of them may be caused by habitat changes or by altered microclimate. For example, the relative abundance of lizard species in areas affected by sand-mining and atmospheric fluoride is caused by habitat changes in polluted landscapes: forest species become less common and lizard species from open areas become more common (Taylor & Fox 2001). Similarly, the altered growth form of trees in industrial barrens (please see color plates 9, 10, 29, 34, 39 and 46 in Appendix II) results from combination of increased light availability and strong wind impact rather than from the toxicity of pollutants (Kozlov 2001b; Kozlov & Zvereva 2007a). The importance of secondary effects is confirmed by the failure of small-scale experiments with industrial pollutants (Nygaard & Abrahamsen 1991; Shevtsova & Neuvonen 1997; Zobel et al. 1999) to mimic some of environmental changes observed around the point polluters, such as a decline of field layer vegetation.

We conclude that pollution ecology suffers from generally poor understanding of cause-and-effect relationships. In many situations, the causal links behind the changes observed around point polluters are far from being transparent due to both the natural variation in study systems (Section 8.4.2) and the importance of indirect effects and secondary impacts acting at different scales. Causality of the effects routinely attributed to pollution need to be explored in greater details.

9.2.4 Linearity of Responses

As previously mentioned (Section 8.4.3), correlation and regression analyses are most frequently used to explore relationships between pollution load and characteristics of terrestrial biota. The wide use of linear statistical models indicates that researchers generally presume that the associations under study are linear or at least monotonic.

There is no theoretical background behind the commonly presumed linearity of biotic responses to pollution. Instead, the experience of toxicology clearly demonstrated that when both small (no effect) and extreme (lethal) concentrations are considered, then the dose–effect relationships are usually approximated by the logistic (S-shaped) function (Streibig et al. 1993). Seven response patterns of species diversity indices along the disturbance gradient (three linear and four non-linear) were identified in experiments with herbaceous vegetation (Li et al. 2004). There are sufficient reasons to believe that the logistic function generally describes association between pollution load and characteristics of biota better than the linear model (Vorobeichik et al. 1994). However, a shortage of information prevents verification of this hypothesis: only a thin fraction of studies conducted around point polluters is based on sufficiently large number of study sites (Fig. 8.1) to allow both (a) accurate approximation of data using non-linear functions and (b) statistical comparison of the fit of non-linear and linear models to the data.

Although assumption on linearity may appear valid only for a relatively minor range of toxicant concentrations, it remains the metric of practical choice when the number of data points is relatively small and visual inspection of data suggests the absence of a dome-shaped or U-shaped pattern. Moreover, the linear correlation is one of the few metrics suitable for meta-analysis (Rosenberg et al. 2000), while there is no commonly accepted way to handle polynomial and other non-linear regressions.

However, when changes along pollution gradients are not monotonic, the use of linear correlation may appear misleading: both dome-shaped and U-shaped patterns may yield zero correlation even in the presence of strong association between pollution load and characteristics of biota. Importantly, some phenomenological models predict the existence of these patterns. In particular, the intermediate disturbance hypothesis proposed by Connell (1978) suggests that communities subjected to moderate level of disturbance may have greater species richness or diversity than both communities existing under benign conditions and those experiencing higher disturbance. In addition, the 'humpback' model predicts the strongest plant-plant facilitation at intermediate levels of stress (Rebele 2000; Maestre & Cortina 2004; Maestre et al. 2005; Gilad et al. 2007), at least when a resource stressor (drought or lack of nutrients) is the most important abiotic force affecting plant-plant interactions (Eränen 2009). On the other hand, survival selection proportional to pollution load may create a U-shaped pattern in some vitality indices: vitality is likely to be lowest at intermediate levels of pollution, where toxicity is high enough to adversely affect fitness, but too low to act as strong selecting agent enhancing population-wide tolerance by elimination of sensitive genotypes.

In our study, the second-degree polynomial regression fitted the data better than linear regression in 64 of 608 statistical tests, i.e., about twice as high as can be expected by chance at the accepted significance level $P = 0.05$. The ratio between dome-shaped and U-shaped patterns did not differ from 1:1 ($G_H = 0.066$, df = 1, $P = 0.79$). These results agree with the estimated frequency (≤5%) of dome-shaped responses to pollution in published data sets on plant diversity (Zvereva et al. 2008) and density of insect herbivores (Zvereva & Kozlov 2009). The frequency of peaked relationships between pollution load and diversity of vascular plants in original data (11.4%) is in agreement with the conclusion (Mackey & Currie 2001; Li et al. 2004) that diversity–disturbance relationships are peaked less frequently than predicted by the intermediate disturbance hypothesis.

We conclude that both dome-shaped and U-shaped dose–response patterns do occur in pollution gradients, but their frequencies are generally low. Therefore, correlation coefficients are likely to produce an unbiased estimates of the strength of relationships between pollution load and the investigated characters of terrestrial biota. However, proportions of dome-shaped and U-shaped patterns changed with investigated characters ($G_H = 31.4$, d.f. = 20, $P = 0.048$), hinting that the accuracy of estimates based on coefficients of linear correlation may vary across our study. More attention should be paid to identification of non-linear effects, in particular in association with exploration of causal relationships behind the observed phenomena (Section 9.2.3).

9.2.5 Gradualness of Responses

Terrestrial habitats that have been damaged the most by industrial pollution are associated with historical smelting sites, such as Trail, Wawa, Sudbery, and Karabash, suggesting that a long period of impact is necessary to cause substantial changes in an ecosystem. In a few well-documented situations, the expansion of industrial barrens lasted for decades (Kozlov & Zvereva 2007a). In combination with observations that ecological damage becomes progressively more severe as distance from the polluter decreases (Freedman 1989) and effects shown in large-scale experiments with both irradiation (Woodwell 1970) and industrial pollutants (Zobel et al. 1999) develop gradually, these data create an impression that the state of ecosystems experiencing pollution impact changes gradually over time.

We had no reason to question the validity of this gradualness assumption until we recognised that correlations between the duration of impact of the pollution and the magnitude of biotic effect occurred less frequently than expected. An analysis of published data only detected two significant relationships: adverse effects of pollution on species richness of woody plants increased with time (Zvereva et al. 2008), while adverse effects on insect performance decreased (Zvereva & Kozlov 2009). Furthermore, the oldest polluters demonstrated the strongest effects on soil microfungi in categorical analyses, but correlations between the magnitude of the effect and duration of impact appeared non-significant for both diversity and abundance characteristics (Ruotsalainen & Kozlov 2006). The failure to model temporal trends using linear functions demonstrates that changes in terrestrial biota around industrial polluters are often non-linear, although generally monotonic.

Catastrophic changes in ecosystems can be seen as extreme cases of non-linear responses to pollution. Changes of some factor may have little effect on an ecosystem until a threshold is reached, but then a shift to an alternative state occurs (Scheffer

& Carpenter 2003). These abrupt changes seem to be more common than earlier thought and evidence of catastrophic regime shifts in ecosystems has been increasingly reported in a variety of aquatic and terrestrial systems (Scheffer & Carpenter 2003; Genkai-Kato 2007).

It is difficult to experimentally prove that a system has multiple stable states (Scheffer & Carpenter 2003; Schroder et al. 2005), and previous observational evidence has mostly concerned aquatic ecosystems (Genkai-Kato 2007). Exploration of polluted ecosystems usually begins when a disturbance is already evident, therefore providing little or no information to judge the applicability of this theory to changes caused by pollution. We have only identified one study that illustrates changes in plant diversity in the first 3 years of a pollution event: near the Puławy nitrogen plant (built in 1966) the number of vascular plant species decreased from presumably 23 (unpolluted control) to 13 in 1968 and two in 1969 (Sokołowski 1971). In contrast, repeated observations conducted around several polluters decades after their commencement revealed only minor changes in species richness with time (Zvereva et al. 2008), suggesting that major changes in diversity are abrupt rather than gradual.

The transition of polluted ecosystems to an alternative state is indirectly confirmed by observations conducted along pollution gradients. In the gradient starting at the copper smelter in Revda, the majority of the study plots were either in their original state or in a relatively stable impact state, suggesting that transition between these two states was very rapid (Vorobeichik 2003b). Similarly, we identified a clear spatial border between industrial barrens and forest ecosystems around Monchegorsk, with the transition zone rarely exceeding 100 m in width (V.E.Z. & M.V.K., personal observations, 2007). However, monitoring the recovery of industrial barrens near Sudbury did not provide support for the catastrophic regime shifts model (Anand et al. 2005).

We conclude that some data contradict an intuitive assumption on gradualness in ecosystem responses to pollution. On the other hand, few observations give support for the hypothesis that some ecosystems shifted to an alternative state following severe impact of industrial pollution and subsequently remained relatively stable in this new state. More long-term observations on both the decline and recovery processes are necessary to distinguish between these two alternatives - gradual changes and catastrophic shifts in polluted ecosystems.

9.3 Changes of Ecosystem Components Along Pollution Gradients: Structure of Phenomenological Model

Building a phenomenological model of ecosystem responses to industrial pollution requires linking observed effects with characteristics of both the polluters and impacted ecosystems. This task is complicated by multiple correlations between the investigated parameters, their generally weak relationships with pollution (Section 8.4.3), and the importance of indirect and secondary effects (Section 9.2.3) that may mask or even counterbalance the direct effects of pollutants. To avoid redundancy, we first identified groups of parameters that demonstrated co-ordinated responses to pollution and selected a set of key parameters for further exploratory analysis. In the second stage, we explored the relationships between these key parameters and the characteristics of the polluters and the impacted ecosystems.

The exploration of the correlation structure of our data using principal component analysis, path analysis or structural equation models was prevented by an incomplete data matrix, resulting from the impossibility of measuring each parameter around all 18 polluters, structure of the data (effect sizes, not primary measurements), and absence of an *a priori* model. Therefore, we chose to identify correlation pleiades, or suites of highly correlated parameters that are relatively independent of other parameters within impacted ecosystems, using a maximum correlation path between our variables (Weldre 1964). The concept of correlation pleiades, suggested by Terentjev (1931), was introduced to the international scientific community by Berg (1960) and was then almost exclusively discussed in studies of plant morphology (Armbruster et al. 1999; Ordano et al. 2008).

The algorithm of the maximum correlation path analysis is straightforward (Weldre 1964; Schmidt 1984). The correlation matrix is searched for the largest (by absolute value) correlation coefficient, which denotes the first link between two characters. Then the correlation matrix is searched for the largest correlation between the two selected characters and all the remaining characters. The procedure is repeated until the last character is attached to the diagram. This algorithm identifies the single correlation path between any two variables, in contrast to diagrams produced by path analysis that allows multiple links between variables.

A stepwise regression analysis was used to search for the associations between changes in the key parameters of the investigated ecosystems, characteristics of the polluters, and the local climate. We used the following explanatory variables: mean January temperature, mean July temperature, annual precipitation (Table 2.2), duration of pollution impact (calculated as the difference between the year when the majority of data were collected [Table 2.25] and the year of the polluter's establishment [Section 2.2.2]), mean annual emissions of SO_2, metals, nitrogen oxides, and fluorine (primarily in the form of HF) during 2000–2004, and maximum documented annual emission of SO_2 since the beginning of pollution. The emission data were extracted from Tables 2.4–2.24.

The maximum correlation path diagram (Fig. 9.1) demonstrated that changes in the measured soil characteristics are more strongly linked with changes in biotic parameters than with each other. Furthermore, changes of soil pH and electrical conductivity, stoniness, and topsoil depth were each explained by different polluter characteristics (Table 9.1). These results indicate that several co-occurring processes modified soil characteristics around industrial polluters. It seems likely that pollution directly affected the soil pH, and changes in the soil pH resulted in a cascading effect on vegetation structure. Modification of vegetation, primarily the decline of top-canopy and understorey plants, caused a decrease in topsoil depth and an increase in stoniness, with a positive feedback to plant performance.

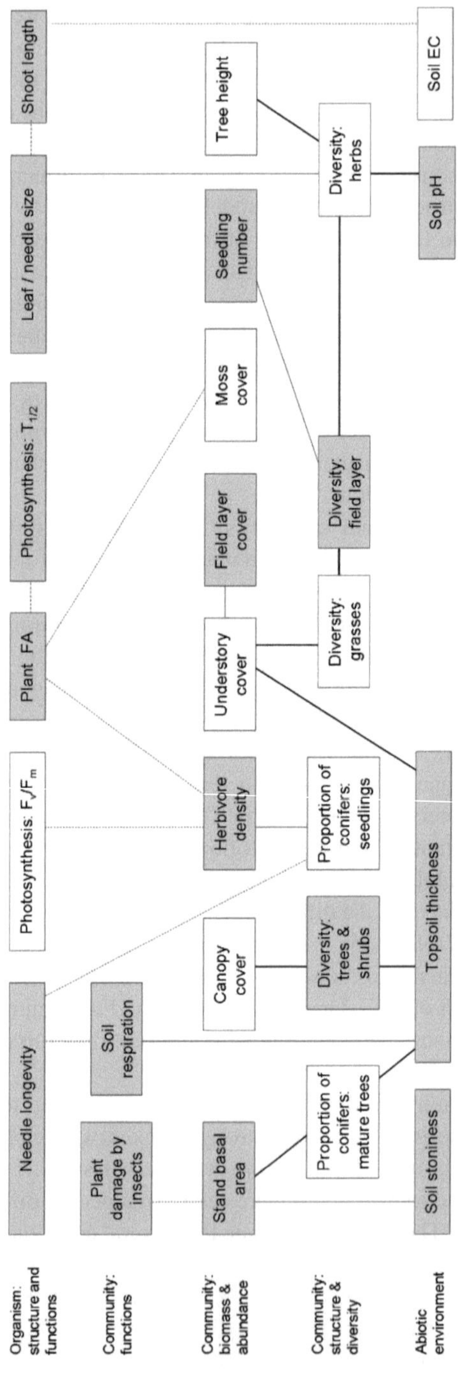

Fig. 9.1 The maximum correlation path diagram showing the strongest correlations between pollution-induced changes in the explored characteristics of terrestrial ecosystems. Thick lines, strong correlations ($|r| > 0.8$); thin lines, moderate correlations ($0.7 < |r| < 0.8$); dotted lines, weak correlations ($|r| < 0.7$). Shaded: characters, for which the sources of variation were explored using stepwise regression analysis (Table 9.1)

Table 9.1 Variables explaining variation in responses of the selected characteristics of terrestrial ecosystems to pollution

Group of characters	Response variable	Explanatory variable	Slope (mean ± S.E.)	Partial R^2	F (d.f.)	P
Abiotic environment	Topsoil depth	Duration of impact	−0.026 ± 0.007	0.70	14.1 (1, 7)	0.0095
	Soil stoniness	Mean emission of SO_2[a]	0.049 ± 0.016	0.39	9.33 (1, 6)	0.03
	Soil pH	Mean emission of metals[a]	−0.206 ± 0.034	0.77	36.0 (1, 11)	<0.0001
Community: structure and diversity	Species richness of field layer vegetation	Mean emission of metals[a]	−0.154 ± 0.024	0.53	11.5 (1, 11)	0.0068
		Mean July temperature	−0.178 ± 0.043	0.30	16.9 (1, 11)	0.0026
	Species richness of woody plants	Mean emission of metals[a]	−0.091 ± 0.042	0.67	4.75 (1, 9)	0.06
Community: biomass and abundance	Cover of field layer vegetation	Duration of impact	−0.016 ± 0.006	0.40	6.66 (1, 10)	0.03
	Density of birch-feeding herbivores	Mean July temperature	−0.084 ± 0.027	0.57	9.39 (1, 8)	0.02
Community: functions	Soil respiration	Mean emission of metals[a]	−0.118 ± 0.028	0.60	17.8 (1, 12)	0.0012
Organism: structure and functions	Rate of photosynthetic reaction ($T_{1/2}$)	Mean July temperature	0.074 ± 0.031	0.38	7.21 (1, 11)	0.02
	Leaf length	Mean emission of metals[a]	−0.065 ± 0.024	0.25	4.30 (1, 12)	0.06
		Mean July temperature	−0.082 ± 0.039	0.20	4.35 (1, 12)	0.06
	Shoot length	Mean emission of SO_2[a]	−0.082 ± 0.029	0.26	4.57 (1, 12)	0.05
	Needle longevity	Mean emission of SO_2[a]	−0.094 ± 0.047	0.47	8.00 (1, 10)	0.02
		Annual precipitation	0.006 ± 0.002	0.26	7.58 (1, 10)	0.02

Stepwise regression analyses included the following explanatory variables: mean January temperature, mean July temperature, annual precipitation, duration of pollution impact, mean annual emissions of SO_2, metals, nitrogen oxides, and fluorine (primarily HF) during 2000–2004, and maximum documented annual emission of SO_2 since the beginning of pollution. None of these variables entered the regression models at $P = 0.06$ for stand basal area, density of woody plant seedlings, plant damage by insect herbivores, and plant fluctuating asymmetry.

[a] Values were log-transformed prior the analysis.

Changes in soil electrical conductivity were not explained by the amount of emissions and were only weakly associated with changes in terrestrial biota.

The lower respiration of polluted soils was primarily (R^2 = 0.60) related to higher metal emissions (Table 9.1). However, it seems likely that decreases in litterfall and soil moisture that occurred following the decline of top-canopy and understorey plants have also contributed to decreased soil respiration. Furthermore, a link between needle longevity and soil respiration (Fig. 9.1) denotes the strong influence of forest canopy processes on the activity of roots and associated organisms (Moyano et al. 2008).

We detected strong to moderate correlations between changes in measured characteristics of plant communities (Fig. 9.1). At the same time, the amount of emissions and duration of impact only explained a small part of changes in plant diversity and productivity (Table 9.1), suggesting that secondary effects modified the trajectories of the successions triggered by pollution (partially described in Section 9.2.3). Also, disturbances other than pollution, the history of which could not be accounted for in the course of our study (such as logging, fires, damage by pests and pathogens, and recreation), may have substantially contributed to modifications of vegetation structure.

Changes in parameters that reflect the state of long-lived woody plants (canopy cover, species richness of trees and large shrubs, stand basal area, and proportion of conifers among top-canopy trees) are only weakly linked to changes in understorey, field layer vegetation and moss cover (Fig. 9.1). Furthermore, no association was detected between changes of top-canopy vegetation and characteristics of polluters. We suggest that the existence of a time lag between the onset of pollution impact and the decline of extant individuals in affected populations (Zverev et al. 2008; Zverev 2009) may mask the association between the characteristics of polluters and changes in populations of long-lived plants.

Although magnitudes of organism-level effects were generally influenced by both the emission of sulphur dioxide and the local climate (Table 9.1), these effects only demonstrated weak links between one another and showed almost no correlation with community-level effects. Especially surprising is the absence of expected associations between changes in photosynthetic processes (measured by the efficiency of photosynthesis and the rate of photosynthetic reaction) and changes in plant growth processes (measured by leaf/needle size and shoot length). Similarly, changes in growth processes showed no strict correspondence with changes in plant abundance or biomass (Fig. 9.1). This pattern supports the conclusion (McClellan et al. 2008) that the results of organism-level studies cannot be directly translated into community-level effects.

Finally, we discovered weak negative correlations between changes in the abundance of birch-feeding insects, plant fluctuating asymmetry, and photosynthesis. At the same time, these indices all showed only weak associations with changes in other ecosystem characteristics. Variation in two of four indices from this group (herbivore density and $T_{1/2}$) is partially explained by mean July temperature, while the two other indices did not depend on either emission of pollutants or climate (Table 9.1). An exploration of the cause-and-effect relationships within this group of parameters requires additional information, since existing data are insufficient to select one of two plausible models. Plant fluctuating asymmetry may increase (Zvereva et al. 1997; Møller & Shykoff 1999; Kozlov 2005c) and photochemical efficiency may decrease (Delaney 2008) following damage by herbivorous insects. On the other hand, low photochemical efficiency and increases in fluctuating asymmetry may indicate that plants experience abiotic stress (Daley 1995; Møller & Shykoff 1999; Nesterenko et al. 2007). Feeding on stressed plants may facilitate some guilds of herbivorous insects (Koricheva et al. 1998), leading to an increase in their population densities. More generally, the correlation structure of our data confirms Vehviläinen et al.'s (2007) conclusion that there are no consistent relationships between the abundance of insect herbivores and the diversity of forest trees.

Thus, we identified several groups of characteristics that showed coordinated responses to pollution. The similarity in responses of individual parameters may result from: (a) the structure of the data, (b) cause-and-effect relationships, and (c) a common cause behind several observed effects. To avoid redundancy, we restricted exploration of the sources of variation in responses to industrial pollution (Sections 9.4.1–9.4.3) to three of four characteristics of soils and to 13 of 22 community- and organism-level characteristics (Table 9.1), which, due to the correlation structure of our data (Fig. 9.1), are sufficiently representative of the entire suite of investigated parameters.

9.4 Sources of Variation in Biotic Responses to Pollution

9.4.1 Variation Related to Emission Sources

The links between the direction and magnitude of changes in different parameters of terrestrial ecosystems around individual polluters and characteristics of these polluters have remained almost unexplored. Even the comparisons between the classes of polluters have only yielded a few intuitive conclusions, like the smaller extent of vegetation damage observed around power plants relative to smelters and fluorine-emitting industries, and the less severe damage of terrestrial ecosystems around sources of fluorine-containing emissions relative to non-ferrous smelters (Bohne 1971; Freedman 1989). This is partially explained by the absence of comparative analyses in the vast majority of these primary studies.

Only four of 1,000 publications from our collection of primary studies (described in Section 8.4.1) compared the effects observed around four polluters, while 91% of publications reported observations conducted around a single polluter. Narrative reviews often considered the effects caused by different pollutants in different chapters (Scurfield 1960a, b, Alstad et al. 1982; Treshow & Andersen 1989; Heliövaara & Väisänen 1993; Flagler 1998). To our knowledge, a statistical comparison between the ecological effects caused by different classes of polluters was performed for the first time in a meta-analysis of soil microfungi data (Ruotsalainen & Kozlov 2006).

Effect sizes were first regressed to the emissions of sulphur dioxide by Zvereva et al. (2008). An analysis of published data demonstrated that the recorded biotic effects generally depend on the type of the polluter, and in some particular situations, also on the duration of the pollution impact (Ruotsalainen & Kozlov 2006; Zvereva et al. 2008; Zvereva & Kozlov 2009; Roitto et al. 2009).

In our study (Chapters 3–7), significant adverse effects of non-ferrous smelters were detected in 54% of individual meta-analyses (Table 9.2), i.e., much more frequently than around aluminium smelters (11%). These results support the conclusion by Freedman (1989) that aluminium industries generally impose weaker effects on terrestrial biota than non-ferrous smelters. However, the generally weak effect of aluminium smelters may be enhanced in some specific situations, especially in warmer and more humid climates (discussed in Section 6.4.3).

The proportion of significant negative effects detected by meta-analyses was highest (62%) among polluters that caused soil acidification. This result, although in agreement with earlier conclusions that acidification causes strong detrimental effects (Abrahamsen 1984; Roem et al. 2002; AMAP 2006), cannot be attributed to acidification alone, since most acidic polluters also emit heavy metals. Similarly, the relatively high (22%) occurrence of negative effects in the absence of changes in soil pH can most probably be explained by the emission of heavy metals by three of five polluters from this category (Harjavalta, Norilsk and Sudbury). Minor effects of Norilsk on soil pH are primarily due to the high buffering capacity of soils (AMAP 2006), while the absence of significant changes in soil pH around Harjavalta and Sudbury resulted from the substantial decline of sulphur dioxide emissions over the past decades. Alkaline polluters affected about the same proportion of variables as neutral polluters, but caused both negative (17%) and positive (8%) effects (Table 9.2).

A stepwise regression analyses conducted to determine the variables associated with the variation in responses of 16 selected ecosystem parameters across the polluters yielded 12 significant models (Table 9.1). Amount of emissions was chosen by the program as explanatory variable in eight of these 12 models. This result seems predictable, since polluters producing more emissions are likely to cause stronger environmental effects. However, in similarly performed analyses of published data, the amount of emissions was not selected in any of the eight significant regression models (Zvereva et al. 2008; Zvereva & Kozlov 2009), although these regression analyses had larger statistical power due to larger sample sizes. We suggest that the absence of association between biotic effects and amount of emissions in earlier analyses may have resulted from research and publication biases that have affected the outcomes of primary studies. More specifically, elimination of small and non-significant effects at both the recording and publication stages (discussed in Section 8.5) may have created a false impression that the effects of pollution are always strong and significant (discussed in Section 9.2.2), and thus the magnitude of effects appeared to be independent of the strength of pollution impacts. On the other hand, emission data for the polluters explored in our study are more complete and correspond better to the timing of data collection than for the polluters, the effects of which were summarised in earlier meta-analyses. The discovery of dose-effect relationships

Table 9.2 Summary of the effects revealed by meta-analyses

			Sources of variation						
			Polluter types		Impact on soil pH			Geographical position	
Group of characters	Character	Overall effect	Non-ferrous smelters	Aluminum smelters	Acidic	Neutral	Alkaline	North	South
Abiotic environment	Topsoil depth	NS	–	NS	–	NS	NS	NS	NS
	Soil stoniness	+	+	+	+	+	+	+	+
	Soil pH	NS	–	+	–	NS	+	NS	NS
	Soil electrical conductivity	NS	NS	NS	NS	NS	+	NS	NS
Community: diversity	Species richness of trees and large shrubs	NS	–	NS	NS	NS	NS	NS	NS
	Species richness of herbs	NS	–	NS	–	NS	NS	NS	NS
	Species richness of grasses	NS	NS	NS	–	NS	NS	NS	NS
	Species richness of field layer vegetation	NS	–	NS	–	NS	NS	NS	NS
	Species richness of all vascular plants	NS	–	NS	–	–	NS	NS	–
	Proportion of conifers in a stand	NS	–	+	–	–	NS	NS	NS
	Proportion of conifers among seedlings	NS	NS	NS	NS	(no data)	NS	NS	NS
Community: biomass and abundance	Cover of top-canopy trees	NS	–	+	–	–	+	NS	NS
	Cover of understorey	NS	–	+	–	NS	+	NS	NS
	Cover of field layer vegetation	–	–	–	–	NS	–	–	NS
	Cover of mosses	–	NS	–	–	NS	–	–	NS
	Stand basal area	–	–	NS	–	–	NS	NS	–
	Stand height	–	–	–	–	–	–	–	–
	Number of seedlings of woody plants	NS	NS	NS	NS	NS	NS	NS	NS
	Density of birch-feeding herbivores	NS	NS	NS	NS	(no data)	NS	NS	–
Community: functions	Soil respiration	–	–	NS	–	NS	NS	–	–
	Plant damage by herbivorous insects	–	NS	NS	–	NS	NS	NS	–
Organism: structure and functions	Efficiency of photosynthesis (F_v/F_m)	NS	NS	NS	NS	NS	NS	NS	NS
	Rate of photosynthetic reaction ($T_{1/2}$)	–	NS	NS	NS	NS	NS	NS	–
	Leaf/needle size of woody plants	NS	–	NS	–	NS	NS	NS	NS
	Shoot length of woody plants	NS	NS	NS	NS	NS	NS	NS	NS
	Needle longevity	–	–	–	–	NS	–	–	NS
	Fluctuating asymmetry of woody plants	NS	NS	NS	NS	NS	NS	NS	NS

Only analyses based on correlation coefficients are included. Plus (+), values increased with pollution; minus (–), values decreased with pollution; NS, values demonstrated no significant changes with pollution.

further stresses the need to provide sufficient information about polluters when describing their impacts on organisms and ecosystems (Section 8.4.6).

Medical studies often link health effects with past air pollution levels (Moffatt et al. 2000; Rosenlund et al. 2006; Kohlhammer et al. 2007). However, the maximum documented annual emissions of SO_2 since the beginning of pollution did not enter any regression model (Table 9.1), while recent emissions of sulphur dioxide and metals entered three and five of 12 significant regression models, respectively (Table 9.1). The average slope of regression models that linked effect size with log-transformed values of annual emissions of heavy metals was twice as high as for emissions of sulphur dioxide (mean ± S.E.: −0.142 ± 0.025 vs. −0.075 ± 0.013, respectively). These results indicate a higher contribution of metal emissions to changes observed around point polluters relative to sulphur dioxide.

The duration of impact entered only two of 12 significant regression models (Table 9.1). The increases in adverse effects on topsoil depth and cover of filed layer vegetation with time are consistent with the effects of the duration of the impact on diversity of vascular plants detected by meta-analysis of published data (Zvereva et al. 2008). The rarity of associations between duration of impact and magnitude of effect among both published and original data may hint at either the non-linearity of responses (Section 9.2.5) or the adaptation to pollution through either physiological acclimation or selection for pollution resistance. The adaptation hypothesis was only confirmed so far for field-collected arthropods, the performance of which improved with the time elapsed since the beginning of the impact (Zvereva & Kozlov 2009).

The relatively low number of data points used in the regression analyses (Table 9.1), as well as the low number of parameters affected by the amount of emissions, resulted in the low accuracy of our threshold emission level estimates. Uncertainties are also associated with the impossibility of separating the effects of sulphur dioxide and metals using our data set, where emissions of these two groups of pollutants strongly correlate to each other ($r = 0.63$, $N = 18$ polluters, $P = 0.0049$). The model calculation suggests that annual emissions that do not exceed 10 t of metals or 350 t of sulphur dioxide are unlikely to cause any changes in terrestrial ecosystems surrounding point polluters. On the other hand, emissions exceeding 390 t of metals or 595,000 t of sulphur dioxide will obviously result in detectable biotic effects. This result agrees with our earlier estimate that polluters emitting <1,500 t of SO_2 annually are unlikely to cause depauperation of plant communities (Zvereva et al. 2008) and further supports the conclusion (Freedman 1989; Kozlov & Zvereva 2007a; Zvereva et al. 2008) that the effects of industries emitting sulphur dioxide but not metals are generally less detrimental than effects of non-ferrous smelters.

The amounts of fluorine-containing emissions were reported for eight of our polluters (six aluminium smelters, the copper smelter at Revda, and the cement plant at Vorkuta), while the amounts of sulphur dioxide released into the ambient air were known for all 18 polluters. Since fluorine is at least as toxic as sulphur dioxide (Bohne 1971; Treshow & Andersen 1989; Flagler 1998; Meldrum 1999), the question arises whether the low number of fluorine-emitting industries in our sample is the primary reason for not detecting associations between fluorine emissions and changes in ecosystem characteristics. To partially test this hypothesis, we calculated the correlations between all investigated parameters and emissions of fluorine across six aluminium plants. A single correlation was found to be significant at the table-wide probability level $P = 0.05$, namely: the needle longevity decreased with increases in fluorine emissions ($r = -0.99$, $N = 4$ polluters). The coefficients of determination (R^2) for sulphur dioxide and fluorine, averaged for these six aluminium smelters across 27 measured parameters, were nearly identical (0.22 and 0.23, respectively). This result is consistent with strong positive correlations between the amount of fluorine and sulphur dioxide released by aluminium smelters ($r = 0.78$, $N = 6$ polluters, $P = 0.07$). Thus, although we do not think that sulphur dioxide is the primary cause of the biotic effects observed around aluminium smelters, still these effects can be predicted on the basis of sulphur dioxide emissions.

Nitrogen oxide emissions did not enter any regression model. This result is in line with the commonly accepted opinion about the low contribution of nitrogen oxides to local effects observed around point polluters (Heliövaara & Väisänen 1993; Flagler 1998; Bytnerowicz et al. 1999). Nitrogen oxides may even act as fertilisers in some situations and actually improve plant growth in soils deficient in nitrogen (Abrahamsen 1984; Treshow 1984). However, excessive emissions of nitrogen-containing dust may cause acute damage, including forest decline around fertilizer producing factories (Sokołowski 1971; Kowalkowskí 1990). Nitrogen oxides are indeed important at regional scales, due to both nitrogen saturation effect and a contribution to ozone formation (Lee 1998; Bytnerowicz et al. 1999; Karnosky et al. 2003a; Hyvönen et al. 2007; Hettelingh et al. 2008).

To conclude, changes in ecosystem parameters around industrial polluters generally depend on both composition and amount of emissions, but are only rarely associated with the time since beginning of pollution. Acidic polluters cause stronger effects on biota than alkaline polluters; the effects of metal-emitting industries are generally more detrimental than the effects of polluters only emitting sulphur dioxide.

9.4.2 Variation Related to Impacted Ecosystems

It became evident long ago that the same pollution loads may cause different effects in different ecosystems and/or regions. Differential sensitivities to pollutants were formalised through the concept of critical loads (Nilsson & Grennfelt 1988). The critical load for a pollutant defines a deposition level below which sensitive parts of ecosystems are not affected. A critical load approach has been widely and successfully used to evaluate the effects of acid precipitation on ecosystems in Europe (Nilsson & Grennfelt 1988; Posch et al. 2001) and have shaped air-pollution control policies since the 1980s (Burns et al. 2008). Reinds et al. (2006) published a preliminary assessment of critical loads for several metal pollutants, including copper and nickel. However, we are not aware of any attempt to link the strength of biotic effects caused by point polluters with the existing estimates of critical loads for the impacted areas.

To ensure the uniformity of data on critical loads, we restricted the following analysis to territories covered by the European assessment programs (Posch et al. 2001; Reinds et al. 2006). We extracted data on the maximum critical load of sulphur, nickel and copper from the maps published in previously mentioned reports, and extrapolated these maps to obtain approximate measures of critical loads for two polluters situated in Ural, i.e. slightly outside the mapped region.

The addition of critical loads of sulphur to the list of explanatory variables did not change the results of the stepwise regression analyses described in Section 9.4.1, while critical loads for nickel entered three regression models. The adverse effects of pollution on soil pH, species richness of field layer vegetation, and leaf/needle size were less expressed in regions with higher critical loads. This result seems to agree with the conclusion that metals contribute to adverse effects more than sulphur dioxide. However, metal emissions were only reported for eight of the 18 polluters included in our analysis, and nickel was the major metal pollutant for only two of these eight polluters. At the same time, the critical loads for copper, which was emitted by a larger number of polluters and in generally larger quantities, entered only one regression model, but with the opposite sign than expected. The adverse effects of pollution on shoot length were stronger ($R^2 = 0.30$) in regions with higher critical loads of copper. This gives us a reason to suspect that the detected correlations with critical loads for metals are spurious. This hypothesis was confirmed by stepwise regression analyses restricted to the eight metal-emitting polluters, which demonstrated a much lower (instead of the expected higher) explanatory power of the critical load for metal pollutants relative to the analyses based on all 18 polluters (data not shown). Our results may hint that uncertainties in input parameters used to calculate the critical loads for Europe are larger than commonly thought.

Another aspect of ecosystem sensitivity to disturbances is linked to ecosystem structure and functional organisation. In particular, relationships between species diversity and ecosystem stability, that have been and still are widely debated, may have direct implications for understanding how communities respond to pollution and other antropogenic stressors (Clements & Newman 2002). While four models reviewed by Peterson et al. (1998) assume a positive relationship between stability and diversity, an opposite pattern was both predicted theoretically (May 1973; Lehman & Tilman 2000) and detected in several experiments (Clements & Newman 2002; Foster et al. 2002).

A meta-analysis of published data (Zvereva et al. 2008) demonstrated that pollution effects on the species richness of vascular plants became more severe with increases in the diversity of the impacted communities. To verify this hypothesis with original data, we conducted a stepwise regression analysis using the mean numbers of the species with the two most distant (unpolluted) sites added to the explanatory variables. A strong negative association ($R^2 = 0.68$) was only detected between the magnitude of the pollution effect and regional diversity for herbs. Among other characteristics of the plant community, only stand basal area showed a moderate negative association ($R^2 = 0.45$) between effect size and regional diversity of vascular plants. These results are in agreement with the overall increase of adverse effects with summer temperatures (Section 9.4.3), which can be seen as a sign of the higher stability of Northern, generally less diverse, ecosystems.

To conclude, little is known about the ecosystem properties that determine their sensitivity to the impacts of industrial pollutants. Critical loads of sulphur, nickel, or copper did not explain the variation in responses of the investigated components of terrestrial ecosystems to pollution, while a lower diversity of plant communities was associated with smaller changes in some characteristics of vegetation around industrial polluters.

9.4.3 Variation Related to Climate

Although environmental pollution is an integral part of global change (Taylor et al. 1994), the vast majority of the research addressing the biotic effects of global change has overlooked the pollution issue (but see Settele et al. 2005). On the other hand, extensive studies on both the distribution of pollutants and the biotic effects of pollution have generally neglected climate effects, even though changes in the toxicity of pollutants as a result of temperature changes were already known several decades ago (Cairns et al. 1975). The interactive effects of air pollution and temperature were documented in a number of studies on plant growth (Alekseev 1991; Shevtsova 1998), different characteristics of aquatic biota (Sokolova & Lanning 2008), and human health (Muggeo 2007; Hu et al. 2008, and references therein).

Recent scientific assessments (Houghton et al. 2001) and innovative integrated projects (European Environment Agency 2003) started to change the situation, bridging research fields that for a long time have developed independently. Simultaneously addressing air pollution and climate change problems would potentially result in better integration of local, national and global environmental policies (Swart et al. 2004; Bytnerowicz et al. 2007).

Our results confirm latitudinal variations in the effects of industrial pollution on biota that were previously revealed in published data on plant diversity (Zvereva et al. 2008). In 8 of 23 individual meta-analyses (Table 9.2), adverse effects were significant only around either northern or southern polluters. This variation can be attributed to variations in both diversity (discussed above, Section 9.4.2) and climate. The discovery of the climatic variation of responses of terrestrial arthropods to industrial pollution, which was consistent with the climatic variations observed in plant communities, allowed us to suggest that the enhancement of pollution effects in warmer and moister climates concerns many aspects of ecosystem structure and functioning (Zvereva & Kozlov 2009). This hypothesis is confirmed by our study of point polluters. Climatic data entered five of 12 significant regression models, consistently demonstrating that adverse effects of pollution on community- and organism-level parameters increased with increases in mean July temperatures (Table 9.1).

The lower sensitivity of high-latitude ecosystems to the impact of industrial pollution, revealed by the analyses of both original and published data, can be explained by several reasons. First, lower temperatures may result in lower stomata opening, decreasing the impact of gaseous pollutants on plants (Bytnerowicz et al. 1999). Second, the shorter vegetation period naturally decreases the exposure to pollutants. Third, high-latitude soils generally contain fewer nutrients, and therefore the adverse effects of toxic pollutants may be counterbalanced to some extent by the fertilising effect of nitrogen oxides (Abrahamsen 1984; Treshow 1984). Finally, higher temperatures and increased precipitation may enhance mobility and increase toxicity of pollutants (Cairns et al. 1975; Klein 1989; Tipping et al. 1999), explaining the generally higher adverse effects seen in a warmer climate.

Only a few attempts have been made to predict the consequences of climate change for polluted areas. At the organism level, the effects of temperature on the sensitivity of Norway spruce and European beech to sulphur dioxide and nitrogen oxides were explored using a model of stomatal conductance. Although the expected responses to temperature increases depend on both species and region, a simulation showed that the sensitivity of trees to air pollution would increase in boreal areas of Europe (Guardans 2002). At the community level, pollution effects on plant diversity are likely to increase in warmer climates (Zvereva et al. 2008). Also the decline of arthropod decomposers and predators and increase in herbivory with increases in pollution become more severe with increases in mean temperature and precipitation (Zvereva & Kozlov 2009). All these effects are likely to have adverse consequences for ecosystem-level processes.

Thus, the effects caused by industrial pollutants critically depend on climate. Although our results are in contrast to the predictions based on the critical loads concept (Posch 2002), they provide support to the hypothesis (Holmstrup et al. 1998) that under a warmer climate, existing pollution loads may become more harmful. This is especially important for Northern Europe and some regions in the USA and Canada, which have been (and in some places still are) affected by emissions from major non-ferrous smelters.

9.5 Exploring Effects of Industrial Pollution: Prospects and Limitations

The need to integrate the knowledge of biotic effects caused by pollution is obvious. Several recent reviews (Percy et al. 1999; Clements & Newman 2002; Adams 2003; Cape et al. 2003; Ferretti et al. 2003; Karnosky et al. 2003c; Percy & Ferretti 2004; Paoletti et al. 2007) have stressed the need to develop pollution ecology through the concerted effort of scientists who monitor the effects of pollution, explore them by means of field and laboratory experiments, and model both distribution of pollutants and ecosystem processes. These reviews also identified critical data gaps and research needs in several specific fields of knowledge.

Over the past decade, the focus of pollution ecology largely shifted from the acute localized effects of industrial pollutants, such as sulphur dioxide and heavy metals, to the regional effects of ozone and nitrogen deposition. In many industrialised countries, 'traditional' pollutants have dropped out of priority lists, particularly in forestry-oriented studies (Ferretti et al. 2003; Paoletti et al. 2007). Under these circumstances, our research may look outdated due to its focus on point polluters. However, the annual symposium of the British Ecological Society in 2008, entitled 'Ecology of Industrial Pollution', attracted delegates from universities, industry, government agencies, museums and consultancy firms, demonstrating the existence of sufficient interest to this problem (Crowden 2008). On the other hand, research funds for studying the effects of 'traditional' pollutants have dwindled considerably in the last decades, leading to the pessimistic conclusion that "progress on these issues will come slowly if at all" (Lovett & Tear 2007).

We are a bit more optimistic. We think that our study not only identified the acute need for a better understanding of the link between sources of industrial pollution and effects on structure and functions of terrestrial ecosystems, but also revealed the possibility of achieving this goal within a reasonable time frame.

First, and most importantly, we demonstrated that it is possible to build a phenomenological model linking changes in key parameters of terrestrial ecosystems with pollution data. However, when building this model, we need to appreciate that the effects caused by pollution are generally low to moderate, and even in the impact zones of large industrial enterprises they can be masked by the effects of both natural variation and other disturbances.

Second, we revealed substantial variation in biotic responses to pollution related to both characteristics of polluters, characteristics of impacted ecosystems, and climate. Our inability to reveal a consistent and significant effect of 'pollution in general' on most of the explored characteristics results not from the absence of responses, but from their diversity, often resulting in a zero overall effect. Thus, rough parameterisation of the most important links within this model is only possible when identified sources of variation are accounted for.

Third, since the links between the organism-level and community- or ecosystem-level changes appeared weak, the impact of pollution on ecosystem-level processes should be explored directly rather than deducted from organism-level studies.

The shortage of funding for field studies of biotic effects caused by industrial pollutants emphasizes the importance of careful selection of the topics of the planned research. In our opinion, the top-priority list includes the following research areas, the exploration of which is of critical importance for large-scale generalizations in the field of pollution ecology:

Study objects:

- Geography: Africa, South-Eastern Asia, South America, Oceania
- Biomes and ecosystems: tropical forests, deserts, grasslands, savannah, tundra
- Industries: power plants, chemical industries, small polluters

Research topics

- Belowground processes
- Changes in species interactions
- Functional diversity of communities

- Spatial structure and demography of populations
- Ecosystem-level effects

We also suggest that pollution ecologists expand the use of the following research approaches:

- Comparisons between polluters, organisms, communities, and ecosystems
- Long-term studies
- Exploration of dose-and-effect and cause-and-effect relationships
- Research synthesis

Finally, pollution ecology cannot develop in isolation. Bridging the gaps between pollution ecology, basic ecology, ecotoxicology, and global change research is critically important for advancing the development of all these disciplines. The experience accumulated during the integrated research of ozone impact on terrestrial biota (Karnosky et al. 2003a, b; Hayes et al. 2007; Marinari et al. 2007) clearly shows the existence of scientific potential for rapid collection of high-quality information, which is necessary to build the predictive models that account for multiple interactions between pollutants, ecosystem properties, and climate.

9.6 Summary

The adverse consequences of pollution impacts on terrestrial ecosystems have been under careful investigation for more than a century. However, we are still far from making large-scale generalizations. This is primarily due to the insufficient quality of many studies, prejudgement, and misinterpretation of the results. Changes occurring in organisms, communities, and ecosystems affected by industrial pollutants are neither general nor uniform. Identification of factors explaining the diversity in responses is crucial for understanding the links between sources of industrial pollution and effects on structure and functions of terrestrial ecosystems. Changes in investigated parameters of biota generally depend on both composition and amount of emissions, but are only rarely correlated with the time since the beginning of pollution. Acidic polluters cause stronger effects on biota than alkaline polluters; effects of metal-emitting industries are more detrimental than the effects of polluters emitting only sulphur dioxide. The detrimental effects are stronger in Southern ecosystems and depend on summer temperatures in such a way that in a warmer climate the existing pollution loads may become more harmful. Information accumulated by pollution scientists allows for the construction of a phenomenological model that links changes in key parameters of terrestrial ecosystems with pollution and climate data, and for a rough parameterisation of the most important links within this model. Uncertainties in this model can be reduced by summarising outcomes of earlier research by means of meta-analysis, searching for factors explaining variation across the primary studies, and designing new projects to cover areas neglected previously by pollution scientists.

Appendix I
List of Abbreviations

CI	Confidence interval
d	Hedge's d measure of effect size (consult Section 2.5.2.2)
DA	Directional asymmetry (consult Section 5.1)
df	Degrees of freedom
EC	Soil electrical conductivity
ES	Effect size (consult Section 2.5.2.2)
FA	Fluctuating asymmetry (consult Section 5.1)
F_v/F_m	The ratio of variable to maximum chlorophyll fluorescence yielded under the artificial light treatment (consult Section 4.1.2.1)
G_H	Heterogeneity G-test checking whether the ratios are all homogeneous across the characters, or across the study sites
ME	Measurement error
pH	A measure of the acidity or basicity of a soil
Q_B	Between-group heterogeneity, used to test whether the categorical variable can account for a significant amount of the variability in the effect sizes
Q_T	Total heterogeneity of a sample, calculated to determine whether a set of effect sizes is homogeneous
RTU	Recognisable taxonomic units (consult Section 7.2.1.2)
$T_{1/2}$	The time needed for the leaf to reach half of its F_m (consult Section 4.1.2.1)
z_r	Effect size based on correlation coefficient (consult Section 2.5.2.2)

Correspondence between vernacular[1] and Latin names of plant species

Black alder	*Alnus glutinosa*
Canoe birch	*Betula papyrifera*
Common birch	*Betula pendula*
Cowberry	*Vaccinium vitis-idaea*
Crowberry	*Empetrum nigrum*
Dwarf birch	*Betula nańa*
European aspen	*Populus tremula*
European beech	*Fagus sylvatica*
Goat willow	*Salix caprea*
Mountain birch	*Betula pubescens* subsp. *czerepanovii*
Norway spruce	*Picea abies* ssp. *abies*
Scots pine	*Pinus sylvestris*
Siberian fir	*Abies sibirica*
Siberian larch	*Larix sibirica*
Siberian spruce	*Picea abies* ssp. *obovata*
Speckled alder	*Alnus incana*
White birch	*Betula pubescens*
Woolly willow	*Salix lanata*

[1] Vernacular names of the most common plants are used through the text to facilitate readers not familiar with plant taxonomy.

Appendix II

Plate 1 Power plant at Apatity, Russia (2008). Photo: V. Zverev

Plate 2 Unpolluted site 2-5, located 5.7 km E of the power plant at Apatity (2008): mixed north taiga forest. Photo: V. Zverev

Plate 3 Polluted site 2-1, located 0.9 km E of the power plant at Apatity (2008): mixed north taiga forest. No visible signs of pollution impact (compare with Plate 2). Photo: V. Zverev

Plate 4 Aluminium smelter at Bratsk, Russia (2003). Photo: V. Zverev

Plate 5 Slightly polluted site 1-4, located 27 km NE of the aluminium smelter at Bratsk (2003): mixed southern taiga forest typical for the study area, without visible signs of pollution impact. Photo: V. Zverev

Plate 6 Polluted site 2-1, located 1.8 km E of the aluminium smelter at Bratsk (2003): declining forest. Note extensive development of shrubs and herbs (compare with Plate 5). Photo: V. Zverev

Plate 7 Nickel smelter and associated industrial enterprises at Harjavalta, Finland (2008). Photo: V. Zverev

Plate 8 Slightly polluted site 2-4, located 4.6 km SE of the smelter at Harjavalta (2008): mixed taiga forest typical for the study area, without visible signs of pollution impact. Photo: V. Zverev

Plate 9 Polluted site 1-1, located 0.9 km NW of the smelter at Harjavalta (2008): industrial barren. Note absence of field layer vegetation (compare with Plate 8), abundance of woody debris, and stunted growth form of Scots pines (all trees are about 50 years old). Photo: V. Zverev

Plate 10 Polluted site 1-1, located 0.9 km NW of the smelter at Harjavalta (2008): industrial barren. Note polycormic growth form of white birch, with numerous dead, partially decayed, trunks. Photo: V. Zverev

Plate 11 Polluted site (additional) located 1.7 km SEE of smelter at Harjavalta (2008): semi-barren landscape. Although mature trees persist on this site, regeneration is prevented by metal toxicity. Photo: V. Zverev

Plate 12 Fertilising factory at Jonava, Lithuania (2005). Photo: V. Zverev

Plate 13 Unpolluted site 2-5, located 18.9 km NW of the fertilising factory at Jonava (2005): Scots pine forest. Photo: V. Zverev

Plate 14 Polluted site 2-1, located 0.8 km W of the fertilising factory at Jonava (2005): Scots pine forest. Extensive development of nitrophilous vegetation (compare with Plate 13) is indicative of high deposition of nitrogen-containing emissions in the past. Photo: V. Zverev

Plate 15 Aluminium smelter at Kandalaksha, Russia (2007). Photo: V. Zverev

Plate 16 Slightly polluted site 1-4, located 6 km N of the aluminium smelter at Kandalaksha (2007): Scots pine forest typical for the study area, without visible signs of pollution impact. Photo: V. Zverev

Plate 17 Polluted site 2-1, located 0.6 km SE of the aluminium smelter at Kandalaksha (2007): Scots pine forest. Crown transparency, absence of Scots pine regeneration, and sparse field layer vegetation are indicative of pollution impact (compare with Plate 16). Photo: V. Zverev

Plate 18 Copper smelter at Karabash, Russia (2007). Photo: V. Zverev

Plate 19 Unpolluted site 1-5, located 32.2 km NE of the copper smelter at Karabash (2007): mixed south taiga forest. Photo: V. Zverev

Plate 20 Polluted site 2-1, located 1.6 km S of the copper smelter at Karabash (2007): industrial barren. Soil organic layer is completely missing. Photo: V. Zverev

Plate 21 Polluted site (additional) located 2 km E of the copper smelter at Karabash (2007): industrial barren. This hill was covered by forest in 1930s. Photo: V. Zverev

Plate 22 Polluted site 2-2, located 3.4 km S of the copper smelter at Karabash (2007): semi-barren landscape. Although soil organic layer persists in some spots, field layer vegetation is absent (compare with Plates 19 and 20). Photo: V. Zverev

Plate 23 Polluted site (additional) located 4.3 km E of the copper smelter at Karabash (2007): semi-barren landscape. Although Scots pines (planted in this heavily eroded area in the early 1990s) show good vitality, natural regeneration is still absent. Photo: V. Zverev

Plate 24 Iron pellet plant at Kostomuksha, Russia (2006). Photo: V. Zverev

Plate 25 Unpolluted site 1-5, located 20.4 km N of iron pellet plant at Kostomuksha (2001): northern taiga forest dominated by Scots pine. Photo: V. Zverev

Plate 26 Polluted site 1-1, located 1 km NW of iron pellet plant at Kostomuksha (2001): northern taiga forest. Crown transparency and absence of lichens on Scots pine trunks (compare with Plate 25) are indicative of pollution impact. Photo: V. Zverev

Plate 27 Copper smelter at Krompachy, Slovakia (2002). Note development of industrial barrens and extensive soil erosion on the hill behind the smelter. Photo: V. Zverev

Plate 28 Slightly polluted site 2-3, located 2.9 km SE of copper smelter at Krompachy (2002): mixed forest typical for the study area, without visible signs of pollution impact. Photo: V. Zverev

Plate 29 Polluted site 2-1, located 1.5 km E of copper smelter at Krompachy (2002): industrial barren. Note heavily eroded soil and stunted growth form of hornbeams Carpinus betulus. Photo: V. Zverev

Plate 30 Nickel-copper smelter at Monchegorsk, Russia (2007). Photo: V. Zverev

Plate 31 Unpolluted site 2-12, located 64 km SE of smelter at Monchegorsk (2007): northern taiga forest dominated by Norway spruce. Photo: V. Zverev

Plate 32 Polluted site (additional) located 8 km NW of smelter at Monchegorsk (2007): dead coniferous stand. This landscape is gradually transforming into industrial barren. Photo: V. Zverev

Plate 33 Polluted site (additional) located 4 km W of smelter at Monchegorsk (2007): dead coniferous stand. Huge amount of dry woody debris makes this landscape prone to fires. Photo: V. Zverev

Plate 34 Polluted site 1-2, located 5 km NW of smelter at Monchegorsk (2007): industrial barren. The upper soil horizons (about 20 cm) have been removed by erosion; stunted mountain birches shown in this photo are at least 50 years old. Photo: V. Zverev

Plate 35 Polluted site 2-3, located 5.7 km S of smelter at Monchegorsk (2007): industrial barren. Note the spatial extent of the disturbed landscape. Photo: V. Zverev

Plate 36 Polluted site (additional) located 2 km W of smelter at Monchegorsk (2007): industrial barren. This landscape has developed from spruce forest (similar to one shown on Plate 31). Photo: V. Zverev

Plate 37 Polluted site 2-4, located 7.5 km S of smelter at Monchegorsk (2007): industrial barren. Due to reduced microbial activity, decomposition of woody debris is very slow and thin branches are preserved for decades. Photo: V. Zverev

Plate 38 Polluted site 2-4, located 7.5 km S of smelter at Monchegorsk (2007): industrial barren. Note numerous dead stems of creeping dwarf shrubs: this area was previously covered by dense matt of crowberry, *Empetrum nigrum* ssp. *hermaphroditum*. Photo: V. Zverev

Plate 39 Polluted site 2-2, located 4.3 km S of smelter at Monchegorsk (2007): industrial barren. Valery Barcan, Deputy Director of the Lapland State Reserve, shows a 'mobile' specimen of mountain birch, with the main root exposed due to extensive soil erosion. Photo: V. Zverev

Plate 40 Slightly polluted site 1-4, located 8.5 km NW of aluminium smelter at Nadvoitsy (2007): mixed mid-taiga forest typical for the study area, without visible signs of pollution impact. Photo: V. Zverev

Plate 41 Polluted site 2-1, located 0.8 km E of aluminium smelter at Nadvoitsy (2007): mixed mid-taiga forest. Although the forest visually resembles unpolluted site (compare with Plate 40), sparse field layer vegetation and absence of epiphytic lichens are indicative of pollution impact. Photo: V. Zverev

Plate 42 Aluminium smelter at Nadvoitsy, Russia (2008). Photo: V. Zverev

Plate 43 Nickel-copper smelter at Nikel, Russia (2006). Photo: V. Zverev

Plate 44 Slightly polluted site 2-4, located 17.1 km S of smelter at Nikel (2006): birch woodlands typical for the study area, without visible signs of pollution impact. Photo: V. Zverev

Plate 45 Polluted site (additional) located 2.3 km NW of smelter at Nikel (2006): forest transforming to industrial barren. Photo: V. Zverev

Plate 46 Polluted site 2-1, located 1.4 km E of smelter at Nikel (2007): industrial barren. Pollution impact and soil erosion have greatly modified growth form of mountain birch. Photo: V. Zverev

Plate 47 Polluted site 2-1, located 1.4 km E of smelter at Nikel (2007): industrial barren. Pollution combined with climatic stress caused branch mortality on willow, *Salix myrsinifolia* ssp. *borealis*, which is one of most tolerant local woody plants. Photo: V. Zverev

Plate 48 Slightly polluted site 2-4, located 31.8 km E of smelter at Norilsk (2002): sub-tundra larch forest typical for the study area, without visible signs of pollution impact. Photo: V. Zverev

Plate 49 Polluted site 2-1, located 4 km NE of smelter at Norilsk (2002): declined larch forest. Presence of extensive field layer vegetation does not allow for classification of this landscape as industrial barren. Photo: V. Zverev

Plate 50 Non-ferrous smelters at Norilsk, Russia (2002). Photo: V. Zverev

Plate 51 Copper smelter at Revda, Russia (2007). Photo: V. Zverev

Plate 52 Unpolluted site 1-5, located 29.7 km W of copper smelter at Revda (2007): south taiga forest. Photo: V. Zverev

Plate 53 Polluted site 2-1, located 1.7 km S of copper smelter at Revda (2007): industrial barren. Pollution continues to cause acute damage of Siberian spruce. Photo: V. Zverev

Plate 54 Polluted site 2-1, located 1.7 km S of copper smelter at Revda (2007): industrial barren. Stumps of conifers confirm that several decades ago this site was covered by mature coniferous forest (compare with Plate 52). Root exposure suggests that, like near Monchegorsk (compare with Plate 34), the upper soil horizons (about 20 cm) have been removed by erosion. Photo: V. Zverev

Plate 55 Aluminium smelter at Straumsvík, Iceland (2007). Photo: V. Zverev

Plate 56 Unpolluted site 1-5, located 5.2 km SW of aluminium smelter at Straumsvík (2007): lava fields covered by specific vegetation (grassy heaths). Photo: V. Zverev

Plate 57 Polluted site 2-1, located 0.7 km NE of aluminium smelter at Straumsvík (2007): grassy heaths. No visible effects of pollution. Photo: V. Zverev

Plate 58 Unpolluted site 1-5, located 43.3 km NW of nickel smelter at Sudbury (2007): mixed forest. Photo: M. Kozlov

Plate 59 Polluted site 2-1, located 1.4 km SW of nickel smelter at Sudbury (2007): industrial barren. Vegetation on this site is recovering after substantial decline of emissions. Photo: M. Kozlov

Plate 60 Superstack of the Copper Cliff nickel smelter at Sudbury, Canada (2007). Photo: M. Kozlov

Plate 61 Aluminium smelter at Volkhov, Russia (2008). Vegetation is not dead (compare with Plate 63); this picture was taken in late autumn. Photo: V. Zverev

Plate 62 Unpolluted site 1-5, located 13.7 km NW of aluminium smelter at Volkhov (2002): mixed forest. Photo: V. Zverev

Plate 63 Polluted site 1-2, located 2 km N of aluminium smelter at Volkhov (2002): secondary vegetation on a bank of the Volkhov river. No visible effects of pollution. Photo: V. Zverev

Plate 64 Cement factory at Vorkuta, Russia (2002). Photo: E. Zvereva.

Plate 65 Power plant at Vorkuta, Russia (2002). Photo: E. Zvereva

Plate 66 Unpolluted site 1-5, located 14 km W of polluters at Vorkuta (2002): shrubby tundra. Photo: E. Zvereva

Plate 67 Polluted site 1-2, located 2 km SW of polluters at Vorkuta (2002): shrubby tundra. No visible effects of pollution. Photo: E. Zvereva

Plate 68 Aluminium smelter at Žiar nad Hronom, Slovakia (2002). Photo: V. Zverev

Plate 69 Unpolluted site 2-5, located 11 km SE of aluminium smelter at Žiar nad Hronom (2002): broadleaved forest dominated by European beech. Photo: V. Zverev

Plate 70 Polluted site 2-1, located 2.6 km NE of aluminium smelter at Žiar nad Hronom (2002): beech forest is only slightly more transparent than in unpolluted sites (compare with Plate 69). Photo: V. Zverev

References

Aamlid D (1992) Air pollution impact on forest and vegetation of eastern Finnmark, Norway. In: Kismul V, Jerre J, Lobersli E (eds) Effects of air pollutants on terrestrial ecosystems in the border area between Russia and Norway. Proceedings from the first symposium, Svanvik, Norway, March 1992. State Pollution Control Authority, Oslo, pp 56–65

Aamlid D, Skogheim I (2001) The occurrence of *Hypogymnia physodes* and *Melanelia olivacea* lichens on birch stems in northern boreal forest influenced by local air pollution. Norsk Geografisk Tidsskrift 55:94–98

Aamlid D, Vassilieva N, Aarrestad PA, Gytarsky ML, Lindmo S, Karaban R, Korotkov V, Rindal T, Kuzmicheva V, Venn K (2000) The ecological state of the ecosystems in the border areas between Norway and Russia. Boreal Environ Res 5:257–278

Abrahamsen G (1984) Effects of acidic deposition on forest soil and vegetation. Phil Trans Royal Soc Lond Ser B – Biol Sci 305:369–382

Adams MB (2003) Ecological issues related to N deposition to natural ecosystems: research needs. Environ Int 29:189–199

Adams WW, Winter K, Lanzl A (1989) Light and the maintenance of photosynthetic competence in leaves of *Populus balsamifera* L. during short-term exposures to high concentrations of sulfur dioxide. Planta 177:91–97

Aerts R (1995) The advantages of being evergreen. Trends Ecol Evol 10:402–407

Agrawal M (2005) Effects of air pollution on agriculture: an issue of national concern. Natl Acad Sci Lett (India) 28:93–106

Ahonen-Jonnarth U, Van Hees PAW, Lundstrom US, Finlay RD (2000) Organic acids produced by mycorrhizal *Pinus sylvestris* exposed to elevated aluminium and heavy metal concentrations. New Phytol 146:557–567

Akselsson C, Ardo J, Sverdrup H (2004) Critical loads of acidity for forest soils and relationship to forest decline in the northern Czech Republic. Environ Monit Assess 98:363–379

Alekseev AS (1991) The radial increment of *Picea abies* under the influence of atmospheric pollution. Botanicheskij Zhurnal [Bot J, St. Petersburg] 76:1498–1503 (in Russian)

Alexeyev VA (1993) Air pollution impact on forest ecosystems of Kola peninsula: history of investigations, progress and shortcomings. In: Kozlov MV, Haukioja E, Yarmishko VT (eds) Aerial pollution in Kola Peninsula. Proceedings of the International Workshop, St. Petersburg, 14–16 April 1992. Kola Science Centre, Apatity, pp 20–34

Alexeyev VA (1995) Impact of air pollution on far north forest vegetation. Sci Total Environ 160/161:605–617

Allen EB, Temple PJ, Bytnerowicz A, Arbaugh MJ, Sirulnik AG, Rao LE (2007) Patterns of understory diversity in mixed coniferous forests of southern California impacted by air pollution. Scient World J 7:247–263

Allen-Gil SM, Ford J, Lasorsa BK, Monetti M, Vlasova T, Landers DH (2002) Heavy metal contamination in the Taimyr Peninsula, Siberian Arctic. Sci Total Environ 301:119–138

Alliende MC (1989) Demographic studies of a dioecious tree. 2. The distribution of leaf predation within and between trees. J Ecol 77:1048–1058

Allum JAE, Dreisinger BR (1987) Remote sensing of vegetation change near INCO's Sudbury mining complexes. Int J Rem Sens 8:399–416

Alpatov YN, Runova EM, Zabotina NN (2001) Application of topological modelling method for investigation of damaged forests. In: Galkin YS, Chernobrovina OK (eds) Mathematical and physical methods in ecology and environmental monitoring. Proceedings of the international conference, Moscow, 23–25 October 2001. Moscow State Forest University, Moscow, pp 367–379 (in Russian)

Alstad DN, Edmunds GF, Weinstein LH (1982) Effects of air pollutants on insect populations. Ann Rev Entomol 27:369–384

AMAP (2006) AMAP assessment 2006: acidifying pollutants, arctic haze, and acidification in the Arctic. Arctic Monitoring and Assesment Programme, Oslo

Ambo-Rappe R, Lajus DL, Schreider MJ (2008) Increased heavy metal and nutrient contamination does not increase fluctuating asymmetry in the seagrass *Halophila ovalis*. Ecol Indic 8:100–103

Amiro BD, Courtin GM (1981) Patterns of vegetation in the vicinity of an industrially disturbed ecosystem, Sudbury, Ontario. Can J Bot 59:1623–1639

An YJ (2006) Assessment of comparative toxicities of lead and copper using plant assay. Chemosphere 62:1359–1365

Anand M, Ma KM, Okonski A, Levin S, McCreath D (2003) Characterising biocomplexity and soil microbial dynamics along a smelter-damaged landscape gradient. Sci Total Environ 311:247–259

Anand M, Laurence S, Rayfield B (2005) Diversity relationships among taxonomic groups in recovering and restored forests. Conserv Biol 19:955–962

Anciaes M, Marini MA (2000) The effects of fragmentation on fluctuating asymmetry in passerine birds of Brazilian tropical forests. J Appl Ecol 37:1013–1028

Andersen JN (ed) (2000) Management of contaminated sites and land in Central and Eastern Europe. Danish Cooperation for Environment in Eastern Europe, Copenhagen

Anderson FK (1966) Air pollution damage to vegetation in Georgetown Canyon, Idaho. M.Sc. thesis, University of Utah, Utah

Andrén H, Delin A, Seiler A (1997) Population response to landscape changes depends on specialization to different landscape elements. Oikos 80:193–196

Andreucci F, Barbato R, Massa N, Berta G (2006) Phytosociological, phenological and photosynthetic analyses of the vegetation of a highly polluted site. Plant Biosyst 140:176–189

Anisimova OA (1989) Xylophagous insects of Pribaikalie coniferous forests weakened by fluorine emissions of aluminium smelters. In: Rozhkov AS, Pleshanov AS (eds) Studies of forest pathology in Baikal region. Siberian Institute of Plant Physiology and Biochemistry, Irkutsk, pp 39–63 (in Russian)

Anonymous (ed) (1998) IUFRO 18th International meeting for specialists in air pollution effects on forest ecosystems. Forest growth responses to the pollution climate of the 21st century. 21–23 September 1998. Heriot-Watt University, Edinburgh

Anonymous (1999) Russia's Karabash: a town with the air of death about it. http://www.bellybuttonwindow.com/1999/russia/serious_soviet_pollu.html. Accessed 20 Jan 2009

Anonymous (ed) (2000) Air pollution, global change and forests in the new Millennium. The 19th International meeting for specialists in air pollution effects on forest ecosystems. 28–31 May 2000. Michigan Technological University, Houghton

Anonymous (ed) (2004) Meeting on 'Forests under changing climate, enhanced UV and air pollution'. 25–30 August 2004, Oulu, Finland. Abstracts. University of Oulu, Oulu

Anonymous (2007) This year Kandalaksha gets a new plan of city development. http://www.hibiny.ru/news/ru/kandalaksha/society/2007/2/8/5352 (in Russian). Accessed 20 Jan 2009

Armbruster WS, Di Stilio VS, Tuxill JD, Flores TC, Runk JLV (1999) Covariance and decoupling of floral and vegetative traits in nine neotropical plants: a re-evaluation of Berg's correlation pleiades concept. Am J Bot 86:39–55

Armentano TV, Bennett JP (1992) Air pollution effects on the diversity and structure of communities. In: Backer JR, Tingey DT (eds) Air pollution effects on biodiversity. Van Nostrand Reinhold, New York, pp 159–176

Armolaitis K (1998) Nitrogen pollution on the local scale in Lithuania: vitality of forest ecosystems. Environ Pollut 102:55–60

Arseneault D, Payette S (1997) Reconstruction of millennial forest dynamics from tree remains in a subarctic tree line peatland. Ecology 78:1873–1883

Arshad MA, Martin S (2002) Identifying critical limits for soil quality indicators in agro-ecosystems. Agric Ecosyst Environ 88:153–160

Atkinson MD (1992) *Betula pendula* Roth (*B. verrucosa* Ehrh.) and *B. pubescens* Ehrh. J Ecol 80:837–870

Auerbach SI (1981) Ecosystem response to stress: a review of concepts and approaches. In: Barrett GW, Rosenberg R (eds) Stress effects on natural ecosystems. Wiley, Chichester, pp 29–41

Augustaitis AA (1989) Peculiarities of above-ground phytomass formation in young pine stands affected by environmental pollution. In: Izrael YuA (ed) Problems of ecological monitoring and ecosystem modelling, vol 12. Gidrometeoizdat, Leningrad, pp 32–51 (in Russian)

Aznar JC, Richer-Lafleche M, Begin C, Marion J (2007) Mining and smelting activities produce anomalies in tree growth patterns (Murdochville, Quebec). Water Air Soil Pollut 186:139–147

Bååth E (1989) Effects of heavy metals in soil on microbial processes and populations (a review). Water Air Soil Pollut 47:335–379

Bach CE (1990) Plant successional stage and insect herbivory – flea beetles on sand-dune willow. Ecology 71:598–609

Baiduzhy A (1994) Nadvoitsy aluminum plant's fluoride pollution decays teeth and causes joint disease. http://www2.fluoridealert.org/Pollution/Aluminum-Industry/Nadvoitsy-Aluminum-Plant-s-Fluoride-Pollution-Decays-Teeth-Causes-Joint-Disease. Accessed 20 Jan 2009

Baikov KS (ed) (2005) Conspectus florae Siberiae. Vascular plants. Nauka, Novosibirsk (in Russian)

Baklanov AA, Rigina O, Rodjushkina IA (1993) Estimation of pollution fallout in surroundings of Apatity. In: Kozlov MV, Haukioja E, Yarmishko VT (eds) Aerial pollution in Kola Peninsula. Proceedings of the International Workshop, St. Petersburg, 14–16 April 1992. Kola Science Centre, Apatity, pp 116–118

Balshem M (1991) Cancer, control, and causality: talking about cancer in a working class community. Am Ethnol 18:152–172

Balster NJ, Marshall JD (2000) Decreased needle longevity of fertilized Douglas-fir and grand fir in the northern Rockies. Tree Physiol 20:1191–1197

Baltensweiler W (1985) Waldsterben – forest pests and air pollution. Zeitschrift für Angewandte Entomologie 99:77–85

Banásová V, Lackovičová A (2004) The disturbance of grasslands in the vicinity of copper plant in the town of Krompachy (Slovenské Rudohorie Mts, NE Slovakia). Bulletin Slovenskej Botanickej Spoločnosti 26:153–161 (in Slovak)

Banásová V, Holub Z, Zelenakova E (1987) The dynamic, structure and heavy metal accumulation in vegetation under long-term influence of Pb and Cu immissions. Ekologia (Bratislava) 6:101–111

Barbault R (2001) Loss of biodiversity. In: Levin SA (ed) Encyclopedia of biodiversity, vol 3. Academic Press, San Diego/London, pp 761–769

Barcan V (2002a) Leaching of nickel and copper from soil contaminated by metallurgical dust. Environ Int 28:63–68

Barcan V (2002b) Nature and origin of multicomponent aerial emissions of the copper-nickel smelter complex. Environ Int 28:451–456

Barcan V, Kovnatsky E (1998) Soil surface geochemical anomaly around the copper-nickel metallurgical smelter. Water Air Soil Pollut 103:197–218

Barecz P (2000) Copper giant hits recovery path. The Slovak Spectator [6 November 2000]. http://www.spectator.sk/articles/view/737/3/. Accessed 20 Jan 2009

BarentsObserver (2007) The factory at Nikel will be closed. Monchegorskij Rabochij 97(10284) [14 August 2007], p 1 (in Russian)

Barkan VS (1993) Measurement of atmospheric concentrations of sulphur dioxide. In: Kozlov MV, Haukioja E, Yarmishko VT (eds) Aerial pollution in Kola Peninsula. Proceedings of the international workshop, 14–16 April 1992, St. Petersburg, Russia. Kola Science Centre, Apatity, pp 90–98

Barker JR, Tingey DT (eds) (1992) Air pollution effects on biodiversity. Van Nostrand Reinhold, New York

Barnes J, Bender J, Lyons T, Borland A (1999) Natural and man-made selection for air pollution resistance. J Exp Bot 50:1423–1435

Barrett GW, Vandyne GM, Odum EP (1976) Stress ecology. BioScience 26:192–194

BBC News (2007) Toxic truth of secretive Siberian city. http://news.bbc.co.uk/2/hi/europe/6528853.stm. Accessed 20 Jan 2009

Beckett PJ (1984) Using plants to monitor atmospheric polltuion. Laurenthian Univ Rev 16:49–66

Beckett PJ (1986) *Pohlia* moss tolerance to the acidic, metal-contaminated substrate of the Sudbury, Ontario, Canada, mining and smelting region. In: Ernst WHO (ed) Environmental contamination. 2nd International Conference, Amsterdam, September 1986. CEP Consultants, Edinburgh, pp 30–32

Begg CB (1994) Publication bias. In: Cooper H, Hedges LV (eds) The handbook of research synthesis. Russel Sage Foundation, New York, pp 399–410

Begg GS, Reid JB, Tasker ML, Webb A (1997) Assessing the vulnerability of seabirds to oil pollution: sensitivity to spatial scale. Colon Waterbird 20:339–352

Bellini P, Baccini M, Biggeri A, Terracini B (2007) The meta-analysis of the Italian studies on short-term effects of air pollution (MISA): old and new issues on the interpretation of the statistical evidences. Environmetrics 18:219–229

Belskii EA, Lyakhov AG (2003) Responses of bird communities of southern taiga in Middle Ural to industrial environmental pollution. Ekologija [Russ J Ecol] 0(3):200–207 (in Russian)

Bennett JP, Resh HM, Runeckle VC (1974) Apparent stimulations of plant growth by air pollutants. Can J Bot 52:35–41

Berg RL (1960) The ecological significance of correlation pleiades. Evolution 14:171–180

Berger R, Katzensteiner K (1994) Mass outbreak of the small spruce sawfly *Pristiphora abietina* (Christ) (Hym., Thentrinidae) in Hausruck. 2. Influence of air pollution. J Appl Entomol 118:253–266

Berlyand ME (ed) (1967) Review of ambient air pollution in cities and industrial centres of the Soviet Union in 1966. Voeikov Main Geophysical Observatory, St. Petersburg (in Russian)

Berlyand ME (ed) (1968) Review of ambient air pollution in cities and industrial centres of the Soviet Union in 1967. Voeikov Main Geophysical Observatory, St. Petersburg (in Russian)

Berlyand ME (ed) (1969) Review of ambient air pollution in 1968 and scientific prognosis of its changes during the nearest 10–15 years in cities and industrial centres of the Soviet Union. Voeikov Main Geophysical Observatory, St. Petersburg (in Russian)

Berlyand ME (ed) (1970) Review of ambient air pollution in cities and industrial centres of the Soviet Union in 1969. Voeikov Main Geophysical Observatory, St. Petersburg (in Russian)

Berlyand ME (ed) (1971) Review of ambient air pollution in cities and industrial centres of the Soviet Union in 1970. Voeikov Main Geophysical Observatory, St. Petersburg (in Russian)

Berlyand ME (ed) (1972) Review of ambient air pollution in cities and industrial centres of the Soviet Union in 1971. Voeikov Main Geophysical Observatory, St. Petersburg (in Russian)

Berlyand ME (ed) (1973) Review of ambient air pollution in cities and industrial centres of the Soviet Union in 1972. Voeikov Main Geophysical Observatory, St. Petersburg (in Russian)

Berlyand ME (ed) (1974) Review of ambient air pollution in cities and industrial centres of the Soviet Union in 1973. Voeikov Main Geophysical Observatory, St. Petersburg (in Russian)

Berlyand ME (ed) (1975) Review of ambient air pollution in cities and industrial centres of the Soviet Union in 1974. Voeikov Main Geophysical Observatory, St. Petersburg (in Russian)

Berlyand ME (ed) (1976) Review of ambient air pollution in cities and industrial centres of the Soviet Union in 1975. Voeikov Main Geophysical Observatory, St. Petersburg (in Russian)

Berlyand ME (ed) (1979) Review of ambient air pollution in cities and industrial centres of the Soviet Union in 1978. Voeikov Main Geophysical Observatory, St. Petersburg (in Russian)

Berlyand ME (ed) (1980) Annual report of ambient air pollution in cities and industrial centres of the Soviet Union in 1979. Voeikov Main Geophysical Observatory, St. Petersburg (in Russian)

Berlyand ME (ed) (1981) Annual report. Ambient air pollution and protection of air basin of cities and industrial centres of the Soviet Union. 1980. Voeikov Main Geophysical Observatory, St. Petersburg (in Russian)
Berlyand ME (ed) (1982) Annual report of ambient air pollution and aerial emissions of pollutants in cities and industrial centres of the Soviet Union. 1981. Voeikov Main Geophysical Observatory, St. Petersburg (in Russian)
Berlyand ME (ed) (1983) Annual report of ambient air pollution and aerial emissions of pollutants in cities and industrial centres of the Soviet Union. 1982. Voeikov Main Geophysical Observatory, St. Petersburg (in Russian)
Berlyand ME (ed) (1984) Annual report of ambient air pollution and aerial emissions of pollutants in cities and industrial centres of the Soviet Union. 1983, vol 2. Voeikov Main Geophysical Observatory, St. Petersburg (in Russian)
Berlyand ME (ed) (1985) Annual report of ambient air pollution and aerial emissions of pollutants in cities and industrial centres of the Soviet Union. 1984, vol 2. Voeikov Main Geophysical Observatory, St. Petersburg (in Russian)
Berlyand ME (ed) (1986) Annual report of ambient air pollution and aerial emissions of pollutants in cities and industrial centres of the Soviet Union. 1985, vol 2. Voeikov Main Geophysical Observatory, St. Petersburg (in Russian)
Berlyand ME (ed) (1987) Annual report of ambient air pollution and aerial emissions of pollutants in cities and industrial centres of the Soviet Union. 1986, vol 1. Voeikov Main Geophysical Observatory, St. Petersburg (in Russian)
Berlyand ME (ed) (1988) Annual report of ambient air pollution and aerial emissions of pollutants in cities and industrial centres of the Soviet Union. 1987, vol 1. Voeikov Main Geophysical Observatory, St. Petersburg (in Russian)
Berlyand ME (ed) (1989) Annual report of ambient air pollution in cities and industrial centres of the Soviet Union. Emissions of pollutants: 1988. Voeikov Main Geophysical Observatory, St. Petersburg (in Russian)
Berlyand ME (ed) (1990) Annual report of ambient air pollution in cities and industrial centres of the Soviet Union. Emissions of pollutants: 1989. Voeikov Main Geophysical Observatory, St. Petersburg (in Russian)
Berlyand ME (ed) (1991) Annual report of ambient air pollution in cities and industrial centres of the Soviet Union. Emissions of pollutants: 1990. Voeikov Main Geophysical Observatory, St. Petersburg (in Russian)
Berlyand ME (ed) (1992) Annual report of ambient air pollution in cities and industrial centres of the Russian Federation. Emissions of pollutants: 1991. Voeikov Main Geophysical Observatory, St. Petersburg (in Russian)
Berlyand ME (ed) (1994) Annual report of ambient air pollution in cities and industrial centres of the Russian Federation. Emissions of pollutants: 1993. Voeikov Main Geophysical Observatory & Research Institute of Ambient Air Protection, St. Petersburg (in Russian)
Berryman AA (1994) Population dynamics: forecasting and diagnosis from time series. In: Leather SR, Watt AD, Mills NJ, Walters KFA (eds) Individuals, populations, and patterns in ecology. Intercept, Andover, pp 119–128
Beyschlag W, Ryel RJ, Dietsch C (1994) Shedding of older needle age classes does not necessarily reduce photosynthetic primary production of Norway spruce: analysis with a 3-dimensional canopy photosynthesis model. Trees – Struct Funct 9:51–59
Bezel VS (1987) Population ecotoxycology of mammals. Nauka, Moscow (in Russian)
Bezel VS (2006) Ecological toxicology: population and ecosystem aspects. Institute of Plant and Animal Ecology, Ekaterinburg (in Russian)
Bezel VS, Bolshakov VN, Vorobeychik EL (1994) Population ecotoxicology. Nauka, Moscow (in Russian)
Bezel VS, Pozolotina VN, Belskii EA, Zhuikova TV (2001) Variation in population parameters: adaptation to toxic environmental factors. Russ J Ecol 32:413–419

Bezuglaya EJ (ed) (1991) Ambient air in the cities of the RSFSR. Annual report on atmospheric contamination in towns of the Russian Federation in 1990. Voeikov Main Geophysical Observatory, St. Petersburg (in Russian)

Bezuglaya EJ, Rastorgueva GP, Smirnova IV (1991) Breathing of an industrial city. Gidrometeoizdat, Leningrad (in Russian)

Biggeri A, Barbone F, Lagazio C, Bovenzi M, Stanta G (1996) Air pollution and lung cancer in Trieste, Italy: spatial analysis of risk as a function of distance from sources. Environ Health Persp 104:750–754

Birdsey RA (2003) Preface. In: Karnosky DF, Percy KE, Chappelka AH, Simpson C, Pikkarainen J (eds) Air pollution, global change and forests in the new millenium. Elsevier, Amsterdam, pp xxi–xxii

Birg VS (1989) Ecophysiological traits of conifer-feeding lepidopteran populations in an idustrially polluted area. Abstract of Ph.D. thesis, Leningrad State University, Leningrad (in Russian)

Biske GS, Nesterenko IM, Potapova OI (eds) (1977) Biological resources of Kostomuksha region: modes of exploitation and protection. Forest Institute, Petrozavodsk (in Russian)

Bjerke JW, Tommervik H, Finne TE, Jensen H, Lukina N, Bakkestuen V (2006) Epiphytic lichen distribution and plant leaf heavy metal concentrations in Russian-Norwegian boreal forests influenced by air pollution from nickel-copper smelters. Boreal Environ Res 11:441–450

Bjorksten TA, Fowler K, Pomiankowski A (2000) What does sexual trait FA tell us about stress? Trends Ecol Evol 15:163–166

Blacksmith Institute (2007) The World's worst polluted places. The top ten (of the dirty thirty). New York, Blacksmith Institute. www.blacksmithinstitute.org/wwpp2007/finalReport2007.pdf. Accessed 20 Jan 2009

Blais JM, Duff KE, Laing TE, Smol JP (1999) Regional contamination in lakes from the Noril'sk region in Siberia, Russia. Water Air Soil Pollut 110:389–404

Bobro M, Hančuľák J, Brehuv J, Stančo P (2000) Heavy metals in the air between Krompachy and Košice. Acta Montanistica Slovaca 5:321–325 (in Slovak)

Bobrova LI, Kachurin MN (1936) Vegetation of Monchetundra. In: Zinzerling YD (ed) Materials on the vegetation of northern and western parts of the Kola Peninsula. Academy of Sciences of the USSR, Moscow/Leningrad, pp 95–121 (in Russian)

Boese SR, Maclean DC, Elmogazi D (1995) Effects of fluoride on chlorophyll *a* fluorescence in spinach. Environ Pollut 89:203–208

Bogacheva IA (1990) Relationships between insect herbivores and plants in subarctic ecosystems. Institute of Plant and Animal Ecology, Sverdlovsk (in Russian)

Bohne H (1971) Changes of the landscape caused by industrial smoke acids. In: Nováková E, Vaněk J, Štěpán J (eds) Bioindicators of landscape deterioration. Terplan, Prague, pp 14–17

Bolger DT, Suarez AV, Crooks KR, Morrison SA, Case TJ (2000) Arthropods in urban habitat fragments in southern California: area, age, and edge effects. Ecol Appl 10:1230–1248

Bolshakov VN, Pyastolova OA, Vershinin VL (2001) Specific features of the formation of animal species communities in technogenic and urbanized landscapes. Russ J Ecol 32:315–325

Boltramovich S, Dudarev G, Gorelov V (2003) The melting iron curtain: a competitive analysis of the northwest Russian metal cluster. The Research Institute of the Finnish Economy, Helsinki

Bouillon D (2003) Preliminary assessment of 100 years of Pb emissions from the Inco Sudbury operations. Internal INCO document. *NV*; cited after Sudbury Area Risk Assessment (2008)

Bouma J (2002) Land quality indicators of sustainable land management across scales. Agric Ecosyst Environ 88:129–136

Boutron CF, Candelone JP, Hong SM (1995) Greenland snow and ice cores – unique archives of large scale pollution of the troposphere of the northern hemisphere by lead and other heavy metals. Sci Total Environ 161:233–241

Boyd R, Barnes S-J, De Caritat P, Chekushin VA, Melezhik VA, Reimann C, Zientek MA (2009) Emissions from the copper-nickel industry on the Kola Peninsula and at Noril'sk, Russia. Atmos Environ 43:1474–1480

Bradshaw AD, McNeilly T (1981) Evolution and pollution. Edward Arnold, London

Brandle M, Amarell U, Auge H, Klotz S, Brandl R (2001) Plant and insect diversity along a pollution gradient: understanding species richness across trophic levels. Biodivers Conserv 10:1497–1511

Braun SD, Jones TH, Perner J (2004) Shifting average body size during regeneration after pollution – a case study using ground beetle assemblages. Ecol Entomol 29:543–554

Bray JR (1964) Primary consumption in three forest canopies. Ecology 45:165–167

Brimblecombe P, Makra L (2005) Selections from the history of environmental pollution, with special attention to air pollution. Part 2: from medieval times to the 19th century. Int J Environ Pollut 23:351–367

Bringmark E, Bringmark L (1995) Disappearance of spatial variability and structure in forest floors – a distinct effect of air pollution? Water Air Soil Pollut 85:761–766

Brooker RW, Callaghan TV (1998) The balance between positive and negative plant interactions and its relationship to environmental gradients: a model. Oikos 81:196–207

Broto VC, Tabbush P, Burningham K, Elghali L, Edwards D (2007) Coal ash and risk: four social interpretations of a pollution landscape. Landscape Res 32:481–497

Brun LA, Le Corff J, Maillet J (2003) Effects of elevated soil copper on phenology, growth and reproduction of five ruderal plant species. Environ Pollut 122:361–368

Bryndina EV (2000) Effects of discharges of a copper smelting plant on communities of wood-decaying fungi of Southern taiga. Sibirskij Ekologicheskij Zhurnal [Siberian Ecol J, Novosibirsk] 0(6):679–684 (in Russian)

Bucha T, Maňkovská B (2002) Change analyses of forest health condition development in Žiar nad Hronom region influenced by aluminium plant. In: Maňkovská B (ed) Long term air pollution effect on forest ecosystems. 20th International meeting for specialists in air pollution effects on forest ecosystems. Book of abstracts. August 30–September 1, 2002. Forest Research Institute, Zvolen, pp 134

Bugrovskii VV, Butusov OB, Orlov DS, Stepanov AM, Sukhanova NI, Chernenkova TV (1985) Satellite monitoring of impact zones of metallurgical industry with considerable chemical emissions. In: Pyavchenko NI (ed) The problem of human impact on environment. Nauka, Moscow, pp 113–123 (in Russian)

Bunce HWF (1979) Fluoride emissions and forest growth. J Air Pollut Cont Assoc 29:642–643

Burns DA, Blett T, Haeuber R, Pardo LH (2008) Critical loads as a policy tool for protecting ecosystems from the effects of air pollutants. Frontiers Ecol Environ 6:156–159

Bussotti F, Ferretti M (1998) Air pollution, forest condition and forest decline in Southern Europe: an overview. Environ Pollut 101:49–65

Bussotti F, Grossoni P, Paoletti E, Raddi R, Tronconi A (eds) (1996) Stress factors and air pollution. 17th International meeting for specialists in air pollution effects on forest ecosystems. Firenze, 14–19 September, 1996. Book of Abstracts. Centro Stampa 2P, Firenze

Bussotti F, Cascio C, Strasser RJ, Schaub M, Gerosa GA (2008) General fature of ozone stress on woody plants, detected by the chlorophyll *a* fluorescence transient (FT). In: Schaub M, Dobbertin MK, Steiner D (eds) Air pollution and climate change at contrasting altitude and latitude. 23rd IUFRO conference for specialists in air pollution and climate change effects on forest ecosystems, Murten, Switzerland, 7–12 September 2008. Abstracts. Swiss Federal Research Institute, Birmensdorf, p 162

Butovsky RO, Gongalsky KB (1999) Morphometric analysis of ground beetles (Coleoptera, Carabidae) in anthropogenic impact bioindication. In: Butovsky RO, Straalen NM van (eds) Pollution-induced changes in soil invertebrate food-webs, vol 2. Vrije Universiteit, Amsterdam, pp 77–87

Butusov OB, Nosova LM, Stepanov AM (1998) Analysis of forest recovery after an indrustry-caused ecological catastrophe by the means of aerial diagnostics. Izvestiya Rossiiskoi Akademii Nauk Seriya Geograficheskaya [Bull Russ Acad Sci Geogr, Moscow] 0(1):90–101 (in Russian)

Bylund H, Tenow O (1994) Long-term dynamics of leaf miners, *Eriocrania* spp., on mountain birch – alternate year fluctuations and interaction with *Epirrita autumnata*. Ecol Entomol 19:310–318

Bytnerowicz A, Godzik S, Poth M, Anderson I, Szdzuj J, Tobias C, Macko S, Kubiesa P, Staszewski T, Fenn M (1999) Chemical composition of air, soil and vegetation in forests of the Silesian Beskid Moutains, Poland. Water Air Soil Pollut 116:141–150

Bytnerowicz A, Omasa K, Paoletti E (2007) Integrated effects of air pollution and climate change on forests: a northern hemisphere perspective. Environ Pollut 147:438–445

Cairns J, Heath AG, Parker BC (1975) Temperature influence on chemical toxicity to aquatic organisms. J Water Pollut Cont Feder 47:267–280

Cairns J, Niederlehner BR (1996) Developing a field of landscape ecotoxicology. Ecol Appl 6:790–796

Calow P, Sibly RM, Forbes V (1997) Risk assessment on the basis of simplified life-history scenarios. Environ Toxicol Chem 16:1983–1989

Cape JN, Fowler D, Davison A (2003) Ecological effects of sulfur dioxide, fluorides, and minor air pollutants: recent trends and research needs. Environ Int 29:201–211

Carlson CE, Bousfield WE, Mcgregor MD (1977) Relationship of an insect infestation on Lodgepole pine to fluorides emitted from a nearby aluminum plant in Montana. Fluoride 10:14–21

Carrier JG, Krippl E (1990) Comprehensive study of European forests assesses damage and economic losses from air pollution. Environ Conserv 17:365–366

Chalcraft DR, Cox SB, Clark C, Cleland EE, Suding KN, Weiher E, Pennington D (2008) Scale-dependent responses of plant biodiversity to nitrogen enrichment. Ecology 89:2165–2171

Chan WH, Lusis MA (1985) Post-superstack Sudbury smelter emissions and their fate in the atmosphere – an overview of the Sudbury environment study results. Water Air Soil Pollut 26:43–58

Chapman PM (2002) Integrating toxicology and ecology: putting the "eco" into ecotoxicology. Marine Pollut Bull 44:7–15

Chekovich KM, Dushatkin AB, Gapon AE, Lomonosov IS, Sidorov II (1993) Ecogeochemistry of cities in regions with modest self-cleaning capacities and modest landscape stability. Surroundings of Norilsk. In: Makarov VN, Marshintsev VK, Fedoseeva VI (eds) Ecogeochemistry of East Siberian cities. Institute of Permafrost Research, Yakutsk, pp 63–67 (in Russian)

Chernenkova TV (1986) Methods of an integrated assessment of the state of forest ecosystems in the impact zone of industrial enterprises. In: Lyubashevsky NM, Sadykov OF (eds) Boundary problems of ecology. Institute of Plant and Animal Ecology, Sverdlovsk, pp 116–127 (in Russian)

Chernenkova TV (1991) Phytoindication of early stages of industrial disturbance of north taiga ecosystems. In: Krivolutsky DA (ed) Bioindication and biomonitoring. Nauka, Moscow, pp 114–120 (in Russian)

Chernenkova TV (2002) Response of forest vegetation to industrial pollution. Nauka, Moscow (in Russian)

Chernenkova TV, Stepanov AM, Gordeeva MM (1989) Changes in structure of forest communities under industrial impact. Zhurnal Obstchei Biologii [J Fund Biol, Moscow] 50:388–394 (in Russian)

Chernenkova TV, Mekhanikova EV, Gusarova AYu (1999) Recovery of forest vegetation following decline of emissions of a metallurgical plant. Lesovedenie [Forestry, Moscow] 0(2):28–34 (in Russian)

Chernenkova TV, Kabirov RR, Mekhanikova EV, Stepanov AM, Gusarova AYu (2001) Recovery of vegetation after a copper smelter shutdown. Lesovedenie [Forestry, Moscow] 0(6):31–37 (in Russian)

Chernousova NF (1990) To the assessment of industrial emission impact on small mammals. In: Vershinin VL (ed) Animals in human-affected landscape. Institute of Plant and Animal Ecology, Sverdlovsk, pp 83–90 (in Russian)

Choi DS, Kayama M, Jin HO, Lee CH, Izuta T, Koike T (2006) Growth and photosynthetic responses of two pine species (*Pinus koraiensis* and *Pinus rigida*) in a polluted industrial region in Korea. Environ Pollut 139:421–432

Chumakov LS (1989) Carabid communities of perennial herb fields under impact of the emissions of chemical industry. Deposited in All-Union Institute for Scientific and Technical Information, No. 4349-B89. Minsk (in Russian)

Cicák A, Mihál I (1996) Health condition and mycoflora of unmixed beech stand in the regions with different immission impact. Lesnicky casopis [Forest J] 42:145–151 (in Slovak)
Clarke GM (1992) Fluctuating asymmetry: a technique for measuring developmental stress of genetic and environmental origin. Acta Zool Fenn 191:31–35
Clements WH (2004) Small-scale experiments support causal relationships between metal contamination and macroinvertebrate community responses. Ecol Appl 14:954–967
Clements WH, Newman MC (2002) Community ecotoxicology. Wiley, New York
Cogbill CV (1977) Effect of acid precipitation on tree growth in eastern North America. Water Air Soil Pollut 8:89–93
Cohen J (1988) Statistical power analysis for the behavioral sciences. Erlbaum, Hillsdale
Coley PD, Barone JA (1996) Herbivory and plant defenses in tropical forests. Ann Rev Ecol Syst 27:305–335
Conn VS, Rantz MJ (2003) Research methods: managing primary study quality in meta-analyses. Res Nurs Health 26:322–333
Connell JH (1978) Diversity in tropical rain forests and coral reefs – high diversity of trees and corals is maintained only in a non-equilibrium state. Science 199:1302–1310
Cook CM, Kostidou A, Vardaka E, Lanaras T (1997) Effects of copper on the growth, photosynthesis and nutrient concentrations of *Phaseolus* plants. Photosynthetica 34:179–193
Corwin DL, Lesch SM (2003) Application of soil electrical conductivity to precision agriculture: theory, principles, and guidelines. Agron J 95:455–471
Costanza R, d'Arge R, de Groot R, Farber S, Grasso M, Hannon B, Limburg K, Naeem S, O'Neill RV, Paruelo J, Raskin RG, Sutton P, van den Belt M (1997) The value of the World's ecosystem services and natural capital. Nature 387:253–260
Costescu LM, Hutchinson TC (1972) The ecological consequences of soil pollution by metallic dust from the Sudbury smelters. In: Environmental progress in science and education. Proceedings of the 18th annual technical meeting, New York, 1–4 May 1972. Institute of Environmental Sciences, New York, pp 540–545
Courtin GM (1994) The last 150 years – a history of environmental degradation in Sudbury. Sci Total Environ 148:99–102
Cox JD, Beckett J (1993) Changes in terrestrial lichen communities along an environmental gradient affected by metals. In: Allan RJ, Nriagu JO (eds) Heavy metals in the environment. 9th International conference. Papers. CEP Consultants, Edinburgh, pp 68–71
Cox R, Percy K, Jensen K, Simpson C (eds) (1996) Air pollution and multiple stresses. Proceedings of the 16th International meeting for specialists in air pollution effects on forest ecosystems, Fredericton, New Brunswick, Canada, 7–9 September 1994. Canadian Forest Service, Fredericton
Cox RM, Hutchinson TC (1980) Multiple metal tolerances in the grass *Deschampsia cespitosa* (L.) Beauv. from the Sudbury smelting area. New Phytol 84:631–647
Cranston PS (1990) Biomonitoring and invertebrate taxonomy. Environ Monit Assess 14:265–273
Crowden A (2008) BES annual symposium 2008: ecology of industrial pollution. Brit Ecol Soc Bull 39:37
Crowther C, Steuart DW (1914) Further studies of the effects of smoke from towns upon vegetation in the surrounding areas. J Agric Sci 6:395–405
Crutsinger GM, Sanders NJ (2005) Aphid-tending ants affect secondary users in leaf shelters and rates of herbivory on *Salix hookeriana* in a coastal dune habitat. Am Midl Nat 154:296–304
Cyr H, Pace ML (1993) Magnitude and patterns of herbivory in aquatic and terrestrial ecosystems. Nature 361:148–150
DalCorso G, Farinati S, Maistri S, Furini A (2008) How plants cope with cadmium: staking all on metabolism and gene expression. J Integr Plant Biol 50:1268–1280
Daley PF (1995) Chlorophyll fluorescence analysis and imaging in plant stress and disease. Can J Plant Pathol 17:167–173
Das S, Sarker M, Mukhopadhyay A (2005) Changing diversity of hymenopteran parasitoids from organically and conventionally managed tea ecosystem of North Bengal, India. J Environ Biol 26:505–509

Dässler H-G (ed) (1976) Einfluß von Luftverunreinigungen auf die Vegetation. Ursachen – Wirkungen – Gegenmaßnahmen. Gustav Fischer, Jena

Dauskevicius V (1984) Impact of industrial pollution on forest plantations. In: Vaicis MV (ed) Impact of industrial pollution on forest ecosystems and measures to increase their resistance. Abstracts of presentations from the All-Union scientific and practical conference, 26–27 June, 1984. Lithuanian Reasearch Institute of Forestry, Kaunas, pp 87–89 (in Russian)

Dauwe T, Janssens E, Eens M (2006) Effects of heavy metal exposure on the condition and health of adult great tits (*Parus major*). Environ Pollut 140:71–78

Davydova ND (2001) Development of technogenic geochemical anomalies in south taiga low-mountain geosystems of Middle Siberia. Geografija i Prirodnye Resursy [Geogr Nat Resour, Irkutsk] 0(2):73–80 (in Russian)

Delaney KJ (2008) Injured and uninjured leaf photosynthetic responses after mechanical injury on *Nerium oleander* leaves, and *Danaus plexippus* herbivory on *Asclepias curassavica* leaves. Plant Ecol 199:187–200

Delgado-Rodríguez M (2006) Systematic reviews of meta-analyses: applications and limitations. J Epidemiol Commun Health 60:90–92

Demyanov VA, Kitsing LI, Yarmishko VT (1996) Effect of industrial pollution on radial increment of *Larix gmelinii* (Pinaceae). Izvestiya Rossiiskoi Akademii Nauk Seriya Biologicheskaya [Bull Russ Acad Sci Biol, Moscow] 0(4):490–494 (in Russian)

Derome J, Lindroos AJ (1998) Effects of heavy metal contamination on macronutrient availability and acidification parameters in forest soil in the vicinity of the Harjavalta Cu-Ni smelter, SW Finland. Environ Pollut 99:225–232

Derome J, Nieminen T (1998) Metal and macronutrient fluxes in heavy metal polluted Scots pine ecosystems in SW Finland. Environ Pollut 103:219–228

Deyeva NM, Maznaja EA (1993) The state of bilberry in polluted and unpolluted forests of the Kola peninsula. In: Kozlov MV, Yarmishko VT, Haukioja E (eds) Aerial pollution in Kola Peninsula. Proceedings of the international workshop, St. Petersburg, Russia, 14–16 April 1992. Kola Science Centre, Apatity, pp 308–312

Didham RK, Norton DA (2006) When are alternative stable states more likely to occur? A reply to T. Fukami and W. G. Lee. Oikos 113:357–362

Ditrikh SA (2001) Basic ecological problems of the Bratsk aluminium plant. Tsvetnye Metally [Nonferr Metals, Moscow] 0(7):55–58 (in Russian)

Divan AM, Oliva MA, Martinez CA, Cambraia J (2007) Effects of fluoride emissions on two tropical grasses: *Chloris gayana* and *Panicum maximum* cv. Coloniao. Ecotoxicol Environ Safety 67:247–253

Dmowski K, Kozakiewicz A, Kozakiewicz M (1998) Small mammal populations and community under conditions of extremely high thallium contamination in the environment. Ecotoxicol Environ Safety 41:2–7

Doak DF, Bigger D, Harding EK, Marvier MA, O'Malley RE, Thomson D (1998) The statistical inevitability of stability-diversity relationships in community ecology. Am Nat 151:264–276

Doak DF, Estes JA, Halpern BS, Jacob U, Lindberg DR, Lovvorn J, Monson DH, Tinker MT, Williams TM, Wootton JT, Carroll I, Emmerson M, Micheli F, Novak M (2008) Understanding and predicting ecological dynamics: are major surprises inevitable? Ecology 89:952–961

Dobbertin M (2005) Tree growth as indicator of tree vitality and of tree reaction to environmental stress: a review. Eur J Forest Res 124:319–333

Dodson SI, Allen TFH, Carpenter SR, Ives AR, Jeanne RL, Kitchell JF, Langston NE, Turner MG (1998) Ecology. Oxford University Press, New York

Doncheva AV (1979) Landscape in zone of industry impact. Lesnaya Promyshlennost, Moscow (in Russian)

Doran JW (2002) Soil health and global sustainability: translating science into practice. Agric Ecosyst Environ 88:119–127

Doran JW, Jones AJ (1996) Methods for assessing soil quality. SSSA Special Publication 49. Soil Science Society America, Madison

Downs TJ, Ambrose RF (2001) Syntropic ecotoxicology: a heuristic model for understanding the vulnerability of ecological systems to stress. Ecosyst Health 7:266–283

Dreisinger BR, McGovern PC (1970) Monitoring atmospheric sulphur dioxide and correlating its effect on crops and forests in the Sudbury area. In: Linzon SN (ed) Impact of air pollution on vegetation. Conference, 7–9 April 1970, Toronto, Ontario. Air Pollution Control Association, Pittsburgh, pp 11–28

Druzhinina OA (1985) Vegetation dynamics in intensively developed regions of the extreme North. In: Chernov YI (ed) Communities of the Far North and man. Nauka, Moscow, pp 205–231 (in Russian)

Dubey B (2004) A model for the effect of time delay on the dynamics of a population living in a polluted environment. J Biol Syst 12:35–43

Duchesne L, Ouimet R, Houle D (2002) Basal area growth of sugar maple in relation to acid deposition, stand health, and soil nutrients. J Environ Qual 31:1676–1683

Dudka S, Adriano DC (1997) Environmental impacts of metal ore mining and processing: a review. J Environ Qual 26:590–602

Dudka S, Poncehernandez R, Hutchinson TC (1995) Current level of total element concentrations in the surface layer of Sudbury's soils. Sci Total Environ 162:161–171

Eeva T, Lehikoinen E (2000) Recovery of breeding success in wild birds. Nature 403:851–852

Eeva T, Tanhuanpää S, Råbergh C, Airaksinen S, Nikinmaa M, Lehikoinen E (2000) Biomarkers and fluctuating asymmetry as indicators of pollution-induced stress in two hole-nesting passerines. Funct Ecol 14:235–243

Eeva T, Belskii E, Kuranov B (2006a) Environmental pollution affects genetic diversity in wild bird populations. Mutat Res – Genet Toxicol Environ Mut 608:8–15

Eeva T, Hakkarainen H, Laaksonen T, Lehikoinen E (2006b) Environmental pollution has sex-dependent effects on local survival. Biol Lett 2:298–300

Efron S (1994) Ecological Russian roulette. Los Angeles Times, 22 November. http://www2.fluoridealert.org/Pollution/Aluminum-Industry/Ecological-Russian-Roulette. Accessed 20 Jan 2009

Elkarmi A, Eideh RA (2006) Allometry of *Urtica urens* in polluted and unpolluted habitats. J Plant Biol 49:9–15

Ellenberg H, Weber HE, Düll R, Wirth W, Werner W, Paulissen D (1992) Zeigerwerte von Pflanzen in Mitteleuropa. Ed. 2. Scripta Geobotanica 18:1–258

Emberson L, Ashmore M, Murray F (eds) (2003) Air pollution impacts on crops and forests, a global assessment. Imperial College Press, London

Emissions of pollutants in Russia (1966–2005): reference used for the serial publications; see Berlyand (1967, 1968, 1969, 1970, 1971, 1972, 1973, 1974, 1975, 1976, 1979, 1980, 1981, 1982, 1983, 1984, 1985, 1986, 1987, 1988, 1989, 1990, 1991, 1992, 1994), Milyaev and Nikolaev (1996), Milyaev et al. (1997a, b, 1998, 1999, 2000, 2001, 2002, 2003), Milyaev and Yasenskij (2004, 2006)

Eränen JK (2008) Rapid evolution towards heavy metal resistance by mountain birch around two subarctic copper-nickel smelters. J Evol Biol 21:492–501

Eränen JK (2009) Evolutionary ecology of mountain birch in subarctic stress gradients: interplay of biotic and abiotic factors in plant-plant interactions and evolutionary processes. Ph.D. thesis, University of Turku, Turku

Eränen JK, Kozlov MV (2006) Physical sheltering and liming improve survival and performance of mountain birch seedlings: a 5-year study in a heavily polluted industrial barren. Restor Ecol 14:77–86

Eränen JK, Kozlov MV (2007) Competition and facilitation in industrial barrens: variation in performance of mountain birch seedlings with distance from nurse plants. Chemosphere 67:1088–1095

Eränen JK, Kozlov MV (2008) Increasing intraspecific facilitation in exposed environments: consistent results from mountain birch populations in two subarctic stress gradients. Oikos 117:1569–1577

Eränen JK, Kozlov MV (2009) Interactions between mountain birch seedlings from differentiated populations in contrasting environments of subarctic Russia. Plant Ecol 200:167–177

Eränen JK, Nilsen J, Zverev VE, Kozlov MV (2009) Mountain birch under multiple stressors – heavy metal resistant populations co-resistant to biotic stress but mal-adapted to abiotic stress. J Evol Biol 22:840–851

Eriksson CP, Holmgren P (1996) Estimating stone and boulder content in forest soils – evaluating the potential of surface penetration methods. Catena 28:121–134

Ermakov AI (2004) Structural changes in the carabid fauna of forest ecosystems under a toxic impact. Russ J Ecol 35:403–408

Ershov SF, Kaitmazov NG, Shur MB, Naftal MN, Kozlov AN, Ilyukhin IV, Tsemekhman LS (2005) Long-term strategy of the Zapolyarnyy branch of Norilsk Nikel Ltd. in sulphur dioxide utilization – the most important step on the way to the global leadership. Tsvetnye Metally [Nonferr Metals, Moscow] 0(10):73–77 (in Russian)

European Environment Agency (2003) Europe's environment: the third assessment. Environmental assessment report No. 10. EEA, Copenhagen

Evdokimova GA (2000) The impact of heavy metals on the microbial diversity of podzolic soils in the Kola peninsula. In: Innes L, Oleksyn J (eds) Forest dynamics in heavily polluted regions. CAB International, Wallingford, pp 67–76

Evdokimova GA, Mozgova NP, Shtina EA (1997) Soil pollution by fluorine and assessment of soil microflora status in the impact area of an aluminium smelter. Pochvovedenie [Soil Sci, Moscow] 0(7):898–905 (in Russian)

Evdokimova GA, Zenkova IV, Mozgova NP, Pereverzev VN (2005) Soil and soil biota under fluorine contamination. Institute of North Industrial Ecology Problems, Apatity (in Russian)

Evdokimova GA, Korneykova MV, Lebedeva EV (2007) Micromycete communities from soils in the impact zone of aluminium plant. Mikologiya i Fitopatologiya [Mycol Phytopathol, St. Petersburg] 41(1):20–28 (in Russian)

Everhart JL, McNear D, Peltier E, van der Lelie D, Chaney RL, Sparks DL (2006) Assessing nickel bioavailability in smelter-contaminated soils. Sci Total Environ 367:732–744

Ewers FW, Schmid R (1981) Longevity of needle fascicles of *Pinus longaeva* (bristlecone pine) and other North American pines. Oecologia 51:107–115

Fabricius C, Burger M, Hockey PAR (2003) Comparing biodiversity between protected areas and adjacent rangeland in xeric succulent thicket, South Africa: arthropods and reptiles. J Appl Ecol 40:392–403

Fabricius KE, De'ath G (2004) Identifying ecological change and its causes: a case study on coral reefs. Ecol Appl 14:1448–1465

Fadeeva MA (1999) Lichens of pine forests of north-west Karelia affected by atmospheric pollution. Abstract of Ph.D. thesis, Komarov Botanical Institute, St. Petersburg (in Russian)

FAO (1992) Mixed and pure forest plantations in the tropics and subtropics. FAO Forestry Paper 103. FAO, Rome

FAO (2006) New_LocClim, Local Climate Estimator Version 1.10. Environment and Natural Resources Service – Agrometeorology Group, FAO/SDRN, Rome, Italy. ftp://ext-ftp.fao.org/SD/Reserved/Agromet/New_LocClim/. Accessed 20 Jan 2009

Fedorets NG, Diakonov VV, Litinsky PY, Shiltsova GV (1998) Contamination of the forested area in Republic of Karelia with heavy metals and sulphur. Forest Research Institute, Petrozavodsk (in Russian)

Feng ZW, Miao H, Zhang FZ, Huang YZ (2002) Effects of acid deposition on terrestrial ecosystems and their rehabilitation strategies in China. J Environ Sci (China) 14:227–233

Feriancová-Masárová Z, Kalivodová E (1965a) Vplyv exhalátov hlinikárne v Žiari nad Hronom na druhové zloženie vtákov v okolí. Biológia (Bratislava) 20:109–121

Feriancová-Masárová Z, Kalivodová E (1965b) Niekoľko poznámok o vplyve fluórových exhalátov v okolí hlinkárne v Žiari nad Hronom na kvantitu hniezdiacich vtákov. Biológia (Bratislava) 20:341–347

Ferretti M, Bucher J, Bytnerowicz A, Prus-Glowacki W, Karnosky DF, Percy KE (2003) State of science and gaps in our knowledge in relation to air pollution. In: Karnosky DF, Percy KE,

Chappelka AH, Simpson C, Pikkarainen J (eds) Air pollution, global change and forests in the new millennium. Elsevier, Amsterdam, pp 437–446
Field CB, Chapin FS, Matson PA, Mooney HA (1992) Responses of terrestrial ecosystems to the changing atmosphere – a resource-based approach. Ann Rev Ecol Syst 23:201–235
Filipov D (2004) Polluted town counts on Putin. The Boston Globe, 13 March 2004. http://www.boston.com/yourlife/health/other/articles/2004/03/13/polluted_town_counts_on_putin/ (in Russian). Accessed 20 Jan 2009
Filser J (2008) Ecotoxicology and ecosystems: relevance, restrictions, research needs. Basic Appl Ecol 9:333–336
Filser J, Koehler H, Ruf A, Rombke J, Prinzing A, Schaefer M (2008) Ecological theory meets soil ecotoxicology: challenge and chance. Basic Appl Ecol 9:346–355
Fimushin BS (1979) Regularities in growth of Scots pine stands and methods of evaluation of damage imposed by industrial emissions in surroundings of Sverdlovsk. Abstract of Ph.D. thesis, Ural Forest Technical Institute, Sverdlovsk (in Russian)
Fisher AEI, Hartley SE, Young M (1999) Behavioural responses of the leaf-chewing guild to the presence of *Eriocrania* mines on silver birch (*Betula pendula*). Ecol Entomol 24:156–162
Fitzgerald GF (1980) Canadian-American arbitration and creation of law – the contribution of the Trail smelter case to the development of international law, including the emerging law of trasboundary air pollution. Etudes Internationales 11:393–419
Flagler RB (ed) (1998) Recognition of air pollution injury to vegetation: a pictorial atlas. Air & Waste Management Association, Pittsburgh
Fomin VV, Shavnin SA (2001) Ecological zoning of forests in areas exposed to industrial air pollution. Russ J Ecol 32:89–93
Fontaine B, Gargominy O, Neubert E (2007) Land snail diversity of the savanna/forest mosaic in Lope National Park, Gabon. Malacologia 49:313–338
Formann AK (2008) Estimating the proportion of studies missing for meta-analysis due to publication bias. Contemp Clin Trial 29:732–739
Foster BL, Smith VH, Dickson TL, Hildebrand T (2002) Invasibility and compositional stability in a grassland community: relationships to diversity and extrinsic factors. Oikos 99:300–307
Fowler D, Cape JN, Coyle M, Flechard C, Kuylenstierna J, Hicks K, Derwent D, Johnson C, Stevenson D (1999) The global exposure of forests to air pollutants. In: Sheppard LJ, Cape JN (eds) Forest growth responses to the pollution climate of the 21st century. Kluwer, Dordrecht, pp 5–32
Fox CA, Kincaid WB, Nash TH, Young DL, Fritts HC (1986) Tree-ring variation in western larch (*Larix occidentalis*) exposed to sulfur dioxide emissions. Can J Forest Res 16:283–292
Fox GA (1991) Practical causal inference for ecoepidemiologists. J Toxicol Environ Health 33:359–373
Freedman B (1989) Environmental ecology. Academic Press, San Diego
Freedman B, Hutchinson TC (1980a) Smelter pollution near Sudbury, Ontario, Canada, and effects on forest litter decomposition. In: Hutchinson TC, Havas M (eds) Effects of acid precipitation on terrestrial ecosystems. Plenum, New York/London, pp 395–434
Freedman B, Hutchinson TC (1980b) Long-term effects of smelter pollution at Sudbury, Ontario, on forest community composition. Can J Bot 58:2123–2140
Freeman DC, Graham JH, Emlen JM (1993) Developmental stability in plants – symmetries, stress and epigenesis. Genetica 89:97–119
Freer-Smith PH (1997) Forest growth and decline: what is the role of air pollutants? In: Yunus M, Iqbal M (eds) Plant response to air pollution. Wiley, Chichester, pp 437–447
Fritze H (1987) The influence of urban air pollution on soil respiration and fungal hyphal length. Ann Bot Fenn 24:251–256
Fritze H (1992) Effects of environmental pollution on forest soil microflora – a review. Silva Fenn 26:37–48
Fritze H, Vanhala P, Pietikainen J, Mälkönen E (1996) Vitality fertilization of Scots pine stands growing along a gradient of heavy metal pollution: short-term effects on microbial biomass and respiration rate of the humus layer. Fresenius J Anal Chem 354:750–755

Frontasyeva MV, Smirnov LI, Steinnes E, Lyapunov SM, Cherchintsev VD (2004) Heavy metal atmospheric deposition study in the South Ural Mountains. J Radioanal Nucl Chem 259:19–26

Führer E (1985) Air pollution and the incidence of forest insect problems. Z Angew Entomol 99:371–377

Fuksman IL, Poikalainen Y, Shreders SM, Kanyuchkova GK, Chinenova LA (1997) Physiological and biochemical indication of the state of Scotch pine under the influence of industrial pollution. Russ J Ecol 28:184–187

Gago C, Marcos MLF, Alvarez E (2002) Aqueous aluminium species in forest soils affected by fluoride emissions from an aluminium smelter in NW Spain. Fluoride 35:110–121

Gangestad SW, Thornhill R (1999) Individual differences in developmental precision and fluctuating asymmetry: a model and its implications. J Evol Biol 12:402–416

Genkai-Kato M (2007) Regime shifts: catastrophic responses of ecosystems to human impacts. Ecol Res 22:214–219

Genoni GP (1997) Towards a conceptual synthesis in ecotoxicology. Oikos 80:96–106

Georgievskij AV (1990) Impact of air pollution on the vegetation of suburban area of Kandalaksha and the Kandalaksha Reserve. Nauchnye Doklady Vysshei Shkoly, Biologicheskie Nauki [Scient Rep High School Biol Sci, Moscow] 0(9):124–133 (in Russian)

Germanova NI, Medvedeva MV (2001) Dynamics of biological activity of forest soil in the impact zone of Kostomuksha ore processing enterprise. In: Krutov VI, Sinkevich SM, Kaibiyainen LK, Fedorets NG (eds) Bioecological aspects of forest ecosystem monitoring in North-West Russia. Institute of Forest, Petrozavodsk, pp 192–202 (in Russian)

Getsen MV (ed) (2000) Ecologo-economic and social problems of the Vorkuta industrial region (search for the solutions and maintenance of sustainability). Committee on Natural Resources of the Republic of Komi, Vorkuta & Syktyvkar (in Russian)

Getsen MV, Stenina A, Patova EN (1994) Algal flora of the Bolshezemelskaya tundra under anthropogenic influence. Nauka, Ekaterinburg (in Russian)

Getsen MV, Stenina AS, Voronin LV (eds) (1996) Bioindication of the environmental state in tundras of Vorkuta. Institute of Biology, Syktyvkar (in Russian)

Gignac LD (1987) La distribution des bryophytes sur des tourbières contaminées par des métaux dans la région de Sudbury, Ontario, Canada. Cryptogamie Bryologie Lichénologie 8:339–351

Gilad E, Shachak M, Meron E (2007) Dynamics and spatial organization of plant communities in water-limited systems. Theor Popul Biol 72:214–230

Gilbert OL (1975) Effects of air pollution on landscape and land-use around Norwegian aluminum smelters. Environ Pollut 8:113–121

Giller KE, Witter E, McGrath SP (1998) Toxicity of heavy metals to microorganisms and microbial processes in agricultural soils: a review. Soil Biol Biochem 30:1389–1414

Gíslason GM, Helgason GV (1989) A literature review on ther environmental impact of air pollution in the surroundings of aluminium smelters, with emphasis on the aluminium smelter in Straumsvík. Preliminary report to the Icelandic Energy Marketing Unit. Reykjavik, Institute of Biology, University of Iceland

Glasov MV (1986) Participation of herbivores in formation of primary production by spruce forests of southern taiga. In: Abaturov VV, Tishkov AA (eds) Herbivores in terrestrial biogeocoenoses. Proceedings of All-Union conference, Valdai, 3–6 June 1984. Nauka, Moscow, pp 58–64 (in Russian)

Glass GV (2000) Meta-analysis at 25. http://glass.ed.asu.edu/gene/papers/meta25.html. Accessed 20 Jan 2009

Glasziou PP, Sanders SL (2002) Investigating causes of heterogeneity in systematic reviews. Stat Med 21:1503–1511

Gleason HA, Cronquist A (1993) Manual of vascular plants of northeastern United States and adjacent Canada. New York Botanical Garden, New York

Goldberg DE, Rajaniemi T, Gurevitch J, Stewart-Oaten A (1999) Empirical approaches to quantifying interaction intensity: competition and facilitation along productivity gradients. Ecology 80:1118–1131

Goldberg IL (1997) Changes in the moss layer of southern taiga dark coniferous forests under conditions of technogenic pollution. Russ J Ecol 28:417–419

Golubeva NI (1991) Accumulation of heavy metals in soils of the city of Apatity. Deposited in the All-Union Institute for Scientific and Technical Information, No. 2194-B91. Moscow (in Russian)

Golutvin GI, Selikhovkin AV, Tokmakov AV (1983) Insects as indicator of gaseous environmental contamination In: Solovyov VA (ed) Ecology and protection of forests: forest pathology and nature conservation. Leningrad State Forest Technical Academy, Leningrad, pp 34–39 (in Russian)

Gomez AA, Swete-Kelly DE, Syers JK, Coughlan KJ (1996) Measuring sustainability of agricultural systems at the farm level. In: Doran JW, Jones AJ (eds) Methods for assessing soil quality. SSSA special publication 49. Soil Science Society of America, Madison, WI, pp 401–410

Gordon AG, Gorham E (1963) Ecological aspects of air pollution from an iron-sintering plant at Wawa, Ontario. Can J Bot 41:1063–1078

Gorham E, Gordon AG (1960a) Some effects of smelter pollution northeast of Falconbridge, Ontario. Can J Bot 38:307–312

Gorham E, Gordon AG (1960b) The influence of smelter fumes upon the chemical composition of lake waters near Sudbury, Ontario, and upon the surrounding vegetation. Can J Bot 38:477–487

Gorshkov SP (1997) Landscape geoecological mapping of the Norilsk area of ecological disaster. In: Gennadiev AN, Milanova EV (eds) Modifications of natural environment: global and regional aspects. Moscow State University, Moscow, pp 148–161 (in Russian)

Gottlieb MS, Shear CL, Seale DB (1982) Lung cancer mortality and residential proximity to industry. Environ Health Persp 45:157–164

Goulden ML, Munger JW, Fan SM, Daube BC, Wofsy SC (1996) Exchange of carbon dioxide by a deciduous forest: response to interannual climate variability. Science 271:1576–1578

Grachev VA (2004) Ecological problems of metallurgical industry and legislation of the Russian Federation. Problemy Regionalnoi Ekologii [Prob Reg Ecol, Moscow] 0(4):98–102 (in Russian)

Graham JH, Freeman DC, Emlen JM (1993) Developmental stability: a sensitive indicator of populations under stress. In: Landis WG, Hughes JS, Lewis MA (eds) Environmental toxicology and risk assessment. American Society for Testing and Materials, Philadelphia, pp 136–158

Greenberg R, Bichier P, Angon AC, MacVean C, Perez R, Cano E (2000) The impact of avian insectivory on arthropods and leaf damage in some Guatemalan coffee plantations. Ecology 81:1750–1755

Grime JP (1973) Competitive exclusion in herbaceous vegetation. Nature 242:344–347

Grime JP (1979) Plant strategies and vegetation processes. Wiley, New York

Grimm NB, Foster D, Groffman P, Grove JM, Hopkinson CS, Nadelhoffer KJ, Pataki DE, Peters DPC (2008) The changing landscape: ecosystem responses to urbanization and pollution across climatic and societal gradients. Frontiers Ecol Environ 6:264–272

Grodziński W, Yorks TP (1981) Species and ecosystem level bioindicators of airborne pollution – an analysis of two major studies. Water Air Soil Pollut 16:33–53

Gruner DS, Smith JE, Seabloom EW, Sandin SA, Ngai JT, Hillebrand H, Harpole WS, Elser JJ, Cleland EE, Bracken MES, Borer ET, Bolker BM (2008) A cross-system synthesis of consumer and nutrient resource control on producer biomass. Ecol Lett 11:740–755

Guardans R (2002) Estimation of climate change influence on the sensitivity of trees in Europe to air pollution concentrations. Environ Sci Policy 5:319–333

Gulyaeva NG, Kalieva OP (2004) Ecological and geochemical evaluation of north taiga landscapes of Karelia and Kola regions. Razvedka i Okhrana Nedr [Explor Prot Earth's Inter] 0(3):82–87 (in Russian)

Gunn JM (ed) (1995) Restoration and recovery of an industrial region – progress in restoring the smelter-damaged landscape near Sudbury, Canada. Springer, New York

Guretzky JA, Moore KJ, Burras CL, Brummer EC (2004) Distribution of legumes along gradients of slope and soil electrical conductivity in pastures. Agron J 96:547–555

Gurevitch J, Hedges LV (1999) Statistical issues in ecological meta-analyses. Ecology 80:1142–1149

Gurevitch J, Hedges LV (2001) Meta-analysis. Combining the results of independent experiments. In: Schneider SM, Gurevitch J (eds) Design and analysis of ecological experiments. Oxford University Press, Oxford, pp 347–369

Hagen SB, Ims RA, Yoccoz NG, Sorlibraten O (2008) Fluctuating asymmetry as an indicator of elevation stress and distribution limits in mountain birch (*Betula pubescens*). Plant Ecol 195:157–163

Haidouti C, Chronopoulou A, Chronopoulos J (1993) Effects of fluoride emissions from industry on the fluoride concentration of soils and vegetation. Biochem Syst Ecol 21:195–208

Hajdúk J (1974) Changes in plant communities in an area affected by fluorine emissions. In: Sobocký E (ed) Les a priemyselne imisie. VEDA, Bratislava, pp 71–79 (in Slovak)

Hajdúk J, Lisiká E (1999) *Cladonia rei* (lichenized Ascomycotina) on heavy metal contaminated habitats near copper smelters at Krompachy (NE Slovakia). Bulletin Slovenskej Botanickej Spoločnosti 21:49–51

Harjavallan kaupunki (1991) Harjavallan ympäristön tila 1990 (Harjavallan Kaupungin Tutkimussarja 1/1991). Harjavallan kaupunki, Harjavalta

Hartsock NJ, Mueller TG, Thomas GW, Barnhisel RI, Wells KL, Shearer SA (2000) Soil electrical conductivity variability. In: Robert PC, Rust RH, Larson WE (eds) Proceedings of 5th international conference on precision agriculture, Bloomington, 16–19 July 2000 [CD-room]. Center for Precision Agriculture, University of Minnesota, St Paul

Haselhoff E, Lindau G (1903) Die Beschädigung der Vegetation durch Rauch. Handbuch zur Erkennung und Beurteilung von Rauchschäden. Gebrüder Borntraeger, Leipzig

Hassan IA (2004) Interactive effects of salinity and ozone pollution on photosynthesis, stomatal conductance, growth, and assimilate partitioning of wheat (*Triticum aestivum* L.). Photosynthetica 42:111–116

Haukioja E, Koponen S, Ojala H (1973) Local differences in leaf consumption by invertebrates in northern Norway and Finland. Rep Kevo Subarct Res Sta 10:29–33

Hayes F, Jones MLM, Mills G, Ashmore M (2007) Meta-analysis of the relative sensitivity of semi-natural vegetation species to ozone. Environ Pollut 146:754–762

Haywood JK (1907) Injury to vegetation and animal life by smelter fumes. J Am Chem Soc 29:998–1009

Hazlett PW, Rutherford GK, Loon GW van (1983) Metal contaminants in surface soils and vegetation as a result of nickel/copper smelting at Coniston, Ontario, Canada. Reclam Reveg Res 2:123–137

Hein L, van Ierland E (2006) Efficient and sustainable management of complex forest ecosystems. Ecol Model 190:351–366

Heino J, Koskenkari T (2004) Industrial ecology and the metallurgy industry. The Harjavalta industrial ecosystem. In: Pongrácz E (ed) Proceedings of the waste minimization and resources use optimization conference, 10 June 2004, University of Oulu, Finland. Oulu University Press, Oulu, pp 143–151

HELCOM (2004) Status of the hot spots in Saint-Petersburg and the Leningrad region. Based on the Fifth HELCOM PITF Regional Workshop, St. Petersburg, Russia, June 2001. (Baltic Sea Environment Proceedings 98). Helsinki, Baltic Marine Environment Protection Commission

Heliövaara K, Väisänen R (1993) Insects and pollution. CRC Press, Boca Raton

Hell P, Stanovský M, Žilinec M (1995) Dentalfluorose des Rehwildes in der Region einer slowakischen Aluminiumfabrik. Zeitschrift für Jagdwissenschaft 41:117–125

Helm B, Albrecht H (2000) Human handedness causes directional asymmetry in avian wing length measurements. Anim Behav 60:899–902

Helmisaari HS (2000) Functioning of forest ecosystems under heavy metal loading. Areas polluted by heavy metals. In: Mälkönen E (ed) Forest condition in a changing environment – the Finnish case. Kluwer, Dordrecht, pp 248–251

Helmisaari HS, Derome J, Fritze H, Nieminen T, Palmgren K, Salemaa M, Vanha-Majamaa I (1995) Copper in Scots pine forests around a heavy-metal smelter in south-western Finland. Water Air Soil Pollut 85:1727–1732

Helmisaari HS, Salemaa M, Derome J, Kiikkila O, Uhlig C, Nieminen TM (2007) Remediation of heavy metal-contaminated forest soil using recycled organic matter and native woody plants. J Environ Qual 36:1145–1153

Helsel DR (1990) Less than obvious – statistical treatment of data below the detection limit. Environ Sci Technol 24:1766–1774
Hettelingh JP, Posch M, Slootweg J (eds) (2008) CCE status report 2008: critical load, dynamic modelling and impact assessment in Europe. Report No. 500090003. Netherlands Environmental Assessment Agency, Bilthoven. http://www.mnp.nl. Accessed 20 Jan 2009
Hewitt JE, Thrush SE, Cummings VJ (2001) Assessing environmental impacts: effects of spatial and temporal variability at likely impact scales. Ecol Appl 11:1502–1516
Hijano CF, Dominguez MDP, Gimenez RC, Sanchez PH, Garcia IS (2005) Higher plants as bioindicators of sulphur dioxide emissions in urban environments. Environ Monit Assess 111:75–88
Hill AB (1965) Environment and disease: association or causation. Proc Royal Soc Med Lond 58:295–300
Hill MO, Preston CD, Roy DB (2004) PLANTATT. Attributes of British and Irish plants: status, size, life history, geography and habitats. Centre for Ecology & Hydrology, Huntingdon
Hobbs NT (2003) Challenges and opportunities in integrating ecological knowledge across scales. Forest Ecol Manage 181:223–238
Hoffman DJ, Rattner BA, Burton GAJ, Cairns JJ (1995) Handbook of ecotoxicology. Lewis, Boca Raton
Høgda KA, Tømmervik H, Solheim I, Marhaud Ø (1995) Use of multitemporal Landsat image data for mapping the effects from air pollution in the Kirkenes-Pechenga area in the period 1973–1994. NORUT-IT Report, IT2039/1–95. University of Tromsø, Tromsø
Holland R (1888) Air pollution as affecting plant life. Air pollution. [Lectures.] No. 9. Manchester & Salford Noxious Vapours Abatement Association, Manchester
Holmstrup M, Petersen BF, Larsen MM (1998) Combined effects of copper, desiccation, and frost on the viability of earthworm cocoons. Environ Toxicol Chem 17:897–901
Honkasalo J (1989) Kolosjoen sulatto. In: Autere E, Liede J (eds). Petsamon Nikkeli: taistelu strategisesta metallista. Vuorimiesyhdistys, Helsinki, pp 74–85
Hose GC, Murray BR, Park ML, Kelaher BP, Figueira WF (2006) A meta-analysis comparing the toxicity of sediments in the laboratory and in situ. Environ Toxicol Chem 25:1148–1152
Houghton JT, Ding Y, Griggs DJ, Noguer M, van der Linden PJ, Dai X, Maskell K, Johnson CA (2001) Climate change 2001: the scientific basis. Third assessment report. WGI Cambridge University Press, Cambridge/New York
Howard DM, Howard PJA (1993) Relationships between CO_2 evolution, moisture content and temperature for a range of soil types. Soil Biol Biochem 25:1537–1546
Hronec O (1996) Exhaláty – Pôda – Vegetácia. TOP s.r.o., Prešov
Hu WB, Mengersen K, McMichael A, Tong SL (2008) Temperature, air pollution and total mortality during summers in Sydney, 1994–2004. Int J Biometeorol 52:689–696
Hume D (2001) Environmental monitoring and fluctuating asymmetry: study design. New Zealand Department of Conservation, Wellington
Hunt M (1997) How science takes stock. Russell Sage Foundation, New York
Hunter MD, Price PW (1998) Cycles in insect populations: delayed density dependence or exogenous driving variables? Ecol Entomol 23:216–222
Hunter MLJ, Gibbs JP (2007) Fundamentals of conservation biology. Blackwell, Cambridge/Oxford
Huopalainen M, Tuittila ES, Vanha-Majamaa I, Nousiainen H, Laine J, Vasander H (2000) The potential of soil seed banks for revegetation of bogs in SW Finland after long-term aerial pollution. Ann Bot Fenn 37:1–9
Hurlbert SH (1984) Pseudoreplication and the design of ecological field experiments. Ecol Monogr 54:187–211
Hursh CR (1948) Local climate in Copper Basin of Tennessee as modified by the removal of vegetation. US Dept Agric Circ 774:1–38
Hustich I (1978) The growth of Scots pine in northern Lapland, 1928–77. Ann Bot Fenn 15:241–252
Hutchinson TC, Whitby LM (1974) Heavy-metal pollution in the Sudbury mining and smelting region of Canada, I. Soil and vegetation contamination by nickel, copper, and other metals. Environ Conserv 1:123–132

Hynninen V (1986) Monitoring of airborne metal pollution with moss bags near an industrial source at Harjavalta, southwest Finland. Ann Bot Fenn 23:83–90

Hyvönen R, Ågren GI, Linder S, Persson T, Cotrufo MF, Ekblad A, Freeman M, Grelle A, Janssens IA, Jarvis PG, Kellomäki S, Lindroth A, Loustau D, Lundmark T, Norby RJ, Oren R, Pilegaard K, Ryan MG, Sigurdsson BD, Strömgren M, van Oijen M, Wallin G (2007) The likely impact of elevated CO_2, nitrogen deposition, increased temperature and management on carbon sequestration in temperate and boreal forest ecosystems: a literature review. New Phytol 173:463–480

Igamberdiev VM, Tereshenkov OM, Kutyev HA, Popova EN (1994) Evaluation of the recent state of the environment: Norilsk industrial region. Narodnoe Khozyaistvo Respubliki Komi [Econ Komi Republic, Vorkuta] 3:54–61 (in Russian)

Innes JL, Haron AH (eds) (2000) Air pollution and the forests of developing and rapidly industrializing countries. CAB International, Wallingford

Innes JL, Oleksyn J (eds) (2000) Forest dynamics in heavily polluted regions. CAB International, Wallingford

Isachenko AG, Kuznetsov VI, Chertov OG, Stchukin YN, Glukhov OD, Gorelik DO, Krasov VI, Milyaev VB, Khvorov GV (1990) Ecological map of the Leningrad Region (The state of the environment). Gortechprogress, Leningrad (in Russian)

Jacobson JS, Hill AC (eds) (1970) Recognition of air pollution injury to vegetation: a pictorial atlas. Air Pollution Control Association, Pittsburgh

Jalkanen R (1995) Needle trace method for retrospective needle retention studies on Scots pine (*Pinus sylvestris* L.). Ph.D. thesis, University of Oulu, Oulu

Jalkanen RE (1996) Needle retention chronology along a pollution gradient. In: Dean JS, Meko DM, Swetnam TW (eds) Tree rings, environment and humanity. Proceedings of the International Conference, Tucson, Arizona, 17–21 May 1994. Radiocarbon, Tucson, pp 419–426

Janssens IA, Lankreijer H, Matteucci G, Kowalski AS, Buchmann N, Epron D, Pilegaard K, Kutsch W, Longdoz B, Grünwald T, Montagnani L, Dore S, Rebmann C, Moors EJ, Grelle A, Rannik Ü, Morgenstern K, Oltchev S, Clement R, Gudmundsson J, Minerbi S, Berbigier P, Ibrom A, Moncrieff J, Aubinet M, Bernhofer C, Jensen NO, Vesala T, Granier A, Schulze ED, Lindroth A, Dolman AJ, Jarvis PG, Ceulemans R, Valentini R (2001) Productivity overshadows temperature in determining soil and ecosystem respiration across European forests. Global Change Biol 7:269–278

Jennions MD, Møller AP (2002) Relationships fade with time: a meta-analysis of temporal trends in publication in ecology and evolution. Proc Royal Soc Lond Ser B – Biol Sci 269:43–48

Jennions MD, Møller AP (2003) A survey of the statistical power of research in behavioral ecology and animal behavior. Behav Ecol 14:438–445

Johansen P, Muir DCG, Law RJ (2000) Contaminants in the Greenland environment – foreword. Sci Total Environ 245:1–2

Johnson AR (2002) Landscape ecotoxicology and assessment of risk at multiple scales. Human Ecol Risk Assess 8:127–146

Jonas T, Rixen C, Sturm M, Stoeckli V (2008) How alpine plant growth is linked to snow cover and climate variability. J Geophys Res – Biogeosci 113:G03013, doi:10.1029/2007JG000680

Jones CC (2001) What causes insect and disease outbreaks on trees? Part 2. Tree Care Ind 12:80–87

Jones ME, Paine TD (2006) Detecting changes in insect herbivore communities along a pollution gradient. Environ Pollut 143:377–387

Juknys R, Vencloviene J, Stravinskiene V, Augustaitis A, Bartkevicius E (2003) Scots pine (*Pinus sylvestris* L.) growth and condition in a polluted environment: from decline to recovery. Environ Pollut 125:205–212

Junttila O, Heide OM (1981) Shoot and needle growth in *Pinus sylvestris* as related to temperature in Northern Fennoscandia. Forest Sci 27:423–430

Kabirov RR (1993) Changes of north forest algaesinusia under human impact in the Kola region. In: Kozlov MV, Haukioja E, Yarmishko VT (eds) Aerial pollution in Kola peninsula. Proceeding of the international workshop, St. Petersburg, 14–16 April 1992. Kola Science Centre, Apatity, pp 268–271

Kaibiyainen LK, Bolondinskii VK, Sofronova GI, Yalynskaya EE (2001) Monitoring of physiological state of woody plants in zones of industrial impact. In: Krutov VI, Sinkevich SM, Kaibiyainen LK, Fedorets NG (eds) Bioecological aspects of forest ecosystems monitoring in North-West Russia. Institute of Forest, Petrozavodsk, pp 100–114 (in Russian)
Kaigorodova SY, Vorobeichik EL (1996) Changes in certain properties of grey forest soil polluted with emissions from a copper-smelting plant. Russ J Ecol 27:177–183
Kalabin GV (1999) Ecological atlas of the Murmansk Region. Institute of North Industrial Ecology Problems and Geographical Faculty of the Moscow State University, Moscow/Apatity (in Russian)
KalaŬP, Nižnanská M, Bevilaqua D, Stašková I (1996) Concentrations of mercury, copper, cadmium and lead in fruiting bodies of edible mushrooms in the vicinity of a mercury smelter and a copper smelter. Sci Total Environ 177:251–258
Kaleta M (1982) Forest ecosystems in area of Slovak smeltings. Čistota ovzdušia 5:1–7 (in Slovak)
Kammer PM, Mohl A (2002) Factors controlling species richness in alpine plant communities: an assessment of the importance of stress and disturbance. Arct Antarct Alp Res 34: 398–407
Kandler O, Innes JL (1995) Air pollution and forest decline in central Europe. Environ Pollut 90:171–180
Karlen DL, Mausbach MJ, Doran JW, Cline RG, Harris RF, Schuman GE (1997) Soil quality: a concept, definition, and framework for evaluation. Soil Sci Soc Am J 61:4–10
Karnosky DF, Percy KE, Thakur RC, Honrath RE (2003a) Air pollution and global change: a double challenge to forest ecosystems. In: Karnosky DF, Percy KE, Chappelka AH, Simpson C, Pikkarainen J (eds) Air pollution, global change and forests in the new millennium. Elsevier, Amsterdam, pp 1–42
Karnosky DF, Percy K, Mankovska B, Prichard T, Noormets A, Dickson RE, Jepsen E, Isebrands JG (2003b) Ozone affects the fitness of trembling aspen. In: Karnosky DF, Percy KE, Chappelka AH, Simpson C, Pikkarainen J (eds) Air pollution, global change and forests in the new millennium. Elsevier, Amsterdam, pp 199–210
Karnosky DF, Percy KE, Chappelka AH, Krupa SV (2003c) Air pollution and global change impacts on forest ecosystems: monitoring and research needs. In: Karnosky DF, Percy KE, Chappelka AH, Simpson C, Pikkarainen J (eds) Air pollution, global change and forests in the new millennium. Elsevier, Amsterdam, pp 447–460
Kataev GD. (1984) Small mammals as indicators of anthropogenic impact in the Kola peninsula. In: Krivolutsky DA (ed) Effects of industrial enterprises on the environment. Abstracts of All-Union seminar, Zvenigorod, 4–6 December 1984. Severtsov Institute of Animal Evolutionary Morphology and Ecology, Pushchino, pp 90–93 (in Russian)
Kataev OA, Golutvin GI, Popovichev BG (1981) Impact of industrial aerial emissions on the state of young Scots pine forests. In: Belov SV (ed) Forestry, silviculture and soil science, vol 10. Leningrad State Forest Technical Academy, Leningrad, pp 40–45 (in Russian)
Keane RE (2008) Biophysical controls on surface fuel litterfall and decomposition in the northern Rocky Mountains, USA. Can J Forest Res 38:1431–1445
Kellerová D (2005) The air pollution in the surroundings of an aluminium plant. Ekologia (Bratislava) 24:108–114
Khantemirova EV (1996) Spruce-fir forests of Middle Ural and industrial pollution. In: Vorobeichik EL, Mukhacheva SV, Mikhailova IN (eds) Problems of general and applied ecology. Materials of the conference of young scientists. Ekaterinburg Publishers, Ekaterinburg, pp 278–290 (in Russian)
Kharuk VI (2000) Air pollution impacts on subarctic forests at Noril'sk, Siberia. In: Innes JL, Oleksyn J (eds) Forest dynamics in heavily polluted regions (IUFRO task force on environmental change. Report No. 1) CAB International, Wallingford, pp 77–86
Kharuk VI, Winterberger K, Tsybulskiy GM, Yakhimovich AP (1995) Satellite data analysis of pollution-induced subtundra forest decline. Issledovanija Zemli iz Kosmosa [Earth Observ Rem Sens, Moscow] 0(4):91–97 (in Russian)
Kharuk VI, Winterberger K, Tsibulskii GM, Yakhimovich AP, Moroz SN (1996) Technogenic disturbance of pretundra forests in Norilsk valley. Russ J Ecol 27:406–410

Kidd NAC (1991) The implications of air pollution for conifer aphid population dynamics – a simulation analysis. J Appl Entomol 111:166–171

Kiikkilä O (2003) Heavy-metal pollution and remediation of forest soil around the Harjavalta Cu-Ni smelter, in SW Finland. Silva Fenn 37:399–415

Kijamov KV (1959) Contamination of ambient air by pulp and paper mill [in Syasstroy] and aluminium plant [in Volkhov]. In: Zhdanov VM (ed) The XIIth All-Union meeting of hygienists, epidemiologists, microbiologists and infectionists, vol 1. Problems of hygiene. State Publishing House of Medical Literature, Moscow, pp 194–196 (in Russian)

Kirtsideli IJ, Vorobeva NI, Tereshenkova OM (1995) Influence of industrial pollution on the communities of soil micromycetes of forest tundra in Taimyr Peninsula. Mikologiya i Fitopatologiya [Mycol Phytopathol, St. Petersburg] 29(4):12–19 (in Russian)

Kitao M, Lei TT, Koike T (1997) Comparison of photosynthetic responses to manganese toxicity of deciduous broad-leaved trees in northern Japan. Environ Pollut 97:113–118

Kitao M, Lei TT, Koike T, Kayama M, Tobita H, Maruyama Y (2007) Interaction of drought and elevated CO_2 concentration on photosynthetic down-regulation and susceptibility to photoinhibition in Japanese white birch seedlings grown with limited N availability. Tree Physiol 27:727–735

Klein DR, Vlasova TM (1992) Lichens, a unique forage resource threatened by air pollution. Rangifer 12:21–27

Klein W (1989) Mobility of environmental chemicals, including abiotic degradation. In: Borudeau P, Haines JA, Klein W, Krishna Murti CR (eds) Ecotoxicology and climate, with special reference to hot and cold climates (SCOPE 38, IPCS Joint Symposia 9). Wiley, Chichester, pp 65–78

Klumpp A, Domingos M, Pignata ML (1999) Air pollution and vegetation damage in South America – state of knowledge and perspectives. In: Agrawal SB, Agrawal M (eds). Environmental pollution and plant responses. Lewis, Boca Raton, pp 111–136

Knierim U, Van Dongen S, Forkman B, Tuyttens FAM, Spinka M, Campo JL, Weissengruber GE (2007) Fluctuating asymmetry as an animal welfare indicator – a review of methodology and validity. Physiol Behav 92:398–421

Knistautas AJ (1982) The major of revealed effects of industrial pollution on populations of forest birds. In: Budris R (ed) Ecological studies and protection of birds in Baltic republics: abstracts of Baltic conference of young ornithologists, dedicated to the centenary of the birth of Prof. T. Ivanauskas. Nature Conservation State Committee of Lithuanian SSR, Kaunas, pp 151–153 (in Russian)

Knistautas AJ (1983) Nesting of the great tit under impact of aerial pollution. Bulleten Moskovskogo Obshchestva Ispytateley Prirody, Serija Biologicheskaja [Bull Moscow Soc Nat Biol Ser] 88(2):17–21 (in Russian)

Kobayashi O, Funada R, Fukazawa K, Ohtani J (1997) A dendrochronological evaluation of the effects of air pollution on the radial growth of Norway spruce. Mokuzai Gakkaishi 43:824–831

Koch R (1880) Investigations into the etiology of traumatic infective deseases. The New Sydenham Society, London

Kočí V (2006) Zelená jezírka, rudá puda a zlutý vzduch. Vesmír 85:26–28

Kohlhammer Y, Rzehak P, Behr J, Wichmann HE, Heinrich J (2007) High prevalence of pneumonia in children of a smelter town. Int J Occup Environ Health 13:167–174

Koleff P, Gaston KJ, Lennon JJ (2003) Measuring beta diversity for presence-absence data. J Anim Ecol 72:367–382

Komarov V (1978) The destruction of nature in the Soviet Union. Pluto, London

Komulainen M, Vieno M, Yarmishko VT, Daletskaja TD, Maznaja EA (1994) Seedling establishment from seeds and seed banks in forests under long-term pollution stress – a potential for vegetation recovery. Can J Bot 72:143–149

Kontrišová O (1980) Meadow associations in an area of action of fluorine-type immissions (Žiar basin). Biologické Práce 26:1–159 (in Slovak)

Koricheva J, Haukioja E (1992) Effects of air pollution on host plant quality, individual performance, and population density of *Eriocrania* miners (Lepidoptera, Eriocraniidae). Environ Entomol 21:1386–1392

Koricheva J, Larsson S, Haukioja E (1998) Insect performance on experimentally stressed woody plants: a meta-analysis. Ann Rev Entomol 43:195–216

Koroleva LI, Shavnin SA (2000) Possibility to indicate the condition of Scots pine by the content and composition of its ester oil in industrially polluted areas of Middle Ural. Tekhnologii Kachestva Zhizni [Qual Life Technol, Ekaterinburg] 0(1):49–56 (in Russian)

Koster M, de Groot A, Vijver M, Peijnenburg W (2006) Copper in the terrestrial environment: verification of a laboratory-derived terrestrial biotic ligand model to predict earthworm mortality with toxicity observed in field soils. Soil Biol Biochem 38:1788–1796

Koutsenogii KP, Smirnova AI, Smolyakov BS, Churkina TV (2002) Application of an integrated approach to evaluate contamination of northern regions by emissions of Norilsk mining and smelting enterprise. In: Izrael YA (ed) Scientific aspects of ecological problems of Russia. Proceedings of All-Russian Conference, Moscow, 13–16 June 2001, vol 2. Nauka, Moscow, pp 229–233 (in Russian)

Kovalenko LA, Babushkina LG, Shebalova NM, Kokovkina TF (1997) Changes in biological activity of litter and soil of pine plantations in an area contaminated with sulfur-containing chemicals and heavy metals. In: Makhnev AK (ed) Biological recultivation of disturbed lands. Proceedings of the International Congress, 26–29 August 1996. Institute of Forest, Ekaterinburg, pp 125–136 (in Russian)

Kovalev BI, Filipchuk AN (1990) Condition of forests in an area of impact by industrial emissions. Lesnoe Khoziaystvo [Forest Manage, Moscow] 0(5):36–38 (in Russian)

Kowalkowskí A (1990) Survival rate of tree and shrub seedlings under extreme environmental nitrogen pollution. Water Air Soil Pollut 54:51–59

Kozlov MV (1984) Pest lepidopterans on birches [in Murmansk region]. Zastchita Rastenij [Plant Prot, Moscow] 0(4):31–32 (in Russian)

Kozlov MV (1985) Changes in phytophagous insect communities affected by airborne pollutants near the nonferrous smelter. In: Kastrel TN (ed) Problems of biosphere (Information bulletin), vol 11. Scientific Board on Biospheric Problems, Academy of Sciences of the USSR, Moscow, pp 76–90 (in Russian)

Kozlov MV (1990) Impact of anthropogenic factors on populations of terrestrial insects. All-Union Institute for Scientific and Technical Information, Moscow (in Russian)

Kozlov MV (1991) Using of moths for bioindication of the state of stands under aerial pollution impact. In: Baranchikov YuN (ed) Resistance of forests to insect impact. Sukachev Forestry Institute, Krasnoyarsk, pp 26–27 (in Russian)

Kozlov MV (1996) Subalpine and alpine assemblages of Lepidoptera in surroundings of a powerful smelter on the Kola Peninsula, NW Russia. Nota Lepidopterologica 18:17–37

Kozlov MV (1997) Pollution impact on insect biodiversity in boreal forests: evaluation of effects and perspectives of recovery. In: Crawford RMM (ed) Disturbance and recovery in Arctic lands: an ecological perspective. Proceedings of the NATO advanced research workshop on disturbance and recovery of Arctic terrestrial ecosystems, Rovaniemi, Finland, 24–30 September 1995. (NATO ASI series. Partnership subseries 2. Environment, vol 25). Kluwer, Dordrecht, pp 213–250

Kozlov MV (1999) Offspring sex ratio in a heavily polluted town. Reprod Toxicol 13:567–568

Kozlov MV (2001a) Developmental stability: an alleged simplicity of methodology (about the manual 'Environmental health: methods of evaluation'). Zapovedniki i Natsionalnye Parki [Nat Reserv Natl Parks, Moscow] 36:23–25 (in Russian)

Kozlov MV (2001b) Snowpack changes around a nickel-copper smelter at Monchegorsk, northwestern Russia. Can J Forest Res 31:1684–1690

Kozlov MV (2002) Changes in wind regime around a nickel-copper smelter at Monchegorsk, northwestern Russia. Int J Biometeorol 46:76–80

Kozlov MV (2003) Density fluctuations of the leafminer *Phyllonorycter strigulatella* (Lepidoptera: Gracillariidae) in the impact zone of a power plant. Environ Pollut 121:1–10

Kozlov MV (2004) Silviculture in polluted areas. In: Burley J, Evans J, Youngquist JA (eds) Encyclopedia of forest science. Elsevier, London, pp 1112–1121

Kozlov MV (2005a) Sources of variation in concentrations of nickel and copper in mountain birch foliage near a nickel-copper smelter at Monchegorsk, north-western Russia: results of long-term monitoring. Environ Pollut 135:91–99

Kozlov MV (2005b) Pollution resistance of mountain birch, *Betula pubescens* subsp. *czerepanovii*, near the copper-nickel smelter: natural selection or phenotypic acclimation? Chemosphere 59:189–197

Kozlov MV (2005c) Contrasting response of mountain birch to damage by *Eriocrania* leafminers in polluted and unpolluted habitats. Can J Bot 83:73–79

Kozlov MV (2006) Severonikel smelter as the model for studies of the impact of industrial pollution on biota: analysis of the accumulated information. In: Evdokimova GA, Vandysh O (eds) Modern ecological problems of the North (To the centenary of the OI Semenov-Tyan-Shanskiy birthday). Proceedings of the international conference, 10–12 October 2006, part 1. Institute of the North Industrial Ecology Problems, Apatity, pp 231–233 (in Russian)

Kozlov MV (2007) Improper sampling design and pseudoreplicated analysis: conclusions by Veličković (2004) questioned. Hereditas 144:43–44

Kozlov MV (2008) Losses of birch foliage due to insect herbivory along geographical gradients in Europe: a climate-driven pattern? Clim Change 87:107–117

Kozlov MV, Barcan V (2000) Environmental contamination in the central part of the Kola Peninsula: history, documentation, and perception. Ambio 29:512–517

Kozlov MV, Haukioja E (1997) Microclimate changes along a strong pollution gradient in northern boreal forest zone. In: Uso JL, Brebbia CA, Power H (eds) Ecosystems and sustainable development (Advances in ecological sciences, vol 1). Computation Mechanics Publishers, Southampton, pp 603–614

Kozlov MV, Hurlbert SH (2006) Pseudoreplication, chatter, and the international nature of science: a response to D.V. Tatarnikov. Zhurnal Obstchei Biologii [J Fund Biol, Moscow] 67(2):145–152 (in Russian)

Kozlov MV, Niemelä P (1999) Difference in needle length – a new and objective indicator of pollution impact on Scots pine (*Pinus sylvestris*). Water Air Soil Pollut 116:365–370

Kozlov MV, Niemelä P (2003) Drought is more stressful for northern populations of Scots pine than low summer temperatures. Silva Fenn 37:175–180

Kozlov MV, Sokolova IV (1984) A method of estimation of birch foliage area and biomass. Lesovedenie [Forestry, Moscow] 0(6):79–83 (in Russian)

Kozlov MV, Whitworth T (2002) Population densities and diversity of Calliphoridae (Diptera) around a nickel-copper smelter at Monchegorsk, Northwestern Russia. Entomol Fenn 13:98–104

Kozlov MV, Zvereva EL (1997) Effects of pollution and urbanization on diversity of frit flies (Diptera: Chloropidae). Acta Oecol – Int J Ecol 18:13–20

Kozlov MV, Zvereva EL (2003) Impact of industrial polluters on terrestrial ecosystems: a research synthesis. In: Honkanen JO, Koponen PS (eds) Sixth Finnish conference of environmental sciences: Proceedings. Finnish Society for Environmental Sciences & University of Joensuu, Joensuu, pp 72–75

Kozlov MV, Zvereva EL (2007a) Industrial barrens: extreme habitats created by non-ferrous metallurgy. Rev Environ Sci Biotechnol 6:231–259

Kozlov MV, Zvereva EL (2007b) Does impact of point polluters affect growth and reproduction of herbaceous plants? Water Air Soil Pollut 186:183–194

Kozlov MV, Haukioja E, Yarmishko VT (eds) (1993) Aerial pollution in Kola Peninsula. Proceedings of the international workshop, St. Petersburg, Russia, 14–16 April 1992. Kola Science Centre, Apatity

Kozlov MV, Haukioja E, Bakhtiarov AV, Stroganov DN (1995) Heavy metals in birch leaves around a nickel-copper smelter at Monchegorsk, Northwestern Russia. Environ Pollut 90:291–299

Kozlov MV, Lvovsky AL, Mikkola K (1996a) Abundance of day-flying Lepidoptera along an air pollution gradient in the northern boreal forest zone. Entomol Fenn 7:137–144

Kozlov MV, Zvereva EL, Selikhovkin AV (1996b) Decreased performance of *Melasoma lapponica* (Coleoptera: Chrysomelidae) fumigated by sulphur dioxide: direct toxicity versus host plant quality. Environ Entomol 25:143–146

Kozlov MV, Wilsey BJ, Koricheva J, Haukioja E (1996c) Fluctuating asymmetry of birch leaves increases under pollution impact. J Appl Ecol 33:1489–1495
Kozlov MV, Byalt VV, Salo J (1998) Species richness of vascular plants in studies of human-induced disturbances in northern coniferous forests: effect of scale. In: Gorshkov VG, Makarieva AM, Kharkina TG (eds) The role of virgin terrestrial biota in the modern processes of global change: biotic regulation of the environment. Proceedings of international seminar, Petrozavodsk, 12–16 October 1998. St. Petersburg Institute of Nuclear Physics, Gatchina, pp 305–307 (in Russian)
Kozlov MV, Haukioja E, Bakhtiarov AV, Stroganov DN, Zimina SN (2000) Root versus canopy uptake of heavy metals by birch in an industrially polluted area: contrasting behaviour of nickel and copper. Environ Pollut 107:413–420
Kozlov MV, Zvereva EL, Niemelä P (2001) Shoot fluctuating asymmetry: a new and objective stress index in Norway spruce (*Picea abies*). Can J Forest Res 31:1289–1291
Kozlov MV, Niemelä P, Mälkönen E (2002) Effects of compensatory fertilization on pollution-induced stress in Scots pine. Water Air Soil Pollut 134:307–318
Kozlov MV, Zvereva EL, Gilyazov AS, Kataev GD (2005a) Contaminated zone around a nickel-copper smelter: a death trap for birds and mammals? In: Burk AR (ed) Trends in biodiversity research. Nova Science, New York, pp 81–99
Kozlov MV, Brodskaya NK, Haarto A, Kuusela K, Schafer A, Zverev V (2005b) Abundance and diversity of human-biting flies (Diptera: Ceratopogonidae, Culicidae, Tabanidae, Simuliidae) around a nickel-copper smelter at Monchegorsk, northwestern Russia. J Vector Ecol 30:263–271
Kozlov MV, Eränen JK, Zverev VE (2007) Budburst phenology of white birch in industrially polluted areas. Environ Pollut 148:125–131
Kozlovich A (2004) Environmental impact from Nadvoitsy aluminium plant. http://www.naturvern.no/data/f/0/69/97/1_2401_0/Background_info_Nadvoicy_eng_yl.pdf. Accessed 20 Jan 2009
Kozlovich A (2006) 'Clean' environment of Nadvoitsy, or how long we will be told lies? http://www.ariston.karelia.org.ru/lb/lb6_1.php (in Russian). Accessed 20 Jan 2009
Kozlowski TT (1980) Impacts of air pollution on forest ecosystems. BioScience 30:88–93
Kozlowski TT, Pallardy SG (1996) Physiology of woody plants. Academic Press, San Diego
Kramer AE (2007) For one business, polluted clouds have silvery linings. The New York Times, 12 June. http://www.nytimes.com/2007/07/12/world/europe/12norilsk.html?_r = 1&ref = world&oref = slogin. Accessed 20 Jan 2009
Krasovskaya TM (1996) Spatio-temporal regularities of heavy metal accumulation in ecosystems of Vorkuta industrial region. In: Getsen MV, Stenina AS, Voronin LV (eds) Bioindication of the environmental state in tundras of Vorkuta. Institute of Biology, Syktyvkar, pp 42–48 (in Russian)
Krater J (2006) Defending the wild in the land of fire and ice. http://www.earthfirstjournal.org/article.php?id = 316. Accessed 20 Jan 2009
Krause GH, Weis E (1988) The photosynthetic apparatus and chlorophyll fluorescence. An introduction. In: Lichtenthaler HK (ed) Applications of chlorophyll fluorescence. Kluwer, Dordrecht, pp 3–11
Kravchenko AN, Thelen KD, Bullock DG, Miller NR (2003) Relationship among crop grain yield, topography, and soil electrical conductivity studied with cross-correlograms. Agron J 95:1132–1139
Krebs CJ (1989) Ecological methodology. Harper & Row, New York
Krebs CJ (2001) Ecology: the experimental analysis of distribution and abundance. Benjamin Cummings, San Francisco
Kremen C, Ostfeld RS (2005) A call to ecologists: measuring, analyzing, and managing ecosystem services. Frontiers Ecol Environ 3:540–548
Krištín A, Žilinec M (1997) Nest box occupancy and breeding success of hole-nesting passerines at various conditions in beech forests. Folia Zool 46:229–241
Kristinsson H (1998) Plant community changes around the aluminium smelter in Straumsvík, Iceland. Náttúrufræðingurinn 67:241–254 (in Icelandic)
Kruglyashov LP (2001) Solving of ecological problems at Kandalaksha aluminium smelter. Tsvetnye Metally [Nonfer Metals, Moscow] 0(1):73–76

Krutov VI, Fedorets NG, Diakonov VV, Litinsky PY, Germanova NI, Fuksman IL (1998) Ecological monitoring of Karelia forests. Inzhenernaya Ekologija [Eng Ecol, Moscow] 0(5):19–29

Kryazheva NG, Chistyakova EK, Zakharov VM (1996) Analysis of development stability of *Betula pendula* under conditions of chemical pollution. Russ J Ecol 27:422–424

Kryuchkov VV, Makarova TD (1989) Aerotechnogenic impact on ecosystems of the Kola North. Kola Science Centre, Apatity (in Russian)

Kryuchkov VV (1993a) Extreme anthropogenic loads and the northern ecosystem condition. Ecol Appl 3:622–630

Kryuchkov VV (1993b) Environment of the central part of Kola peninsula: relief, climate, soil, vegetation, emission sources. In: Kozlov MV, Haukioja E, Yarmishko VT (eds) Aerial pollution in Kola Peninsula. Proceedings of the International Workshop, April 14–16, 1992, St. Petersburg. Kola Science Centre, Apatity, pp 12–15

Kryuchkov VV (1993c) Degradation of ecosystems around the "Severonikel" smelter complex. In: Kozlov MV, Haukioja E, Yarmishko VT (eds) Aerial pollution in Kola Peninsula: Proceedings of the International Workshop, April 14–16, 1992, St. Petersburg. Kola Science Centre, Apatity, pp 35–46

Kubin E, Lippo H, Poikolainen J (2000) Heavy metal loading. In: Mälkönen E (ed) Forest condition in changing environment – the Finnish case. Kluwer, Dordrecht, pp 60–71

Kubota Y, Hara T (1996) Recruitment processes and species coexistence in a sub-boreal forest in northern Japan. Ann Bot 78:741–748

Kucherov SE, Muldashev AA (2003) Radial growth of Scots pine in the vicinity of Karabash copper smelter. Lesovedenie [Forestry, Moscow] 0(2):43–49 (in Russian)

Kucken DJ, Davis JS, Petranka JW, Smith CK (1994) Anakeesta stream acidification and metal contamination – effects on a salamander community. J Environ Qual 23:1311–1317

Kulfan J, Zach P, Šušlík V, Zelinková D, Anderson J (2002) Is abundance of the moth *Bucculatrix ulmella* affected by immissions? In: Maňkovská B (ed) Long term air pollution effect on forest ecosystems. 20th International meeting for specialists in air pollution effects on forest ecosystems. Book of abstracts. August 30–September 1, 2002. Forest Research Institute, Zvolen, p 137

Kuliev AN (1977) Impact of dust on tundra vegetation in surroundings of Vorkuta. In: Syroechkovskij EE (ed) Scientific basis of nature protection. Central Research Laboratory of Nature Conservation, Moscow, pp 22–28 (in Russian)

Kuliev AN (1979) Investigation of dust impact on tundra vegetation in surroundings of Vorkuta. In: Syroechkovskij EE (ed) Impact of human activity on natural ecosystems. Central Research Laboratory of Nature Conservation, Moscow, pp 60–66 (in Russian)

Kuliev AN, Lobanov VA (1978) Distribution of dust contaminants in surroundings of Vorkuta. In: Syroechkovskij EE (ed) Changes of natural environment under human activity. Central Research Laboratory of Nature Conservation, Moscow, pp 83–89 (in Russian)

Kunilov VY (1994) Geology of the Noril'sk region: the history of discovery, prospecting, exploration and mining of the Noril'sk deposits. In: Lightfoot PC, Naldrett AJ (eds) Proceedings of the Sudbury – Noril'sk Symposium, vol 5. Ontario Geological Survey, Ontario, pp 203–216

Laaksovirta K, Silvola J (1975) Effect of air pollution by copper, sulphuric acid and fertilizer factories on plants at Harjavalta, W. Finland. Ann Bot Fenn 12:81–88

Lackovičová A (1995) Epiphytic lichen diversity in the Krompachy region. In: Topercer J (ed) Diverzita rastlinstva Slovenska. Zborník referátov zo VI Zjazdu Slovenskej Botanickej Spoločnosti pri SAV. Blatnica, 6–10 Juna 1994. Slovenská botanická spoločnosť pri SAV v Bratislave, Nitra, pp 158–163 (in Slovak)

Lackovičová A, Martiny E, Pišút I, Streško V (1994) Element content of the lichen *Hypogymnia physodes* and spruce needles in the industrial area of Rudnany and Krompachy (NE Slovakia). Ekologia (Bratislava) 13:415–423

Lairand NI, Lovelius NV, Yatsenko-Khmelevsky AA (1979) Effect of anthropogenic influences on the increment of *Pinus sylvestris* (Pinaceae) in the Bratsk region. Botanicheskij Zhurnal [Bot J, Leningrad] 64:1187–1195 (in Russian)

Lamppu J (2002) Scots pine needle longevity and other shoot characteristics along pollution gradients. Ph.D. thesis, University of Oulu, Oulu

Lamppu J, Huttunen S (2001) Scots pine needle longevity and gradation of needle shedding along pollution gradients. Can J Forest Res 31:261–267

Lamppu J, Huttunen S (2003) Relations between Scots pine needle element concentrations and decreased needle longevity along pollution gradients. Environ Pollut 122:119–126

Larsson S (1989) Stressful times for the plant stress – insect performance hypothesis. Oikos 56:277–283

Lavelle P, Decaens T, Aubert M, Barot S, Blouin M, Bureau F, Margerie P, Mora P, Rossi JP (2006) Soil invertebrates and ecosystem services. Eur J Soil Biol 42:S3–S15

Law BE, Ryan MG, Anthoni PM (1999) Seasonal and annual respiration of a ponderosa pine ecosystem. Global Change Biol 5:169–182

Law BE, Falge E, Gu L, Baldocchi DD, Bakwin P, Berbigier P, Davis K, Dolman AJ, Falk M, Fuentes JD, Goldstein A, Granier A, Grelle A, Hollinger D, Janssens IA, Jarvis P, Jensen NO, Katul G, Mahli Y, Matteucci G, Meyers T, Monson R, Munger W, Oechel W, Olson R, Pilegaard K, Paw KT, Thorgeirsson H, Valentini R, Verma S, Vesala T, Wilson K, Wofsy S (2002) Environmental controls over carbon dioxide and water vapor exchange of terrestrial vegetation. Agric Forest Meteorol 113:97–120

Lazareva IP, Vassilieva ES, Vuorimaa TA, Kabanova VS (1988) Aerial industrial contamination of forest ecosystems in Karelia. Forest Institute, Petrozavodsk (in Russian)

Leamy L (1999) Heritability of directional and fluctuating asymmetry for mandibular characters in random bred mice. J Evol Biol 12:146–155

Lebedev A, Sinikova N, Nikolaeva S, Poliakova O, Khrushcheva M, Pozdnyakov S (2003) Metals and organic pollutants in snow surrounding an iron factory. Environ Chem Lett 1:107–112

Lebedeva EV, Lugauskas AY (1985) Effects of industrial pollution on soil micromycetes. Mikologiya i Fitopatologiya [Mycol Phytopathol, St. Petersburg] 19(1):16–19 (in Russian)

LeBlanc F, Rao DN, Comeau G (1972) The epiphytic vegetation of *Populus balsamifera* and its significance as an air pollution indicator in Sudbury, Ontario. Can J Bot 50:519–528

Lechowicz MJ (1987) Resource allocation by plants under air pollution stress: implications for plant-pest-pathogen interactions. Bot Rev 53:281–300

Lee JA (1998) Unintentional experiments with terrestrial ecosystems: ecological effects of sulphur and nitrogen pollutants. J Ecol 86:1–12

Lefohn AS, Husar JD, Husar RB (1999) Estimating historical anthropogenic global sulfur emission patterns for the period 1850–1990. Atmos Environ 33:3435–3444

Lehman CL, Tilman D (2000) Biodiversity, stability, and productivity in competitive communities. Am Nat 156:534–552

Lehvavirta S, Rita H (2002) Natural regeneration of trees in urban woodlands. J Veg Sci 13:57–66

Leimu R, Koricheva J (2004) Cumulative meta-analysis: a new tool for detection of temporal trends and publication bias in ecology. Proc Royal Soc Lond Ser B – Biol Sci 271:1961–1966

Lennon JJ, Koleff P, Greenwood JJD, Gaston KJ (2001) The geographical structure of British bird distributions: diversity, spatial turnover and scale. J Anim Ecol 70:966–979

Lepedus H, Cesar V, Ljubesic N (2005) Photosystem II efficiency, chloroplast pigments and fine structure in previous-season needles of Norway spruce (*Picea abies* L. Karst.) affected by urban pollution. Period Biol 107:329–333

Levula T (1993) Metsien terveydentila Harjavallan ympäristössä. Metsätutkimuslaitoksen Tiedonantoja 470:54–66

Li J, Loneragan WA, Duggin JA, Grant CD (2004) Issues affecting the measurement of disturbance response patterns in herbaceous vegetation – a test of the intermediate disturbance hypothesis. Plant Ecol 172:11–26

Lichtenthaler HK, Rinderle U (1988) The role of chlorophyll fluorescence in the detection of stress conditions in plants. CRC Crit Rev Anal Chem 19:S29–S85

Light RJ, Pillemer DB (1984) Summing up: the science of reviewing research. Harvard University Press, Cambridge

Likens GE, Bormann FH (1974) Acid rain – serious regional environmental problem. Science 184:1171–1179

Lincoln DE, Mooney HA (1984) Herbivory on *Diplacus aurantiacus* shrubs in sun and shade. Oecologia 64:173–176

Linzon SN (1958) The influence of smelter fumes on the growth of white pine in the Sudbury region. Ontario Department of Lands and Forests and Ontario Department of Mines, Ontario

Linzon SN (1966) Damage to eastern white pine by sulfur dioxine, semimature tissue needle blight, and ozone. J Air Pollut Cont Assoc 16:140–144

Linzon SN (1986) Effects of gaseous pollutants on forests in eastern North America. Water Air Soil Pollut 31:537–550

Liu N, Peng CL, Lin ZF, Lin GZ, Pan XP (2007) Effects of simulated SO_2 pollution on subtropical forest succession: toward chlorophyll fluorescence concept. Pak J Bot 39:1921–1935

Lloyd OL, Lloyd MM, Holland Y, Lyster WR (1984) An unusual sex ratio of births in an industrial town with mortality problems. Brit J Obstet Gynaecol 91:901–907

Lloyd OL, Smith G, Lloyd MM, Holland Y, Gailey F (1985) Raised mortality from lung cancer and high sex ratios of births associated with industrial pollution. Brit J Ind Med 42:475–480

Löffler A (1983) Regressionsanalytische Bewertung der Auswirkungen von Exhalaten auf die Reaktion der Waldböden in dem Imissionsgebiet von Žiar nad Hronom. Acta Instituti Forestalis Zvolen 6:235–249

Long RP, Davis DD (1999) Growth variation of white oak subjected to historic levels of fluctuating air pollution. Environ Pollut 106:193–202

Lorentzen Y (2005) "Her i byen er alt i orden". Nat Miljø 0(3):14–16

Lortie CJ, Callaway RM (2006) Re-analysis of meta-analysis: support for the stress-gradient hypothesis. J Ecol 94:7–16

Lortie CJ, Brooker RW, Kikvidze Z, Callaway RM (2004) The value of stress and limitation in an imperfect world: a reply to Körner. J Veg Sci 15:577–580

Lovett GM, Tear T (2007) Effects of atmospheric deposition on biological diversity in the eastern United States. Report published by The Institute of Ecosystem Studies and the Nature Conservancy. http://www.ecostudies.org/reprints/Effects_of_atmospheric_deposition_on_biodiversity.pdf. Accessed 20 Jan 2009

Ludwig W (1932) Das Rechts-Links Problem im Tierreich und beim Menschen. Springer, Berlin

Lugo AE (1978) Stress and ecosystems. In: Thorp JW, Gibbons JW (eds) Energy and environmental stress in aquatic ecosystems. DOE Symposium Series (CONF.-771114), Oak Ridge, pp 61–101

Łukaszek M, Poskuta JW (1998) Development of photosynthetic apparatus and respiration in pea seedlings during greening as influenced by toxic concentration of lead. Acta Physiol Plant 20:35–40

Lukin NA, Gavrilova NA (2004) Strategy of environmental protection from contaminants at NAS [Nadvoitsy Aluminium Smelter] – SUAL. Tsvetnye Metally [Nonferr Metal, Moscow] 0(8):91–93 (in Russian)

Lukina NV, Nikonov VV (1999) Pollution-induced changes in soils subjected to intense air pollution. In: Nikonov VV, Koptsik GN (eds) Acidic deposition and forest soils. Kola Science Centre, Apatity, pp 79–126 (in Russian)

Lukina NV, Suhareva TA, Isaeva LG (2005) Technogenic digressions and recovery successions in northern taiga forests. Nauka, Moscow (in Russian)

Lukyanova LE (1987) Assessment of reproductive parameters of bank voles inhabiting territories of high and background pollution level in Southern Ural. In: Vershinin VL, Krinitsin SV (eds) Proceedings of regional scientific and practical conference "Ecological systems of Ural: investigation, protection, exploitation", Sverdlovsk, 16–21 March 1987. Regional Committee of Scientific and Technical Societies, Sverdlovsk, pp 32 (in Russian)

Lukyanova LE (1990) Study of ecological parameters of small mammals in human-affected areas. In: Vershinin VL (ed) Animals in human-affected landscape. Institute of Plant and Animal Ecology, Sverdlovsk, pp 55–60 (in Russian)

Lumme I, Arkhipov V, Fedorets N, Mälkönen EE (1997) The state of pine stands in Karelian Isthmus–South-east Finland and in Kostomuksha-Kainuu area. Concluding Report on the Russian-Finnish Research Project. Metsäntutkimuslaitoksen Tiedonantoja 665:1–75

Lyubashevsky NM, Tokar VI, Shcherbakov SV (1996) Industrial fluorine in forest ecosystems of Ural and Siberia. Institute of Plant and Animal Ecology, Ekaterinburg (in Russian)

Mackey RL, Currie DJ (2001) The diversity-disturbance relationship: is it generally strong and peaked? Ecology 82:3479–3492

MacMillan D (2000) Smoke wars: Anaconda copper, Montana air pollution, and the cours, 1820–1924. Montana Historical Society Press, Helena

Macnair MR (1997) The evolution of plants in metal-contaminated environments. In: Bijlsma R, Loeschchke V (eds). Environmental stress, adaptation and evolution. Birkhäuser Verlag, Basel, pp 2–24

Maestre FT, Cortina J (2004) Do positive interactions increase with abiotic stress? – a test from a semi-arid steppe. Proc Royal Soc Lond Ser B – Biol Sci 271:S331–S333

Maestre FT, Valladares F, Reynolds JF (2005) Is the change of plant-plant interactions with abiotic stress predictable? A meta-analysis of field results in arid environments. J Ecol 93:748–757

Magnússon SH (2002) Heavy metal deposition in the vicinity of the aluminium smelter in Straumsvik, Iceland. Report NÍ-02010. Náttúrufræðistofnun Íslands, Reykjavik

Magnússon SH, Thomas B (2007) Heavy metals and sulphur in mosses around the aluminium smelter in Straumsvik in 2005. Report NÍ-07003. Náttúrufræðistofnun Íslands, Reykjavik

Magurran AE (1988) Ecological diversity and its measurement. Chapman & Hall, London

Makra L, Brimblecombe P (2004) Selections from the history of environmental pollution, with special attention to air pollution. Part 1. Int J Environ Pollut 22:641–656

Makunina GS (2002) Degradation and chemical properties of soils in the industrially altered area of Karabash. Pochvovedenie [Soil Sci, Moscow] 0(3):368–376 (in Russian)

Maleque MA, Ishii HT, Maeto K (2006) The use of arthropods as indicators of ecosystem integrity in forest management. J Forest 104:113–117

Maňkovská B (1979) Pollution of spruce *Picea abies* Karst by emissions of F, As, Pb, Cd, and S from an aluminum plant. Biológia (Bratislava) 34:563–570

Maňkovská B (1984) The effects of atmospheric emissions from the Krompachy, Nižná Slaná, Rudňany iron ore mines on forest vegetation and soils. Ekologia (Bratislava) 3:331–344

Maňkovská B (2001) Long-term ecological research in Žiarska area. Ekologia (Bratislava) 20:122–127

Maňkovská B (ed) (2002) Long term air pollution effect on forest ecosystems. 20th International meeting for specialists in air pollution effects on forest ecosystems. Book of abstracts. August 30–September 1, 2002. Forest Research Institute, Zvolen

Maňkovská B (2003) Long term effects of pollutants on forest vegetation in Central Spis region. In: Mining and the Environment III Conference, Sudbury 2003. Centre for Environmental Monitoring, Laurentian University, Sudbury, pp 1–5

Maňkovská B (2004) Biomonitoring of pollution load lowering with change of the aluminium production technology. Ekologia (Bratislava) 23:169–183

Maňkovská B, Kohút R (2002) 40 years of the research of air pollutants effect on forest ecosystem in Žiarska valley. ENTERPRISE, Banská Bystrica

Maňkovská B, Steinnes E (1995) Effects of pollutants from an aluminum reduction plant on forest ecosystems. Sci Total Environ 163:11–23

Margalef R (1968) Perspectives in ecological theory. University of Chicago Press, Chicago

Marinari S, Calfapietra C, De Angelis P, Mugnozza GS, Grego S (2007) Impact of elevated CO_2 and nitrogen fertilization on foliar elemental composition in a short rotation poplar plantation. Environ Pollut 147:507–515

Markert B, Herpin U, Berlekamp J, Oehlmann J, Grodzinska K, Maňkovská B, Suchara I, Siewers U, Weckert V, Lieth H (1996) A comparison of heavy metal deposition in selected Eastern European countries using the moss monitoring method, with special emphasis on the 'Black Triangle'. Sci Total Environ 193:85–100

Markow TA, Clarke GM (1997) Meta-analysis of the heritability of developmental stability: a giant step backward – comment. J Evol Biol 10:31–37

Martinez-Carrasco R, Perez P, Morcuende R (2005) Interactive effects of elevated CO_2, temperature and nitrogen on photosynthesis of wheat grown under temperature gradient tunnels. Environ Exp Bot 54:49–59

Mastauskis MM (1987) Economical importance of pests and diseases of Scots pine in the impact area of a nitrogen fertilizer plant. In: Becius P (ed) Investigation and modelling of forest growth in polluted environment. Lithuanian Agricultural Academy, Kaunas, pp 29–34 (in Russian)

Matt GE (2003) Will it work in Münster? Meta-analysis and the empirical generalization of causal relationships. In: Schulze R, Holling H, Bohning D (eds) Meta-analysis: new developments and applications in medical and social sciences. Hogrefe & Huber, Cambridge, pp 113–139

Matthews RA, Landis WG, Matthews GB (1996) The community conditioning hypothesis and its application to environmental toxicology. Environ Toxicol Chem 15:597–603

Matveeva EM, Kovalenko TE, Gruzdeva LI, Ieshko EP (2001) Soil nematodes as indicators of industrial pollution. In: Ieshko EP (ed) Eco-parasitological research of animals and plants in the European North. Institute of Biology, Petrozavodsk, pp 69–77 (in Russian)

Maxwell K, Johnson GN (2000) Chlorophyll fluorescence – a practical guide. J Exp Bot 51:659–668

May RM (1973) Stability and complexity in model ecosystems. Princeton University Press, Princeton

McCall J, Gunn J, Struik H (1995) Photo interpretive study of recovery of damaged lands near the metal smelters of Sudbury, Canada. Water Air Soil Pollut 85:847–852

McClellan K, Altenburger R, Schmitt-Jansen M (2008) Pollution-induced community tolerance as a measure of species interaction in toxicity assessment. J Appl Ecol 45:1514–1522

McClenahen JR (1978) Community changes in a deciduous forest exposed to air pollution. Can J Forest Res 8:432–438

McDonnell MJ, Pickett STA (1990) Ecosystem structure and function along urban rural gradients – an unexploited opportunity for ecology. Ecology 71:1232–1237

McEnroe NA, Helmisaari HS (2001) Decomposition of coniferous forest litter along a heavy metal pollution gradient, south-west Finland. Environ Pollut 113:11–18

McNaughton SJ, Oesterheld M, Frank DA, Williams KJ (1989) Ecosystem-level patterns of primary productivity and herbivory in terrestrial habitats. Nature 341:142–144

Medina MH, Correa JA, Barata C (2007) Micro-evolution due to pollution: possible consequences for ecosystem responses to toxic stress. Chemosphere 67:2105–2114

Meldrum M (1999) Toxicology of hydrogen fluoride in relation to major accident hazards. Regul Toxicol Pharmacol 30:110–116

Menshchikov SL (1992) Impact of aerial industrial pollution on forest-tundra ecosystems. In: Makhnev AK, Koltunov EV (eds) Industrial impact on forest communities and problems of their restoration and conservation. Nauka, Ekaterinburg, pp 81–86 (in Russian)

Menshchikov SL, Ivshin AP (2006) Regularities in transformation of sub-tundra and taiga forests under impact of aerial industrial pollution. Botanical Garden, Ekaterinburg (in Russian)

Menzel A, Sparks TH, Estrella N, Koch E, Aasa A, Ahas R, Alm-Kübler K, Bissolli P, Braslavská O, Briede A, Chmielewski FM, Crepinsek Z, Curnel Y, Dahl Å, Defila C, Donnelly A, Filella Y, Jatcza K, Måge F, Mestre A, Nordli Ø, Peñuelas J, Pirinen P, Remišová V, Scheifinger H, Striz M, Susnik A, Van Vliet AJH, Wielgolaski FE, Zach S, Zust A (2006) European phenological response to climate change matches the warming pattern. Global Change Biol 12:1969–1976

Merilä J, Björklund M (1995) Fluctuating asymmetry and measurement error. Syst Biol 44:97–101

Merino A, Real C, Rodríguez-Guitián MA (2008) Nutrient status of managed and natural forest fragments of *Fagus sylvatica* in southern Europe. Forest Ecol Manage 255:3691–3699

Mihál I, Buňinová K (2005) Species diversity, abundance and dominance of macromycetes in beech forest stands. J Forest Sci 51:187–194

Mikhailova IN (2007) Populations of epiphytic lichens under stress conditions: survival strategies. Lichenologist 39:83–89

Mikhailova IN, Vorobeichik EL (1999) Dimensional and age structure of populations of epiphytic lichen *Hypogymnia physodes* (L.) Nyl. under conditions of atmospheric pollution. Russ J Ecol 30:111–118

Mikhailova TA (1997) Ecological and physiological state of forests affected by industrial emissions. Abstract of Doctoral thesis, Irkutsk State University, Irkutsk (in Russian)

Mikhailova TA (2003) Impact of industrial emissions on the forests of Baikal territory. Geografiya i Prirodnye Resursy [Geogr Nat Resour, Irkutsk] 0(1):51–59 (in Russian)

Mikhailova TA, Berezhnaya NS (2000) Assessment of the pine forest state under long-lasting impact of the emissions from Irkutsk aluminium plant. Geografiya i Prirodnye Resursy [Geogr Nat Resour, Irkutsk] 0(1):43–50 (in Russian)

Milchunas DG, Lauenroth WK (1995) Inertia in plant community structure – state changes after cessation of nutrient-enrichment stress. Ecol Appl 5:452–458

Miller PR, McBridge JR (1975) Effects of air pollution in forests. In: Mudd JB, Kozlowski TT (eds) Responses of plant to air pollution. Academic Press, New York, pp 196–235

Milyaev VB, Nikolaev VD (eds) (1996) Annual report of ambient air pollution in cities and industrail centres of the Russian Federation. Emissions of pollutants: 1994. Voeikov Main Geophysical Observatory & Research Institute of Ambient Air Protection, St. Petersburg (in Russian)

Milyaev VB, Yasenskij AN (eds) (2004) Annual report on emissions of pollutants into ambient air in cities and regions of the Russian Federation: 2003. Research Institute of Ambient Air Protection, St. Petersburg (in Russian)

Milyaev VB, Yasenskij AN (eds) (2005) Annual report on emissions of pollutants into ambient air in cities and regions of the Russian Federation: 2004. Research Institute of Ambient Air Protection, St. Petersburg (in Russian)

Milyaev VB, Yasenskij AN (eds) (2006) Annual report on emissions of pollutants into ambient air in cities and regions of the Russian Federation in 2005. Asterion, St. Petersburg (in Russian)

Milyaev VB, Nikolaev VD, Yasenskij AN (eds) (1997a) Annual report on emissions of pollutants into ambient air in cities and regions of the Russian Federation: 1995. Research Institute of Ambient Air Protection & Voeikov Main Geophysical Observatory, St. Petersburg (in Russian)

Milyaev VB, Nikolaev VD, Yasenskij AN (eds) (1997b) Annual report on emissions of pollutants into ambient air in cities and regions of the Russian Federation: 1996. Research Institute of Ambient Air Protection & Voeikov Main Geophysical Observatory, St. Petersburg (in Russian)

Milyaev VB, Nikolaev VD, Yasenskij AN (eds) (1998) Annual report on emissions of pollutants into ambient air in cities and regions of the Russian Federation: 1997. Research Institute of Ambient Air Protection, St. Petersburg (in Russian)

Milyaev VB, Nikolaev VD, Yasenskij AN (eds) (1999) Annual report on emissions of pollutants into ambient air in cities and regions of the Russian Federation: 1998. Research Institute of Ambient Air Protection, St. Petersburg (in Russian)

Milyaev VB, Nikolaev VD, Yasenskij AN (eds) (2000) Annual report on emissions of pollutants into ambient air in cities and regions of the Russian Federation: 1999. Research Institute of Ambient Air Protection, St. Petersburg (in Russian)

Milyaev VB, Nikolaev VD, Yasenskij AN (eds) (2001) Annual report on emissions of pollutants into ambient air in cities and regions of the Russian Federation: 2000. Research Institute of Ambient Air Protection, St. Petersburg (in Russian)

Milyaev VB, Nikolaev VD, Yasenskij AN (eds) (2002) Annual report on emissions of pollutants into ambient air in cities and regions of the Russian Federation: 2001. Research Institute of Ambient Air Protection, St. Petersburg (in Russian)

Milyaev VB, Nikolaev VD, Yasenskij AN (eds) (2003) Annual report on emissions of pollutants into ambient air in cities and regions of the Russian Federation: 2002. Research Institute of Ambient Air Protection, St. Petersburg (in Russian)

Ministry of the Environment of Finland (1991a) Reduction of the environmental effects of Syasstroy pulp and paper mill and Volkhov aluminium factory. (Environmental priority action programme for Leningrad, Leningrad region, Karelia and Estonia. Pre-feasibility study No. 7). Plancenter, Helsinki

Ministry of the Environment of Finland (1991b) Reduction of the environmental effects of Segezha pulp and paper milll and Nadvoitsy aluminium factory. (Environmental priority action programme for Leningrad, Leningrad region, Karelia and Estonia. Pre-feasibility study No. 11). Plancenter, Helsinki

Ministry of the Environment of the Slovak Republic (2004) State of the Environment Report of the Slovak Republic, 2003. http://www.sazp.sk/slovak/periodika/sprava/sprava_2003_en/index.html. Accessed 20 Jan 2009

Miroevskij GP, Demidov KA, Dubrovskij VL, Karasev YA, Goncharov AV, Tsemekhman LS (2001) State and prospects of solution of the environmental protection problems at the Severonikel smelting complex. Tsvetnye Metally [Nonferr Metals, Moscow] 0(2):80–84

Moffatt S, Phillimore P, Hudson E, Downey D (2000) "Impact? What impact?" Epidemiological research findings in the public domain: a case study from north-east England. Social Sci Med 51:1755–1769

Mohammed GH, Binder WD, Gillies SL (1995) Chlorophyll fluorescence – a review of its practical forestry applications and instrumentation. Scand J Forest Res 10:383–410

Mokany K, Raison RJ, Prokushkin AS (2006) Critical analysis of root: shoot ratios in terrestrial biomes. Global Change Biol 12:84–96

Mokrotovarova OI (2003) Long-term dynamics of air pollution in industrial centres of the Murmansk region. In: Kashulin NA, Vandysh OI (eds) Kola Peninsula on the threshold of the third millenium: ecological problems. Kola Science Centre, Apatity, pp 5–13 (in Russian)

Molau U (2003) Work package 10: Assess the sensitivity of tree seedling establishment and growth to presence of tree canopy. In: Huntley B (ed) DART: Final report, vol 1, pp 197–205. http://www.dur.ac.uk/DART/PDF_Documents/Final_Report_-_Volume_1_-_Scientific_Report.pdf. Accessed 20 Jan 2009

Møller AP, Jennions MD (2001) Testing and adjusting for publication bias. Trends Ecol Evol 16:580–586

Møller AP, Shykoff JA (1999) Morphological developmental stability in plants: patterns and causes. Int J Plant Sci 160:S135–S146

Møller AP, Swaddle JP (1997) Asymmetry, developmental stability, and evolution. Oxford University Press, Oxford

Monge-Corella S, García-Pérez J, Aragonés N, Pollán M, Pérez-Gómez B, López-Abente G (2008) Lung cancer mortality in towns near paper, pulp and board industries in Spain: a point source pollution study. BMC Public Health 8:288

Moretti M, Duelli P, Obrist MK (2006) Biodiversity and resilience of arthropod communities after fire disturbance in temperate forests. Oecologia 149:312–327

Morison J, Morecroft M (eds) (2006) Plant growth and climate change. Blackwell, Oxford

Morshina TG (1986) Local environmental contamination by fluorine. In: Zyrin NG, Malakhov SG, Stasjuk NV (eds) Impactive contamination of soils by metals and fluorine. Gidrometeoizdat, Leningrad, pp 128–136 (in Russian)

Moskalenko NG (1965) Vegatation cover in surroundings of Norilsk. Botanicheskij Zhurnal [Bot J, Leningrad] 50:829–837 (in Russian)

Motulsky H, Christopoulos A (2004) Fitting models to bological data using linear and nonlinear regression: a practical guide to curve fitiing. Oxford University Press, Oxford

Mowat FS, Bundy KJ (2002) A mathematical algorithm to identify toxicity and prioritize pollutants in field sediments. Chemosphere 49:499–513

Moyano FE, Kutsch WL, Rebmann C (2008) Soil respiration fluxes in relation to photosynthetic activity in broad-leaf and needle-leaf forest stands. Agric Forest Meteorol 148:135–143

Muggeo VMR (2007) Bivariate distributed lag models for the analysis of temperature-by-pollutant interaction effect on mortality. Environmetrics 18:231–243

Mukatanov AH, Shigapov ZH (1997) Soil evaluation in Scots pine forests at the Karabash monitoring station. Problemy Ekologii Juzhnogo Urala [Prob South Ural Ecol, Chelyabinsk] 0(3):44–46 (in Russian)

Mukhacheva SV (1996) Ecotoxicological peculiarities and structure of small mammal communities in industrial pollution gradient. Abstract of Ph.D. thesis, Institute of Plant and Animal Ecology, Ekaterinburg (in Russian)

Mukhacheva SV (2007) Spatiotemporal population structure of the bank vole in a gradient of technogenic environmental pollution. Russ J Ecol 38:178–184

Mumtaz MM, Poirier KA, Colman JT (1997) Risk assessment for chemical mixtures: fine-tuning the hazard index approach. J Clean Technol Environ Toxicol Occup Med 6:189–204

Munton D (2002) Fumes, forests and further studies: environmental science and policy inaction in Ontario. J Can Stud 37:130–163

Murtaugh PA (2002) Journal quality, effect size, and publication bias in meta-analysis. Ecology 83:1162–1166

Murtha PA (1972) SO_2 forest damage delineation on high altitude photographs. In: White D (ed) Proceedings of the first Canadian Symposium on remote sensing, vol 1. Department of Energy, Mines and Resources, Ottawa, pp 71–82
Nagamori M, Watanabe Y, Hase T, Kurata Y, Ono Y, Kawamura K (2007) A simple and convenient empirical survey method with a soil electrical conductivity meter for incineration residue-derived soil contamination. J Mater Cycles Waste Manage 9:90–98
Nahmani J, Lavelle P (2002) Effects of heavy metal pollution on soil macrofauna in a grassland of Northern France. Eur J Soil Biol 38:297–300
National Research Council of Canada (1939) Effect of sulphur dioxide on vegetation, prepared for the Associate Committee on Trail Smelter Smoke. (Publ. No. 815). National Research Council of Canada, Ottawa
Neal C (2002) Interception and attenuation of atmospheric pollution in a lowland ash forested site, Old Pond Close, Northamptonshire, UK. Sci Total Environ 282:99–119
Negretov PI (1977) How Vorkuta began. Soviet Studies 29:565–575
Nekrasova LS (1993) Impact of copper smelting on soil mesofauna. Ekologija (Russian Journal of Ecology) 0(5):83–85 (in Russian)
Nesatyy VJ, Suter MJF (2008) Analysis of environmental stress response on the proteome level. Mass Spectrom Rev 27:556–574
Nesterenko TV, Tikhomirov AA, Shikhov VN (2007) Chlorophyll fluorescence induction and estimation of plant resistance to stress factors. Zhurnal Obstchei Biologii [J Fund Biol, Moscow] 68:444–458 (in Russian)
Nesterenko VS (1997) Karabash technogenic system (evaluation of environmental situation and search for remediation measures). Problemy Ekologii Juzhnogo Urala [Prob South Ural Ecol, Chelyabinsk] 0(3):4–14 (in Russian)
Newman EI (1993) Applied ecology. Blackwell, Oxford
Newman JR (1977) Sensitivity of house martin, *Delichon urbica*, to fluoride emissions. Fluoride 10:73–76
Newman JR (1979) Effects of industrial air pollution on wildlife. Biol Conserv 15:181–190
Newman JR, Schreiber RK (1984) Animals as indicators of ecosystem responses to air emissions. Environ Manage 8:309–324
Newman JR, Schreiber RK, Novakova E (1992) Air pollution effects on terrestrial and aquatic animals. In: Barker JR, Tingey DT (eds) Air pollution effects on biodiversity. Van Nostrand Reinhold, New York, pp 177–233
Newson M (1995) The Earth as output: pollution. In: Johnston RJ, Taylor PJ, Watts MJ (eds) Geographies of global change. Blackwell, Oxford/Cambridge, pp 333–353
Nieminen TM (2004) Effects of soil copper and nickel on survival and growth of Scots pine. J Environ Monit 6:888–896
Nieminen TM (2005) Response of Scots pine (*Pinus sylvestris* L.) to a long-term Cu and Ni exposure. Metsäntutkimuslaitoksen Tiedonantoja 942:1–63
Nieminen TM, Derome J, Helmisaari HS (1999) Interactions between precipitation and Scots pine canopies along a heavy-metal pollution gradient. Environ Pollut 106:129–137
Nieminen TM, Ukonmaanaho L, Shotyk W (2002) Enrichment of Cu, Ni, Zn, Pb and As in an ombrotrophic peat bog near a Cu-Ni smelter in southwest Finland. Sci Total Environ 292:81–89
Nilsson J, Grennfelt P (1988) Critical loads for sulphur and nitrogen. NORD 1988:97. Nordic Council of Ministers, Copenhagen
Nilsson S, Blauberg K, Samarskaia E, Kharuk V (1998) Pollution stress of Siberian forests. In: Linkov I, Wilson R (eds) Air pollution in the Ural Mountains. Kluwer, Dordrecht, pp 31–54
Niskavaara H, Reimann C, Chekushin V (1996) Distribution and pathways of heavy metals and sulphur in the vicinity of the copper-nickel smelters in Nikel and Zapoljarnij, Kola peninsula, Russia, as revealed by different sample media. Appl Geochem 11:25–34
Nöjd P, Reams GA (1996) Growth variation of Scots pine across a pollution gradient on the Kola Peninsula, Russia. Environ Pollut 93:313–325
Norilsk Nikel (2007) Annual report 2006. http://www.norilsknickel.ru/upload/report2006_eng.pdf. Accessed 20 Jan 2009

Norin BN, Yarmishko VT (eds) (1990) Effects of industrial atmospheric pollution on pine forests of Kola peninsula. Komarov Botanical Institute, Leningrad (in Russian)

Norokorpi Y, Frank H (1993) Effect of stand density on damage to birch (*Betula pubescens*) caused by phytotoxic air pollutants. Ann Bot Fenn 30:181–187

Novoselov VN (2002) Karabash: years and fortunes. Kniga, Chelyabinsk (in Russian)

Nriagu JO, Pacyna JM (1988) Quantitative assessment of worldwide contamination of air, water and soils by trace metals. Nature 333:134–139

Nriagu JO, Wong HKT, Lawson G, Daniel P (1998) Saturation of ecosystems with toxic metals in Sudbury basin, Ontario, Canada. Sci Total Environ 223:99–117

Nygaard PH, Abrahamsen G (1991) Effects of long-term artificial acidification on the ground vegetation and soil in a 100 year old stand of Scots pine (*Pinus sylvestris*). Plant Soil 131:151–160

Odasz-Albrigtsen AM, Tømmervik H, Murphy P (2000) Decreased photosynthetic efficiency in plant species exposed to multiple airborne pollutants along the Russian-Norwegian border. Can J Bot 78:1021–1033

Odum EP (1969a) Fundamentals of ecology. Saunders, Philadelphia

Odum EP (1969b) Strategy of ecosystem development. Science 164:262–270

Odum EP (1985) Trends expected in stressed ecosystems. BioScience 35:419–422

Odum EP, Finn JT, Franz EH (1979) Perturbation theory and the subsidy stress gradient. BioScience 29:349–352

Opekunova MG, Arestova IY (2005) Bioindicators of aerial industrial contamination of Kostomuksha. In: Bolshakov VN (ed) Basic and applied ecology: urbanisation problems. Materials of international scientific and applied conference, Ekaterinburg, 3–4 February 2005. Ural University, Ekaterinburg, pp 250–253 (in Russian)

Öquist G (1987) Environmental stress and photosynthesis. In: Biggins J (ed) Progress in Photosynthesis Research, vol 4. Martinus Nijhoff, Dordrecht, pp 1–10

Ordano M, Fornoni J, Boege K, Dominguez CA (2008) The adaptive value of phenotypic floral integration. New Phytol 179:1183–1192

Ormerod SJ, Pienkowski MW, Watkinson AR (1999) Communicating the value of ecology. J Appl Ecol 36:847–855

Ornelas JF, Gonzalez C, Jimenez L, Lara C, Martinez AJ (2004) Reproductive ecology of distylous *Palicourea padifolia* (Rubiaceae) in a tropical montane cloud forest. II. Attracting and rewarding mutualistic and antagonistic visitors. Am J Bot 91:1061–1069

Orwin KH, Wardle DA, Greenfield LG (2006) Context-dependent changes in the resistance and resilience of soil microbes to an experimental disturbance for three primary plant chronosequences. Oikos 112:196–208

Otnyukova TN (1997) Early diagnostics of atmospheric pollution by the condition of reindeer lichen. Doklady Rossiyskoy Akademii Selskokhozyaystvennykh Nauk [Proc Russ Acad Agric Sci, Moscow] 0(3):21–22 (in Russian)

Ouzounidou G, Moustakas M, Symeonidis L, Karataglis S (2006) Response of wheat seedlings to Ni stress: effects of supplemental calcium. Arch Environ Contam Toxicol 50:346–352

Owen RD, Mcbee K (1990) Analysis of asymmetry and morphometric variation in natural populations of chromosome-damaged mice. Texas J Sci 42:319–332

Ozolincius R, Stakenas V, Serafinaviciute B (2005) Meteorological factors and air pollution in Lithuanian forests: possible effects on tree condition. Environ Pollut 137:587–595

Pacyna JM, Pacyna EG (2001) An assessment of global and regionalemissions of trace metals to the atmosphere from anthropogenic sources worldwide. Environ Rev 9:269–298

Padgett PE, Kee SN (2004) Impacts of air pollution on forest ecosystems. In: Burley J, Evans J, Youngquist JA (eds) Encyclopedia of forest sciences. Elsevier, Amsterdam, pp 132–139

Palmer AR (1994) Fluctuating asymmetry analyses: a primer. In: Markov TA (ed) Developmental instability: its origins and evolutionary implications. Kluwer, Dordrecht, pp 335–364

Palmer AR (1996) Waltzing with asymmetry. BioScience 46:518–532

Palmer AR (1999) Detecting publication bias in meta-analyses: a case study of fluctuating asymmetry and sexual selection. Am Nat 154:220–233

Palmer AR (2000) Quasireplication and the contract of error: lessons from sex ratios, heritabilities and fluctuating asymmetry. Ann Rev Ecol Syst 31:441–480

Palmer AR, Strobeck C (1986) Fluctuating asymmetry – measurement, analysis, patterns. Ann Rev Ecol Syst 17:391–421

Palmer AR, Strobeck C (2003) Fluctuating asymmetry analyses revisited. In: Polak M (ed) Developmental instability: causes and consequences. Oxford University Press, Oxford, pp 279–319

Pankakoski E, Koivisto I, Hyvärinen H (1992) Reduced developmental stability as an indicator of heavy metal pollution in the common shrew *Sorex araneus*. Acta Zool Fenn 191:137–144

Pankhurst C, Doube B, Gupta VVSR (1997) Biological indicators of soil health. CAB International, Wallingford

Paoletti E, Bytnerowicz A, Andersen C, Augustaitis A, Ferretti M, Grulke N, Günthardt-Goerg MS, Innes J, Johnson D, Karnosky D, Luangjame J, Matyssek R, McNulty S, Müller-Starck G, Musselman R, Percy K (2007) Impacts of air pollution and climate change on forest ecosystems – emerging research needs. Scient World J 7:1–8

Parker ED, Forbes VE, Nielsen SL, Ritter C, Barata C, Baird DJ, Admiraal W, Levin L, Loeschke V, Lyytikainen-Saarenmaa P, Hogh-Jensen H, Calow P, Ripley BJ (1999) Stress in ecological systems. Oikos 86:179–184

Parkinson KJ (1981) An improved method for measuring soil respiration in the field. J Appl Ecol 18:221–228

Parsons PA (1992) Fluctuating asymmetry – a biological monitor of environmental and genomic stress. Heredity 68:361–364

Patterson WA, Olson JJ (1983) Effects of heavy metals on radicle growth of selected woody species germinated on filter paper, mineral and organic soil substrates. Can J Forest Res 13:233–238

Pavlov IN (2006) Woody plants under industrial pollution. Buryat Science Centre, Ulan-Ude (in Russian)

Pen-Mouratov S, Shukurov N, Steinberger Y (2008) Influence of industrial heavy metal pollution on soil free-living nematode population. Environ Pollut 152:172–183

Pensa M, Liblik V, Sellin A (2000) Growth and needle retention of Scots pine trees in the region of oil shale industry. Oil Shale 17:154–167

Pensa M, Liblik V, Jalkanen R (2004) Temporal changes in the state of a pine stand in a bog affected by air pollution in northeast Estonia. Water Air Soil Pollut 159:87–99

Pensa M, Jalkanen R, Liblik V (2007) Variation in Scots pine needle longevity and nutrient conservation in different habitats and latitudes. Can J Forest Res 37:1599–1604

Percy K, Bucher J, Cape J, Ferretti M, Heath R, Jones HE, Karnosky D, Matyssek R, Müller-Starck G, Paoletti E, Rosengren-Brinck U, Sheppard L, Skelly J, Weetman G (1999) State of science and knowledge gaps with respect to air pollution impacts on forests: reports from concurrent IUFRO 7.04.00 working party sessions. Water Air Soil Pollut 116:443–448

Percy KE, Ferretti M (2004) Air pollution and forest health: toward new monitoring concepts. Environ Pollut 130:113–126

Peterson G, Allen CR, Holling CS (1998) Ecological resilience, biodiversity, and scale. Ecosystems 1:6–18

Petrov I (2004) Ecological policy [of the Kola Metallurgical Company] as a business card. Monchegorsk Worker [Monchegorskij Rabochij] 9 September 2004:2 (in Russian)

Phoenix GK, Hicks WK, Cinderby S, Kuylenstierna JCI, Stock WD, Dentener FJ, Giller KE, Austin AT, Lefroy RDB, Gimeno BS, Ashmore MR, Ineson P (2006) Atmospheric nitrogen deposition in world biodiversity hotspots: the need for a greater global perspective in assessing N deposition impacts. Global Change Biol 12:470–476

Pimenov A (2003) Estimation of the stability of the development of birch (*Betula pubescens*) as a component of an integrated estimation of a natural complex. In: Heikkilä R, Lindholm T (eds) Biodiversity and conservation of boreal nature. Proceedings of the 10 years anniversary symposium of the Nature Reserve Friendship. (The Finnish Environment 405). Ministry of the Environment, Helsinki, pp 287–289

Pitelka LF, Raynal DJ (1989) Forest decline and acidic deposition. Ecology 70:2–10

Poikolainen J, Lippo H (1995) The effects of the emissions of the Kostomuksha mining complex on the chemical composition of deposition and soil water in the surrounding pine forests. Water Air Soil Pollut 85:1689–1694

Polak M (2003) Developmental instability: causes and consequences. Oxford University Press, Oxford

Polak T, Rock BN, Campbell PE, Soukupova J, Solcova B, Zvara K, Albrechtova J (2006) Shoot growth processes, assessed by bud development types, reflect Norway spruce vitality and sink prioritization. Forest Ecol Manage 225:337–348

Pollution Probe (2003) Sulphur dioxide and toxic metal emissions from smelters in Ontario. Pollution Probe, Toronto

Polyakov VI, Ivanov VA, Soldatov VA, Abaimov AP, Zubareva ON, Lenkova TL, Polyakova GG (2005) Stand biomass in sub-tundra forests of Southern Taymyr under industrial pollution impact. In: Pleshanov AS, Pomazkina LV (eds) All-Russian conference 'Natural and antropogenic dynamics of terrestrial ecosystems', dedicated to memory of outstanding explorer of Siberian forests, Anatoly Sergeevich Rozhkov (1925–2005). Irkutsk, 11–15 October 2005: Materials. Irkutsk Polytechnic University, Irkutsk, pp 297–300 (in Russian)

Pook EW, Gill AM, Moore PHR (1998) Insect herbivory in a *Eucalyptus maculata* forest on the south coast of New South Wales. Aust J Bot 46:735–742

Popova N (2007) In the town of Nikel the rain is making holes in umbrellas and raincoats. Argumenty Nedeli [Newspaper, Moscow] 19 July 2007 (in Russian)

Popovichev BG, Golutvin GI (1983) Natural forest regeneration in areas affected by industrial aerial emissions. In: Belov SV (ed) Forestry, silviculture and soil science. Leningrad State Forest Technical Academy, Leningrad, pp 67–73 (in Russian)

Posch M (2002) Impacts of climate change on critical loads and exceedances in Europe. Environ Sci Policy 5:307–317

Posch M, de Smet PAM, Hettelingh JP, Downing RJ (eds) (2001) Modelling and mapping of critical thresholds in Europe: status report 2001. RIVM Report No. 259101010. Coordination Center for Effects, National Institute for Public Health and the Environment, Bilthoven

Potapova I, Markkanen S-L (2003) Air pollution of the environment caused by the Kostomuksha ore-dressing mill. In: Heikkilä R, Lindholm T (eds) Biodiversity and conservation of boreal nature – proceedings of the 10 years anniversary symposium of the Nature Reserve Friendship (The Finnish Environment 485). Ministry of the Environment, Vantaa, pp 282–286

Poutanen P, Kuisma M (1998) Puoli vuosisataa kuparia ja nikkeliä. Outokummun Harjavallan tehtaat. Gummerus Kirjapaino Oy, Jyväskylä

Pozniakov VYa (1993) The "Severonikel" smelter complex: history of development. In: Kozlov MV, Haukioja E, Yarmishko VT (eds) Aerial pollution in Kola Peninsula. Proceedings of the international workshop, St. Petersburg, 14–16 April 1992. Kola Science Centre, Apatity, pp 16–19

Pozniakov VYa (1999) Severonikel (History pages of the enterprise). Ruda i Metally, Moscow (in Russian)

Prescott CE, Parkinson D (1985) Effects of sulfur pollution on rates of litter decomposition in a pine forest. Can J Bot 63:1436–1443

Prokacheva VG, Usachev VF, Chmutova NP (1992) The areas of chronic contamination around towns and along roads in republics and districts of the Russian Federation. The State Hydrological Institute, St. Petersburg (in Russian)

Purvis OW, Chimonides PJ, Joness GC, Mikhailova IN, Spiro B, Weiss DJ, Williamson BJ (2004) Lichen biomonitoring near Karabash Smelter Town, Ural Mountains, Russia, one of the most polluted areas in the world. Proc Royal Soc Lond Ser B – Biol Sci 271:221–226

Qaderi MM, Kurepin LV, Reid DM (2006) Growth and physiological responses of canola (*Brassica napus*) to three components of global climate change: temperature, carbon dioxide and drought. Physiol Plant 128:710–721

Rabitsch WB (1997) Levels of asymmetry in *Formica pratensis* Retz. (Hymenoptera, Insecta) from a chronic metal-contaminated site. Environ Toxicol Chem 16:1433–1440

Rachwal L, Wit-Rzepka M (1989) Responses of birches to pollution from a copper smelter. Part II. Results of field experiment. Arboretum Kórnickie 34:185–205

Raguotis AD (1989) Impact of industrial emissions on soil microflora in forests of extreme North. In: Kozubov GM (ed) Ecology of northern forests. Abstracts of presentations of the 1st all-

union meeting, 2–7 October 1989, vol 2. Institute of Biology, Komi Science Centre, Syktyvkar, pp 45–46 (in Russian)
Rajput M, Agrawal M (2005) Biomonitoring of air pollution in a seasonally dry tropical suburban area using wheat transplants. Environ Monit Assess 101:39–53
Rämö K, Kanerva T, Nikula S, Ojanpera K, Manninen S (2006) Influences of elevated ozone and carbon dioxide in growth responses of lowland hay meadow mesocosms. Environ Pollut 144:101–111
Rapport DJ, Whitford WG (1999) How ecosystems respond to stress – common properties of arid and aquatic systems. BioScience 49:193–203
Rapport DJ, Regier HA, Hutchinson TC (1985) Ecosystem behavior under stress. Am Nat 125:617–640
Rebele F (2000) Competition and coexistence of rhizomatous perennial plants along a nutrient gradient. Plant Ecol 147:77–94
Rebristaya OV (1977) Flora of the eastern part of the Bolshezemelskaya tundra. Nauka, Leningrad (in Russian)
Reimann C, De Caritat P, Niskavaara H, Finne TE, Kashulina G, Pavlov VA (1998) Comparison of elemental contents in O- and C-horizon soils from the surroundings of Nikel, Kola Peninsula, using different grain size fractions and extractions. Geoderma 84:65–87
Reinds GJ, Groenenberg JE, de Vries W (2006) Critical loads of copper, nickel, zink, arsenic, chromium and selenium for terrestrialecosystems at a European scale. A preliminary assessment. Alterra-rapport 1355. Alterra, Wageningen
Rennenberg H, Herschbach C, Polle A (1996) Consequences of air pollution on shoot-root interactions. J Plant Physiol 148:296–301
Renner E (2002) The Black Triangle area – fit for Europe? Numerical air quality studies for the Black Triangle area. Ambio 31:231–235
Riemer J, Whittaker JB (1989) Air pollution and insects herbivores: observed interactions and possible mechanisms. In: Bernays EA (ed) Insect-plant interactions. CRC Press, Boca Raton, pp 73–105
Rigina O, Kozlov MV (2000) The impacts of air pollution impact on the northern taiga forests of the Kola peninsula, Russian Federation. In: Innes JL, Oleksyn J (eds) Forest dynamics in heavily polluted regions. CAB International, Wallingford, pp 37–65
Rintakoski H (2004) Harjavalta puhdistunut teollisuuden saasteista. Turun Sanomat 26 August 2004. http://www.turunsanomat.fi/kotimaa/?ts = 1,3:1002:0:0,4:2:0:1:2004–08–26,104:2:246660,1:0:0:0:0:0:. Accessed 20 Jan 2009
Roem WJ, Klees H, Berendse F (2002) Effects of nutrient addition and acidification on plant species diversity and seed germination in heathland. J Appl Ecol 39:937–948
Roitto M, Kozlov MV (2007) Impact of point polluters on growth and reproduction of vascular plants – a meta-analysis. In: Xiang H, Akieh MN, Vuorio A-M, Jokinen T, Sillanpää M (eds) Eighth Finnish conference of environmental sciences. Proceedings. Mikkeli, 10–11 May 2007. Finnish Society for Environmental Sciences & University of Kuopio, Kuopio, pp 119–122
Roitto M, Kozlov MV, Zvereva EL (2009) Impact of point polluters on growth and reproduction of vascular plants (submitted)
Rombke J, Jansch S, Junker T, Pohl B, Scheffczyk A, Schallnass HJ (2006) Improvement of the applicability of ecotoxicological tests with earthworms, springtails, and plants for the assessment of metals in natural soils. Environ Toxicol Chem 25:776–787
Rooney CP, Zhao FJ, McGrath SP (2006) Soil factors controlling the expression of copper toxicity to plants in a wide range of European soils. Environ Toxicol Chem 25:726–732
Rosenberg MS, Adams DC, Gurevich J (2000) MetaWin: statistical software for meta-analysis, Version 2.0. Sinauer, Sunderland
Rosenlund M, Berglind N, Pershagen G, Hallqvist J, Jonson T, Bellander T (2006) Long-term exposure to urban air pollution and myocardial infarction. Epidemiology 17:383–390
Roth LH, Selwyn B, Holguin A, Christensen BL (1982) Principles of epidemiology. Academic Press, New York
Royama T (1992) Analytical population dynamics. Chapman & Hall, London

Rozhkov AS, Mikhailova TA (1989) The effect of fluorine-containing emissions on conifers. Nauka, Novosibirsk (in Russian)

Rozhkov AS, Mikhailova TA (1992) The effect of fluorine-containing emissions on conifers. Springer, Berlin/New York

Rühling A, Tyler G (1973) Heavy metal pollution and decomposition of spruce needle litter. Oikos 24:402–416

Runova EM (1999) Impact of industrial pollution on Priangarye forests. Bratsk Industrial Institute, Bratsk (in Russian)

Ruohomäki K, Kaitaniemi P, Kozlov M, Tammaru T, Haukioja E (1996) Density and performance of *Epirrita autumnata* (Lepidoptera: Geometridae) along three air pollution gradients in northern Europe. J Appl Ecol 33:773–785

Ruotsalainen AL, Kozlov MV (2006) Fungi and air pollution: is there a general pattern? In: Rhodes D (ed) New topics in environmental research. Nova Science, Hauppauge, NY, pp 57–103

Ruotsalainen AL, Markkola A, Kozlov MV (2007) Root fungal colonisation in *Deschampsia flexuosa*: effects of pollution and neighbouring trees. Environ Pollut 147:723–728

Ruotsalainen AL, Markkola A, Kozlov MV (2009) Mycorrhizal colonisation of mountain birch (*Betula pubescens* ssp. *czerepanovii*) along three environmental gradients: does life in harsh environments alter plant-fungal relationships? Environ Monit Assess 148:215–232

Rusek J, Marshall VG (2000) Impacts of airborne pollutants on soil fauna. Ann Rev Ecol Syst 31:395–423

Rustad LE, Campbell JL, Marion GM, Norby RJ, Mitchell MJ, Hartley AE, Cornelissen JHC, Gurevitch J (2001) A meta-analysis of the response of soil respiration, net nitrogen mineralization, and aboveground plant growth to experimental ecosystem warming. Oecologia 126:543–562

Ryan MG, Law BE (2005) Interpreting, measuring, and modeling soil respiration. Biogeochemistry 73:3–27

Ryser P, Sauder WR (2006) Effects of heavy-metal-contaminated soil on growth, phenology and biomass turnover of *Hieracium piloselloides*. Environ Pollut 140:52–61

Saarinen T (1993) Chlorophyll fluorescence, and nitrogen and pigment content of Scots pine (*Pinus sylvestris*) needles in polluted urban habitats. Ann Bot Fenn 30:1–7

Saarinen T, Liski J (1993) The effect of industrial air pollution on chlorophyll fluorescence and pigment contents of Scots pine (*Pinus sylvestris*) needles. Eur J Forest Pathol 23:353–361

Sabová M, Valocká B (1996) Communities of soil nematodes on the localities polluted with acid and alkaline emissions. Helminthologia 33:27–29

Sage RF, Kubien DS (2007) The temperature response of C_3 and C_4 photosynthesis. Plant Cell Environ 30:1086–1106

Sahi SV, Israr M, Srivastava AK, Gardea-Torresdey JL, Parsons JG (2007) Accumulation, speciation and cellular localization of copper in *Sesbania drummondii*. Chemosphere 67:2257–2266

Salemaa M, Uotila T (2001) Seed bank composition and seedling survival in forest soil polluted with heavy metals. Basic Appl Ecol 2:251–263

Salemaa M, Vanha-Majamaa I, Derome J (2001) Understorey vegetation along a heavy-metal pollution gradient in SW Finland. Environ Pollut 112:339–350

Salminen J, Haimi J (1999) Horizontal distribution of copper, nickel and enchytraeid worms in polluted soil. Environ Pollut 104:351–358

Sandermann H (2004) Molecular ecotoxicology: from man-made pollutants to multiple environmental stresses. In: Sandermann H (ed) Molecular ecotoxicology of plants (Ecological studies, vol 170). Springer, Berlin, pp 1–16

Sandermann H, Wellburn A, Heath RL (1997) Forest decline and ozone: a comparison of controlled chamber and field experiments. Springer, Berlin/London

Sapelnikova II (1997) Experience of application of 'Biotest' methods to analyse developmental stability of *Betula pendula*. In: Emets VM (ed) Problems of conservation and evaluation of natural ecosystems and objects. Materials of scientific and applied conference devoted to the 70th anniversary of the Voronezh Biosphere State Reserve. Voronezh, 8–11 September 1997. BIOMIK, Voronezh, pp 94–95 (in Russian)

SAS Institute (2007) SAS/Stat. User's guide, Version 9.1. SAS Institute, Cary

Sataeva LV (ed) (1991) Annual report. Contamination of soils in Russian Federation by industrial toxicants in 1990. Institute of Experimental Meteorology, Obninsk (in Russian)
Sataeva LV (ed) (1992) Annual report. Contamination of soils in Russian Federation by industrial toxicants in 1991, part 1. Institute of Experimental Meteorology, Obninsk (in Russian)
Savchenko VA (1998) Ecological problems of Taymyr. Russian Engineering Academy, Moscow (in Russian)
Saxe H, Ellsworth DS, Heath J (1998) Tree and forest functioning in an enriched CO_2 atmosphere. New Phytol 139:395–436
Sayed OH (2003) Chlorophyll fluorescence as a tool in cereal crop research. Photosynthetica 41:321–330
Sazonova TA, Pridacha VB, Terebova EN, Shreders SM, Kolosova SV, Talanova TYu (2005) Morphophysiological response of Scots pine to industrial pollution. Lesovedenie [Forestry, Moscow] 0(3):11–19 (in Russian)
Scale PR (1982) The effects of emissions from an iron-sintering plant in Wawa, Canada on forest communities. M.Sc. thesis, University of Toronto, Toronto
Scheffer M, Carpenter SR (2003) Catastrophic regime shifts in ecosystems: linking theory to observation. Trends Ecol Evol 18:648–656
Scheidegger C, Mikhailova I (2000) Umweltforschung – Flechten als Bioindikatoren für die Luftverschmutzung im Ural: Eindrücke von einem gemeinsamen Forschungsprojekt. In: Landolt R (ed) Forum für Wissen. Eidgenössische Forschungsanstalt WSL, Birmensdorf, pp 55–59
Scheiner SM, Willig MR (2008) A general theory of ecology. Theor Ecol 1:21–28
Schindler DW (1987) Detecting ecosystem responses to anthropogenic stress. Can J Fish Aquat Sci 44:6–25
Schmidt VM (1984) Mathematic methods in botany. Leningrad State University, Leningrad (in Russian)
Schoettle AW (1990) The interaction between leaf longevity and shoot growth and foliar biomass per shoot in *Pinus contorta* at two elevations. Tree Physiol 7:214
Schowalter TD, Hargrove WW, Crossley DA (1986) Herbivory in forested ecosystems. Ann Rev Entomol 31:177–196
Schroder A, Persson L, De Roos AM (2005) Direct experimental evidence for alternative stable states: a review. Oikos 110:3–19
Schroeter SC, Dixon JD, Kastendiek J, Smith RO (1993) Detecting the ecological effects of environmental impacts: a case study of kelp forest invertebrates. Ecol Appl 3:331–350
Schubert R (ed) (1985) Bioindikation in terrestrichen Ökosystemen. G. Fischer, Stuttgart
Schulz H, Härtling S (2003) Vitality analysis of Scots pines using a multivariate approach. Forest Ecol Manage 186:73–84
Schulz H, Huhn G, Härtling S (1998) Response of sulphur- and nitrogen-containing compounds in Scots pine needles. In: Hüttl RF, Bellmann K (eds) Nutrients in ecosystems: changes of atmospheric chemistry and effects on forest ecosystems. Kluwer, Dordrecht, pp 37–63
Schweingruber FH, Voronin V (1996) Eine dendrochronologisch-bodenchemische Studie aus dem Waldschadengebiet Norilsk, Sibirien und die Konsequenzen für die Interpretation grossflächiger Kronentaxationsinventuren. Allgemeine Forst und Jagdzeitung 167:53–68
Scurfield G (1960a) Air pollution and tree growth. Forest Abstr 21:339–347
Scurfield G (1960b) Air pollution and tree growth. Part II. Forest Abstr 21:517–528
Seguin V, Gagnon C, Courchesne F (2004) Changes in water extractable metals, pH and organic carbon concentrations at the soil-root interface of forested soils. Plant Soil 260:1–17
Selikhovkin AV (1986) Seed weevil (*Apion simile*) under impact of severe air pollution. In: Solovjev VA (ed) Ecology and forest protection. Ecology of forest animals. Leningrad State Forest Technical Academy, Leningrad, pp 48–51 (in Russian)
Selikhovkin AV (1988) Impact of industrial aerial pollution on herbivorous insects. In: Narchuk EP (ed) Lectures on 39th annual meeting in memory of N. A. Cholodkovsky. Nauka, Leningrad, pp 3–42 (in Russian)

Selikhovkin AV (1992) Adaptations of moth (Lepidoptera) populations to aerial pollutants. Entomologicheskoe Obozrenie [Entomol Rev, St. Petersburg] 71:22–27 (in Russian)

Selikhovkin AV (1995) Population dynamics of Microlepidoptera under impact of aerial pollution. Trans St-Petersburg Forest Tech Acad 2:26–38 (in Russian)

Selikhovkin AV (1996) Population dynamics of Microlepidoptera under impact of aerial pollution (continuation). Trans St-Petersburg Forest Tech Acad 4:27–38 (in Russian)

Settele J, Hammen V, Hulme P, Karlson U, Klotz S, Kotarac M, Kunin W, Marion G, O'Connor M, Petanidou T, Peterson K, Potts S, Pritchard H, Pysek P, Rounsevell M, Spangenberg J, Steffan-Dewenter I, Sykes M, Vighi M, Zobel M, Kuhn I (2005) ALARM: Assessing LArge-scale environmental Risks for biodiversity with tested Methods. GAIA – Ecol Persp Sci Soc 14:69–72

Shavnin SA, Fomin VV, Marina NV (1997) Application of the generalized state index determination to ecological monitoring of forests in polluted areas. In: San José R, Brebbia CA (eds) Measurements and modelling in environmental pollution. First international conference computational mechanics publications, Southampton/Boston, pp 399–407

Shavnin SA, Sveshnikov VA, Kirpichnikova TV (1999) Integrated comparative assessment of the condition of young pine and spruce trees in the air pollution zone of Revda-Pervouralsk industrial center. Defektoskopiya [Russ J Nondest Test, Ekaterinburg] 0(9):79–84 (in Russian)

Shaw AJ (ed) (1990) Heavy metal tolerance in plants: evolutionary aspects. CRC Press, Boca Raton

Shebalova NM, Babushkina LG (1999) Forest soil of Scots pine forests growing in industrially polluted areas: bioindicators of pollution. Ural State Forest Engineering Academy, Ekaterinburg (in Russian)

Shelukho VP (2002) Xylophages of pine under chronic alkalyne pollution impact. In: Demakov YuP (ed) Ecological basis for rational forest use in Middle Volga area. Proceedings of scientific and practical conference, April 9–12, 2001. Mari State Technical University, Yoshkar-Ola, pp 30–33 (in Russian)

Sherzhukova LV, Krivtsova AN, Meluzova MI, Mishalenkova YN (2002) Estimation of developmental stability of small-leaved ime on reserved and urbanized territories. Ontogenez [Onthogeny, Moscow] 33(1):16–18 (in Russian)

Shevchenko V, Lisitzin A, Vinogradova A, Stein R (2003) Heavy metals in aerosols over the seas of the Russian Arctic. Sci Total Environ 306:11–25

Shevtsova A (1998) Response of subarctic dwarf shrubs to climate change and air pollution. Ph.D. thesis, University of Turku, Turku

Shevtsova A, Neuvonen S (1997) Responses of ground vegetation to prolonged simulated acid rain in sub-arctic pine-birch forest. New Phytol 136:613–625

Shiltsova GV, Lastochkina VG (2004) Chemical composition of atmospheric depositions in the impact zone of Kostomuksha iron ore processing industry (Northern Karelia). Institute of Forest, Petrozavodsk (in Russian)

Shinozaki K, Yoda K, Hozumi K, Kira T (1964a) A quantitative analysis of plant form-the pipe model theory. I. Basic analysis. Jpn J Ecol 14:97–105

Shinozaki K, Yoda K, Hozumi K, Kira T (1964b) A quantitative analysis of plant form – the pipe model theory. II. Further evidence of the theory and its application in forest ecology. Jpn J Ecol 14:133–139

Shukurov N, Pen-Mouratov S, Steinberger Y (2006) The influence of soil pollution on soil microbial biomass and nematode community structure in Navoiy Industrial Park, Uzbekistan. Environ Int 32:1–11

Sigal LL, Suter GW (1987) Evaluation of methods for determining adverse impacts of air pollution on terrestrial ecosystems. Environ Manage 11:675–694

Siikamäki P, Lammi A (1998) Fluctuating asymmetry in central and marginal populations of *Lychnis viscaria* in relation to genetic and environmental factors. Evolution 52:1285–1292

Simachev IV, Vaganov EA, Vysotskaya LG (1992) Dendroclimatological analysis of larch growth variation in the impact area of Norilsk mining and smelting complex. Geografiya i Prirodnye Resursy [Geogr Nat Resour, Irkutsk] 0(3):136–142 (in Russian)

Simonsen V, Laskowski R, Bayley M, Holmstrup M (2008) Low impact of metal pollution on genetic variation in the earthworm *Dendrobaena octaedra* measured by allozymes. Pedobiologia 52:51–60

Sinadsky JV (1973) Birch, its pests and diseases. Nauka, Moscow (in Russian)

Singh JS, Gupta SR (1977) Plant decomposition and soil respiration in terrestrial ecosystems. Bot Rev 43:499–528

Sinkevich SM (2001) Dynamics of tree growth in zones of industrial emissions. In: Krutov VI, Sinkevich SM, Kaibiyainen LK, Fedorets NG (eds) Bioecological aspects of forest ecosystems monitoring in North-West Russia. Institute of Forest, Petrozavodsk, pp 130–143 (in Russian)

Sitch S, Smith B, Prentice IC, Arneth A, Bondeau A, Cramer W, Kaplan JO, Levis S, Lucht W, Sykes MT, Thonicke K, Venevsky S (2003) Evaluation of ecosystem dynamics, plant geography and terrestrial carbon cycling in the LPJ dynamic global vegetation model. Global Change Biol 9:161–185

Slavenene LV, Vaychis MV, Onyunas VM (1987) To the question of chemical content of individual components of forest biogeocoenoses subjected to industrial pollution. In: Antanaitis VV, Juknys RA, Ignataviciene IA, Mastauskis MM, Zingeris MM (eds) Biomonitoring of forest ecosystems. Materials of the International Seminar, Vilnius, 17–18 March 1987. Lithuanian Agricultural Academy, Kaunas, pp 171–174

Slovak Environmental Agency (1997) State of the Environment Report of the Slovak Republic, 1996. http://www.sazp.sk/slovak/periodika/sprava/sprava97eng/index.html. Accessed 20 Jan 2009

Slovak Environmental Agency (1998) State of the Environment Report of the Slovak Republic, 1997. http://www.sazp.sk/slovak/periodika/sprava/sprava98eng/index.html. Accessed 20 Jan 2009

Smith B, Prentice IC, Sykes MT (2001) Representation of vegetation dynamics in the modelling of terrestrial ecosystems: comparing two contrasting approaches within European climate space. Global Ecol Biogeogr 10:621–637

Smith WH (1974) Air pollution – effects on the structure and function of the temperate forest ecosystem. Environ Pollut 6:111–129

Smith WH (1981) Air pollution and forests: interactions between air contaminants and forest ecossystems. Springer, New York

Smith WH (1992) Air pollution effects on ecosystem processes. In: Backer JR, Tingey DT (eds) Air pollution effects on biodiversity Van Nostrand Reinnhold, New York, pp 234–260

Snel JFH, Vankooten O, Vanhove LWA (1991) Assessment of stress in plants by analysis of photosynthetic performance. Trends Anal Chem 10:26–30

Sobocky E (1977) Management in the polluted area Žiar nad Hronom. Final research report. Forest Research Institute, Zvolen (in Slovak)

Sokolova IM, Lanning G (2008) Interactive effects of metal pollution and temperature on metabolism in aquatic ectotherms: implications of global climate change. Clim Res 37:181–201

Sokołowski AW (1971) Wpływ na roślinność leśną zanieczyszczeń powietrza emitowanych przez Zakłady Azotowe w Puławah. Sylwan 115:47–56

Sokov MK, Rozhkov AS (1975) Dynamics of increment reduction in coniferous trees under impact of aerial emissions of aluminum plants. In: Rozhkov AS (ed) Effects of anthropogenic and biotic factors on conifers (Pathology and resistance). Siberian Institute of Plant Physiology and Biochemistry, Irkutsk, pp 9–60 (in Russian)

Solovieva N, Jones VJ, Appleby PG, Kondratenok BM (2002) Extent, environmental impact and long-term trends in atmospheric contamination in the Usa basin of east-European Russian arctic. Water Air Soil Pollut 139:237–260

Song F, Eastwood AJ, Gilbody S, Duley L, Sutton AJ (2000) Publication and related biases. Health Technol Assess 4:1–115

Soulé ME (1991) Conservation – tactics for a constant crisis. Science 253:744–750

Southwood TRE (1978) Ecological methods, with particular reference to the study of insect populations. 2nd ed. Chapman & Hall, London

Spurgeon DJ, Lofts S, Hankard PK, Toal M, McLellan D, Fishwick S, Svendsen C (2006) Effect of pH on metal speciation and resulting metal uptake and toxicity for earthworms. Environ Toxicol Chem 25:788–796

Staley JT, Hartley SE (2002) Host-mediated effects of feeding by winter moth on the survival of *Euceraphis betulae*. Ecol Entomol 27:626–630

Stallings JH (1964) Phosphorus and water pollution. J Soil Water Conserv 22:228–231

Stapel JO, Waters DJ, Ruberson JR, Lewis WJ (1998) Development and behavior of *Spodoptera exigua* (Lepidoptera: Noctuidae) larvae in choice tests with food substrates containing toxins of *Bacillus thuringiensis*. Biol Cont 11:29–37

Stebel K, Christinsen G, Derome J, Grekelä I (2007) State of the environment in the Norwegian, Finnish, and Russian border area. (The Finnish Environment 6/2007). Lapland Regional Environmental Centre, Rovaniemi

Štefančík I (1995) Forest damage in the region of Žiarska basin. Lesnícky Casopis [Forest J] 51:361–373

Stenström E (1991) The effects of flooding on the formation of ectomycorrhizae in *Pinus sylvestris* seedlings. Plant Soil 131:247–250

Stepanov AM, Kabirov RR, Chernenkova TV (1992) Integrated ecological assassment of industrial impact on ecosystems of Southern taiga. Institute of Forest Resources, Moscow (in Russian)

Stern DI (2006) Reversal of the trend in global anthropogenic sulfur emissions. Global Environ Change – Human Policy Dimensions 16:207–220

Stewart-Oaten A, Bence JR (2001) Temporal and spatial variation in environmental impact assessment. Ecol Monogr 71:305–339

Stjernquist I, Nihlgard B, Filiptchouk AN, Strakhov VV (1998) Soil and forest vitality as affected by air pollutants on the Kola Peninsula. Chemosphere 36:1119–1124

Stoklasa J (1923) Die Beschädigungen der Vegetation durch Rauchgase und Fabriksexhalationen. Urban & Schwarzenberg, Berlin

Stolte KA, Duriscoe DM, Cook ER, Cline SP (1992) Methods of assessing responses of trees, stands and ecosystems to air pollution. In: Olson RK, Binkley D, Böhm M (eds) The response of western forests to air pollution. Springer, New York, pp 259–330

Strand M (1993) Photosynthetic activity of Scots pine (*Pinus sylvestris* L.) needles during winter is affected by exposure to SO_2 and NO_2 during summer. New Phytol 123:133–141

Streibig JC, Rudemo M, Jensen JE (1993) Dose-response curves and statistical models. In: Streibig JC, Kudsk P (eds) Herbicide bioassays. CRC Press, Boca Raton, pp 29–55

Sudbury Area Risk Assessment (2008) Sudbury soils study, vol 1. Background, study organization and 2001 soil survey. www.sudburysoilstudy.com. Accessed 20 Jan 2009

Sujetovinė G, Stakėnas V (2007) Changes in understorey vegetation of Scots pine stands under the decreased impact of acidifying and eutrophying pollutants. Baltic Forest 13:190–196

Šušlík V, Kulfan J (1993) The caterpillars of *Lepidoptera* butterflies as the indicators and pests of beech in the immission region of aluminium works near Žiar nad Hronom. Lesnícky Casopis [Forest J] 39:387–394

Susser M (1986) The logic of Sir Karl Popper and the practice of epidemiology. Am J Epidemiol 124:711–718

Sutherland EK, Martin B (1990) Growth response of *Pseudotsuga menziesii* to air pollution from copper smelting. Can J Forest Res 20:1020–1030

Sutinen ML, Raitio H, Nivala V, Ollikainen R, Ritari A (1996) Effects of emissions from copper-nickel smelters on the frost hardiness of *Pinus sylvestris* needles in the subarctic region. New Phytol 132:503–512

Svoboda L, Zimmermannová K, Kalač P (2000) Concentrations of mercury, cadmium, lead and copper in fruiting bodies of edible mushrooms in an emission area of a copper smelter and a mercury smelter. Sci Total Environ 246:61–67

Swart R, Amann M, Raes F, Tuinstra W (2004) A good climate for clean air: linkages between climate change and air pollution – an editorial essay. Clim Change 66:263–269

Symeonides C (1979) Tree ring analysis for tracing the history of pollution – application to a study in northern Sweden. J Environ Qual 8:482–486

Syroid NA (1987) Impact of gaseous waste of industrial enterprises on plant communities of surrounding areas. In: Karnovich VN (ed) Problems of investigation and conservation of nature in White Sea area. Murmansk City Publishers, Murmansk, pp 139–147 (in Russian)

Talloen W, Lens L, Van Dongen S, Matthysen E (2008) Feather development under environmental stress: lead exposure effects on growth patterns in Great Tits *Parus major*. Bird Study 55:108–117

Tarchevskij VV (1964) Impact of smoke and gases emitted by industrial enterprises of Ural region on vegetation. In: Tarchevskij VV (ed) Plants and industrial environment. Ural State University, Sverdlovsk, pp 5–69 (in Russian)

Tarko AM, Bykadorov AV, Kryuchkov VV (1995) Modeling of air pollution impacts on forest ecosystems in region. Doklady Rossiiskoi Akademii Nauk [Russ Acad Sci Dokl, Moscow] 341:571–573 (in Russian)

Tatarnikov DV (2005) On methodological aspects of ecological experiments (comments on M.V. Kozlov publication). Zhurnal Obstchei Biologii [J Fund Biol, Moscow] 66:90–93 (in Russian)

Taylor GE, Johnson DW, Andersen CP (1994) Air pollution and forest ecosystems – a regional to global perspective. Ecol Appl 4:662–689

Taylor JE, Fox BJ (2001) Assessing the disturbance impact on vegetation and lizard communities of fluoride pollution interacting with fire and mining in eastern Australia. Austral Ecol 26:321–337

Teale EW (1951) The murder of a landscape. Nat Hist 60:352–356

Terentjev P (1931) Biometrical tests on the morphological characteristics of *Rana ridibunda* Pall. (Amphibia, Salientia). Biometrika 23:23–51

Terrin N, Schmid CH, Lau J (2005) In an empirical evaluation of the funnel plot, researchers could not visually identify publication bias. J Clin Epidemiol 58:894–901

Tesche M, Feiler SE (eds) (1992) Air pollution and interactions between organisms in forest ecosystems. 15th International meeting for specialists in air pollution effects on forest ecosystems. 9–11 September 1992, Tharandt/Dresden, Germany. Technische Universität Dresden, Dresden

Thompson K (2000) The functional ecology of soil seed banks. In: Fenner M (ed) Seeds: the ecology of regeneration in plant communities. CAB International, Wallingford, pp 215–235

Thormar H, Jóhannesson T (1981) Flúor og brennisteinstvíoxid í lofti vid álverid í Straumsvík. Tímarit um lyfjafræði 16:11–19

Tikkanen E, Niemelä I (eds) (1995) Kola Peninsula pollutants and forest ecosystems in Lapland. Final report of the Lapland Forest Damage Project. Finnish Forest Research Institute, Rovaniemi

Tilman D (1999) Ecology – diversity and production in European grasslands. Science 286:1099–1100

Timofeeva O (2005) Forests will grow on industrial barren. Monchegorskij Rabochij [Monchegorsk Worker, Monchegorsk], 8 September 2005 pp 2 (in Russian)

Tipping E, Woof C, Rigg E, Harrison AF, Ineson P, Taylor K, Benham D, Poskitt J, Rowland AP, Bol R, Harkness DD (1999) Climatic influences on the leaching of dissolved organic matter from upland UK moorland soils, investigated by a field manipulation experiment. Environ Int 25:83–95

Tómasson T, Thormar H (1998) Pollution from the aluminium smelter in Straumsvík. Náttúrufræðingurinn 67:233–240 (in Icelandic)

Tømmervik H, Johansen BE, Pedersen JP (1995) Monitoring the effects of air pollution on terrestrial ecosystems in Varanger (Norway) and Nikel-Pechenga (Russia) using remote sensing. Sci Total Environ 161:753–767

Tømmervik H, Johansen ME, Pedersen JP, Guneriussen T (1998) Integration of remote sensed and in-situ data in an analysis of the air pollution effects on terrestrial ecosystems in the border areas between Norway and Russia. Environ Monit Assess 49:51–85

Tømmervik H, Hogda KA, Solheim L (2003) Monitoring vegetation changes in Pasvik (Norway) and Pechenga in Kola Peninsula (Russia) using multitemporal Landsat MSS/TM data. Rem Sens Environ 85:370–388

Treshow M (1984) Air pollution and plant life. Wiley, Chichester

Treshow M, Anderson FK (1989) Plant stress from air pollution. Wiley, Chichester

Trubina MR (2002) Plant communities of different landscapes under long-lasting disturbance. In: Vasilyev AS (ed) Ecological problems of mountain territories. Materials of the international scientific conference, 18–20 June 2002. Akademkniga Publishers, Ekaterinburg, pp 240–244 (in Russian)

Trueman JWH, Cranston PS (1994) An evaluation of some methods of rapid biodiversity assessment for estimating arthropod diversity. A report to the Department of the Environment, Sport and Territories. CSIRO Division of Entomology, Canberra

Tuma I (2003) A study of the effect of pollution and climate impact on plant growth based on field transplantation (A phytometer method). Ekologia (Bratislava) 22:253–267

Tuovinen JP, Laurila T, Lattila H, Ryaboshapko A, Brukhanov P, Korolev S (1993) Impact of the sulphur dioxide sources in the Kola Peninsula on air quality in northernmost Europe. Atmos Environ Part A – Gen Top 27:1379–1395

Turchin P (1990) Rarity of density dependence or population regulation with lags. Nature 344:660–663

Turner AP. (1994) The responses of plants to heavy metals. In: Ross SM (ed) Toxic metals in soil-plant systems. Wiley, Chichester, pp 153–187

Tutin TG (ed) (1964) Flora Europaea, vol 1. Cambridge University Press, Cambridge

Tutin TG (ed) (1968) Flora Europaea, vol 2. Cambridge University Press, Cambridge

Tutin TG (ed) (1972) Flora Europaea, vol 3. Cambridge University Press, Cambridge

Tutin TG (ed) (1976) Flora Europaea, vol 4. Cambridge University Press, Cambridge

Tutin TG (ed) (1980) Flora Europaea, vol 5. Cambridge University Press, Cambridge

Tutubalina OV, Rees WG (2001) Vegetation degradation in a permafrost region as seen from space: Noril'sk (1961–1999). Cold Reg Sci Technol 32:191–203

Tyler G (1978) Leaching rates of heavy metal ions in forest soil. Water Air Soil Pollut 9:137–148

Udachin V, Williamson BJ, Purvis OW, Spiro B, Dubbin W, Brooks S, Coste B, Herrington RJ, Mikhailova I (2003) Assessment of environmental impacts of active smelter operations and abandoned mines in Karabash, Ural Mountains of Russia. Sustain Dev 11:133–142

Ugrjumov BI, Runova EM, Nezhevets GP (1996) Aspects of forest exploatation in areas affected by industrial emissions. Bratsk Industrial Institute, Bratsk (in Russian)

Uhlig C, Salemaa M, Vanha-Majamaa I, Derome J (2001) Element distribution in *Empetrum nigrum* microsites at heavy metal contaminated sites in Harjavalta, western Finland. Environ Pollut 112:435–442

Underwood AJ (1989) The analysis of stress in natural populations. Biol J Linn Soc 37:51–78

UN-ECE (2006) Manual on methods and criteria for harmonized sampling, assessment, monitoring and analysis of the effects of air pollution on forests. UNEP/UN-ECE, Strasbourg

Uri ND, Lewis JA (1998) The dynamics of soil erosion in US agriculture. Sci Total Environ 218:45–58

US Geological Survey (1999) Primary aluminium plants worldwide – 1998. Part II – Summary. Reston

Utriainen MA, Karenlampi LV, Karenlampi SO, Schat H (1997) Differential tolerance to copper and zinc of micropropagated birches tested in hydroponics. New Phytol 137:543–549

Vacek S, Bastl M, Lepš J (1999) Vegetation changes in forests of the Krkonoše Mts. over a period of air pollution stress (1980–1995). Plant Ecol 143:1–11

Vajda A, Venäläinen A (2005) Feedback processes between climate, surface and vegetation at the northern climatological tree-line (Finnish Lapland). Boreal Environ Res 10:299–314

Valentin DW, Soule ME, Samollow P (1973) Asymmetry analysis in fishes – possible statistical indicator of environmental stress. Fish Bull 71:357–370

Valkama J, Kozlov MV (2001) Impact of climatic factors on the developmental stability of mountain birch growing in a contaminated area. J Appl Ecol 38:665–673

Valladares GR, Hartley SE (1994) Effects of scale on detecting interactions between *Coleophora* and *Eriocrania* leaf-miners. Ecol Entomol 19:257–262

Van Dongen S (2006) Fluctuating asymmetry and developmental instability in evolutionary biology: past, present and future. J Evol Biol 19:1727–1743

Van Dongen S, Molenberghs G, Matthysen E (1999) The statistical analysis of fluctuating asymmetry: REML estimation of a mixed regression model. J Evol Biol 12:94–102

Van Kooten O, Snel JFH (1990) The use of chlorophyll fluorescence nomenclature in plant stress physiology. Photosynth Res 25:147–150

Vanhala P (2002) Seasonal variation in the soil respiration rate in coniferous forest soils. Soil Biol Biochem 34:1375–1379

Vanhala P, Tamminen P, Fritze H (2005) Relationship between basal soil respiration rate, tree stand and soil characteristics in boreal forests. Environ Monit Assess 101:85–92

Vanhala PT, Ahtiainen JH (1994) Soil respiration, ATP content, and photobacterium toxicity test as indicators of metal pollution in soil. Environ Toxicol Water Qual 9:115–121

Varaksin SG, Kuznetsova GV (2008) Assessment of ground vegetation in the zones under technogenic impact in the suburbs of Norilsk city. Sibirskij Ekologicheskij Zhurnal [Siberian J Ecol, Novosibirsk] 0(4):655–659 (in Russian)

Vasfilov SP (2002) Impact of aerial pollution on Scots pine. Botanical Garden, Ekaterinburg (in Russian)

Vasilenko VN, Nazarov IM, Fridman SD, Belikova TV, Dlikman IF (1991) Atmospheric depositions of pollutants in the USSR, part 1. Data of 1987–1989. Gidrometeoizdat, Moscow (in Russian)

Vehviläinen H, Koricheva J, Ruohomäki K (2007) Tree species diversity influences herbivore abundance and damage: meta-analysis of long-term forest experiments. Oecologia 152:287–298

Veličković M (2007a) Measures of the developmental stability, body size and body condition in the black-striped mouse (*Apodemus agrarius*) as indicators of a disturbed environment in northern Serbia. Belg J Zool 137:147–156

Veličković M (2007b) Sampling designs, pseudoreplication and a good practice in modern science: a response to Mikhail V. Kozlov desultoriness, and recommendations to environmental scientists. Hereditas 144:45–47

Veličković M, Perisic S (2006) Leaf fluctuating asymmetry of common plantain as an indicator of habitat quality. Plant Biosyst 140:138–145

Venediktov PS, Volgin SL, Kazimirko YV, Krendeleva TE, Kukharskikh GP, Makarova VV, Lavrukhina OG, Pogosyan SI, Yakovleva OV, Rubin AB (1999) Control of the physiological state of green plantations in urban ecosystems. Biofizika 44:1037–1047

Veselkin DV (2004a) Anatomical structure of ectomycorrhiza in *Abies sibirica* Ledeb. and *Picea obovata* Ledeb. in forest ecosystems affected by emissions from a copper smelter. Russ J Ecol 35:71–78

Veselkin DV (2004b) Influence of heavy metal and SO_2 pollution on ectomycorrhizas of *Picea obovata* and *Abies sibirica*. Mikologiya i Fitopatologiya [Mycol Phytopathol, St. Petersburg] 38(1):20–26 (in Russian)

Vike E (1999) Air-pollutant dispersal patterns and vegetation damage in the vicinity of three aluminium smelters in Norway. Sci Total Environ 236:75–90

Virtanen T, Mikkola K, Patova E, Nikula A (2002) Satellite image analysis of human caused changes in the tundra vegetation around the city of Vorkuta, north-European Russia. Environ Pollut 120:647–658

Vlasova TM, Filipchuk AN (1990) Selection of bioindicators for local monitoring of northern forests under aerial industrial pressure. In: Anonymous (ed) International Symposium 'Northern forests: state, dynamics, human impact'. Arkhangelsk, 16–26 July 1990, vol 4. The USSR State Forest Committee, Moscow, pp 6–17 (in Russian)

Vorobeichik EL (1995) Changes in thickness of forest litter under chemical pollution. Russ J Ecol 26:252–258

Vorobeichik EL (1997) On the methods for measuring forest litter thickness to diagnose the technogenic disturbances of ecosystems. Russ J Ecol 28:230–234

Vorobeichik EL (1998) Populations of earthworms (Lumbricidae) in forests of the Middle Urals in conditions of pollution by discharge from copper works. Russ J Ecol 29:85–91

Vorobeichik EL (2002) Changes in the spatial structure of the destruction process under the conditions of atmospheric pollution of forest ecosystems. Biol Bull 29:300–310

Vorobeichik EL (2003a) Response of forest litter and soil biota to toxic pollution. Lesovedenie [Forestry, Moscow] 0(2):32–42 (in Russian)

Vorobeichik EL (2003b) Ecological standardization of pollution loads on terrestrial ecosystems. Abstract of Doctoral thesis, Institute of Plant and Animal Ecology, Ekaterinburg, Russia (in Russian)

Vorobeichik EL (2005) Pollution ecology in Institute of Plant and Anumal Ecology. In: Smirnov NG (ed) Ural ecological scientific school: milestones of establishment and development. Institute of Plant and Animal Ecology, Ekaterinburg, pp 175–217 (in Russian)

Vorobeichik EL (2007) Seasonal changes in the spatial distribution of cellulolytic activity of soil microflora under conditions of atmospheric pollution. Russ J Ecol 38:398–407

Vorobeichik EL, Khantemirova EV (1994) Reaction of forest phytocenoses to technogenic pollution – dose-effect dependences. Russ J Ecol 25:171–180

Vorobeichik EL, Sadykov OF, Farafontov MG (1994) Ecological standardization of industrial pollution of terrestrial ecosystems (local scale). Nauka, Ekaterinburg (in Russian)

Voroney RP (2006) The soil habitat. In: Paul EA (ed) Soil microbiology, ecology and biochemistry. Academic Press, Amsterdam, pp 25–52

Wagner M (2004) The roles of seed dispersal ability and seedling salt tolerance in community assembly of a severely degraded site. In: Temperton VM, Hobbs RJ, Nuttle T, Halle S (eds) Assembly rules and restoration ecology: briging gap between theory and practice. Island Press, Washington, pp 266–284

Wagner M, Heinrich W, Jetschke G (2006) Seed bank assembly in an unmanaged ruderal grassland recovering from long-term exposure to industrial emissions. Acta Oecol 30:342–352

Walker TR, Young SD, Crittenden PD, Zhang H (2003a) Anthropogenic metal enrichment of snow and soil in north-eastern European Russia. Environ Pollut 121:11–21

Walker TR, Crittenden PD, Young SD (2003b) Regional variation in the chemical composition of winter snow pack and terricolous lichens in relation to sources of acid emissions in the Usa river basin, northeast European Russia. Environ Pollut 125:401–412

Walker TR, Habeck JO, Karjalainen TP, Virtanen T, Solovieva N, Jones V, Kuhry P, Ponomarev VI, Mikkola K, Nikula A, Patova E, Crittenden PD, Young SD, Ingold T (2006) Perceived and measured levels of environmental pollution: interdisciplinary research in the subarctic lowlands of northeast European Russia. Ambio 35:220–228

Walsh B (2007) The World's most polluted places: Norilsk, Russia. Time Magazine, 12 September 2007. http://www.time.com/time/specials/2007/article/0,28804,1661031_1661028_1661022,00.html. Accessed 20 Jan 2009

Waring RH (1987) Characteristics of trees predisposed to die. BioScience 37:569–574

Weldre SR (1964) On the correlation structure of external morphological characters of toad-headed agama *Phrynocephalus mystaceus* (Pallas 1776). In: Terentjev PV (ed) Using of matematical methods in biology, part 3. Leningrad State University, Leningrad, pp 75–85 (in Russian)

Weltje L (1998) Mixture toxicity and tissue interactions of Cd, Cu, Pb and Zn in earthworms (Oligochaeta) in laboratory and field soils: a critical evaluation of data. Chemosphere 36:2643–2660

Wentzel KF, Ohnesorge B (1961) Zum Auftreten von Schadinsekten bei Luftverunreinigung. Untersuchungen über Bezienhungen zwischen Raucheinfluss und einer local engbegrenzten Massenvermehrung der Kleinen Fichtenblattwespe *Pristiphora abietina* (Christ). Forstarchiv 32:177–186

West DC, McLaughlin SB, Shugart HH (1980) Simulated forest response to chronic air-pollution stress. J Environ Qual 9:43–49

West PW (2003) Tree and forest measurement. Springer, Berlin/London

Whittaker RH, Bormann FH, Likens GE, Siccama TG (1974) Hubbard Brook ecosystem study – forest biomass and production. Ecol Monogr 44:233–252

Wilcke W, Kaupenjohann M (1998) Heavy metal distribution between soil aggregate core and surface fractions along gradients of deposition from the atmosphere. Geoderma 83:55–66

Wilcke W, Zech W, Kobza J (1996) PAH-pools in soils along a PAH-deposition gradient. Environ Pollut 92:307–313

Wilcke W, Guschker C, Kobza J, Zech W (1999) Heavy metal concentrations, partitioning, and storage in Slovak forest and arable soils along a deposition gradient. J Plant Nutr Soil Sci 162:223–229

Wilcke W, Totsche KU, Korber M, Kobza J, Zech W (2000) Fluoro-mobilization of metals in a Slovak forest soil affected by the emissions of an aluminum smelter. J Plant Nutr Soil Sci 163:503–508

Williams FLR, Lawson AB, Lloyd OL (1992) Low sex ratios of births in areas at risk from air pollution from incinerators, as shown by geographical analysis and 3-dimensional mapping. Int J Epidemiol 21:311–319
Williams FLR, Ogston SA, Lloyd OL (1995) Sex ratios of births, mortality, and air pollution: can measuring the sex ratios of births help to identify health hazards from air pollution in industrial environments? Occup Environ Med 52:164–169
Williams MR, Abbott I (1991) Quantifying average defoliation using leaf-level measurements. Ecology 72:1510–1511
Williamson BJ, Udachin V, Purvis OW, Spiro B, Cressey G, Jones GC (2004) Characterisation of airborne particulate pollution in the Cu smelter and former mining town of Karabash, south Ural Mountains of Russia. Environ Monit Assess 98:235–259
Williamson JR, Neilsen WA (2003) The effect of soil compaction, profile disturbance and fertilizer application on the growth of eucalypt seedlings in two glasshouse studies. Soil Till Res 71:95–107
Wilsey BJ, Saloniemi I (1999) Leaf fluctuating asymmetry in tree-line mountain birches, *Betula pubescens* ssp. *tortuosa*: genetic or environmentally influenced? Oikos 87:341–345
Winterhalder K (2000) Landscape degradation by smelter emissions near Sudbury, Canada, and subsequent amelioration and restoration. In: Innes JL, Oleksyn J (eds) Forest dynamics in heavily polluted regions. CAB International, Wallingford, pp 87–119
Winterhalder K (2002) Reclamation of smelter-damaged lands. In: Richard I, Barnhisel RI, Darmody RG, Daniels WL (eds) Reclamation of drastically disturbed lands (Agronomy Monograph, no. 41). American Society of Agronomy, Madison, pp 819–853
Winterhalder K, Lautenbach WE, Beckett PJ (2001) The Sudbury regional land reclamation program: a case study. In: Burley JB (ed) Environmental design for reclaming surface mines. The Edwin Mellen Press, Lewiston, pp 286–295
Wolf A, Kozlov MV, Callaghan TV (2008) Impact of non-outbreak insect damage on vegetation in northern Europe will be greater than expected during a changing climate. Clim Change 87:91–106
Wolff EW, Suttie ED, Peel DA (1999) Antarctic snow record of cadmium, copper, and zinc content during the twentieth century. Atmos Environ 33:1535–1541
Wołk A (1977) Microclimate alterations in relation to the extent of forest destruction in vicinity of the nitrogen plant at Puławy. Sylwan 121(7):33–46 (in Polish)
Wolterbeek HT, Bode P, Verburg TG (1996) Assessing the quality of biomonitoring via signal-to-noise ratio analysis. Sci Total Environ 180:107–116
Wong MH (1978) An ecological survey of effect of sulfur dioxide emitted from an acid work factory. Bull Environ Contam Toxicol 19:715–723
Wood CW, Nash TNI (1976) Copper smelter effluent effects on Sonoran desert vegetation. Ecology 57:1311–1316
Woodwell GM (1967) Radiation and patterns of nature. Science 156:461–470
Woodwell GM (1970) Effects of pollution on structure and physiology of ecosystems. Science 168:429–433
Wotton DL, Hogan GD (1981) The effects of atmospheric emissions from the Thompson mining and smelting industry on forest vegetation and soils. Manitoba Department of Consumer and Corporate Affairs and Environment, Winnipeg
Wulff A, Ahonen J, Karenlampi L (1996) Cell ultrastructural evidence of accelerated ageing of Norway spruce needles in industrial areas. New Phytol 133:553–561
Yan ND, Leung B, Keller W, Arnott SE, Gunn JM, Raddum GG (2003) Developing conceptual frameworks for the recovery of aquatic biota from acidification. Ambio 32:165–169
Yarmishko VT (1993) The analysis of morphological structure and crown state of *Pinus sylvestris* L. under different levels of air pollution in the Kola peninsula. In: Kozlov MV, Haukioja E, Yarmishko VT (eds) Aerial pollution in Kola Peninsula. Proceedings of international workshop, St. Petersburg, 14–16 April 1992. Kola Science Centre, Apatity, pp 236–251
Yarmishko VT (1997) Scots pine and aerial pollution in the European North. St. Petersburg University, St. Petersburg (in Russian)

Yezerinac SM, Lougheed SC, Handford P (1992) Morphological variability and enzyme heterozygosity – individual and population-level correlations. Evolution 46:1959–1964

Ympäristöministeriö (2007) Suomen alueelle Venäjältä kohdistuvat rajat ylittävät ympäristöuhat. Karri Eloheimo, Helsinki. http://www.ymparisto.fi/download.asp?contentid = 64243&lan = FI. Accessed 20 Jan 2009

Yusupov IA, Luganskij NA, Zalesov SV (1999) The state of young plantations of Scots pine affected by aerial pollution. Ural State Forest Academy, Ekaterinburg (in Russian)

Zaguralskaya LM (1997) Microbiological monitoring of forest ecosystems under industrial impact in northern taiga. Lesovedenie [Forestry, Moscow] 0(5):3–12 (in Russian)

Zainagutdinova EM (2003) Study of industrial emission impact on soil-dwelling larvae of click beetles (Elateridae). In: Dolgin MM (ed) Invertebrate animals diversity in the North. 2nd International Conference, March 17–22, 2003. Abstracts. Institute of Biology, Syktyvkar, p 28 (in Russian)

Zakharov VM (1990) Analysis of fluctuating asymmetry as a method of biomonitoring at the population level. In: Krivolutsky D (ed) Bioindications of chemical and radioactive pollution. Mir, Moscow, pp 187–198 (in Russian)

Zakharov VM, Clarke GM (1993) Biotest: a new integrated approach for assessing the condition of natural environments. Moscow Affiliate of the International Biotest Foundation, Moscow (in Russian)

Zakharov VM, Chubinishvili AT, Baranov AS, Borisov VI, Chistyakova EK, Dmitriev SG, Kryazheva NG, Krysanov EY, Pronin AV, Valetsky AV (2001) The health of the environment: experience of assessment. Center for the Health of the Environment, Moscow (in Russian)

Zhirov VK, Kashulin PA, Kuzmin AV, Kostyuk VI (1993) Modification of free radical oxidation activity and content of pigments in plants affected by aerial emission. In: Kozlov MV, Haukioja E, Yarmishko VT (eds) Aerial pollution in Kola peninsula. Proceedings of the international workshop, St. Petersburg, 14–16 April. Kola Science Centre, Apatity, pp 211–214

Zhuikova TV, Bezel VS, Pozolotina VN (2001) Demographic structure of *Taraxacum officinale* s. l. Populations under toxic environmental contamination. Botanicheskij Zhurnal [Bot J, St. Petersburg] 86:103–112

Zlobin V (2002) Nadvoitsy Aluminum Works: 100 mln for ecology. Severny Kurier 51, 11 April 2002. http://svanhovd.no/abstracts/ab_2002/04_2002_1.pdf. Accessed 22 Dec 2007

Zobel M, Pilt I, Moora M, Pärtel M, Liira J (1999) Small-scale dynamics of plant communities in an experimentally polluted and fungicide-treated subarctic birch-pine forest. Acta Oecol 20:29–37

Zubareva ON, Skripal'shchikova LN, Greshilova NV, Kharuk VI (2003) Zoning of landscapes exposed to technogenic emissions from the Norilsk mining and smelting works. Russ J Ecol 34:375–380

Zverev VE (2009) Mortality and regeneration of mountain birch around the nickel-copper smelter following substantial emission decline: results of fifteen years of monitoring. Russ J Ecol 40 (in press)

Zverev VE, Zvereva EL, Kozlov MV (2008) Slow growth of *Empetrum nigrum* in industrial barrens: combined effect of environmental disturbance and age of extant plants. Environ Pollut 156:454–460

Zvereva EL (1994) Effects of industrial pollution on fly communities (Diptera, Brachycera). Entomologicheskoje Obozrenie [Entomol Rev, St. Petersburg] 73:45–57 (in Russian)

Zvereva EL, Kozlov MV (2000a) Effects of air pollution on natural enemies of the leaf beetle *Melasoma lapponica*. J Appl Ecol 37:298–308

Zvereva EL, Kozlov MV (2000b) Pollution suppresses delayed inducible resistance in boreal willow *Salix borealis*. Ecol Lett 3:85–89

Zvereva EL, Kozlov MV (2000c) Bud removal alleviates negative effects of pollution on quality of *Salix borealis* for insect herbivore. Basic Appl Ecol 1:171–176

Zvereva EL, Kozlov MV (2001) Effects of pollution-induced habitat disturbance on the response of willows to simulated herbivory. J Ecol 89:21–30

Zvereva EL, Kozlov MV (2004) Facilitative effects of top-canopy plants on four dwarf shrub species in habitats severely disturbed by pollution. J Ecol 92:288–296

Zvereva EL, Kozlov MV (2005) Growth and reproduction of dwarf shrubs, *Vaccinium myrtillus* and *V. vitis-idaea*, in a severely polluted area. Basic Appl Ecol 6:261–274

Zvereva EL, Kozlov MV (2006a) Consequences of simultaneous elevation of carbon dioxide and temperature for plant-herbivore interactions: a metaanalysis. Global Change Biol 12:27–41

Zvereva EL, Kozlov MV (2006b) Top-down effects on population dynamics of *Eriocrania* miners (Lepidoptera) under pollution impact: does an enemy-free space exist? Oikos 115:413–426

Zvereva EL, Kozlov MV (2007) Facilitation of bilberry by mountain birch in habitat severely disturbed by pollution: importance of sheltering. Environ Exp Bot 60:170–176

Zvereva EL, Kozlov MV (2009) Responses of terrestrial arthropods to industrial pollution: a meta-analysis. Environ Sci Pollut Res (DOI 10.1007/s11356-009-0138-0)

Zvereva EL, Kozlov MV, Neuvonen S (1995) Population density and performance of *Melasoma lapponica* (Coleoptera, Chrysomelidae) in surroundings of smelter complex. Environ Entomol 24:707–715

Zvereva EL, Kozlov MV, Haukioja E (1997a) Stress responses of *Salix borealis* to pollution and defoliation. J Appl Ecol 34:1387–1396

Zvereva EL, Kozlov MV, Niemelä P, Haukioja E (1997b) Delayed induced resistance and increase in leaf fluctuating asymmetry as responses of *Salix borealis* to insect herbivory. Oecologia 109:368–373

Zvereva EL, Kozlov MV, Kruglova OY (2002) Colour polymorphism in relation to population dynamics of the leaf beetle, *Chrysomela lapponica*. Evolut Ecol 16:523–539

Zvereva EL, Toivonen E, Kozlov MV (2008) Changes in species richness of vascular plants under the impact of air pollution: a global perspective. Global Ecol Biogeogr 17:305–319

Zwolinski J (1994) Rates of organic-matter decomposition in forests polluted with heavy metals. Ecol Eng 3:17–26

Zyrin NG, Gorbatov VS, Sadovnikova LK, Tananykina EF, Obukhov AI, Reshetnikov SI, Pershina NZ (1986) Mid-Ural industrial region. In: Zyrin NG, Malakhov SG, Stasjuk NV (eds) Contamination of soils by metals and fluorine. Gidrometeoizdat, Leningrad, pp 66–90 (in Russian)

Geo Index

Organism Index

Subject Index

L

M

N

O

MIX
Papier aus verantwortungsvollen Quellen
Paper from responsible sources
FSC® C105338

If you have any concerns about our products,
you can contact us on
ProductSafety@springernature.com

In case Publisher is established outside the EU,
the EU authorized representative is:
Springer Nature Customer Service Center GmbH
Europaplatz 3, 69115 Heidelberg, Germany

Printed by Libri Plureos GmbH
in Hamburg, Germany